Rubber

TROPICAL AGRICULTURE SERIES

The Tropical Agriculture Series, of which this volume
forms part, is published under the editorship of Gordon
Wrigley

ALREADY PUBLISHED

Tobacco *B C Akehurst*
Sugar-cane *F Blackburn*
Tropical Grassland Husbandry *L V Crowder and H R Chheda*
Sorghum *H Doggett*
Sheep Production in the Tropics and Sub-Tropics *Ruth M Gatenby*
Rice *D H Grist*
The Oil Palm *C W S Hartley*
Cotton *John M Munro*
Cattle Production in the Tropics volume 1 *W J A Payne*
Spices Vols 1 & 2 *J W Purseglove* et al.
Tropical Fruits *J A Samson*
Bananas *R H Stover and N W Simmonds*
Agriculture in the Tropics *C C Webster and P N Wilson*
Rubber *C C Webster and W J Baulkwill*
Oilseed Crops *E A Weiss*
An Introduction to Animal Husbandry in the Tropics
  G Williamson and W J A Payne
Cocoa *G A R Wood and R A Lass*
Coffee *G Wrigley*

Rubber

C C Webster

*formerly Director of the Rubber Research
Institute of Malaysia*

and

W J Baulkwill

*formerly of the Rubber Research Institute
of Malaysia*

Longman
Scientific &
Technical

Copublished in the United States with
John Wiley & Sons, Inc., New York

Longman Scientific & Technical,
Longman Group UK Limited,
Longman House, Burnt Mill, Harlow,
Essex CM20 2JE, England
and Associated Companies throughout the world.

Copublished in the United States with
John Wiley & Sons, Inc., 605 Third Avenue, New York,
NY 10158

First published 1989

British Library Cataloguing in Publication Data
Webster, C. C.
    Rubber. — (Tropical agriculture series).
    1. Rubber
    I. Title II. Baulkwill, W. J.   III. Series
    633.8'952   TS1890
ISBN 0-582-40405-3

Library of Congress Cataloguing in Publication Data
Rubber/[edited by] C. C. Webster and W. J. Baulkwill.
    p. cm. — (Tropical agriculture series)
    Bibliography: p.
    Includes index.
    1. Hevea. 2. Rubber. I. Webster, C. C. (Cyril Charles). 1909–
II. Baulkwill, W. J. (William John), 1919–.   . III. Series.
SB291.H4R78 1989
633.8'952–dc19
ISBN 0-470-21027 (Wiley, USA only)

Set in Linotron 202 10/11pt Times

Produced by Longman Singapore Publishers (Pte) Ltd.
Printed in Singapore

Contents

NOTE:

The author has endeavoured to ascertain the accuracy of the statements in this book. However, facilities for determining such accuracy with absolute certainty in relation to every particular statement have not necessarily been available. Therefore, the reader is always advised to seek an independent expert opinion before implementing in practice any particular pesticide technique or method described herein.

Acknowledgements

The editors gratefully acknowledge the assistance received from the following persons and organizations.

Dr J. W. Blencowe, who gave much help in preparing the reference list.

Mr Philip Watson, Statistician, and Dr P. Jumpasut, Economist, of the International Rubber Study Group, London, for their advice on the section on economic forecasts.

The Director and library staff of the Royal Botanic Gardens, Kew, for the provision of documents and for kindly reading the manuscript on the domestication of *Hevea brasiliensis*.

The Director of the Rubber Research Institute of Malaysia for permission to reproduce a number of tables and figures and, especially, our thanks are due to two members of the staff of the Institute, Mr J. C. Rajarao and Mr Hoh Lian Yong, who kindly prepared a number of photographs, including most of those illustrating diseases and pests of the rubber tree.

We are indebted to the following for permission to reproduce copyright material:

Food and Agriculture Organization of the United Nations for tables 4.5, 4.6 from *World Soil Resources Report* No. 45: 'Report on the Ad Hoc Consultation on Land Evaluation', Rome 6–8 January 1975, and table 7.7 from Note No. 131 of *UNDP/FAO Project INS/72/004*, by M.E.B. Reed; the Incorporated Society of Planters for fig. 4.2 by Pushparajah from *The Planter* 1983, and table 8.23 by Ling and Mainstone from *The Planter* 1983; Institut de Recherches sur le Caoutchouc for fig. 4.4 by Omont from *Le caoutchouc naturel*, M. Compagnon 1986, Editions G.P. Maisonneuve et Larose amd table 4.4 from *Série Agronomique Physiologie* 1976; International Potash Institute, Berne for table 8.13 by D. Liang from pp. 1–5 of *Potash Review* Subject 27, No. 5 1983; International Rubber Study Group, London for table 1.10 from table

4.4 of *Rubber Statistical Bulletin*, Feb. 1986; Johns Hopkins University Press for fig. 1.7 from fig. 2.4 by Grilli, Agostini & Hooft-Welvaars from *World Bank Staff Occasional Papers* No. 30; Longman Group (U.K.) Ltd. and the author for fig. 2.1 from *Tropical Crops* by John Purseglove; Oxford University Press (East Asia regional office) for an extract and tables 1.3 & 7.13 from *The Natural Rubber Industry: Its Development, Technology and Economy in Malaysia* by C. Barlow; Royal Botanic Gardens, Kew for fig. 1.3; Rubber Manufacturers Association Inc. for an extract from *The Green Book*; Rubber Research Institute of Malaysia for figs. 2.2, 2.3, 2.4, 4.5, 4.8, 4.9, 6.5, 6.6, 6.10, 6.12, 6.13, 8.1, 8.2, 8.5, 9.1, 9.2, 9.3, 9.4, 9.5, 9.6, 9.7, 9.9, 9.10, 9.12, 9.13, 9.15, 10.19, 10.20 and tables 3.1, 3.2, 3.3, 3.4, 3.5, 3.6, 4.2, 4.7, 4.8, 4.9, 4.10, 4.11, 6.1, 6.2, 6.3, 6.4, 6.5, 6.6, 6.7, 7.1, 7.2, 7.3, 7.4, 7.5, 7.6, 7.9, 7.10, 7.11, 7.12, 7.14, 7.15, 8.1, 8.2, 8.3, 8.4, 8.5, 8.6, 8.7, 8.8, 8.9, 8.11, 8.12, 8.14, 8.15, 8.16, 8.17, 8.18, 8.19, 8.20, 8.21, 8.22, 8.24, 8.25, 8.26, 8.27, 8.28, 8.29, 11.6, 12.1, 12.2 from various sources; the author, Dr. A. N. Strahler for fig. 4.1 from fig. 13.5 of *Physical Geography* 3rd Edition 1969; the editor of *Sudhevea*, Brazil for fig 4.7 by K. H. Chee; University of Miami Academic Services for table 4.3 from *Hevea, Thirty Years of Research in the Far East* by M. J. Djikman 1951.

Whilst every effort has been made to trace the owners of copyright material, in a few cases this has proved impossible and we take this opportunity to offer our apologies to any copyright holders whose rights we may have unwittingly infringed.

List of contributors

W. J. Baulkwill
formerly of the Rubber Research Institute of Malaysia

J. W. Blencowe
formerly Deputy Director of the Rubber Research Institute of Malaysia and subsequently Project Manager, FAO rubber development projects, Thailand

A. Johnston
formerly Director of the Commonwealth Mycological Institute, Kew, England

J. E. Morris
Consultant, formerly of the Rubber Research Institute of Malaysia

The late E. C. Paardekooper
formerly tree crops specialist with the FAO-World Bank cooperative programme

N. W. Simmonds
formerly Director of the Scottish Plant Breeding Station. Consultant to the Rubber Research Institute of Malaysia

G. A. Watson
Consultant. Formerly Head of the Soils Division of the Rubber Research Institute of Malaysia

C. C. Webster
formerly Director of the Rubber Research Institute of Malaysia

Chapter 1

The history of natural rubber production

W. J. Baulkwill

Organized production of natural rubber to meet industrial needs is now 150 years old and was preceded by many centuries of localized rubber use in the tropics. Over these years the once simple activity of gathering wild rubber to make a few useful or ceremonial objects has developed into the extensive and complex plantation industry of today. Our purpose here is to show how this development has resulted from the technological innovations and socio-economic changes of the nineteenth and twentieth centuries. Attention is focused throughout this chapter on the responses of natural rubber producers to demand trends and on the consequently changing fortunes of the production industry itself. Some emphasis is also placed on the increasingly specific requirements of the manufacturer and, in more recent times, the relative capacities of natural and synthetic rubbers to meet them. If a 'lesson' for our industry emerges from this chapter, it is that efficient production is necessary but not enough: anticipation of future demand and knowledge of the manufacturer's technical requirements are no less needed.

Four phases of the natural rubber production history are described here: (1) the pre-industrial but often sophisticated use of rubber from wild plants in tropical America; (2) the organized gathering and export of wild rubber in response to demand from nineteenth-century manufacturers; (3) the establishment of cultivated *Hevea* plantations as the era of the motor vehicle began, and the cultivated crop's domination of the market in the first half of the

	Thousand tonnes					*Million tonnes*			
Pre-*1820*	*1830*	*1850*	*1880*	*1910*	*1940*	*1950*	*1970*	*1980*	Forecast *2000*
?	0.02	0.4	17	101	1	2.5	9	12.5	20±
Phase 1 pre-industrial	Phase 2 wild natural rubber for industry			Phase 3 cultivated natural rubber		Phase 4 natural rubber and synthetic rubber			

present century; (4) the emergence of synthetic rubber in large quantities in the Second World War and its subsequent capture of two-thirds of the market. Though of equal historical interest the successive phases are most unequal *in terms of quantities of rubber consumed.*

1.1 Early history: to 1830*

1.1.1 Rubber as a religious substance: Amerindian civilization

The importance of rubber to the Aztec and Mayan civilizations of Mexico and Yucatan was reported by numerous commentators from de Sahagun (1530) to de Torquemada (1615) and is described by Schurer (1957b, 1958). Rubber, *olli* or *ulli*, symbolizing blood or life-force, was burnt or paraded in the form of figurines, or paintings on bark-paper, at times when human sacrifices were made to propitiate the gods. The ceremonial and social importance of ball-games and the remarkable elasticity of the solid rubber balls were also recorded. The balls were valuable and, in their thousands, were used to pay taxes to rulers (Schurer 1957b). The source of rubber was less reported, but Hernandez (1570–77) made an accurate drawing of the Mexican rubber tree – later called *Castilla elastica*. As the time passed, Spanish colonists in Mexico found another use for the magic substance; the practice of waterproofing textiles with several layers of latex began in the seventeenth century and was reported in detail by Cervantes (1794) who described the preparation of rubber-coated capes, boots and even carriage hoods. There is only a little evidence for the pre-industrial use of rubber in continents other than America.

1.1.2 From curiosity to commodity: 1700–1830

Despite the long list of historical records which we can now assemble on tropical American rubber, the substance aroused no permanent interest in Europe until it caught the attention of a group then rapidly acquiring esteem and influence – the scientists of the mid-eighteenth century. At that time there were two distinct, geographically separated, rubber-making crafts: waterproofing in Mexico and the production of hollow, moulded objects in the Amazon-Guianas region. It was in the latter region exclusively that several French scientists embarked, in ignorance of earlier accounts, on the rediscovery and reappraisal of rubber (Schultes 1977a; Schurer 1956, 1957a). Among these scientists, the astronomer-geographer de La Condamine (1745), in his book on the Amazon, described

---

* See Appendix 1.1 (p. 539) for further bibliographic detail on early history.

how Amerindians in Ecuador and Brazil made torches, boots, bottles and syringes, employing an elastic resin or gum called *caoutchouc*, 'weeping wood', from the sap of a tree called *hhévé*. The syringes were used as ceremonial and medicinal squirts and enemas. Fresneau (1755), an engineer who worked in French Guiana, published a rough drawing of an *Hevea* species, describing how it was tapped and how objects were made by smoke-drying layers of latex on to clay formers (Schurer 1956, 1960a). Aided by de La Condamine, Fresneau was to become the first enthusiast for rubber as a material with practical possibilities in Europe.

In Europe the only way to use the rubber objects coming from America seemed to be to dissolve them, so restoring, as it were, the latex phase. In the period 1750–1800 a number of scientists (see Appendix 1.1 (p. 539)), mainly in France, and inspired by Fresneau, set about discovering solvents and then using the solution to make the flexible tubes – catheters – which physicians needed for the relief of bladders and similar medical purposes. The solvents found in the laboratory were turpentine, rectified ether and rectified petroleum. The tubes were made by applying rubber solution to wax formers, or to tubes made of textile, or by 'welding' spirals of rubber around glass rods.[1] In addition, around 1790, European scientists re-invented the rubber syringe, a device long known to the Amazonian Indians. In the novel sphere of aerostatics, rubber solution was applied to the silk fabric of the hydrogen balloons that made successful ascents in France in 1783.

Despite these advances, as the nineteenth century began, only a few minor uses had been found for rubber outside the narrow domain of physics, chemistry and medicine. In England rubber had been used to erase pencil marks since 1770, hence its popular name, and in Paris some elastic garters were manufactured in 1803. A few patents were filed for waterproofing with rubber solution, but with no immediate commercial result. For the next 20 years, it seemed, in modern terms, that research had entirely outstripped development.

The first factories of the Industrial Revolution 1800–30

The first rubber manufacturers were concerned with clothing. They either cut thin rubber strips to make elasticated garters, gloves, etc., or applied rubber solution to cloth, to make waterproofed garments (Porritt 1926). Thomas Hancock in 1820 in England and Charles Macintosh in 1823 in Scotland both made waterproofed fabrics, Macintosh using his unique fabric–rubber–fabric sandwich and his patented coal-tar-derived naphtha as the rubber solvent. Hancock also devised processes for cutting thin strips of thread from large sheets and for rubberizing fabric belting. In 1820 he invented the revolutionary *masticator* which shredded and compressed solid rubbers and scraps to a warm, homogenous, mouldable or rapidly soluble mass – this was the technical advance that made modern rubber processing possible.

1.2 The response of supply to demand, 1830–1914

1.2.1 Industry accelerates demand

The early part of this period saw the 'take-off' of industrial development in Europe and the USA in which all economic activities were intensified, fed by raw materials from overseas, including rubber. Barker and Holt (1939) described the early rubber industry at Boston USA, starting with the large-scale importation of rubber shoes fabricated in Brazil and made to fit wooden lasts built to clients' measurements. In 1836 Edwin Chaffee made a significant step forward in the United States when he invented the steam-heated *mixing mill and calender*, an economic machine which forced a soft, warm film of rubber on to cloth, so dispensing with the use of solution in making waterproof garments. An early rubber clothing business in New England boomed, but collapsed in 1837 because the cloth cracked in winter and became tacky and smelly in summer. Only the discovery of vulcanization could solve this problem.

The introduction of practical methods of *vulcanization* during 1838–44 by the two giants of the rubber industry, Thomas Hancock and Charles Goodyear, changed the nature of rubber and thus created a new demand: the incorporation of sulphur, plus the admixture of lead oxide and the use of heat, produced a rubber which was proof against hot- and cold-air temperatures and resistant to melting. Ebonite or vulcanite was made similarly but with more sulphur and greater heat – and quickly produced a crop of new products: furniture, dental plates, insulators, and in 1884, a fountain pen. The important process of *cold* vulcanization by dipping thin rubber articles in sulphur chloride solution was first patented by Andrew Parkes in England in 1846.

The Great Exhibitions of London in 1851 and of Paris in 1855 were showplaces of the new vulcanized industry where ebonite furniture, odourless rubberized garments, elastic fabrics, prototype solid rubber tyres, vulcanized rubber shoes and airbeds fought for attention. At both exhibitions Charles Goodyear won the highest possible awards.

From the mid-century, the explosive growth of the industrial world, and its population, increased the demand for rubber, from a recorded 391 t in 1850 to 119,224 t in 1914 (see Table 1.1). By 1880 the main applications of rubber were essentially the same as those so well described and illustrated by Thomas Hancock in his *Personal Narrative* of 1857 – surgical appliances, footwear, waterproof clothing and sheeting, inflatables of all kinds, elasticated fabrics, hoses, belting and engine components – but the engineering uses had been vastly increased by the proliferation of factories, houses, public services, the growth of the steam transport network and latterly electric lighting and telephones. As an electrical insulant, rubber was not always the winner: gutta-percha was the choice for submarine cables, and waxed paper and impregnated cotton for many purposes, but rubber insulation conquered the indoor cable in the 1890s (Porritt 1926). All these applications were supported by patient research into rubber properties, and improvements in the existing processes – milling, compounding, calendering, extrusion and vulcanization – the latter much improved from 1905 by the use of organic accelerators which reduced the time taken and the amount of sulphur required. An economically important development was the reclaimed rubber industry using acid (1858) and alkali (1899) processes. In the acute shortage of 1917, reclaim provided 36 per cent of USA rubber consumption (Barker & Holt 1940).

For ordinary citizens, and soon for the rubber industry, the most startling changes came in the period 1885–95 with the appearance of 'safety' chain-driven bicycles and motor vehicles powered by internal combustion petrol engines, soon using adaptations of Dunlop's pneumatic rubber tyre. After 1900, the automobile tyre industry began to create a demand for the raw material which 10 years later the wild-rubber industry would be unable to satisfy. Ford's model-T motor car was first made in 1908 and *15 million* were produced by 1927.

Table 1.1 *Recorded consumption of rubber 1830–1914 (t)*

Date	USA	UK	Russia	Germany	France	Others	Total
1830	n.a.	23	—	—	—	—	23
1840	n.a.	312	—	—	—	—	312
1850	n.a.*	391	—	—	—	—	391
1860	762*	2,187	—	—	—	—	2,949
1870	4,365	7,779	—	—	—	—	12,144
1880	8,239	8,615	—	—	—	—	16,854
1890	15,582	13,412	—	—	—	—	28,994
1900	20,634	11,159	4,293	8,652	2,520	6,200	53,458
1910	42,887	20,782	7,349	13,951	3,800	12,181	100,950
1914	63,630	18,868	11,830	8,574 (6 mths)	4,447	11,875†	119,224

Sources: 1825–1900 McFadyean (1944); 1900–14 Barker and Holt (1938).
* Records of US manufactures (Barker and Holt 1939) indicate consumption of *c.* 1,000 for 1850 and *c.* 1,800 for 1860.
† Includes Canada 1,000, Japan 1,000.

1.2.2 The statistics of demand and supply

Data for consumption to 1914 are shown in Table 1.1 and for production in Fig. 1.1.

Rubber demand was modest until around 1850 but accelerated thereafter. The sharpest rise (doubling) between 1900 and 1914, was

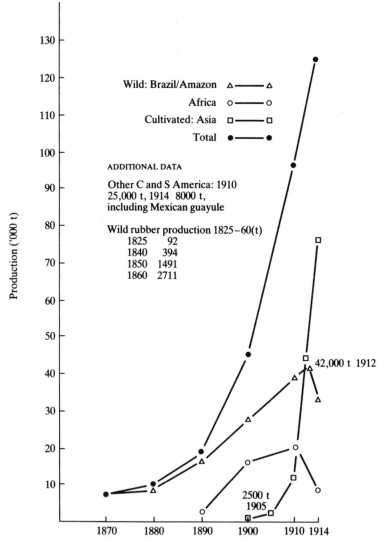

Fig. 1.1   Production of wild and cultivated rubber to 1914 (*Sources:* to 1900 Wallace 1952, Barker and Holt 1939; 1900–14 Barker and Holt 1938, McFadyean 1944)

a consequence of the automobile tyre industry and was most marked in the USA.

Brazilian exports of rubber coped with the rising demand to about 1880 when a determined world-wide quest for new sources of wild rubber began. At about the same time *Hevea brasiliensis* from the Amazon was successfully domesticated in Ceylon (Sri Lanka) and Malaya, leading to commercial plantings by 1895. Wild rubber, including that from non-*Hevea* spp., reached a peak of production in 1912 when Asian plantation rubber was already impinging on the market, but was surpassed by cultivated rubber in 1914 and thereafter fell into decline.

1.2.3 Finding the raw material: from wild to cultivated rubber

The supply of wild rubber

The wild rubber exported from Brazil (Fig. 1.2) came, with few exceptions, either from the *Hevea* spp. in the Amazon basin or from *Castilla ulei* which closely resembles *C. elastica*, the historic source of rubber in Mexico, and had a wider distribution than *Hevea*. Serious production of *Castilla* rubber (caucho) began in Peru in 1882 (Schurtz *et al.* 1925) and later accounted for almost one-fifth of Amazonian production. The distribution of the *Hevea* spp. and the quality of their rubbers are described in Chapter 2. Because of its concentration in the more accessible areas, *H. brasiliensis* supplied the bulk of wild *Hevea* rubber in the nineteenth century.

Production from Amazonian Hevea

Early production was from the lower Amazon (Pará State) but, as demand grew, rubber gathering expanded westwards first into the Amazonas region, then Peru (1850s) and Bolivia (1880s). The Matto Grosso and Acre areas were late-comers (1900).

The earliest exports in the 1820s were in the form of figurines, boots, playballs, syringes and bottles, but later solid balls weighing 30–60 kg became the standard product. The exporters at Pará and Manaus financed middlemen to purchase rubber for them from the wild-rubber estates. The owner, or lessee, of the estate or *seringal* contracted out rubber production to the tappers to whom he supplied food and other goods on credit, deducting the cost when paying the tapper for his rubber. This commonly resulted in the tapper being deeply in debt to his employer at the end of a 5–6 month season. (Woodroffe & Hammel-Smith 1915; Schurtz *et al.* 1925).

The trees, which were scattered in the forest and were mostly 24–30 m high with a girth of about 2 m, were tapped in tasks of 100–150 trees, the trees of each task being linked by a loop-shaped trail starting and ending at the tapper's hut. Each tapper had two tasks which he tapped on *alternate days* during a tapping season

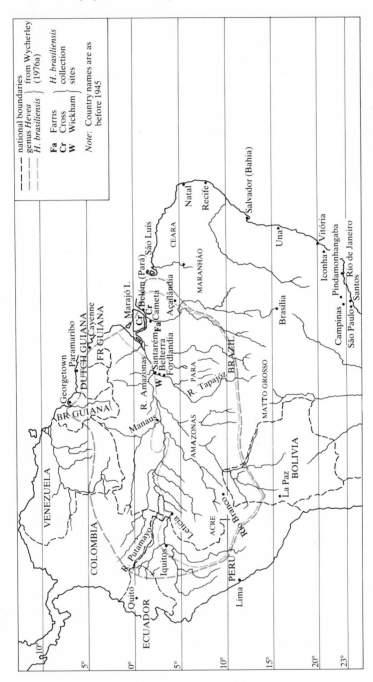

Fig. 1.2 The Amazon basin showing the distribution of *Hevea brasiliensis* and other *Hevea* spp. and approximate sites of the historical collections of *H. brasiliensis* intended for domestication

which, being limited by rain and floods, lasted for 5–7 months, depending on locality, between May and November (Montenegro 1908; Woodroffe & Hamel-Smith 1915; Schurz *et al.* 1925; Weinstein 1983). The arduous life of the tapper or *seringueiro* has been described by many observers including Cross (Kew Gardens Correspondence 1877a), Wickham (1908), Akers (1914), Woodruffe and Hamel-Smith (1915) and Schurz *et al.* (1925). Starting before dawn, he used a small hatchet to make 2–10 small upward incisions into the bark of each tree in a horizontal line about 2 m above the base and then affixed tin or earthenware collection cups below each incision. After a day's rest the trees would be tapped 5 cm lower, any scrap being collected at the same time. A common fault was excessively deep tapping which could kill the trees.

On completing his tapping round, the tapper retraced his steps (*running* according to Cross) to collect the latex in a bucket. In the afternoon he coagulated his latex in an unhealthy smoke-laden hut by dipping a paddle into the latex and rotating it over a hot stove, usually fuelled with the nuts of the uricuri palm, *Attalea excelsa*. Repetition of the process produced the well-cured balls of rubber which, when classified as 'fine hard Para', secured a premium over the plantation product until the 1930s (Burkill 1935). According to Schurz *et al.* (1925) the average season's yield from tapping 300 trees was 200 kg, but could be 360–900 kg in good virgin stands.

Other wild rubber — the non-Heveas

As demand and prices rose around 1898, a frenetic search for wild rubbers outside Brazil was set in motion, notably in Africa, Central and South America, and India (Fig. 1.1).

The very numerous wild species that produced rubber of diverse quality have been described by Schidrowitz and Dawson (1952), Polhamus (1957, 1962) and for peninsular Malaysia by Burkill (1935). Of these, the best known are the already mentioned *Castilla* spp. producing 'caucho' rubber in Central America and the Amazon Basin, *Manihot* spp. producing Ceara or maniçoba rubber in South America, *Funtumia elastica* in Africa, *Ficus elastica* or rambong in Asia, and several tropical vines among which *Landolphia* spp. in Africa are perhaps the best known. Three species of Compositae have yielded enough rubber to merit domestication: *Parthenium argentatum* (guayule) of Central America, *Taraxacum* (kok-saghyz) of southern USSR, *Solidago* spp. (golden rod) of North and South America, the first being the highest yielding (National Academy of Sciences 1977).

The twentieth-century searchers for wild rubber used all the species listed above and many more. There was large-scale destruction of the botanical resources, and brutal exploitation of gatherers in the Congo, the Upper Amazon and elsewhere (Appendix 1.1).

At the height of the last of these public scandals (Putamayo, Peru, 1910) the press was turning its attention to a new rubber sensation – the boom in the dividends and shares of the companies that held rubber plantations in Asia.

The domestication of rubber: 1870–1914

The domestication of *Hevea brasiliensis* in the East is the most spectacular event in rubber history, for, in the space of only 40 years, a novel plantation industry of some 900,000 ha, or $2\frac{1}{4}$ million acres, was created to meet the new industrial demand (Table 1.2).

Table 1.2 *Area and output of cultivated Hevea in 1900–14(ha and t of dry rubber)*

Date	Ceylon	India/Burma	Malaya (and Labuan)	British N. Borneo/ Brunei/ Sarawak	Netherlands East Indies	Indochina	Total ha (t)
1900							
ha	400	—	2,400	—	—	—	2 800
(t)	(4)	—	—	—	—	—	(4)
1905							
ha	26,700	2,800	18,600	c. 800	3,200	—	52,000
(t)	(70)	—	(110)	—	—	—	(180)
1910							
ha	104,400	20,200	218,900	8,500	99,600	3,600	455,200
(t)	(1,550)	(140*)	(5,800)	(60)	(3,050)	(180*)	(10,780)
1914							
ha	117,300	41,700	472,700	20,600	245,200	16,200	913,700
(t)	(15,580)	(1,370*)	(47,180)	(900)	(9,110)	(190)	(74,330)

Sources: Figart (1925) and * McFadyean (1944). Total tonnage data differ slightly from those in Fig. 1.1

Domestication was chronicled and analysed in great detail by 1912, Petch (1914), also by Wickham (1908), Wright (1912), Ridley (1912, 1955), and the subject has been reviewed more recently by Wycherley (1959, 1968a,b), Drabble (1973) and Schultes (1977a). The principal sources of information are the archives and publications of the botanical gardens involved, notably at Kew, Singapore and in Ceylon.

Factors favouring cultivated rubber

The successful transfer of *Hevea brasiliensis* to Asia and the subsequent establishment of commercial rubber plantations there, in response to the rising demand for the raw material, were the result of the political forethought and practical abilities of a few people assisted by many favourable circumstances: the ecological suitability for *Hevea* of several Asian countries in the tropical rainforest belt (see Ch. 4); the prior establishment of botanical gardens around the world; and not least the availability of capital and labour. Malaya, where production was soon to be the highest, was the best example of favourable circumstances: all those mentioned above were present, plus a developed transport system related to the tin industry, expanses of available and accessible land, and around 1900 a declining coffee industry, soon replaced by *Hevea*. The Netherlands East Indies (NEI), now Indonesia, with more crop options, was slower to adopt the new plantation crop (Allen & Donnithorne 1957; Chan 1967; Jackson 1968; Drabble 1973; Barlow 1978; Andaya & Andaya 1982). The Malayan authorities also assisted rubber planting by release of land at low rentals and by loans to planters. There was some support for research on rubber in Malaya, Ceylon and NEI and also important schemes to assist immigration, and in NEI transmigration, of plantation workers.

Domestication: the sequence of events

Four gifted men played the leading roles in the domestication of *Hevea brasiliensis*. They were Clements Markham of the India Office, Joseph Hooker, Director of Kew Gardens, Henry Wickham, planter, rubber trader and naturalist, and Henry Ridley, protégé of Hooker and from 1888 Director of Singapore Botanic Gardens. Kew Gardens played a special part in the domestication of wild plants, for it was here that planting materials of potential economic value were assembled from abroad, propagated and then distributed to other botanical gardens around the world.

Markham provided the political initiative, his grand design being the development of profitable crops in India and Ceylon. During 1860–64, he had already successfully collaborated with the botanist R. Spruce and a Kew gardener R. M. Cross in domesticating South American *Cinchona* spp. in India (Kew Gardens correspondence 1860). In 1872 he commissioned James Collins, an early proponent of 'the acclimatization of the different species of *Hevea* . . . in such of our Eastern provinces as will be found best suited', to report on the rubber-yielding plants of the world (Collins 1869, 1872). Then in 1873, working with Collins and with Hooker at Kew, Markham arranged the purchase of 2000 *Hevea brasiliensis* seeds collected by Mr Farris at Cameta near the port of Pará (Belém) — but of these only 12 of the resultant seedlings survived at Kew, of which 6 were sent to the Botanic Gardens Calcutta and there failed, while 6 were used to produce cuttings at Kew (Royal Botanic Gardens, Kew 1914). Despite this setback, Markham commissioned two further collections: the first by Henry Wickham, then in Brazil, who (after an initial failure in 1875) successfully brought 70,000 seeds from the Rio Tapajóz region of the Upper Amazon to Kew, arriving in June 1876 (see Figs 1.2 and 1.3); the second by Cross who brought over

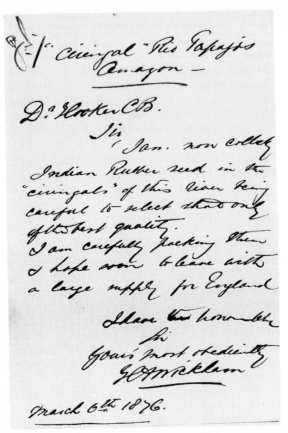

Fig. 1.3   Henry Wickham announces dispatch to Kew of *Hevea brasiliensis* seeds from the Rio Tapajós region, Upper Amazon, Brazil (From Kew Gardens Correspondence 1876b)

1000 plants from the Pará/Marajó Island region of the Lower Amazon, the plants arriving at Kew in November of the same year (Royal Botanic Gardens, Kew 1914). A much-embroidered story has persisted that Wickham smuggled out his *Hevea* seeds, but in fact the export was officially authorized, as was later confirmed by the Brazilian Government (Wycherley 1968a).

The germination of over 2000 of the Wickham seeds at Kew and the successful transfer of the seedlings to Ceylon in miniature glasshouses or 'Wardian cases', as well as the subsequent maturing and flowering of the trees and distribution of the resultant seed, was well documented, but the fate of Cross's plants is obscure, as can be seen from the succession of events outlined below.

14 June 1876   Seventy thousand Wickham Upper Amazon
                    Hevea brasiliensis seeds arrived at Kew and
                    2700 germinated (Kew Gardens Correspon-
                    dence 1876a, b).

8 August 1876   One thousand nine hundred and nineteen of
                    the Wickham seedlings sent to the Botanical
                    Gardens Ceylon and 90 per cent survived
                    (Kew Gardens Correspondence 1876c).

30 August 1876   Eighteen of the Wickham seedlings sent to
                    Bogor Botanic Gardens, NEI, where 2
                    survived, and 50 to Singapore where probably
                    none survived (Royal Botanic Gardens, Kew
                    1914).

23 November 1876   One thousand and eighty Cross Lower
                    Amazon plants arrived in poor condition at
                    Kew, where 400 were retained at the Botanic
                    Gardens and 680 by a neighbouring nursery-
                    man, the respective numbers of surviving
                    plants being 12 and 14 (Kew Gardens Corre-
                    spondence 1877a; Royal Botanic Gardens,
                    Kew 1878, 1898, 1914).

11 June 1877   Twenty-two *Hevea* plants, not specified as
                    Wickham or Cross, sent from Kew to Singa-
                    pore, where they were successfully raised and
                    distributed in Malaya. They are thought to be
                    the prime source of the '1000 tappable trees'
                    found by Ridley in Malaya in 1888 (Petch
                    1914; Ridley 1955; Wycherley 1968a, b).

14 September 1877   One hundred *Hevea* plants specified as Cross
                    material sent to Ceylon Botanic Gardens.
                    Subsequent fate unknown (Kew Gardens
                    Correspondence 1877b; Petch 1914).

1883   Trees of Wickham origin in Ceylon seeded.
                    Seeds and seedlings distributed worldwide,
                    including Singapore.

1888   Ridley began experiments on the tappable
                    trees he found in Malaya (Kew Gardens
                    Correspondence 1907; Wycherley 1968a, b).
                    There were also experiments in Ceylon.

Since there is thus no positive evidence that Cross (or Farris)
material survived in the East, modern Asian planting material up
to the 1980s is commonly called the 'Wickham genetic base' (Ch. 3).
This implies that most modern genetic material comes from a
narrow base of a few trees that grew in the Upper Amazon and does

not contain material from the Lower Amazon swamplands. Nevertheless, some small admixture of Cross genetic material cannot be entirely ruled out, for two reasons: the 22 plants sent to Singapore on 11 June 1877 were unspecified and the fate of the 100 Cross plants sent to Ceylon on 14 September 1877 is unrecorded. Before 1914, and less commonly thereafter, some admixture of Cross material in the Asian stock of *Hevea* was considered quite possible (Wright 1912; Petch 1914; Figart 1925 p. 91; Ridley 1955, writing in 1930). However, in 1914 Colonel D. Prain, Director of Kew, in a letter to J. S. Cramer in Java (Royal Botanic Gardens, Kew 1914) extinguished for many people the 'Cross infiltration' thesis[2]: 'Whether a single plant brought home by Cross ever became fit to send to Asia I do not know. I cannot find any entry in our archives which could be so interpreted.' Petch (1914) quickly entered an addendum to his paper on this subject conceding 'the entire absence of Cross material is possible but not proved'. Burkill (1935), in his authoritative work, said that no Cross material was known to have survived. Since then, 'infiltration by Cross' has never been seriously raised, but a few doubts must remain – failing further evidence about the provenance of the 22 and fate of the 100 *Hevea* plants sent from Kew to the East in 1877.

Whatever the source of the planting materials, they were soon brought into cultivation in Ceylon and Malaya and a little later in NEI. By 1892 the production of good quality rubber was reported from Ceylon and the first 120 ha of *Hevea* were planted then (Kew Gardens Correspondence 1892). In Malaya, Ridley (1955) reported; 'In 1895 I induced a Chinaman Mr Tan Chay Yan of Malacca and Messrs R. C. M. and D. C. P. Kindersley in the Federated Malay States to plant rubber on a moderately large scale.' Within a few years extensive planting of *Hevea brasilensis* was under way.

After the first plantings of *Hevea* rubber in 1895–1905, capital began to be attracted to the East by high prices and dividends. At the peak prices, dividends of over 300 per cent were paid and 30 per cent was frequently attained (Allen & Donnithorne 1957; Drabble 1973; Barlow 1978). The Singapore import-export companies, known as agency houses, which had long supplied secretarial/management services to plantations, now became active in raising London capital to form new rubber plantation companies in Malaya, Ceylon and NEI; the local expertise of the agencies being a guarantee of success.

Hevea was not the only candidate for domestication among the many wild rubbers. Between 1870 and 1914 four species were given extensive trials: *Ficus elastica* (Asia); *Castilla elastica* (Mexico, Ceylon, West Indies and elsewhere); Ceara rubber *Manihot glaziovii* (Tanganyika and Asia); *Funtumia elastica* (Africa). None of these plants, however, was sufficiently productive to compete with *Hevea*.

Early experimental work with domesticated Hevea

As the seedlings from Kew matured in the Ceylon Botanic Gardens, the first test tapping was done there (Drabble 1973) and other experiments followed. Parkin (1900, 1910) worked on latex flow, defined 'wound response' and used acetic acid to coagulate latex. Leguminous cover crops were tried, especially as green manure (Wright 1912). Petch (1921) worked on rubber diseases and eventually wrote the standard work on this subject.

Most progress in the period 1890–1914 was made in Malaya – under the inspiration of H. N. Ridley from Kew, who worked as Director of the Singapore Botanic Gardens from 1888–1911. The research advances and practical influence of Ridley and his innovative associates in Malaya have been clearly reviewed by Wycherley (1959, 1968a,b) and described and documented by Drabble (1973).

By 1897 Ridley had replaced Brazilian-style tapping – multiple incisions which damaged the cambium and eventually made trees untappable — by an entirely new system of excision, that is, removing at each tapping thin slivers of bark from the lower edge of a sloping cut (using a chisel or knife, later a gouge), much as is done today. The initial Ridley tapping pattern, much modified as time went on, was a 'herringbone' of sloping incisions with a central vertical channel for latex offtake. Thus, in essence, Ridley invented the tapping method still in commercial use, one which ensures a smooth and easily collectable flow of latex, preserves the life of the tree and allows the bark to renew itself smoothly, so permitting efficient tapping of the renewed bark.

A long list of additions to rubber knowledge followed from Ridley and his associates during the period up to the rubber boom of 1909–10: bark renewal after excision tapping, value of early-morning tapping, comparison of daily and alternate daily tapping (the latter preferred), best age of opening trees (then six years), influence of girth and direction of cut on yield, value of contour planting, use of smokehouses, coagulation of latex in shallow trays, and, towards 1909, identification and treatment of diseases. Ridley was cautious about overtapping and the early official recommendations of his era would have restricted tapping to one or two months a year for each tree. The planters were more reckless, reducing rest periods to two months and, during the rubber boom, to zero (Drabble 1973 p. 8). One aspect of planting changed little – the fashion of clean-weeding, contrary to Ridley's ideas, became more and more entrenched and remained the vogue into the 1930s.

The earliest plantings made in the period 1895–1905 were extremely dense, around 1000/ha, but in 1906 Ridley, observing the relation between girth and yield, recommended sparser stands of 370/ha (Wycherley 1959). Actual estate stands in Malaya dropped to around 270/ha in 1910 (Drabble 1973) and were not increased until the 1920s. The first Malayan plantations yielded some 390 kg/ha of dry rubber[3] (Schultes 1977a) and average yields of mature trees in 1929 were not much higher (Bauer 1948). Work on breeding for higher yields, which started in NEI in 1912, is described later in section 1.5.1.

The *Hevea* plantation industry in the East: 1900–14

Cultivated *Hevea* rubber production began in Ceylon around 1900 (Table 1.2). By 1910 the rubber boom was approaching its peak and by 1914 the basic pattern of Eastern plantations, for many years to come, had been established: there were large *Hevea* areas in Malaya and NEI, 473,000 ha and 245,000 ha respectively; smaller areas in Ceylon and India and expanding ones in Indo-China and what is now eastern Malaysia. Within NEI, the Javanese hectarage was surpassed in 1914 by that of Sumatra, where far more land was available (Dijkman 1951).

Estates and smallholdings

Estates accounted for nearly all the *Hevea* area in the early days. For NEI in 1914, Figart (1925) recorded that only 8100 ha were held by Asians. For Malaya he obtained more detailed records:

European and Asian holdings, Malaya, 1907–14 (ha)

	1907	1910	1914
European	68,000	152,000	272,000
Asian	800	66,000	201,000

These figures for Asian-owned holdings are a good guide to the exceedingly rapid growth of Malayan smallholder rubber planting from 1907 to 1914, although the Asian holdings also included an unknown, but limited, number of small estates.

The growth of estate production depended on hired labour not readily available in most of the countries concerned, notably Malaya, Sumatra, Ceylon and parts of Indo-China, but this short-coming was quickly remedied by the recruitment of immigrant labour. Drabble (1973) has described the steady growth of the estate labour force in the Federated Malay States, from 58,000 in 1907 to 161,000 in 1914. This force was almost entirely immigrant and came mainly from south India, but also from China and Java. The vital role of immigrant estate labour is discussed again in section 1.3, 1914–45.

Table 1.3 *Total production and consumption of rubber, 1914 and 1920–46 ('000 t dry rubber)*

Year	Production			Consumption			Production–consumption balance	
	Natural rubber	Synthetic rubber	Total	Natural rubber	Synthetic rubber	Total	Natural rubber	Synthetic rubber
1914	125	—	125	121	—	121	+4	—
1920	358	—	358	301	—	301	+57	—
1921	307	—	307	282	—	282	+25	—
1922*	409	—	409	408	—	408	+1*	—
1923*	411	—	411	452	—	452	−41*	—
1924*	430	—	430	471	—	471	−41*	—
1925*	537	—	537	563	—	563	−26*	—
1926*	633	—	633	550	—	550	+83*	—
1927*	619	—	619	603	—	603	+16*	—
1928*	666	—	666	693	—	693	−27*	—
1929	874	—	874	816	—	816	+58	—
1930	838	—	838	721	—	721	+117	—
1931	813	—	813	690	—	690	+123	—
1932	721	—	721	703	—	703	+18	—
1933	865	—	865	833	—	833	+32	—
1934†	1049	—	1049	930	—	930	+119†	—
1935†	843	—	843	952	—	952	−109†	—
1936†	880	—	880	1052	—	1052	−172†	—
1937†	1185	4	1189	1109	n.a.	1109	+76†	n.a.
1938†	886	6	892	948	n.a.	948	−62†	n.a.
1939†	1006	24	1030	1119	n.a.	1119	−113†	n.a.
1940†	1417	44	1461	1127	n.a.	1127	+290†	n.a.
1941†	1504	78	1582	1259	74	1333	+245†	+4
1942†	650	122	772	778	115	893	−128†	+7
1943†	472	356	828	625	297	922	−153†	+59
1944†	366	916	1282	394	749	1143	−28†	+167
1945	254	880	1134	267	879	1146	−13	+1
1946	851	820	1671	564	928	1492	+287	−108

Source: Barlow (1978)

Note: Figures for eastern Europe were not included in western data before 1962; SR was produced there before 1937. Total NR tonnage data differ slightly from those in Fig. 1.4a.

* Restriction years, Stevenson Restriction Scheme.

† Restriction years, International Rubber Regulation Agreement.

1.3 Demand and supply in war and peace: 1914–45

1.3.1 Demand

There was an underlying upward trend in consumption of natural rubber throughout this period, despite a minor depression in 1921–22 and a major one in 1929–34 (Table 1.3). Demand, led by the motor vehicle tyre industry, primarily in the USA but increasingly elsewhere, rose from 121,000 t in 1914 to 1,127,000 t in 1940, thus on average more than doubling every decade.

Table 1.4 gives an idea of the broad categories of uses found for rubber in industrialized countries in 1941. Tyres dominated consumption, accounting normally for 70 per cent or more in the USA, but this was also a time when new uses, industrial and domestic, were constantly being found. This level of quantity and diversity was achieved through ceaseless scientific, technological and managerial progress in the 1920s, 1930s and war years, described in detail by the authors of the *History of the Rubber Industry*, edited by Schidrowitz and Dawson (1952).

Table 1.4   *USA rubber consumption 1940–41 ('000 t)*

Tyres and tyre products		*Non-tyre products*	
Tyres	416	Mechanical goods (belting hoses, tubes, etc.)	68
Repair materials and retreads	18	Latex foam	15
Inner tubes	56	Footwear, heels, soles	40
	490	Insulation	22
		Waterproof fabric	10
		Miscellaneous	49
			204
Grand total 694,000 t		*Percentage tyre products* 70%	

Source: United States Department of Commerce (1952) *Materials survey: rubber*

During this period there were interesting changes in the geography of demand; there was continued market domination by the USA whose share never fell below 50 per cent, but also a spread of consumption beyond USA and western Europe to Canada, Australia, Japan, India and Brazil. In the Second World War, when natural rubber production was reduced to a third of its peace-time level following the occupation of Malaya and NEI in 1942, the deficiency was made up by synthetic rubber (Table 1.3): mainly styrene-butadiene, produced primarily in Germany, USA and USSR. The synthetic rubbers are discussed in detail in section 1.4 covering post-war history.

1.3.2 Supply

The response to demand of the eastern producers of 'plantation rubber', or cultivated *Hevea brasiliensis*, was swift. Table 1.3 shows how natural rubber production rose from 125,000 t in 1914 to 1,504,000 t in 1941, thus by a factor of 12. The period 1914–45 saw the decline of wild rubber into insignificance, except in times of war, and the precipitate rise of cultivated rubber, whose success brought first over-production and then complex schemes to restrict output and maintain prices. Restriction intensified rivalries on two fronts: between the colonial 'duopolists', Britain and the Netherlands, for the greater share of *Hevea* output; and between the smallholders and the estates who, both under pressure, solved their problems in very different ways (Whitford 1931; Bauer 1948 p. 211).

In the absence of accurate predictions of demand, the producers relied on price as their guide and there was frequent overshooting and occasional undershooting of the demand target as shown in Table 1.3. Attempts to remedy price falls by restriction of supplies eventually succeeded, despite severe competition between producer countries for a share of the market. The restrictions and their effects are described in some detail because of their historic and current interest to producers.

Restriction of supply: the Stevenson Scheme and the International Rubber Agreement

Restriction of rubber supplies should be seen against the political and economic background of the 1920s and 1930s: an underlying upward trend in consumption which included rubber (Table 1.3), depressions in demand in 1920–22 and 1929–33, and a chronic oversupply of raw materials leading to price instability and inadequate income for producers. Numerous, sometimes desperate, attempts were made to raise commodity prices by restricting supplies of, for example: coffee, sugar, tea, wheat, tin. On the other hand, there were suggestions that producers and users should *jointly* control supplies. Sir Josiah Stamp, Director of the Bank of England, in his introduction to *The World's Struggle for Rubber 1905–1931* (Laurence[4] 1931) a powerful American polemic against rubber restriction 'for British Imperial advantage', emphasized (my italics) '*the possibilities of world planning and supply-control which are now filling our minds as the next step in large-scale industry and international relations*'.

The Stevenson Scheme 1922–28: an exercise in intervention

As prices collapsed, Sir James Stevenson's Committee in Britain devised a scheme of compulsory restriction by Malaya, Ceylon and the NEI. The NEI, unable to guarantee control of its smallholder production, refused to join. The other two countries went ahead.

Winston Churchill, Colonial Secretary, gave other reasons for the scheme besides price maintenance – to sustain the Empire and to help repayment of Britain's dollar war debt to the USA (Laurence 1931).

Restriction of supplies was achieved by imposing on the estates quotas which were *percentages of their previous production* called 'standard production'. In additional a pivotal price was fixed to give a 'reasonable' profit; only if this price were exceeded would the percentage and therefore total production be increased.

For smallholders, the standard was assessed from the age of their trees. Export by estates was checked through an 'export ledger' system. Smallholders were allotted coupons to accompany their produce. The 'exportable percentage' allowed was 60 per cent of the standard and later 50 per cent. As prices rose sharply the percentage was increased to 100 per cent in 1926. This over-optimistic move caused excess production and a price collapse in 1928 – at this stage the scheme was abandoned for a number of reasons. The British has lost an economic battle with the Dutch in NEI, whose share of production rose from 26 to 38 per cent in 1922–27 while Malaya's fell from 53 to 38 per cent. On the political front, Americans had complained of colonialist impositions, as had the legislature in Ceylon (Laurence 1931, quoting Hansard).

As derestriction took effect, production, especially by small-holders, assumed its seemingly natural course – perilously upward.

The International Rubber Regulation Agreement (IRRA) 1934–44

Prices crashed to ruinous levels in 1931–33, to around 3d. per pound on average and touching $1\frac{5}{8}$d., well below the Malayan estate production cost of 6d. per pound (Barlow 1978).

The Rubber Growers' Association moved towards renewed control in 1930, knowing that NEI now had to be included. Because of NEI objections that smallholder production was uncontrollable the IRRA was not agreed until 7 May 1934. It was operated, from 1 June, by representatives of the United Kingdom, NEI, Burma, Malaya, Ceylon, India, Sarawak, French Indo-China and Siam, with an advisory panel of manufacturers.

Following the Stevenson Scheme model, quotas were set according to a *percentage of previous production* for Malaya and Ceylon; but for NEI, Siam and Sarawak, and other countries, quotas were set *by negotiation*. This time no pivotal prices were fixed. New planting was forbidden (except for experimental purposes) to major producing countries up to 1938, whereafter 5 per cent new planting was permitted. Replanting was restricted to 20 per cent per annum of the existing agreed acreage but was unre-stricted from 1938. An innovation for most countries was an export tax or 'cess' to cover costs which included *statistical surveys* and *research*. Stocks held in producing countries were limited by the Agreement.

As expected, it was almost impossible to restrict output by NEI smallholders. A special export tax failed to restrain them. In 1935 the NEI quota was therefore raised and the resultant stimulating effect on NEI output can be appreciated in Fig. 1.4. NEI production rose to equal Malaya's by 1941.

In both Malaya and NEI the output of smallholders was badly underestimated and Bauer has pointed out that this constituted unfair protection of the commercial plantations. Malayan small-holders were allotted 38.9 per cent of national production in 1934 but their actual share had been 45 per cent (see Bauer 1948 and Table 1.7).

Exportable percentages, after a running-in period, were adjusted quarterly in full statistical knowledge of rates of absorption, accumulated stocks, and prices. Thus the initial percentage in 1934 was 100 per cent, reduced to 70–60 per cent in 1935. A short-lived boom in 1937 caused a temporary increase in the percentage. When prices sagged, as in 1938, the percentage was lowered to 45 per cent. As war-induced demand grew in 1941 the percentage was raised to 120 per cent.

The scheme seemed to work well, with an average deviation from permissible exports of only +0.4 per cent (McFadyean 1944). Price stability was achieved. The IRRA was made less necessary by the war, but was doggedly applied until 1944.

Effectiveness of the schemes

There is substantial agreement on the merits and demerits of the schemes between Bauer (1948) and Barlow (1978) who quoted other commentators. I have resumed the main arguments below.

1. The Stevenson Scheme (1922–28) raised prices, but with violent fluctuation which drew intense criticism from manufacturers. Essentially the scheme had to be dropped because it did not include a major producer, NEI.
2. The International Rubber Regulation Agreement (1934–44) worked smoothly and kept prices stable while producers lowered their costs. But Malaya lost further ground to NEI.
3. The schemes were unfair to smallholders, reducing their share of production.
4. The schemes sheltered high-cost estates and must therefore have delayed (though they did not prevent) introduction of efficient production methods.

Side-effects of restriction

1. The USA was extremely resistant to colonial control of raw materials, especially as demonstrated in the Stevenson Scheme.

Consequently, the USA made determined but not immediately rewarding attempts to 'grow its own rubber'; examples were plantations by Firestone in Liberia, Goodyear in the Philippines and Ford in Brazil.

2. The same desire to escape reliance on Malaya and NEI, increased by expectation of war, led the USA, USSR and Germany to hasten research on *synthetic* rubber which emerged in quantity in 1941 (Table 1.3).

3. A similar effect was produced by the development of reclaimed rubber in the 1920s and thereafter, especially in the USA.

Tonnage: the struggle for the lion's share of supply

By 1914 Malaya, with all the advantages described, had leapt ahead, producing 47,000 t of dry rubber from 473,000 ha or over a million acres, thus eclipsing the wild rubber industry of the Amazon. Figure 1.4a shows how, thereafter, the two leading producers, Malaya and

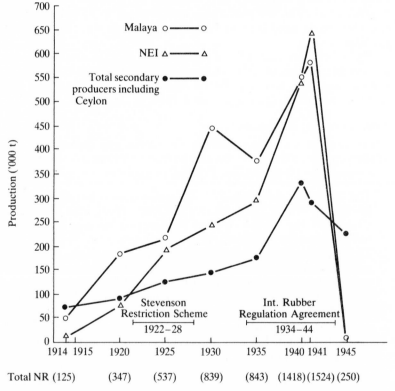

Fig. 1.4a Trends in natural rubber production 1914–45: the two main producers (*Sources:* McFadyean 1944; *Rubber Statistical Bulletin*)

NEI (including Dutch Borneo), met demand. Both restriction schemes improved the position of NEI, which finally overhauled Malayan output in 1941. Lesser producers together made a very important contribution (Fig. 1.4b), Ceylon being particularly consistent. Late but important arrivals were Indo-China, Siam, and (shown in Fig 1.4b under 'Other Asia') British North Borneo, Sarawak and Brunei. Elsewhere in Asia, India and Burma were early producers, but Indian annual production did not rise above the 20,000 t level until the 1950s.

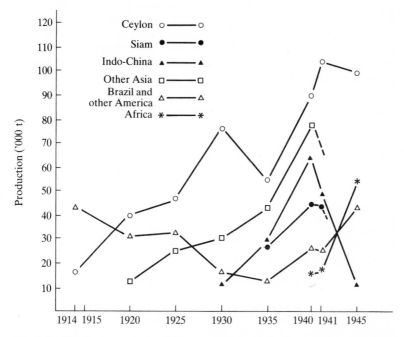

Fig. 1.4b Trends in natural rubber production 1914–45: secondary producers (*Sources:* McFadyean 1944; *Rubber Statistical Bulletin*)

Brazilian wild rubber, still highly regarded for its quality, reached a very low ebb of production in the 1930s, but again passed the 20,000 t level in the war years when other Latin American countries, notably Bolivia, Ecuador and Mexico, similarly raised production. In Brazil (Fig. 1.2), Ford's huge concessions for cultivated *Hevea* (at Fordlandia, Pará State, in 1927 and at Belterra in 1940) suffered severely from attacks of South American Leaf Blight, as did similar ventures elsewhere in Latin America (see Chs 4 and 10).

African production, mainly of wild rubber, passed its peak of

20,000 t in 1910 and sagged to around 3000 t in 1922. The early German plantings of Ceara rubber, *Manihot glaziovii*, mainly in Tanganyika, were soon abandoned and the pre-war plantings of *Hevea* elsewhere in Africa were not generally successful in a commercial sense (Whitford & Anthony 1926). The African revival, when it came in the late 1930s, was mainly due to the efforts of the Firestone Company in Liberia; the American company started operations in 1924 and by 1941 was producing 8,000 t/an. of which half was exported as latex. Similar, but smaller-scale, efforts were made by other companies in the Belgian Congo, Nigeria, the Cameroons and in French west and equatorial Africa. During the Second World War, wild rubber was again exported, the main sources being *Funtumia elastica, Landolphia* sp., *Clitandra* spp. supplemented, in Tanganyika, by derelict *Manihot* plantations (Wren 1947). An outline of African production, wild and cultivated, is given in Table 1.5.

Table 1.5   *An outline of wild and cultivated African production, 1914–46*

	('000 t)		
	1914	*1940*	*1946*
Liberia	—	7.34	21.0
Nigeria	0.17	2.95	11.6
Belgian Congo	2.28	1.25	5.0
French West Africa	1.07	0.61	n.a.
Cameroons*	0.34	1.68	n.a.
Other Africa[†]	3.60	2.52	10.3
	7.46	16.35	47.9

Sources: 1914, 1940 McFadyean (1944); 1946 International Rubber Study Group (1974)
* German 1914; French Mandate from 1922
[†] 1914, Angola and numerous other countries; 1940–46, main contributors Gold Coast, French Equatorial and West Africa

Planted areas: growth and restriction for smallholders and estates

The relative shares of estates and smallholdings, in the growing total area up to the Second World War, are outlined in Table 1.6. The data on areas, especially for the earlier period, were not totally reliable,[5] and there was special doubt about the area under small-holdings in NEI. Nevertheless, allowing for these reservations on data, a clear picture emerges.

While 1914–40 production rose by a factor of 12, the total planted area increased by a factor of 4 from 911,000 ha to 3,580,000 ha. During the Stevenson Scheme restriction period NEI's hectarage 'caught up' with Malaya's and thereafter their positions

Table 1.6 *Planted area of estates and smallholdings in the leading Asian producing countries 1914–40 ('000 ha), with percentage share of smallholdings*

	Malaya			NEI			Ceylon			All Asian countries		
	Estates	Smallholdings	Total	Estates	Smallholdings	Total	Estates	Smallholdings	Total	Estates	Smallholdings	Total
1914	272	200 (42%)	472	225	20 (8%)	243	91	26 (22%)	117	654	257 (28%)	911
1923	579	380 (39%)	959	381	123 (24%)	504	127	52 (29%)	179	1111	623 (36%)	1734
1929	739	463 (39%)	1202	546	727 (57%)	1273	114	163 (59%)	277	1410	1673 (54%)	3083
1934	813	516 (39%)	1329	599	695 (54%)	1294	140	103 (42%)	243	1763	1627 (48%)	3390
1940	842	546 (39%)	1389	633	729 (53%)	1360	145	113 (44%)	258	1851	1727 (48%)	3580

Other areas ('000 ha)

	Sarawak	British North Borneo	Indo-China	Siam	India/Burma
1929	105	48	98	60	68
1940	96	53	133	169	99

Sources: Malaya 1921–40, Barlow (1973); – Years 1914, 1923, Figart (1925); – 1934, 1940 (Bauer 1948) and von Saher and Verhaar (1979).
Note: The years are selected for availability and comparability of the statistics.

were roughly stabilized under the IRRA regime. From 1934 onwards there was little expansion of total planted hectarage. Malaya's smallholders held 42 per cent of national rubber area by 1914, but were reduced to 39 per cent in the restriction period. In NEI, smallholders advanced to 57 per cent of the national area during the Stevenson Scheme but fell back to 53 per cent during the later IRRA restrictions.

Production and yield of smallholdings and estates

Separate production figures for estates and smallholdings in this period are not readily accessible; they are available for Malaya and Indonesia only from 1929 (Table 1.7). The general picture, *for the period 1929–39* is one of static total estate production and declining smallholder output in Malaya while in NEI there was a fluctuating increase of total output, and an increase in the smallholders' share. In Malaya the recovery of rubber production began only with the onset of war, and was destined to be short-lived (Table 1.7).

The state of the natural-rubber-producing industry 1914–45

From a statistical point of view, rubber succeeded equally on highly capitalized estates and tiny family holdings. What human experiences lay beneath these statistics?

Smallholdings

The smallholders of Asia rivalled the estates in yield as well as in planted area, for in 1929 Malayan smallholdings were found in a survey to be *outyielding* estates (Malaya, Department of Agriculture 1934; Bauer 1948; and see Table 1.7). In fact, a distinctive form of smallholder exploitation, totally different from that on estates, had developed. It was carefully described for the US Department of Commerce by Figart (1925) and for the Rubber Manufacturers Association, New York, by Whitford (1929–34). Figart's maps of Asian producer countries are shown in Fig. 1.5.

   In *Malaya* villagers held a few hectares of rubber interspersed with, or adjoining, their fruit orchards or coconut groves, while outside the villages small capitalists bought somewhat larger blocks of land. Their system was one of minimum investment and maximum flexibility. On areas planted at 500–1000 trees/hectare, very much more densely than on estates, the villager adjusted tapping intensity to prices and to the seasonal labour demands of other crops. A proportion of the trees could be rested at will. In addition, compared with estates, the smallholders' densely planted more heavily shaded and little weeded holdings were less prone to soil erosion and the spread of lalang (*Imperata cylindrica*). The system worked and survived as the production figures have shown. Figart (1925) and also Whitford (1931 pp. 12,33), quoting the *India*

Table 1.7 *Production 1929–40 by estates and smallholdings in Malaya and NEI ('000 t dry rubber) with yields for Malaya (kg/mature ha)*

Year	Malaya						NEI			
	Estates		Smallholdings		Total	Smallholdings	Estates	Smallholdings	Total	Smallholdings
	'000 t	kg/ mature ha	'000 t	kg/ mature ha	'000 t	% of total production	'000 t	'000 t	'000 t	% of total production
1929	248.9	459	215.4	567	464.3	46	154.2	108.6	262.8	41
1930	237.8	424	218.4	562	456.2	48	153.6	90.5	244.1	37
1931	240.8	416	198.1	497	438.9	45	165.8	88.7	254.5	35
1932	241.8	408	179.8	434	421.6	43	150.9	61.4	212.3	29
1933	242.8	394	222.5	519	465.3	48	172.2	115.5	287.7	40
1934	264.2	358	220.5	443	484.7	45	192.8	185.9	378.7	49
1935	243.9	327	135.1	287	379.0	36	154.9	144.9	299.8	48
1936	234.7	306	131.0	269	365.7	36	161.7	151.4	313.1	48
1937	317.0	418	189.0	368	506.0	37	245.1	208.5	453.6	46
1938	247.9	322	114.8	226	362.7	32	175.0	146.6	321.6	45
1939	246.9	323	116.8	218	363.7	32	198.1	184.9	383.0	48
1940	336.9	463	216.1	426	553.0	39	282.7	266.3	549.0	49

Sources: Malaya, Barlow (1978), using Malayan official statistics, – NEI, Creutzberg (1975), quoted in von Saher and Verhaar (1979)

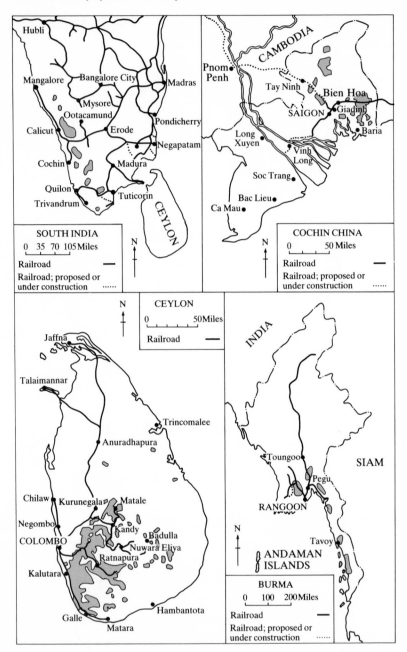

Fig. 1.5a   Rubber plantation, areas of some producing countries in 1925 (based on maps in 'The plantation rubber industry of the Middle East', (Figart 1925)

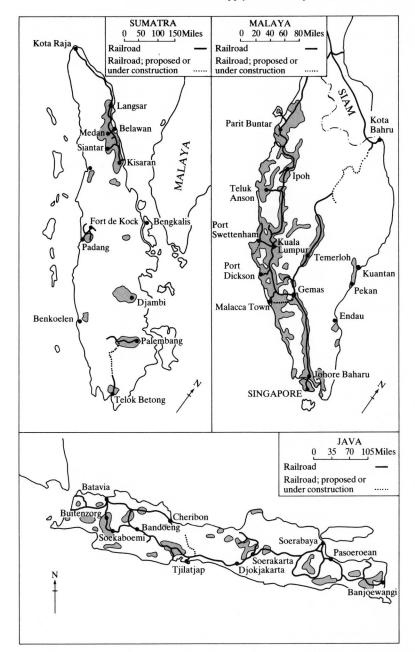

Fig. 1.5b   Rubber plantation areas of some producing countries in 1925 (continued)

Rubber Journal (Anon 1931), made it clear that plantation companies were aware of the smallholders' advantages, both in Malaya and NEI.

For *NEI*, interesting variations on the smallholder theme were described in Figart's survey (1925). In Palembang, Sumatra, the system closely resembled the Malayan one but was combined with forest gathering of gutta-percha and the gum-bearing wood jelutong. Land was free of charge and the Palembang Government gave technical advice to smallholders. In Djambi, further north in Sumatra, a sparse population became full-time rubber cultivators, employing non-family labour. Here, the wet, unrolled, alum-coagulated sheet, with a high content of foreign matter, was preserved under water and shipped to Singapore. When depressions came, these rubber-dependent producers of Djambi were unlikely to cut production – the reverse was more probable. NEI smallholders' production was not confined to Sumatra; in 1933 about 40 per cent of it came from *Dutch Borneo* (Whitford 1934), but very little from *Java*.

Estates

Estates, with their orderly appearance and disciplined labour forces, were similar everywhere and were controlled by similar European or American investors, the majority British or Dutch. Additional uniformity was imposed by the Agency Houses (mostly British and working from Singapore, see p. 14) who worked for the rubber companies and employed visiting agents to advise estates on management procedures. There was a small proportion, not easily quantifiable, of Asian-owned estates.

In the mid-1920s an estate typically averaged less than 500 ha in Malaya and Sumatra (Figart 1925; Barlow 1978). It was usually arranged in several blocks according to time of planting and contained a central processing factory. The trees were usually tapped alternate daily and the tapping pattern was commonly either a 'V-cut' or the familiar 'half-spiral' of today. Most estates were rigorously clean-weeded and had some root disease. Typically the end-product would be 'thin pale crepe' or ribbed smoked sheet. The manager would have one or two assistants, and returned to Europe on leave at five-yearly intervals.

Modern estate managers may be surprised by the extent of the imaginative literature devoted to their (then expatriate) life-style as it existed in the first half of this century. Among major writings were the short stories of Somerset Maugham, the planter Henri Fauconnier's *Malaisie* (The Soul of Malaya, 1931) awarded the *Prix Goncourt*, Pierre Boulle's *Le Sacrilège Malais* (Sacrilege in Malaya, 1959) and Vicki Baum's *Weeping Wood* (1943). In these writings the emphasis is on the uniqueness of the planter's life rather than everyday practicalities, but Boulle's comic description of time and motion studies on a rubber estate is an exception. Austrian-born Vicki Baum was fascinated by the romance, enterprise, and sometimes suffering, associated with the pioneer rubber industry. In contrast with his fictional counterpart, the real-life manager of the depression years must have been obsessed by business and career problems. If he kept his job, he was commonly asked to manage twice as much as before and lose a third of his salary. Price depressions and restriction brought every manner of cost reduction: abandonment of clean-weeding, reduction of maintenance, introduction of tapping rests and selective tapping. Commonly the least productive areas were put out of tapping completely. All this meant that labour would be reduced by a quarter or more.

Estate labour

Estate labour in the inter-war period has been described by Bauer (1948) for Malaya and in less detail for other countries; also by Barlow (1978). A feature of estate labour in Asia was that it was largely *immigrant* (or migrant) and the conditions of immigrant labour were roughly parallel in the different countries. The one major producing area using local village labour was *Java*.

In *Malaya*, as in *Ceylon*, labour from a relatively poor part of Asia, south India, was the backbone of the plantations. This was a flexible labour supply; it came when wanted, accepted downwardly adjustable wages, could be returned unwanted when prices fell – despite protests from the Indian Government. Some end-of-year employment levels of immigrant and other estate workers in Malaya are shown below. When prices collapsed in 1932 there were wholesale dismissals, repatriation and, additionally, a legal 40 per cent reduction in estate wages, to match a fall in cost of living (Table 1.8). In 1934 profits and labour demand rose and immigration recommenced.

In *Malaya*, as elsewhere, the estate labourer's health was poor in the early part of this period, when death rates were high,

Table 1.8 *Workers employed on Malayan estates*

	Indian	Chinese	Others	Total	
1929	205,000	42,000	11,000	258,000	all estates, mainly rubber
1932	104,000	35,000	6,000	145,000	
1933	111,000	39,000	10,000	160,000	
1934	179,000	86,000	37,000	302,000	rubber estates
1940	218,000	88,000	45,000	351,000	

Source: Bauer (1948)
Note: from 1934, a much improved recording system was introduced, and separate data were given for rubber estates.

especially in Negri Sembilan and Pahang. This high mortality was ascribed by contemporary commentators to the high incidence of malaria in areas of felled jungle (Figart 1925).

Death rates per thousand of estate workers in Malaya and Sumatra

	Federated Malay States (Indians)	*Sumatra*
1911	49–195 (not averaged)	46
1916	22	17
1917	22	14
1922	17	7

Source: Figart (1925)

In *Sumatra* a similar labour situation prevailed (Whitford 1934). Transmigrant labour was brought from Java, initially on an indenture system. In both Malaya and Sumatra there were typical twentieth-century enactments to protect workers from abuses. Malaya had requirements for estate-related hospitals (1909) and workers' living space (1916), a labour code (1923) and standard wages (1927). In Sumatra and Ceylon most of these provisions were paralleled.

Statistics of the industry

Especially in connection with restriction schemes, attention was focused on world markets and national outputs. The Rubber Growers' Association (RGA) collected statistics for issue in its *Bulletin* from 1924. In 1934 the work was transferred to the International Rubber Regulation Committee (IRRC) and from 1944, on the Committee's demise, to the newly formed International Rubber Study Group whose statistical surveys have become an essential element in rubber production and have furnished much of the statistical information discussed in this book. Statistical sources for the industry to 1945 are outlined in Appendix 1.1 (p. 540).

Research and development

Important investigations continued during this time both at government-sponsored institutes and on certain estates; major advances during the whole post-1914 period are discussed in section 1.5.

1.4 A share of the market: natural and synthetic rubber after 1945

1.4.1. Consumption

As shown by Table 1.9 total consumption of rubber by manufacturers has increased eightfold from 1946 to 1984, reflecting the

Table 1.9 *Consumption of rubber 1946 and 1984*

	USA	W. Europe	E. Europe	Japan	China	Others	Total ('000 t)
1946	1,056	268	175	20	10	118*	1,647
	64%	16%	10.5%	1.2%	0.5%	7%	
1984	2,793	2,661	3,005	1,440	620	2596†	13,115
	21%	20%	23%	11%	5%	19%	

Sources: International Rubber Study Group (1974, 1984); *Rubber Statistical Bulletin* (Sept. 1985)
* Largest consumer: Canada 40.
† Largest consumers: Brazil 312, Canada 310, India 270.

growth of the world economy and the geographical extension of the motor vehicle and tyre industries, from the USA and Western Europe to new areas of expansion around the world (Grilli *et al.* 1980).

Average growth of total rubber consumption (Table 1.10) for 1948–77 was high, 6.3 per cent per annum (Grilli *et al.* 1980), but since the oil crisis of the 1970s, and during the subsequent economic recession, growth has slackened. For 1985–2000, average annual growth rates of only 2.4–3.4 per cent are currently forecast, though with higher rates in newly industrializing countries (Jumpasut 1985).

Table 1.10 *Consumption and production of dry rubber 1946–85, with forecasts to 2000 ('000 t)*

Year	Consumption			Production			
	NR	SR	Total	NR	SR	Total	NR%
1946	590	1,057	1,647	867	970	1,837	47
1948	1,405	690	2,093	1,547	745	2,292	67
1950	1,707	835	2,542	1,887	787	2,674	70
1955	1,927	1,522	3,450	1,950	1,542	3,492	55
1960	2,115	2,347	4,462	2,035	2,447	4,482	45
1965	2,445	3,740	6,185	2,475	3,795	6,270	39
1970	2,990	5,635	8,625	3,125	5,890	9,015	35
1975	3,370	7,030	10,400	3,330	6,850	10,180	33
1980	3,760	8,685	12,445	3,845	8,645	12,490	31
1984	4,235	8,880	13,115	4,250	8,950	13,200	32
1985(est.)	4,350	9,005	13,355	4,300	8,985	13,285	33
Forecasts							
1990	5,236	9,981	15,217	4,890	—	15,217*	34
2000	6,267	13,048	19,315	6,286	—	19,315*	32

Sources: 1946–80 International Rubber Study Group (1974, 1984); 1981–85 *Rubber Statistical Bulletin*. Forecasts with various economic scenarios are discussed at the end of this chapter; those quoted here are from Jumpasut (1985) with a lower growth scenario (see Table 1.14)
* Consumption forecast.

Major uses of rubber: the importance of tyres for natural rubber

The main categories of uses today closely resemble those shown for 1940–41 in Table 1.4. Vehicle tyres have continued to dominate the picture, in 1986 absorbing over 50 per cent of all rubber consumed, but with wide regional variations. A further 10 per cent, approximately, is considered to be absorbed by other motor vehicle parts. Tyres are especially important for natural rubber, accounting for some 70 per cent of total NR output (International Rubber Study Group 1984). Within these categories of rubber goods, there have been radical changes: more rigorous quality specifications and novel methods of fabrication, especially in the construction of tyres. These changes and their influence on demand for SR and NR are discussed in section 1.4.3 on SR/NR competition.

1.4.2 Production of natural rubber

Data for overall NR production after 1945 were shown in Table 1.10, while country-by-country output is shown in Fig. 1.6 and very briefly resumed below for the six leading producers. Malaysia and Indonesia have continued to lead production, but apart from a brief period in the 1950s, Indonesian output has lagged behind Malaysia's. In recent years Thailand, India and China have made the most rapid progress.

Natural rubber production ('000 t)

	Malaysia	Indonesia	Thailand	China	India	Sri Lanka	World total
1945	438	178	25	n.a	16	96	867
1960	765	620	171	n.a	25	99	2035
1970	1269	815	287	n.a	90	159	3125
1984	1529(36%)	1115(26%)	629	190	184	142	4250

Sources: International Rubber Study Group (1974, 1984); *Rubber Statistical Bulletin* (1985)

Synthetic rubber had supplied three-quarters of the market at the end of the Second World War, but Malaysia and Indonesia rapidly re-established pre-war production levels and enabled natural rubber to raise its share of consumption from 36 per cent in 1946 to 67 per cent in 1950. Thereafter the oil-based SR industry, as described below, exploited new technical capabilities to increase its share at a rate that alarmed NR producers and caused them to revise their methods drastically.

Obstacles to natural rubber productivity: an excess of old trees

In the 1950s cultivated rubber began to feel the effects of age. The

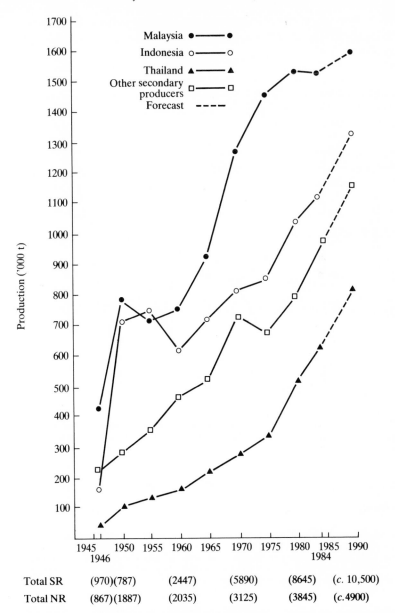

Total SR	(970)(787)	(2447)	(5890)	(8645)	(c. 10,500)
Total NR	(867)(1887)	(2035)	(3125)	(3845)	(c.4900)

Fig. 1.6a Natural rubber production 1946–84: the three major producers, with forecasts to 1990 (*Sources:* International Rubber Study Group 1974, 1984, *Rubber Statistical Bulletin* 1980–85; forecasts 1990, Jumpasut 1985)

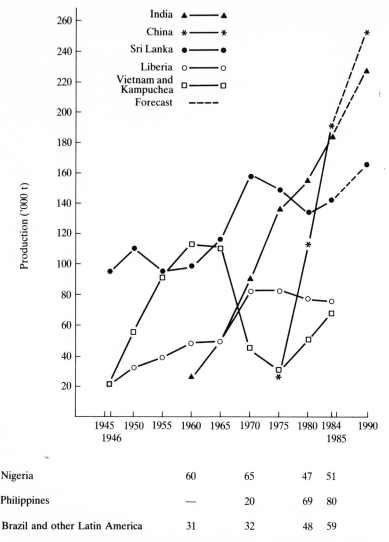

Nigeria	60	65	47 51
Philippines	—	20	69 80
Brazil and other Latin America	31	32	48 59

Fig. 1.6b   Natural rubber production 1946–84: secondary producers, with forecasts to 1990 (*Sources:* International Rubber Study Group 1974, 1984; *Rubber Statistical Bulletin* 1980–85; forecasts 1990, Jumpasut 1985

trees were often some 30 years old (50 per cent of estates in Malaya as late as 1954) and mostly of low-yielding stock. In Malaya calculated yields of mature trees actually fell from 1949–52: from 558 to 500 kg/ha/an. on estates and 432 to 389 kg/ha/an. on smallholdings (Barlow 1978 p. 444). Serious replanting with improved stock began in Malaya and Ceylon in the mid-1950s, with substantial government assistance and with excellent results (see later sections and Ch. 12).

Price difficulties for natural rubber producers

There were the usual sharp fluctuations of NR price partly due to the inability of perennial crop plantations to adjust short-term output in response to changes of demand (Goering 1982). Synthetic rubber's relative price stability (see Fig. 1.7) helped it to penetrate the market. Although there were peak natural rubber prices above trend, including the Korean War period 1950–51, the underlying movement of NR price was *downward*, under the pressure of increasingly low-cost SR, until the oil crisis of the early 1970s forced the SR, and therefore the NR, price to rise.

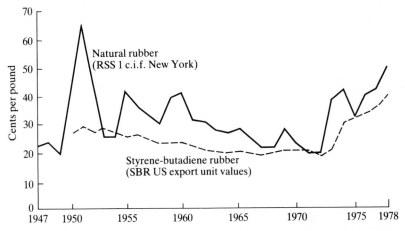

Fig. 1.7 Price trends of synthetic and natural rubber 1947–78 (After Grilli *et al.* 1980, Fig. 2–4)

International control of production and prices found little favour in the early post-war years, though individual producers and consumers[6] did accumulate stocks. More recent price fluctuations prompted prolonged discussions between producers and consumers resulting, in 1979, in the conclusion of the International Natural Rubber Agreement. Under this agreement the International Natural Rubber Organisation, based on Kuala Lumpur, can accumulate a buffer stock of up to 500,000 t in producing and consuming countries by purchases and sales at intervention prices related to a reference price. So far this has not been entirely successful in achieving the overall objectives of smoothing out price fluctuations and providing a minimum-price 'safety net' for smallholders. Following a fall in the price of NR in 1980–81, a buffer stock of

270,000 t was purchased. This stabilized the price within the agreed band for a time, but subsequently, despite an increase in the consumption of NR and despite further buying to increase the buffer stock to over 400,000 t, prices have continued to fall until March 1987, since when market conditions have provoked a sharp price rise and some disposal of stocks.

1.4.3 Competition between synthetic and natural rubber

In 1946 only four synthetic rubbers were important (see Table 1.11): the general purpose styrene-butadiene (SBR), accounting for over 85 per cent of all SR use in the USA, and three more specialized rubbers, IIR (butyl rubber), CR (Neoprene) and NBR (the nitrile group). In the next four years SBR lost ground to the restored supply of NR, which by 1950 had captured 67 per cent of the total market because of its lower cost and superiority for tyre making.

Percentage consumption of synthetic and natural rubber 1946–84

	1946	1950	1960	1965	1970	1975	1980	1983	1984
NR	36	67	47	39	34	32	30	32	32
SR	64	33	53	61	66	68	70	68	68

Sources: International Rubber Study Group (1974, 1984); *Rubber Statistical Bulletin* (1985)

After this brief post-war resurgence of NR the tide turned again; SR claimed over 40 per cent of consumption in 1951 and 70 per cent in 1978–79. The advance of SR was due to technical developments in the 1950s, new synthetics in the 1960s and the inability of the NR producers at the time to produce enough rubber at the right price. Changes in production technology improved SBR as a tyre rubber by giving it more resistance to abrasion and ageing with less heat build-up, while oil-extension gave better road grip and lower costs (McHale 1964; Allen 1972). From 1955 to 1970 SBR had a distinct price advantage (Fig. 1.7).

Of the new synthetics, *cis* 1,4 polyisoprene, IR or 'synthetic natural', almost replicated NR qualities but was not generally an important competitor because of its higher cost. A more serious threat came from polybutadiene, BR, which, when blended with SBR in tyres, improved resilience and reduced abrasion and heat build-up. Synthetic rubber also had economic advantages in nearness to industrial markets, and vertical integration of petrochemical/rubber-fabricating enterprises.

A further, important advantage of SR was that it was a clean, standardized, technically specialized, ready-to-mix product, packed

Table 1.11 Qualities of the principal general and special purpose synthetic rubbers compared with natural rubber

Code/Name	Chemical structure/composition	Characteristics	Examples of uses
*NR natural rubber	Cis 1,4 polyisoprene with carbonyl group occurring on molecule chain; plus 'impurities' (see Ch. 11)	Good dynamic qualities. Good tear strength and low heat build-up. Good green strength (cohesiveness) and tack (mixing capacity)	Various. Tyre carcasses, esp. radial. Heavy duty and high-speed tyres involving heat build-up
IR, polyisoprene	Cis 1,4 polyisoprene sold in 'high-cis' 95–96% pure and 'low-cis' forms	Similar to NR but in comparison has low strength at high temperatures and lacks green strength and tack	Can be substituted for NR but lack of green strength is a tyre-fabricating defect
*SBR, styrene-butadiene	Styrene-butadiene copolymer	Resembles NR but inferior in tear strength, heat build-up (greater than NR), resistance to fatigue, green strength. Superior in resilience, flexing strength, anti-abrasion qualities, resistance to wet road skidding	Various. Especially used in radial tyre treads in combination with BR (see below) which reduces defects
BR, polybutadiene	1,4 polybutadiene	When blended with SBR raises resilience and shock elasticity; and reduces abrasion, heat build-up	Mainly for tyre compounding with SBR
*IIR, butyl rubber	Isobutylene/butadiene copolymer	Resistance to permeation by gas and to ageing. Inferior resilience	Inner tubes of tyres. Gas proofing
*CR, Neoprene	Polychloroprene	Many properties similar to NR. Resistance to: flame, ozone, sunlight ageing, oil	Bridge pads. Industrial uses. Electric motor mountings. Belting. Cable. Adhesives
*NBR, nitrile rubbers	Butadiene/acrylonitrile copolymers	High resistance to oil, flame	Oil seals. Oil hoses
EPR, ethylene/propylene rubbers	Ethylene/propylene copolymers and terpolymers	Resistance to: heat, ageing, ozone. Copolymers hard to vulcanize	Automotive components. Cables

*In commercial use before 1960s.
Other elastomers and elastomer-like materials include: silicone rubber, resists chemicals and low temperatures; polyurethane/polyesters for synthetic foams; thermoplastics; thermoplastic rubber, easily moulded, numerous applications; polyvinyl chloride (PVC), for flooring, footwear, etc; EVAC ethylene-vinyl-acetate, oil and heat resistance; thiokol, a polysulphide, oil resistance.

in relatively small, easily handled, polythene-wrapped bales. By contrast, most NR was graded on visual features that bore little relation to technical properties and was marketed in large bales requiring treatment before being put into the manufacturer's mixer. Moves to counter these advantages of SR originated chiefly, but not entirely, in Malaysia where, as described in Chapter 11, commercial production of block rubbers developed rapidly and where the Standard Malaysian Rubber Scheme, which provided for grading on technical specifications and for standard methods of packing, was introduced in 1965. Later, similar schemes in other producing countries enabled NR to meet SR competition with a uniform product of defined characteristics.

The competitive situation was also markedly changed in 1971–74 by the quadrupling of the price of oil, on which SR is 70 per cent cost-dependent. This favourable situation for NR, described by Goering (1982), has since been maintained (although a sustained fall in oil prices could modify the situation) and has been further reinforced by the gradual adoption of radial tyres, which require NR in the side-walls and also to ensure adhesion between steel and rubber, and hence use 20–30 per cent more NR than non-radials. The overall effect of these changes has been that NR'S share of total consumption has steadied at an average of just over 30 per cent.

The geography of synthetic rubber: consumption and production

The advance of SR *consumption* was geographically uneven; SR began its post-war career in the USA and quite separately in the USSR – by 1960 SR accounted for nearly 70 per cent in the USA, 60 per cent in Eastern Europe and much less elsewhere. Since then, Eastern Europe has increasingly dispensed with NR, while the difference in percentage NR consumption between the USA and other industrialized countries has diminished. In 1984 the percentage of SR absorbed in some of the principal regions of consumption was: eastern Europe 87, USA 73, EEC 68, Japan 64, China 36 (*Rubber Statistical Bulletin* 1985). *Production* of SR, after initial concentration in the USA and USSR has now taken root in some 30 countries, including some newly industrializing ones, notably Brazil, Mexico, South Korea, and Taiwan.

The choice: synthetic or natural rubber?

Natural rubber's share of the market has stabilized over the last decade and has been predicted to remain fairly constant at around 30 per cent until the year 2000 (Grilli *et al.* 1980; Goering 1982; Smit 1982, 1984; Jumpasut 1985). The share is determined by technical and economic factors, whose anticipated interplay in practice is outlined at the end of this chapter in section 1.6 'Forecasts'.

Technical factors: tyres and non-tyres
Natural rubber's technical place in the world now relies primarily on its superior qualities over SBR for the fabrication of radial tyre carcasses, as opposed to treads. Is NR's role assured as provider of non-tread tyre components? The answer on economic and technical grounds is 'yes', but with reservations related to the technical potential of SBR and to the possibility of radical technical innovations in future. Buckler *et al.* (1980) have outlined experiments aimed at introducing the missing 'green strength' (cohesiveness during tyre fabrication) to SBR by providing cross-links between molecular chains either chemically or by irradiation, thus allowing substitution of SR for NR. Radical future developments could generate new kinds of tyres or tyre-making processes, with unpredictable effects on the role of NR or of rubber itself.

Beyond the domain of tyres, which occupies over 50 per cent of rubber consumption, exists the more shadowy (less recorded or analysed) region of non-tyres, where plastics, SR and NR compete and often replace one another. It is clear, however, that a proportion of the non-tyre sector, perhaps nearly a third, must be occupied by special synthetic rubbers, notably some of those used in automotive parts, while the remainder may be occupied by NR or SR, depending on cost. In effect, the portion of the non-tyre sector that uses general-purpose rubbers is being gradually reduced as the demand increases for more precisely specified products like *flexible, oil-resistant* pipelines for offshore oilfields, *flame-proof* conveyor belts, and so on. The more specialized SRs (all SRs other than SBR, BR and IR) occupied only 9 per cent of the total rubber market in 1965, but 15 per cent in 1979 (Bachman 1980). The further development of modified, special-purpose NRs could help to arrest this process.

Economic determinants
The growth of the world economy, GDP or GNP, determines total demand for rubber, and accelerated growth can influence NR : SR market share, since NR production may fail to keep pace with demand. A second, related, economic factor is oil price which, relayed through SR price, encourages or discourages NR production. The third, vital, economic factor affecting NR's share is the level of NR production itself (largely determined by previous planting). Allen, Thomas and Sekhar (1974) and Allen (1979) have claimed that NR has the 'techno-economic potential to take well over 40 per cent of the market'.

Regional and national differences affecting the NR share of rubber consumption
While technical and economic factors are the overall determinants

of the NR share of total consumption, deviations from this average share are very great, so that in forecasting NR consumption it is essential to know the 'preferences' for NR and SR in the consuming countries. In broad outline, these preferences (Smit 1984; Jumpasut 1985), depend on:

1. The growth of the motor vehicle industry in each country.
2. The relatively high proportion of commercial vehicles in developing and newly industrializing countries; commercial vehicle tyres absorb more rubber and proportionately more NR than do private vehicle tyres.
3. The proportion of tyre and non-tyre manufactures in a given country, since tyres consistently consume more NR than non-tyres.
4. The (gradually diminishing) national differences in the use and production of radial tyres.
5. Influence on consumption of NR production in some manufacturing countries such as India or China, or of strategic SR production in others. Such influences can result from political decisions to support or create domestic industries. Examples are the deliberate use of a high proportion of NR in India (79 per cent in 1984) and the reverse in Eastern Europe (13 per cent).

Conclusion

A picture has emerged in which, assuming modest world economic growth and assuming that NR retains its place in radial tyres, the SR industry seems unlikely to reduce the NR share of around 30 per cent of consumption (see section 1.6.2, production forecasts). Nevertheless, it would seem unwise for natural rubber producers to ignore certain unfavourable possibilities for NR, however remote they may seem:

1. Improvement of processing qualities of SBR which could make it more suitable for radial tyre carcasses.
2. Extension of the role of synthetic special-purpose rubbers.
3. A long-term decline in oil prices which could reduce the price of SR relative to NR (but nevertheless could also increase the world economic growth rate and therefore total rubber consumption).

1.4.4 The state of the natural-rubber-producing industry from 1945

Structure

The industry consists mainly of estates, defined as being over 40 ha, and traditional peasant smallholdings, which are usually of only a few hectares. In addition to these two, clearly defined major sectors, there has been since the late 1950s a considerable development of a hybrid type of enterprise in the form of government-sponsored land settlement schemes, where the settlers are provided with a range of facilities but the establishment and maintenance of their smallholdings is subject to some degree of control by the supervisors of the schemes. These schemes, which combine the size, investment and management style of estates with traditional-style individual or group land ownership, are described in Chapter 12. A less common variant of such schemes is the nuclear estate, in which an estate provides central processing facilities and other services to surrounding smallholdings.

The areas of estates and smallholdings are shown in Table 1.12, the figures for land settlement schemes being included under

Table 1.12   *Area under plantation rubber ('000 ha)*

Territory	Year	Estates*	Smallholdings	High-yielding	Ordinary	Grand total
Pen. Malaysia†	1982	465.5	1227.5	460.3	5.2	1693.0
Sarawak§	1982	2.1	201.6	101.8	101.9	203.7
Sabah§	1982	14.3	94.9	91.0	18.2	109.2
Indonesia	1977	465.6	1862.0	—	—	2327.5
Sri Lanka	1975	105.6	122.0	172.4	55.2	227.6
Thailand	1979	75.9	1442.1	356.7	1161.3	1518.0
Vietnam	1983	—	—	—	—	115.2
Kampuchea	1984	—	—	—	—	19.0
India	1983	65.6	266.1	299.7	32.0	331.7
Burma	1984	—	—	—	—	76.0
China	1982	—	—	—	—	453.0
Philippines	1983	57.7	—	—	—	57.7
Papua NG	1970	13.7	—	—	—	13.7
Brazil	1965	9.8	10.2	—	—	20.0
Mexico	1982	—	—	—	—	8.2
Zaire	1959	67.6	25.4	—	—	93.0
Liberia	1973	76.7	43.1	—	—	119.8
Nigeria	1982	—	—	—	—	185.0
Cameroon UR	1982	—	—	—	—	36.0
Central African Rep.	1965	—	—	—	—	1.2
Ivory Coast	1982	—	—	—	—	41.5
Ghana	1965	6.6	5.1	9.3	2.3	11.6
Total		1426.7	5300.1			7661.6

Source: *Rubber Statistical Bulletin* (Feb. 1986). Estate areas refer broadly to holdings of 40 ha (100 acres) or over
* Editor's note: separate estate/smallholder areas not given for Ivory Coast, Cameroon, Nigeria, Mexico, China, Vietnam and Kampuchea. The estate areas shown here occupy 18.6% of the grand total.
† 'High yield' and 'Ordinary' figures refer to estates only.
§ 'Smallholdings' figures include land schemes.

smallholdings. The estate area has declined in all the main producing countries, falling from an estimated 1,850,000 ha in 1940 (Table 1.6) to 1,430,000 recorded in 1984, or from over 50 per cent of the total of 3,580,000 ha of 1940 to barely 19 per cent of the more recent total of around 7,660,000 ha. Estates now predominate in only a few countries, notably in Liberia, Zaire, Ivory Coast and in the Philippines where the recently planted rubber is all on estates.

Estates

Probably more than three-quarters of the total world estate area is under 'large' estates of over 400 ha and ranging up to 12,000 ha (Barlow 1983). These are generally well supplied with capital and constitute the professionally managed, high-input and high-yield sector of the industry. In some countries (e.g. Malaysia, Sri Lanka) almost all large estates are now planted with high-yielding trees capable of giving at least 1000–1500 kg/ha, but in Indonesia a considerable proportion remains under low-yielding trees. Most of these large estates are no longer foreign-owned. In Indonesia, Sri Lanka and Vietnam many private estates have been nationalized, while in Malaysia local ownership has been achieved through the acquisition of majority shareholdings by government or by Malaysian citizens. There is also a considerable number of small estates, mostly ranging from 50 to 300 ha, which are owned and run by private companies. The capital invested in these is relatively small, the level of purchased inputs and the yields obtained are commonly lower than on the large estates, and the quality of rubber produced is poorer (Barlow 1983).

By providing employment, earning foreign currency, paying export duty and other taxes and by leading the way in innovation, the efficient estates have made a major contribution to national wealth and development. They still continue to do so, although their area under rubber and the number of workers employed on them has been declining. In peninsular Malaysia, rubber estate labour (50 per cent female) has fallen from 298,100 in 1947 to 197,000 in 1973 and 128,000 in 1984. As efficiency has improved and wages have increased, there has been a gradual increase in estate hectarage per worker, which was 3.1 in 1973 and 3.5 in 1984.

Smallholdings

The traditional smallholder section has also advanced, largely thanks to government intervention to raise returns through replanting or new planting with high-yielding trees and the use of better processing methods. As replanting involves a loss of income during the five to six years when the young trees cannot be tapped, some governments have made substantial payments per hectare to encourage smallholders to replace their old, low-yielding trees and have also provided advisory services (see Ch. 12). Despite these efforts and the resultant increase in yields, individually held small-holdings still make much less use than do the estates of techniques that involve extra outlay, notably yield stimulation and fertilizer application.

The socio-economic value of strengthening smallholder communities by increasing their income is emphasized by the governments of most countries with large rural populations (Johnston & Kilby 1975; Goering 1982; Barlow 1978, 1983). Rubber has played a large part in development plans aimed at raising rural incomes and benefiting national economies, especially because it can utilize moderately fertile, well-drained, gently sloping land less suitable for many food crops (Hansell 1981; Goering 1982; and Ch. 4). The merits and demerits of various types of government rehabilitation and development schemes are discussed in Chapter 12. In addition to investment in rubber developments by the governments of the main producing countries, finance has also been provided from international agencies. For example, during 1971–80 the World Bank assisted with 23 projects concerned with replanting or new planting of rubber estates and smallholdings in Cameroon, Ivory Coast, Liberia, Indonesia and Thailand (Goering 1982).

Performance of *Hevea* plantations, by country[7]

Malaysia

Malaysia shows an increasing predominance of smallholdings due to a decline of estate rubber areas, following some fragmentation and considerable replanting with more profitable oil-palm and other crops, while the smallholder area (including settlement schemes) has been greatly extended (Lim 1986). Peninsular Malaysian production, planted areas and yields are shown in Table 1.13.

Since the late 1970s, the decline of estate production and slower increase of smallholder output has been due to increased planting of oil-palms and cocoa instead of rubber on estates and settlement schemes, and to weak prices and higher labour costs, together with labour shortage in some areas.

East Malaysian production has declined: 60,000, 73,000 and 43,000 t in 1955, 1974 and 1984 respectively.

The relatively high performance of peninsular Malaysian plantations since the 1950s has been discussed and explained frequently (Barlow 1978; Smit 1980, Andaya & Andaya 1982). Apart from geographical and infrastructural advantages, the dominant factors have been institutional (well-organized and well-funded research programmes, advisory services and replanting/new planting/marketing organizations). Performance is reflected in the high yields of both estates and smallholdings (Table 1.13) compared with those of other

Table 1.13   *Areas, production and average yield of dry rubber per planted hectare of Malayan estates and smallholdings 1955–84*

	Estates			Smallholdings			
	ha	t	(kg/ha)	ha	t	(kg/ha)	Total area ('000 ha)
1955	816,000	358,000	(439)	722,000	291,000	(403)	1,538
1971	632,000	662,000	(1,048)	1,087,000	609,000	(560)	1,719
1976	533,000	652,000	(1,123)	1,148,000	885,000	(771)	1,681
1983	465,000*	540,000	(1,161)	1,228,000*	983,000§	(800)	1,693
1984	451,400†	508,000	(1,125)	1,249,300†	978,000§	(783)	1,700
1985**	432,000	499,000	(1,155)	1,238,000	920,000	(743)	1,670

Sources: 1955–76 Smit (1982); 1982–84 *Rubber Statistical Bulletin*
* Areas are 1982 figures from *Rubber Statistical Bulletin.*
† Areas from *Annual Report of the Rubber Research Institute of Malaysia* (1985).
§ Of which over 100,000 t from FELDA settlement schemes (see text).
** 1985 data are from Lim (1986) quoting official sources.

major producing countries. Average yields in kg per tapped ha in 1985 reported by Lim (1986) were 1490 and 920 for estates and smallholdings respectively.

Serious replanting of estates and smallholdings began in 1953, and today estates have nearly 100 per cent high-yielding material. Replanting was undertaken by the Rubber Industries Replanting Board which for smallholders became in 1973 the Rubber Industry Smallholders Development Authority (RISDA). The Federal Land Consolidation and Rehabilitation Authority (FELCRA) consolidates and replants derelict developments of the past. From 1956 *new planting* (settlement) schemes for smallholders were undertaken by the Federal Land Development Authority (FELDA). The Malaysian Rubber Development Corporation (MARDEC), with its processing centres, has since 1971 existed to produce high-quality rubber and concentrated latex from smallholder output. The Malaysian Rubber Research and Development Board (1983) has summarized smallholder replanting and new planting progress to 1980 ('000 ha):

Year	Individual holdings		Group planting schemes			Total
	Replanted*	Not yet replanted	FELDA	FELCRA	State schemes†	
1980	612.3	190.0	168.9	41.8	192.7	1205.7

* Includes 6000 ha of mini-estates, 1980.
† Block replanting, other than FELDA and FELCRA, organized by individual states and regional development agencies.

The organization of smallholder production in Malaysia and other producer countries is described and discussed in Chapter 12.

Indonesia

Briefly outlined below, estate and smallholder area and production show the decline in estate *vis-à-vis* smallholder area and the considerable advance of smallholder production, but these figures conceal sharp annual fluctuations in smallholder output.

	Estates		Smallholdings	
	ha	*t*	*ha*	*t*
1955	—	266,000	—	483,000
1964	507,000	223,000	1,599,000	425,000
1978	441,000	269,000	1,871,000	633,000
1984	n.a.	304,000	n.a.	810,000

Replanting schemes started later than in Malaysia (about 1965) and have been less successful than elsewhere in raising yields. Estates yields have been modest in the past, but are much better on public than on private estates. Public estates achieved a recorded average yield of 865 kg/tapped ha in 1985. Yield on smallholdings is only 300–400 kg/ha but plans have been developed for replanting, new planting and improved techniques (Indonesia Department of Trade and Cooperatives 1981). In particular, as explained in Chapter 12, the Directorate-General of Estates is operating a plan for block replanting and new planting of smallholdings which reached a level of 30,000 ha/an. in 1983.

Thailand

In Thailand, almost entirely under smallholdings, a drive to extend and improve smallholder rubber (see Ch. 12) also brought better yields, from 162 kg/ha/an. in 1966 to 330 kg/ha/an. in 1978, as well as bringing Thailand's rubber area to a level similar to Malaysia's. Recorded average yield per tapped ha was 671 kg in 1985.

Sri Lanka

Despite the small area of land available (decreased by urbanization) and the limited size of the smallholdings, mostly under 1 ha, Sri Lanka has increased output by some 50 per cent since 1945, thanks to a replanting programme that started in the mid-1950s. Estates occupy 46 per cent of the total area. Recorded average yield of all holdings was 818 kg/tapped ha in 1985.

India

Indian production has grown from a mere 25,200 t in 1960 to

183,000 t in 1984, mainly achieved by new planting. Three-quarters of the present rubber area is occupied by smallholders. Recorded national average yield in 1985 was higher than most at 902 kg/tapped ha. A special feature is that the whole product is absorbed by India's rubber manufacturing industry, which uses a very high proportion of natural rubber.

Africa

Africa contributed 5 per cent of world production in 1984, over 200,000 t, made up as follows for 1984:

	tonnes		tonnes
Liberia	76,000	Cameroon	17,600
Nigeria	51,000	Ghana	1,500
Ivory Coast	35,500	Central African Republic	1,200
Zaire	18,000		
		Total	200,800

Cameroon had an average yield of 890 kg/tapped ha in 1982. The Ivory Coast is a latecomer, its rubber area mainly under estates and its average yield, at 1240 kg/tapped ha, is one of the highest; quality is also remarkable, the total output being sold as technically classified rubber.

Brazil

In Brazil, since the 1930s, production has been around 20,000 t/an., mostly from wild sources. Large *Hevea* estates at Fordlandia, 1927–45, and Belterra, 1939–42, (Fig. 1.2) were abandoned because of the depredations of *Microcyclus ulei* (South American leaf blight or SALB).

The recent increase of output to 36,000 t in 1984 has been achieved through the efforts of SUDHEVEA, a government agency responsible for new planting — 16,000 ha distributed over Amazonian states during 1973–82 (Ribeiro 1980; de Barros *et al.* 1983). Since 1982 plantings have been extended to drier, more southerly states where the incidence of SALB is likely to be lower. SALB in Brazil is discussed in Chapters 3 (section 3.5.1.1), 4 and 10.

Other America

Production has increased over recent years, in 1984 reaching 23,000 t. Mexico has a programme to produce 45,000 t of *Hevea* rubber, and smaller quantities of guayule, by 1990 (Sanchez & Rivera 1980).

China

China, like India, has expanded rubber production very rapidly in recent years, to 190,000 t in 1984, so becoming the fourth highest producer in the world. The selection and breeding of wind-fast and cold-endurant clones, and the development of practices and choice of sites to avoid the ill effects of wind and cold (see Ch. 4), have been discussed by Huang and Zheng (1983). The best areas for rubber are Hainan Island and the Xishuangbanna Prefecture of Yunnan Province.

1.5 Research and development in natural rubber production

Research has enabled the plantation industry to become increasingly science-based and, latterly, has made a notable contribution to the success of NR in retaining a share of the market, by making reduced production costs possible and by modifying NR to meet the needs of manufacturers. Only a brief summary of some major advances since 1914 is given here and more information on most topics will be found in the relevant chapters. Ridley's very early research, crucial for the foundation of the industry, has been described in connection with the domestication of *Hevea* (p. 15).

Between 1914 and 1940 the Dutch workers in NEI were leaders in almost all fields of biological and agronomic research. The main experiment stations were the Proefstation voor Rubber at Buitenzorg (now Bogor) in Java and the Proefstation der Algemeen Vereeniging van Rubberplanters ter Oost Kust van Sumatra (AVROS) near Medan. An account of this work has been given by Dijkman (1951). In Malaya and Ceylon, investigations on rubber were carried out by the Departments of Agriculture until the formation of the Rubber Research Institutes in 1925 and 1927 respectively. Since 1945 the greatest contribution has come from the RRIM and the research units of major plantation companies in Malaysia, but important work has also continued in Indonesia, Ceylon (Sri Lanka), India, Thailand, Vietnam, Ivory Coast and China.

In the earlier years evaluation and testing of the product was chiefly at the Imperial Institute, London, and the government rubber service (Rijksrubberdienst), Delft. Later on, work on the chemistry and physics of natural rubber and on its manufacturing technology and uses was undertaken by Rubber Stichting, Amsterdam, by the Natural (now Malaysian) Rubber Producers' Research Association (MRPRA) founded 1938 in England and by the Institut Français du Caoutchouc founded in Paris in 1936, this last institute being currently linked with the Institut des Recherches sur le Caoutchouc en Afrique in the Ivory Coast. The importance of such institutions located in industrial countries should be stressed, for it is they (notably MRPRA in recent decades) who have worked with manufacturers to adapt NR to novel technologies, such as injection moulding, and have also developed new uses for NR, especially engineering applications which now absorb some 100,000 t/an. of natural rubber.

The coordination of *all* NR producer research, and consumption-related research on the producer's behalf, is clearly desirable and was made possible by the foundation of the International Rubber Research and Development Board reconstituted under its present name in 1960, but with a history dating from 1937. It currently comprises representatives of research institutes in 11 producer countries. The Board is especially active today in the fields of breeding/selection, physiology, exploitation, South American leaf blight, liquid natural rubber, industrial applications of NR, and personnel training.

1.5.1 Selection and breeding (Ch. 3)

Selection, breeding and clonal propagation of elite trees have been of the first importance and have been estimated by some to account for 70 per cent of all improvement in yield. Following observations during 1912–14 by Cramer in NEI, Lock in Ceylon and Whitby in Malaya on the very variable yield of unselected seedling trees, Cramer began selection in Java and initially recommended the use of open-pollinated seed from the best yielders in the older seedling plantations. Clonal propagation began after van Helden perfected a bud-grafting technique in 1916 and breeding began in NEI after Maas had worked out the technique for hand cross-pollinating. According to Dijkman (1951), there were clones in NEI in the 1930s yielding over 1700 kg/ha in the ninth year of tapping.

In Malaya, H. Gough began selection and multiplying of Prang Besar clones in the early 1920s and planted his first clonal seed garden in 1928. Breeding began at the RRIM in 1928 and, apart from the war years, has continued there and at Prang Besar. This has resulted in a steady improvement in yield; pre-war clones exceeded 2500 kg/ha and more recently Sekhar and Pee (1980) have reported clones reaching 5000 kg/ha on experimental plots.

Although the original introduction of seed into the East was subsequently supplemented by several further small importations, no serious effort was made to widen the genetic base until 1981. In that year the International Rubber Research and Development Board, acting for the producing countries, organized an expedition to collect *Hevea* germ plasm in various parts of Brazil. The material so obtained has been established in Brazil, Malaysia and the Ivory Coast for evaluation, conservation and, eventually, distribution to other countries (Ong *et al.* 1983).

1.5.2 Propagation (Ch. 6)

Since its introduction in 1917, the only significant innovation in budding techniques has been green budding on 2–6-month-old seedlings which was announced from North Borneo by Hurov (1961). Evidence of stock influence on scion was obtained from experiments started in NEI and Malaya in the 1930s, but so far there has been no significant production and use of superior rootstocks, although some monoclonal seedling stocks improved yields of certain scions in experiments. Stocks raised from cuttings in the 1960s were not root-firm, but clonal stocks may soon be produced by tissue culture. Double working, or crown budding, was first tried out in Java in 1926. As a result of trials in Malaya in the 1960s and 1970s crown budding has been found useful in some areas to confer wind and disease resistance on high-yielding clones lacking these qualities.

1.5.3 Fertilizers, cover plants, weed control (Chs. 7, 8)

The laborious, erosion-provoking practice of clean-weeding, previously considered essential on estates in NEI and Malaysia, went out of fashion as labour forces were reduced in the 1930s depression. For a time the use in young rubber of a ground cover of controlled natural regeneration was advocated (Haines 1934) but this was soon replaced by the use of leguminous cover crops which were shown to improve tree growth and lessen the need for nitrogenous fertilizers (see Ch. 7). For many years weed control was done by slashing or hoeing, but after the war the use of chemicals soon became general on estates. Much experimental work has been done from the 1950s onwards to test new herbicides and more refined methods of application.

Before the 1930s there was little use of fertilizers on the estates, where the trees were mostly low-yielding and often planted on virgin land. Nevertheless, significant responses to N or NPK had earlier been demonstrated in NEI, where some experiments were started in 1918 (Grantham 1924). Over 50 experiments were reported by Schmöle (1926) on estates in Sumatra and 58 in Java by Vollema (1931). Few properly planned experiments were done in Malaya in the early years. Haines and Flint (1931) reported 14 Malayan experiments which gave variable and inconclusive results. The first RRIM experiment, started in 1928, showed no significant responses from the immature rubber. Several series of experiments, started on estates in 1930–31 by Dunlop and Imperial Chemical Industries, showed widespread economic responses to N, and in some cases to NPK (Haines & Guest 1936). Modern fertilizer recommendations are based on information obtained over the years by field and pot experiments with fertilizers and cover crops, soil and leaf analyses and soil surveys (Ch. 7).

1.5.4 Exploitation (Ch. 9)

By far the most important advance has been the introduction of yield stimulants (Chapman 1951; Baptist & de Jonge 1955) which are capable of giving sustained yield increases varying greatly with clone but averaging about 30 per cent. Investigations of the anatomy of bark and latex vessels in NEI (Dijkman 1951), of the ultracytology of latex vessels (Andrews & Dickenson 1961), and of the mechanism of latex flow and vessel plugging (e.g. Boatman 1966), provided much new knowledge which helped in the formulation of new exploitation systems, but did not result in any major change in tapping technique. However, the availability of yield stimulants permitted the recent intoduction of puncture and 'micro-X' tapping (Tupy 1973g; Hashim & P'ng 1980). The most recent innovation, still in the experimental stage, is the semi-automation of tapping to reduce the need for skilled labour (Rubber Research Institute of Malaysia 1980a).

1.5.5 Diseases and pests (Ch. 10)

Following early investigations by Ridley in Singapore, work on pests and diseases continued in the main producing countries and comprehensive accounts of these maladies were produced by Petch (1921) in Ceylon, Steinmann (1925) in NEI and Sharples (1936) in Malaya. Knowledge of the pathogenic fungi and their life cycles was built up, but fully effective control measures often had to await the new chemical products of the post-war years. Considerable losses from root disease in the 1930s could have been reduced by control measures then known, but it was not until the early 1960s that the RRIM produced a comprehensive control system which, if properly implemented, could eliminate root disease in replantings.

1.5.6 Processing (Ch. 11)

In the inter-war period steady improvements in factory processing on estates enhanced the quality and lowered the cost of the main products – sheet and crepe. As new manufacturing uses were found for latex in the 1920s, processes for reducing its water content by evaporation, creaming or centrifuging were developed and in addition improved methods of preserving the concentrate were introduced. A number of methods for modifying natural rubber to meet manufactures' needs were devised but the demand for most of these modified rubbers has been relatively small. Pre-vulcanized 'superior processing' (SP) rubbers with improved extrusion and calendering properties have been made in small tonnages since

1955. Others have been: methyl methacrylate (MG) graft rubber, impact- and abrasion-resistant; oil-extended and carbon-black-containing rubbers; constant viscosity (CV) rubbers not subject to hardening in storage (Sekhar 1961). More recently, thermoplastic, epoxidized and other modified rubbers have been under study (Campbell *et al*, 1978; Baker *et al*. 1985; Sekhar & Pee 1980).

The two most important post-war advances occurred in the early 1960s. As already mentioned earlier in this chapter (p. 40), these were the development of block rubber processes and the introduction of grading to technical specifications to replace the old visual grading. These two developments improved the ability of NR to compete with SR and today international production of all technically specified rubbers, predominantly made in Malaya and Indonesia, exceeds 50 per cent of the NR total.

1.5.7 Conclusion

Since 1914, research and development, and the control of output, have enabled the NR industry to survive economic depression and, latterly, competition from synthetic materials. Today, refined methods of forecasting the consumption and production of elastomers make it more feasible than in the past to plan future NR output at levels that are likely to match demand.

1.6 Forecasts of consumption and production

We have seen the adverse effects of over- and underproduction on the competitive position of the NR industry in the past. Planners of future NR output now benefit from frequently revised forecasts of future consumption, production and price movements, based on estimates of economic growth and industrial demand and on detailed information about the capacities of the SR and NR industries. The available data on rubber demand and supply, the nature of NR supply, and NR's future share of the market given different economic scenarios, have been discussed comprehensively by Smit (1984) and are also outlined by Jumpasut (1985). Despite the usefulness of such econometric forecasts their inherent weakness must be kept in mind; essentially they extrapolate from past events and cannot allow for future technological innovations that could, for example, alter the relative market shares held by NR and SR (see p. 41 and Allen 1972, p. 171). Forecasts must be adjusted as new data become available and for this reason interested readers should consult the long- and short-term predictions revised regularly (at two-year intervals) by the International Rubber Study Group.

1.6.1 Consumption forecasts

The starting point for forecasting rubber consumption is usually the current prediction of national and world economic growth rates (GNP or GDP) with which industrial growth rates, vehicle and tyre consumption, petroleum prices (and therefore SR prices) will be linked.

As the economic future is invariably uncertain, economists commonly predict at least two GNP/GDP growth rates, sometimes combined with alternative predictions of energy supplies and prices.

Total rubber demand is considered to be largely predetermined by national income and to be relatively insensitive to expected variations in supplies and prices (Smit 1982 p. 346) because rubber typically forms a small part of the price of the end-product, such as a motor vehicle, and because SR over-capacity well into the future means that growth in rubber demands can easily be met.

Once the predicted economic growth rates are known, estimates are made of *tyre* and *non-tyre* production in each rubber-consuming country, with subdivisions into NR and SR shares determined by the technical factors and national preferences already described, (p. 41–2), and by relative NR : SR prices. The SR prices are derived from the future price of general-purpose SR (derived from petroleum prices) and the predicted surplus of NR supply over demand, but not the supply of SR which has excess capacity. At high NR prices the supply of IR (polyisoprene, or 'synthetic natural', see Table 1.11) would become an influential factor, while at very high prices even the supply of guayule rubber could become important (Bennett 1980). The modelling of the world's rubber economy ends with a summation of the world's NR and SR consumption for specific future years, with variations determined by the different economic growth scenarios. Models used for such calculations of consumption have been described by Grilli *et al.* (1980) for the World Bank, by Smit (1984) of the Economical and Social Institute, Amsterdam, and by Jumpasut (1985) for the International Rubber Study Group.

1.6.2 Production forecasts

Because *Hevea* is a slow-maturing crop and a long-term investment, prediction of production depends less on immediate reaction to demand and price than on the long-term situation of the crop in each producer country – the planted and replanted hectarage and its division between estates and smallholdings; estimated yields per ha (widely divergent between countries); age and yield status of trees; costs of production; alternative crops; and finally the perceived benefit of rubber to the economy, which influences government planning of development.

The final prediction of NR production is derived from calculated potential output (primarily from Indonesia, Malaysia and Thailand) and predicted price, which depends on the demand-supply position already explained. There is not much apparent disagreement between current forecasts of NR production (Table 1.14) given moderate economic growth. They are, however, much more modest than the forecasts made in the 1975–80 period which were based on relatively high presumed economic growth rates. Smit (1982 p. 244) reviewed forecasts of that period, among which those of the International Rubber Study Group, World Bank and FAO were

Table 1.14 *Examples of 1982–85 forecasts of total rubber and NR consumption and NR production ('000 t), with 1983–85 data for comparison*

	Jumpasut (1985) for IRSG				MRRDB** (1983)	
Year and economic growth rate*	Consumption Total	NR	(%)	Production NR	Total consumption	NR potential supply (%)
1983–85 DATA						
Actual 1983[†]	12,310	3,985	(32.4)	4,025		
Actual 1984[†]	13,115	4,235	(32.3)	4,250		
Estimate 1985[§]	13,355	4,350	(32.6)	4,300		
FORECASTS						
1990						
A (low)	15,217	5,236	(34.4)			
B (standard)				4,890	15,500	c.5,000
C (high)	16,282	5,603	(34.4)		±1,500	(32.3)
1995						
A (low)	17,142	6,081	(35.5)			
B (standard)				5,477		
C (high)	19,065	6,764	(35.5)			
2000						
A (low)	19,315	6,267	(32.4)			
B (standard)				6,286	18,000	c.6,000
C (high)	22,358	7,254	(32.4)		±3,000	(33)

* The scenarios are: Jumpasut A scenario 1, lower growth, lower NR consumption/stock ratio, lower oil price; C scenario 5, higher growth, lower NR consumption/stock ratio, lower oil price. (There are other Jumpasut scenarios, considered less likely to occur, where higher NR consumption/stock ratios combined with higher oil prices induce higher relative NR prices and a *lower* NR market share.)
[†] *Rubber Statistical Bulletin.*
[§] Jumpasut (1985).
** Malaysian Rubber Research and Development Board.
Note: for other forecasts, also with low, standard and high predicted growth scenarios, see also Smit (1984, 1985).

prominent; the earlier estimates typically showed NR approaching 6 million tonnes by 1990, 5–10 years earlier than has generally been projected more recently.

More recent estimates of production for 1990–1995–2000 are illustrated in Table 1.14. If restricted or moderate economic growth prevails, output of NR is expected to be 4.5 to 5 million tonnes in 1990 and about 6 million tonnes in 2000, thus meeting over 30 per cent of total consumption, although with some (on the whole less likely) economic growth scenarios NR's share could fall well below this level.

Notes

1. Thomas Hancock in 1824 found a simpler method – joining the freshly cut edges of thin sheet (Schurer 1959b). Eventually the rubber-tube problem was solved by the invention in 1845 by H. Bewley of the gutta-percha extruder, an idea later extended to rubber (Stern 1982).
2. Prain in the same letter similarly ruled out Cramer's suggestion that the Farris material from the Lower Amazon might have survived either in the East or at Kew.
3. Throughout this book the quoted weights of the plantation product refer to dry rubber unless otherwise stated. Quoted weights of wild rubber included some moisture and varying quantities of impurities.
4. James Cooper Laurence, Dean of the University of Minnesota. Former President of Faultless Rubber Company. Persuasive writer against economic controls when exercised by producers only.
5. In Malaya there was land registration for smallholders who were (and are) defined as holding under 100 acres or 41.4 ha. Elsewhere some early estimates of small-holder hectarage were back-calculations from output. Some official statistics recorded, not size of holding but 'European' or 'Asiatic' ownership, here very roughly equated with 'estates' and 'smallholdings' respectively.
6. One consumer, USA, stockpiled 1,250,000 t by 1954, and disposed of it by 1974; the effects are discussed by Barlow (1978).
7. Except where otherwise stated, 'recorded average yields' of dry rubber are those supplied by the International Rubber Study Group, London, in January 1987, and were the most recent figures available at that time.

Chapter 2

The botany of the rubber tree

C. C. Webster and E. C. Paardekooper

2.1 The genus *Hevea*

The genus *Hevea* is a member of the family Euphorbiaceae and comprises 10 species, of which the Para rubber tree, *H. brasiliensis*, is the only one planted commercially.

2.1.1 Area of origin

Species of the genus occur in the wild in an area which covers the whole of the Amazon basin and extends southwards into the foothills of the Matto Grosso region of Brazil and northwards into the upper part of the Orinoco basin, the lower slopes of the Guiana Highlands and parts of the lowlands of the Guianas. This large area covers parts of Brazil, Bolivia, Peru, Colombia, Ecuador, Venezuela, French Guiana, Surinam and Guyana – see Fig. 1.2.

Within this region there are two main types of climate. Most of the lowlands of the northern part of the Amazon basin, the upper Orinoco and the Guianas have a wet equatorial climate with rainfall considerably in excess of evaporation and generally more than 2000 mm/an. There is no dry season, only brief, less wet periods, and monthly temperature means are about 28 °C throughout the year. A monomodal, tropical monsoon type of climate, with one dry season of 3–5 months during the year, occurs in the foothill regions of the Amazon basin towards the Andes in the west and in the Matto Grosso in the south, but only in the latter area is the rainfall less than 1500 mm/an. A climate transitional between the two already mentioned, with annual precipitation exceeding evaporation, but with a somewhat dry spell of 1–3 months, occurs in a large area along the lower Amazon below Manaus, in the coastal area south of the delta and in the lowlands along the southern tributaries of the Amazon as far west as the Rio Madeira near Porto Vello.

2.1.2 Features common to all *Hevea* species

All the species examined are diploid with $2n = 36$ (Majumder 1964a; Ong 1979) and all can be crossed inter-specifically by artificial pollination. Some natural crossing occurs in the wild. The following features are common to all species. The leaves are trifoliate with long petioles bearing nectaries where they give rise to the entire, pinnately veined leaflets, which are folded back on emergence but subsequently assume various positions from reclinate through horizontal to nearly erect. Separate male and female flowers are borne in the same inflorescences with the females at the ends of the main branches of the panicles. Flowers of both sexes have a 5-lobed calyx borne on a basal disc of 5 free or united glands, but no petals. Male flowers have 5–10 stamens with their filaments united into a column and their anthers sessile. The ovary is 3-celled with one ovule per cell. The fruit is a trilocular capsule, usually containing 3 seeds, which, in all species except *H. spruceana* and *H. microphylla*, dehisces explosively to scatter the seeds. All species contain latex in all parts of the plant.

2.1.3 Species other than *H. brasiliensis*

Schultes has described the members of the genus in a number of papers, among which two (1970, 1977b) cover all the then known species and provide good illustrations. The following notes briefly mention the habitats and distinctive features of the species other than *H. brasiliensis*, which is subsequently described in more detail.

H. benthamiana Mueller–Argoviensis is a tree growing to a height of up to 27 m, usually swollen at the base of the trunk. Leaflets have a reddish-brown pubescence on the lower surface, are horizontal to slightly reclinate when mature, and the leaves are shed before the appearance of the inflorescences. The tree grows in periodically flooded alluvial areas along rivers, mainly in the northwestern parts of the Amazon basin and in the upper Orinoco basin. It gives a modest yield of pure white latex which contains rubber of as good a quality as some grades from *H. brasiliensis*.

H. camargoana N. C. Bastos; N. A. Rosa and C. Rosario, the most recently identified species of the genus, has only been collected on the island of Marajó in the Amazon delta. It is a small tree growing to a height of 2–12 m in woodlands fringing streams and swamps. Flowers of both sexes have a pink or red coloration at the base. It produces a small amount of white latex. There is evidence of natural hybridization with *H. brasiliensis* which grows in the same area (Pires 1981).

H. camporum Ducke is a dwarf species growing to no more than 2 m and found only in the region of the headwaters of the rivers Marmellos, Manicoré and Cucieré, where it grows in savannas. It produces a white latex.

H. guianensis Aublet, the type species described by Aublet from French Guiana in 1775, is a tree growing to 30–35 m with a cylindrical trunk, usually without branches for half its length, and bearing a compact crown. It is the only species having conspicuously erect mature leaves; these have obovate leaflets and mostly persist until the appearance of the inflorescences. It gives a small amount of yellowish latex which yields a rather inferior rubber. This species is found throughout the geographic range of the genus up to an altitude of 1100 m, usually growing on well-drained sites.

H. guianensis Aublet var. *lutea* (Spruce ex Benth.) Ducke and R. E. Schultes is similar to the above but is usually taller and has broadly lanceolate leaflets which are often pubescent along the midvein.

H. guianensis Aublet var. *marginata* (Ducke) Ducke is a much smaller tree than the two preceding, with strongly marginate leaves; it occurs in scattered, isolated localities and yields very little latex.

H. microphylla Ule is a tree growing up to 18 m with a slender trunk, slightly swollen at the base, having reddish bark and carrying a sparse and lax crown. The leaves fall before the appearance of the inflorescences. The female flowers have a much swollen, bell-shaped receptacle, which is lacking in other species. The pyramidal capsule is yellow with green stripes and a red tip when ripe; it has leathery, not woody, valves which open slowly and do not eject the seed. The white, watery latex contains little rubber. The trees grow along river banks subject to flooding and are abundant in the Rio Negro area of Brazil and Colombia.

H. nitida Mart. ex Mueller–Argoviensis var. *nitida* usually forms a small to medium-sized tree, but occasionally grows as tall as 27 m. It has a cylindrical trunk with dark red bark. The reclinate leaves are bright green with a very shiny upper surface and tend to persist after the inflorescences have appeared. When mature, the fruit is purplish with a red tip. The white, tacky latex contains much resin, but little rubber, and acts as an anticoagulant when added to the latex of *H. guianensis, H. benthamiana* or *H. brasiliensis*. This species occurs sporadically over a wide area in the western half of the Amazon basin, usually growing in forests on well-drained soils but sometimes found on poorly drained sites.

H. nitida Mart. var. *toxicodendroides* (R. E. Schultes and Vinton) is so named because its glossy trifoliate leaves are suggestive of the poison ivy of North America (*Rhus toxicodendron*). It is a curious, dwarf, shrubby variety, not exceeding 2 m in height, found growing in isolated sites on hills in Colombia, but maintaining its dwarf, shrubby habit when grown on good soil under high rainfall (Schultes 1977b). Its thick, white latex yields a fairly good quality rubber, in contrast to *H. nitida*, above.

H. pauciflora (Spruce ex Benth.) Mueller–Argoviensis var. *pauciflora* is a large tree growing to over 27 m and having a cylindrical trunk. with dark brown bark. The large, blunt-tipped leaflets become leathery when mature and persist until after the appearance of inflorescences. Mature capsules are purplish with reddish tips and contain relatively large seeds. The white latex is very resinous, rapidly oxidizes and contains little rubber. This variety of the species occurs relatively rarely in the Rio Negro area, usually on well-drained sites.

H. pauciflora (Spruce ex Benth.) Mueller–Argoviensis var. *coriacea* Ducke is a smaller tree than that just described above, from which it differs in leaf texture, bark colour and smaller seed size. It is much the commoner of the two varieties, being sporadically distributed in the western and south-western parts of the Amazon basin in Brazil, Colombia and Peru and also in Venezuela and the Guianas. It usually grows on well-drained slopes of low hillsides. It was formerly known as *H. confusa*.

H. rigidifolia (Spruce ex Benth.) Mueller–Argoviensis is a medium-sized tree, usually growing to 12–18 m, with a slender trunk having greyish-tawny bark. The leaves are very thick and leathery, marginate, reclinate and persisting until the appearance of the inflorescences. The white, resinous latex does not yield a rubber of commercially acceptable quality. This is a relatively rare species restricted to the upper Rio Negro basin where it is usually found on well-drained sites.

H. spruceana (Benth.) Mueller–Argoviensis [syn. *H. discolor* (Spruce ex Benth.) Muel–Arg.] grows to as much as 27 m and has a trunk which is much swollen at the base and carries a dense crown. The leaves are reclinate, pubescent on the underside and persist until the appearance of the inflorescences. The flowers are large and reddish-purple; with those of *H. camargoana*, they contrast with the flowers of all other species of the genus which are green or yellow. The capsules and seeds are the largest in the genus and the capsule opens slowly, usually not ejecting the seeds. The white, watery latex yields little rubber. The tree grows on low, periodically flooded river banks and islands along the Amazon and its tributaries from the borders of Peru and Colombia to the delta.

2.2 The rubber tree, *Hevea brasiliensis* (Willd. ex A. de Juss) Mueller–Argoviensis

In South America this species occurs naturally over about half the range of the genus. It mainly occupies the region south of the Amazon, extending to the Acre, Matto Grosso and Parana areas of Brazil and into parts of Bolivia and Peru, but it is also found north of the Amazon to the west of Manaus as far as the extreme south of Colombia. It is usually found on well-drained sites but can grow in areas subject to brief, shallow flooding. According to Schultes (1977b), geographical or ecological strains can be recognized within the species, but it is almost impossible botanically to delimit them because there are all shades of inter-gradation. However, there seem to be significant differences between some strains in the amount and quality of the latex that they produce.

The seed which formed the foundation of the rubber industry of the East was collected by Wickham in a very small area at the confluence of the rivers Tapajóz and Amazon, near the town of Santarém, and was clearly not a representative sample of the germplasm of the species. Indeed, it has been suggested that it might have included seed from inter-specific hybrids, but it has now been reliably established that it was genetically pure *H. brasiliensis*. However, Schultes (1977b) has pointed out that the trees in the Santarém locality did not produce the best quality rubber, which was known as 'Acre fino' and was collected from trees growing on well-drained soils in the south-western part of the Amazon basin.

Traditionally the Amazonian tappers distinguished between three types of tree according to the colour of the hard bark beneath the periderm, claiming that those with black inner bark yielded more latex and better quality rubber than those with white or red. Schultes (1977b) studied this in an area of south-eastern Colombia and confirmed the existence of trees of the three bark types, with associated differences in leaf shape and seed size, which he usually found under different ecological conditions. The black bark type, which occurred in the areas most subject to flooding, had more narrowly lanceolate leaves, smaller seeds and thicker, softer bark than the other two and yielded a thicker latex containing more rubber. The other two types grew on better-drained sites, subject only to light and transient flooding, and had harder, thinner bark which gave a more watery latex.

Hevea brasiliensis is a quick-growing, erect tree with a straight trunk and bark which is usually grey and fairly smooth, but varies somewhat in both colour and surface. It is the tallest species of the genus and in the wild trees may grow to over 40 m and live for over 100 years, but in plantations they rarely exceed 25 m because growth is reduced by tapping and because they are usually replanted after 25–35 years when yields fall to an uneconomic level. On both estates and smallholdings rubber trees are either grown as clones of buddings on seedling rootstocks or as seedlings, the latter nowadays being raised from 'clonal' seed produced by natural crossing between several selected clones planted in an isolated seed garden. Details about these planting materials are given in Chapter 6 and Appendix 2.1.

Fig. 2.1 *Hevea brasiliensis*. A, shoot with dehiscing fruit ($\times\frac{1}{2}$); B, inflorescence ($\times\frac{1}{2}$); C, male flower cut open ($\times3$); D, female flower in longitudinal section ($\times3$); E1, E2, fruits ($\times\frac{1}{4}$); F, seed ($\times\frac{1}{2}$) (After Purseglove 1968, Fig. 25)

2.2.1 Seed

The seeds are large (usually weighing between 3.5 and 6.0 g) and ovoid in shape with the ventral surface slightly flattened. The seed coat, or testa, is hard and shiny, brown or grey-brown with numerous darker mottles or streaks on the dorsal surface, but few or none on the ventral side. It is possible to identify the female parent of a seed by its markings and its shape because the testa is maternal tissue and the shape of the seed is determined by the pressure exerted by the fruit capsule during its development. These features provide a reliable means of identifying clones of buddings and the female parent of clonal seed. The hilum can be seen as a shallow, approximately circular depression on the ventral surface and the micropyle is adjacent to it. A papery integument lines the inside of the testa and encloses the endosperm, which fills the seed. The embryo is situated in the middle of the endosperm with the radicle pointing towards the micropyle. The two white, veined cotyledons are pressed against the endosperm and enclose the plumular end of the axis of the embryo. The endosperm, which forms 50–60 per cent of the weight of the seed, contains a semi-drying oil which can be used as a rather poor substitute for linseed oil. Unless special precautions are taken (see section 6.2.2) the seeds lose viability and deteriorate rapidly on storage, evolving hydrocyanic acid as a result of enzymatic hydrolysis of the cyano-genetic glucoside, linamarin, which they contain.

2.2.2 Germination

Germination is hypogeal and mostly occurs within 3–25 days after sowing. The radicle breaks through the testa at the hilar depression and very soon produces a ring of primordia which rapidly grow out as lateral roots. Further development of the radicle is briefly delayed until the laterals have grown about 2 cm, after which it grows rapidly to form the primary tap-root. The cotyledons remain permanently within the seed but their stalks elongate and emerge with the plumule between them. Initially, the emerging plumule is bent like an inverted U, but it soon withdraws its tip from within the seed, straightens up and grows vigorously. It produces the first pair of leaves about eight days after germination has started. Subsequently an internode is grown and the first flush of three leaves is produced above it. At the same time further lateral roots have grown out below the first ring of laterals on the primary tap-root, which is continuing rapid growth and is well provided with root hairs near its tip (Gomez 1982).

2.2.3 Periodic growth

Growth in length of the stem is discontinuous, being characterized by rapid elongation of an internode towards the end of which a cluster of leaves is produced, followed by a rest period at a node during which scale leaves develop around the terminal bud. By repetition of this sequence, leaves are produced in tiers, storeys or whorls separated by lengths of bare stem. (It should be noted that the term 'node' is not here employed according to strict botanical usage, but designates a position on the stem where the terminal bud passed through a resting stage, while 'internode' refers to a section of stem grown between two resting stages.) Young scions of buddings elongate internodes for 2–3 weeks and rest at nodes for about the same period; vigorous clones take less time than weaker ones to grow internodes, but both spend a similar time resting. Although the elongation of stems is intermittent, their girth increases continuously.

2.2.4 Leaves

On emergence of the spirally arranged, trifoliate leaves, the laminae hang downwards approximately parallel to the petioles and are reddish or bronze in colour. As they grow older they become green and make an increasing angle with the petioles until they reach an angle of about 180°, in which position they then remain until they senesce. The mature laminae are shiny dark green on their upper surfaces and a paler, glaucous green below. The leaves are glabrous with long petioles (about 15 cm) which bear three extra-floral nectaries at the point where they give rise to the leaflets. Nectar is only secreted on the new flush of leaves during flowering. The leaflets, which have short petiolules, are elliptic or obovate with the base acute and the apex acuminate; they have entire margins and pinnate venation.

The upper epidermis and palisade parenchyma of the leaflets are single layers of cells below which there are several layers of spongy parenchyma and the single layer of the lower epidermis. The latter has many ridge-like appendages and a reticulate cuticle (Rao 1963). Stomata are only present in the lower epidermis. Senanayake and Samaranayake (1970) examined 25 clones and found that these showed a wide variation in stomatal density from 22,000 to 38,000 per cm^2, but they found no significant relationship between stomatal density and latex yield. Gomez and Hamzah (1980) investigated variation in leaf morphology and anatomy in 11 clones. They found significant differences between clones in stomatal density (which ranged from 28,000 to 37,000 per cm^2), cell number in the upper epidermis, thickness of palisade and spongy layers and of leaf, and mean number of cells per unit length of palisade and spongy sections. There were no significant differences between clones as to mean surface area per leaflet and no obvious relations between any of the features examined and the yield or vigour of the clones.

The latex vessels of the leaves are mainly confined to the phloem but some make their way up through the palisade cells and end blindly at the epidermis. By the time that the leaves attain their final green colour and posture the latex vessels are completely blocked at the base of the petioles by callose which is deposited at the time of secondary thickening of the petioles (Spencer 1939). This must prevent the transport of latex or photosynthate from the leaves to the trunk in the latex vessels.

In order to characterize the physiological age of young leaves, Samsuddin *et al.* (1978) developed a 'leaf blade class' (LBC) concept based on the gradual change in the angle which the laminae of growing leaves make with the petioles. For determination of the LBC, the maximum angle of 180° was divided into 9 parts of 20°, with LBC 1 corresponding to an angle of 0°–20° and LBC 9 to 160°–180°, the measurement recorded being the angle which the middle leaflet made with the petiole. It was observed that the rate of increase in the angle was linearly related to the age of the leaf in days. In one trial the time taken by five clonal seedling families to reach LBC 9 ranged from 32.5 to 37.0 days with a mean of 35.0 days. Thus, the LBC can be used to indicate the physiological age of the leaf up to about 35 days, when it is considered to be morphologically and physiologically mature. However, the relation between LBC and age in days is influenced by environmental factors, such as water and nutrient supply, relative humidity and temperature, and can only be established by experiment in the relevant locality.

Photosynthetic rate

Samsuddin and Impens (1979a, b) measured changes in the net photosynthetic rate with leaf age, using attached leaves of clonal seedlings and buddings grown in controlled environment chambers, and found that the changes were similar with all the clones tested. Photosynthesis did not start for at least a week after leaf emergence; the rate then increased up to LBC 6–8 (about 23–30 days from emergence) and thereafter fell slowly to what appeared to be a constant rate at 50–60 days. The mean maximum photosynthetic rate of four clones was 0.55 mg CO_2/m/s, falling to 0.40 at 50 days. As these observations were continued for only 2 months, they could not indicate further changes that might occur during the period of nearly a year that the leaves remain on the trees, but the authors suggested that the maximum photosynthetic rate is probably sustained for only a short period during the life of the leaf.

These experiments in controlled environment appeared to show differences between clones in net photosynthetic rate measured at LBC 9 (Samsuddin & Impens 1978a, b; 1979c). But measurements made on leaves (fully exposed to the sun) of 20 *H. brasiliensis* clones and 5 other *Hevea* spp. in the field in Malaysia failed to establish significant differences between clones or species. Maximum photosynthetic rates recorded for the clones in the field, ranging from 0.36–1.14 mg $CO^2/m/s$, were higher than those recorded under controlled environment (Ceulemans *et al.* 1984).

There has been some investigation of the possibility of using photosynthetic rate as a criterion in early selection for yield. Samsuddin *et al.* (1985) made measurements in a hand-pollinated seedling population, derived from a breeding programme and grown in a nursery, but found no significant correlation between photosynthetic rate and nursery yield (early test tapping) or girth measurements. On the other hand, measurements of photosynthetic rate made on mature leaves of young buddings of 23 clones grown in a controlled environment chamber were found to be significantly and positively correlated with the mean yields over five years of tapping of the same clones in a number of large-scale field trials in Malaysia (Samsuddin *et al.* 1987). However, as the correlation coefficient was low (0.469) and only significant at p = 0.05, these workers concluded that further investigations are needed before it can be decided whether the determination of photosynthetic rates in growth chambers is likely to be useful as a criterion for early culling of low yielders in a breeding programme. Although the leaf area index of most commercial clones in mature stands may be similar, it is quite possible that differences between clones in crown architecture, and in the partition of assimilates between growth and latex production, may preclude any direct relationship between the photosynthetic rate of single leaves and the yield of mature trees.

Wintering

Trees older than 3 or 4 years are subject to 'wintering', which is the term used to describe the annual shedding of senescent leaves which renders the trees wholly or partly leafless for a short period. Leaf fall is normally followed within 2 weeks by the terminal buds bursting and by the expansion of new leaves within a further week. Latex yields usually fall slightly at the onset of leaf fall and are more markedly reduced during refoliation. Wintering is induced by a season of drier weather and is much influenced by the rainfall occurring at this time. Where there is a marked dry season the duration of wintering tends to be short; pretty well all the leaves fall in a short time, refoliation is rapidly completed before wetter weather resumes and the reduction in yield is usually not great. On the other hand, where there is no pronounced dry season during the year, but only a season of lower and more infrequent rainfall, leaf fall occurs gradually, trees are never completely leafless, production of new leaves is slower and yield reduction greater. Furthermore, as explained on p. 136, where refoliation is not completed before the arrival of wetter weather, leaf diseases and pests are liable to damage the young foliage, causing secondary leaf fall and more prolonged yield depression. In places where there is more than one drier season during the year there may be a second, usually partial, leaf change.

There are marked differences between clones in wintering behaviour. A few tend to shed and replace part of their foliage simultaneously over a relatively long period and may thus show no very obvious signs of wintering, while at the other extreme some become completely leafless for a time. The majority are intermediate between these extremes. Clones also vary considerably in the extent to which they suffer yield depression during refoliation.

2.2.5 Flowers

Inflorescences are borne in the axils of the basal leaves of the new shoots that grow out after wintering. As with wintering, there may be two flowering seasons during the year. For example, over much of Malaysia the main flowering season occurs in the period February to April, following wintering in January and February, and there is a lesser flowering season during September to October after secondary wintering in August to September.

The inflorescence is a many-branched, shortly pubescent panicle bearing flowers of both sexes. The larger female flowers are borne at the end of the central axis and main branches of the inflorescence with the smaller and much more numerous male flowers on other parts of the panicle. Flowers of both sexes are shortly stalked, scented, and have a greenish-yellow, bell-shaped calyx with 5 triangular lobes, but no petals. The male flowers, about 5 mm long, have a slender, white staminal column bearing 10 very small, sessile anthers in 2 rings of 5, one above the other. The female flowers are about 8 mm long and have a green, disc-like base surmounted by a 3-celled, shortly pubescent, conical ovary with 3 short, white, sticky, sessile stigmas. Flowering usually lasts about two weeks on any one tree. In an inflorescence, some male flowers open first and fall after one day; then the female flowers open and remain open for 3–5 days, after which the remainder of the male flowers open. Thus there is incomplete protandry, and a relatively large proportion of cross-pollination usually occurs in a plantation.

2.2.6 Pollination

Pollination is mainly by insects. Wind appears to play little or no part since no pollen was collected in spore traps placed within 15 m of heavily flowering trees, and inflorescences enclosed in insect-proof bags did not set seeds. In Malaysia over 30 species of insects have been seen to visit the flowers, but it is likely that pollination is almost entirely effected by midges and, to a minor extent, by thrips. In Puerto Rico, Brazil and Malaysia, Ceratopogonid midges, which are very small, hairy, and have a capacity for sustained flight, were found to be the most important pollinators, while several species of thrips, which are not very active fliers, probably play a minor role (Warmke 1951, 1952; Rao 1961).

The pollen grains are triangular, measuring about 35–40 h on each side, and their surface is sticky. The percentage viability of pollen grains can be as high as 90, but on average is only about 50 (Ghandimathi & Yeang 1984). In tests on artificial media, Majumder (1964b) found no differences between a large number of clones in percentage germination of healthy, well-filled pollen grains. He observed that pollen taken from male flowers during, or soon after, rain gave a smaller germination percentage than pollen from dry flowers.

2.2.7 Fruiting

Fertilization takes place within 24 hours after pollination (Majumder 1964b), and unfertilized female flowers quickly wither and fall. Only a small proportion of the female flowers set fruit and the majority of the latter are soon shed. Warmke (1951, 1952) reported that in Puerto Rico 5 per cent or less of the female flowers matured fruit after natural pollination in the field, but in Malaysia Rao (1961) found it to be only 0.3–1.6 per cent. Warmke did not think that pollination was inadequate in Puerto Rico. Observations by Ghandimathi and Yeang (1984) on the *artificial* pollination technique used by plant breeders in Malaysia indicate that it may not transfer pollen satisfactorily and that this may partly account for the low fruit set usually obtained. They noted that mature fruits containing less than three seeds are rare, suggesting that all three ovules of the female flower must form seeds for the fruit to develop to maturity. If so, then for successful fruit development pollen must germinate on all three stigmas of a flower, but in experiments on the artificial pollination technique they found that 70 per cent of the flowers had no pollen at all on at least one stigma.

Clones vary greatly in fertility, some being almost sterile and others prolific. A few clones are highly self-fertile, but most set more fruit when cross-pollinated than when selfed. There is no evidence of self-incompatibility. All the clones so far found to be largely or completely male sterile have also proved to have very low female fertility.

The mature fruit is a large 3-lobed capsule, 3–5 cm in diameter, having a woody endocarp and a thin, leathery mesocarp and containing 3 seeds. The fruit reaches its maximum size in about 12 weeks and the endocarp becomes woody by about 16 weeks. The testa and endosperm of the seed are formed early, the latter being mature by about 19–20 weeks, by which time the embryo, which is the last component to develop, is fully formed and the cotyledons are adpressed to the endosperm. The moisture content of the capsule wall is at a maximum of about 70 per cent in the 16th week, thereafter declining slowly until the 18th week and then more rapidly (Husin *et al.* 1981). At 20–24 weeks after pollination the ripe, dry capsules dehisce explosively, the endocarp splitting into 6 pieces and being dispersed with the seeds to a distance of about 15 m from the tree.

2.2.8 Root system

Trees grown as seedlings, or as buddings on seedling rootstocks, develop a strong tap-root and extensive lateral roots, the whole root system forming about 15 per cent of the total dry weight of a mature tree. In a study of rooting habit on a range of soils, which revealed no marked differences between trees of 9 clonal seedling families, it was found that on deep soils without impediments to root growth trees 3 years old had tap-roots about 1.5 m long and laterals 6–9 m long, while at 7–8 years the tap-roots were about 2.4 m and the laterals over 9 m (Rubber Research Institute of Malaya 1958). The laterals normally extend well beyond the spread of the branches so that in plantations at the usual spacings they commonly grow through the adjacent planting rows. The roots of neighbouring trees intermingle and some may become grafted together. The major lateral roots almost invariably arise from the tap-root in a whorl within 30 cm of the soil surface and grow horizontally, or only slightly downwards. Further laterals are commonly produced at a depth of 40–80 cm, but do not extend horizontally as far as those nearer the surface. All the laterals ultimately give rise to unsuberized, yellow-brown roots of about 1 mm diameter, possessing root hairs and known as feeder roots since they are mainly responsible for absorption of nutrients and water. While the feeder roots are mostly in the top 30 cm of the soil, a proportion arise from the deeper laterals and there is no reason to believe that these are less efficient absorbers than those nearer the surface.

An investigation of the distribution of the feeder roots of clonal seedling trees aged 1–22 years planted in rows 6.1 m apart on a range of soils showed that up to 3 years after planting the roots were concentrated near the trunks; at 4 years the feeder roots of trees in adjacent rows met; and at 5–7 years the feeder root density in the centre of the inter-rows was significantly greater than close to the trees. After the roots of neighbouring trees had met, ramification occurred nearer the trunk with the result that in the mature plantation there was little variation in the concentration of feeder roots across the inter-rows, except where the roots branched prolifically on entering a patch of particularly well-aerated, moist or nutrient-rich soil (Rubber Research Institute of Malaya 1958).

While the development of the root system always follows the general pattern described above, considerable variations occur due to soil type, soil aeration and moisture content, cultivation, nature of the ground vegetation and mode of fertilizer application (see pp. 148, 246 and 343). Soong (1976) investigated the influence of several factors on the density and distribution of feeder roots to a depth of 45 cm by augur sampling the roots of mature trees of 4 clones, all budded on Tjir 1 clonal seedlings and growing on 7 different soil series. He found that feeder root development was markedly influenced by the scion clone; for example, the vigorous clone RRIM 605 had about 80 per cent more feeder roots by weight than the slower-growing RRIM 513. Soil texture had a marked effect. On sandy soils the weight of feeder roots was significantly greater than on clayey soils, probably due to the plant's reaction to the lower moisture retention, or better aeration, of the former soils. Over a range of soil types, feeder root densities were positively and significantly correlated with fine sand content and negatively correlated with clay content. However, this did not apply on some clayey soils which possessed a good structure due to their high sesquioxide content. On most soils about 50 per cent of the feeder roots in the 0–45 cm layer were in the top 7.5 cm and the proportion decreased rapidly with depth, only about 10 per cent occurring between 30 and 45 cm below the surface. Exceptions to this were found on soils with a compact or poorly structured sub-surface layer, where 70 per cent or more of the feeder roots could occur in the top 7.5 cm, and also on uncompacted soils with little or no profile differentiation, where distribution was fairly uniform throughout the 0–45 cm layer.

The growth of feeder roots was found to vary seasonally, being at its maximum in the surface 7.5 cm of soil at the time of refoliation after wintering and at a minimum during the period of senescence before leaf fall. In the 7.5–45 cm soil layer the lowest feeder root density occurred at the same time as in the surface soil but peak root development was about three months later than it was in the upper layer.

The presence of mycorrhizae on the roots of rubber trees was recorded by Park in Ceylon in 1928. In Malaysia, the occurrence of endotropic, vesicular arbuscular mycorrhizae of the *Endogone* type on the roots was found to be general on rubber trees of all ages and on a variety of soils (Rubber Research Institute of Malaya 1954a; Wastie 1965). Spores of several species of endomycorrhizal fungi have been identified in soil samples from rubber plantations examined by Jayaratne (1982) in Sri Lanka and by Ikram and Mahmud (1984) in Malaysia. It is not known whether the mycorrhizae are of significance in the nutrition of the tree. Young rubber trees grow well from sterilized seed in sterile sand if supplied with the requisite nutrients, but mycorrhizae may be of value on soils high in organic matter or of low phosphate content.

2.2.9 Juvenile and mature characteristics

When grown from seed the rubber tree passes through a juvenile stage to a mature stage, the latter normally beginning with the formation of the branches. Throughout its life a seedling tree exhibits certain juvenile characteristics, in that its bark is somewhat rough, and with increasing height the trunk tapers, the thickness of the bark decreases and the number of latex vessel rings therein declines. The 'mature-type' buddings used commercially are formed by budding seedling rootstocks with buds which are mature stage tissue, being taken from high on the stem or from the branches of a seedling tree, or after passage through several bud generations. The scion does not pass through a juvenile phase and therefore grows into a trunk which lacks juvenile characteristics; it does not taper but is almost cylindrical, the bark is smoother than that of a seedling and both its thickness and the number of latex vessel rings within it remain virtually constant with increasing height. Juvenile-type buddings, which can only be made by taking buds from low down on the stem of a young seedling and which are not used commercially, are intermediate between seedlings and mature-type buddings, but more closely resemble seedlings.

There is an important difference in the behaviour of cuttings made from juvenile or mature tissues. Cuttings taken from the base of the stems of young seedlings can be rooted without difficulty and grow into seedling-like plants with normal tap-roots. Cuttings from higher up the stem or from branches of seedlings, or from the scions of budded plants, require special treatment to induce rooting and produce lateral roots but no tap-roots. This matter is discussed further in Chapter 6.

2.2.10 Growth studies in the plantation

Templeton (1968, 1969a) studied the growth before and during tapping of two clones, RRIM 501 and RRIM 513, budded in the field on seedling stocks planted at the normal spacing of 9.15 × 2.44 m (30 × 8 ft), giving 444 trees per hectare. Only the pre-tapping period, which extended to 81 months after budding, is considered here, a discussion of the growth and partition of assimilates of trees in tapping being given in Chapter 9. The determination of the dry weight of the different parts of the plants, including the roots, was done by destructive sampling of complete trees at 9, 15, 21, 27, 39, 55, 63 and 81 months after budding.

The total dry weight per tree increased approximately exponentially up to 39 months from budding, after which the rate was slower. Trees of RRIM 501 exceeded 300 kg within 81 months, a figure agreeing well with that obtained by Shorrocks (1965a) for the same clone. The rate of girth increment rose to a maximum of 1.0 cm/month between 27 and 39 months and then declined. The relative growth rate (RGR–rate of increase in dry weight per unit of dry matter present per unit of time) declined steadily from 0.04 g/g/week at 9 months to 0.005 at 81 months.

The leaf area ratio (LAR–leaf area per unit of dry weight) was naturally low at first with a small scion on a one-year-old rootstock, but rose to 12 cm^2/g by 9 months from budding and remained at about this level until 39 months, after which it declined steadily, reflecting the increasing proportion of plant weight in the non-photosynthetic tissues of trunk and branches. The leaf area index (LAI–area of leaf laminae per unit area of ground) increased rapidly to reach a maximum of 5.8 between 50 and 60 months, when a complete canopy over the ground was achieved, and continued at this level up to 81 months. The net assimilation rate (NAR–rate of increase in dry weight per unit area of leaf) declined slowly from 0.0032 g/cm^2/week at 9 months to 0.0013 at 81 months due to increased self-shading of the foliage as the LAI increased, resulting in a lower photosynthetic rate per unit leaf area. The crop growth rate (CGR–rate of dry matter production per unit area of ground per year, which is proportional to LAI × NAR) increased to a peak of 35.5 t/ha/an. at 55 months, after which it declined. The cumulative dry weight for clone RRIM 501 at 81 months was 135 t/ha, which is close to the figure for the same clone quoted by Shorrocks (1965a).

Templeton considered that the RGR, LAR, NAR and CGR would all decline slowly after canopy closure, but that the LAI, which had levelled off at 5.8 after 63 months, would remain near this value for many years. However, Shorrocks (1965a), who reported a similar LAI value (6.3 at 6 years) for RRIM 501 to that of Templeton, found that the LAI rose to 14 at 10 years and was still about 9 at 24 years. Premature senescence and fall of lower leaves is frequently observed in plantations with dense canopies. It is also common for some of the lower branches of mature stands to die back and be shed. Both these features suggest that the LAI is high enough for self-shading to reduce the light intensity to below the compensation point low down in the canopy. Templeton's data show that the CGR of RRIM 501 fell off after 4–5 years, while the LAI rose to 5.8 at 5 years and remained at this level. Similarly, Shorrocks' figures indicate that the CGR fell from the 6th to the 10th year, while the LAI rose over the same period. Thus, there is some evidence to suggest that the LAI may be above the optimum for dry matter production (but not necessarily for latex production) over much of the life of the plantation.

Templeton (1969a) estimated the efficiency of the rubber tree from his maximum recorded CGR of 35.5 t/ha/an. of dry matter which, at 4800 cal/g, is equivalent to 1704×10^8 cal/ha. The average solar radiation in Malaysia is 420 cal/cm^2/day and assuming 40 per cent utilization of this energy for photosynthesis (including 10 per cent loss of radiation energy to non-photosynthetic pigments), the available energy amounts to about $61,320 \times 10^8$ cal/ha/an., so that the efficiency of utilization of solar radiation by the closed canopy of the rubber plantation is about 2.8 per cent.

The growth and partition of assimilates of trees in tapping is discussed in Chapter 9, but it may be noted here that tapping retards dry weight increment to an extent that varies widely between clones.

2.2.11 Bark anatomy

This section briefly summarizes the main structural features of the bark of the mature rubber tree, most of which were described by Bryce and Campbell in 1917. A more detailed review of current knowledge of the anatomy of the bark and other organs of *Hevea* has been provided by Gomez (1982). As with all woody dicotyledons, in the trunk of a mature rubber tree a central cylinder of wood is surrounded by bark, the two tissues being separated by a thin layer of vascular cambium which generates xylem tissues on the inside and phloem tissues on the outside. The wood (xylem) consists mainly of fibres, tracheids and vessels, the latter two being responsible for the upward transport of water and nutrients taken up from the soil by the roots.

In the virgin bark three concentric layers can be distinguished, as illustrated in the three-dimensional diagram of Fig. 2.2. The soft bark, closest to the cambium, mainly comprises concentric cylinders of sieve tubes and phloem parenchyma alternating with thinner cylinders of laticifers. The former are concerned with the transport and storage of assimilates originating in the leaves; the latter hold the rubber-containing latex. These tissues are traversed horizontally

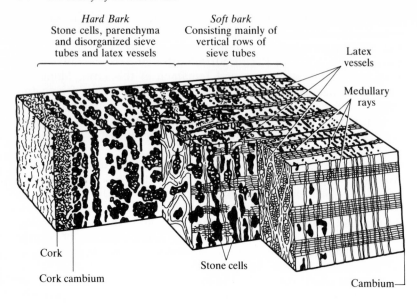

Note: The sieve tubes which make up the bulk of the soft bark, and parenchyma in both soft and hard bark, are not shown in the diagram.
The black areas are clusters of stone cells.

Fig. 2.2   Three-dimensional diagram of the bark of *Hevea brasiliensis*

at intervals by the thin plates of the medullary rays which originate in the cambium and run radially through the stele, petering out in the older xylem and phloem tissues. Their main function is the radial transport of synthesized food materials. The second layer, the hard bark, also contains sieve tubes and laticifers, but both become progressively disorganized and non-functional with increasing distance from the cambium. In this layer many of the parenchyma cells have become sclerified to form hard stone cells. The outermost layer, the periderm, consists of the external covering of cork cells (phellum), beneath which is the cork cambium (phellogen) which generates cork cells on the outside and phelloderm, a tissue resembling cortical parenchyma, on the inside.

The latex vessel system

The laticifers, within which rubber is formed and stored, occur in concentric cylinders because they are differentiated by the cambium at fairly regular intervals, normally at the rate of 1.5–2.5 rings per annum. In each ring the individual vessels are fairly close together and there are frequent connections between them, so that they form a cylindrical meshwork of tubes inter-weaving between the medullary rays. A feature of practical importance is that anastomoses between vessels in adjacent rings are rare and, consequently, those vessels that are not cut in tapping cannot contribute to the latex flow. As the tree grows, the latex vessel cylinders successively produced by the cambium are gradually pushed outwards and the increasing circumference of the tree subjects them to a tangential pull which results in the intrusion of other tissue between the vessels. The older vessels located in the hard bark are therefore disrupted and lose their function.

The vessels are aligned at an angle to the vertical axis of the trunk and generally incline from lower left to upper right as one faces the tree, but in a few clones the inclination is in the opposite direction. The angle of deviation to the vertical is a clonal characteristic and usually ranges between 2.1° and 7.1° (Gomez & Chen 1967). Because of this alignment tapping cuts are normally made to slope from high left to low right at an angle of about 30°, as this cuts more vessels and gives a higher yield than if the cut is made to slope in the opposite direction.

Both the thickness of the virgin bark and the number of latex vessel rings therein increase with the age of the tree. The number of latex vessel rings is primarily a clonal characteristic, but the frequency with which they are initiated by the cambium also depends on the growth rate of the tree, which is influenced by such factors as planting density and nutrient status, as well as by the clone. Paardekooper (unpublished) and Gomez *et al.* (1972) found the rate of vessel ring formation to be almost constant throughout the life of the tree. Although, as will be discussed later, tapping markedly depresses the girthing of the tree, it does not effect the rate at which latex vessels are formed, so that the number of rings increases approximately linearly with age. As might be expected, the correlation between bark thickness and number of latex vessel rings is much higher within a clone than between clones. As a result of selection for high yield the total number of latex vessel rings at a given age, or the number per millimetre of bark thickness, is considerably greater in modern clones than in older planting material. Gomez *et al.* (1972) found a mean of 25.6 rings for 112 clones aged 8½ years, whereas Bryce and Gadd (1923) reported a mean of 11.25 rings for seedlings at a height of 61 cm. In seedling trees bark thickness and the number of latex vessel rings decrease with increasing height up the slightly conical trunk.

The distribution of the latex vessel rings in the bark has been studied by Paardekooper (unpublished) and Gomez *et al.* (1972). In young mature trees the consecutive rings are closest together adjacent to the cambium and become more widely separated further away from it. With 112 clones 'aged 8½ years a mean of 40 per cent of the rings lay within 1 mm of the cambium, but the proportion in this zone varied with the clone from 20 to 55 per cent. As the tree grows older the concentration of the rings near the cambium is diminished and a higher percentage of the total number is found outside the first millimetre until in trees aged 25 years about 75 per cent of the rings is fairly uniformly distributed through the innermost 5 mm of bark. Because of this distribution of the rings in young mature trees and the fact that there are few or no connections between rings it is necessary to tap as close to the cambium as practicable in order to maximize yields. However, tapping in normal practice is not closer than 1 mm to the cambium because damage to the cambium would prevent bark renewal of a quality suitable for future tapping. Gomez *et al.* (1972) found that tapping young mature trees to this depth would leave an average of 40 per cent of the rings uncut but that this proportion would decrease as the trees grew older, until at 32 years old it would be only 8–13 per cent.

The density of the vessels within the rings, i.e. the number of vessels per millimetre of ring, varies between clones, but is always greater in the rings close to the cambium than in those farther away from it. The mean diameter of the vessels also varies with clone and ranged from 21.6 to 29.7 μm in 8 clones examined by Gomez *et al.* (1972).

The amount of rubber-containing tissue per unit volume of bark, which is clearly an important yield determinant, depends on the number of latex vessel rings, the number of vessels per millimetre of ring and the mean diameter of the vessels. It has been found that the number of latex vessel rings is the most important factor and is usually significantly and positively correlated with yield. High correlation coefficients, often in excess of 0.8, have been found for trees within a clone, while between clones the coefficients have usually been between 0.35 and 0.57 (Narayanan *et al.* 1973; Narayanan *et al.* 1974a; Gomez 1982). The diameter of the vessels and their density within the rings appear to have only a minor influence on yield and taking these parameters into account does not improve the correlation coefficients.

While the latex vessels are of primary importance in rubber formation and storage, the surrounding cells must provide them with the required assimilates to regenerate the latex withdrawn through tapping. Gunnery (1935) and Fernando and Tambiah (1970a) observed that high-yielding clones tended to have sieve tubes of wider diameter than those of low yielders, but in a more extensive study of 80 clones Narayanan *et al.* (1974a) found no significant correlation between sieve tube diameter and yield.

By severing much of the phloem tissue, the tapping cut must interfere to some extent with the downward transport of assimilates, but although regular tapping depresses the rate of girthing of the tree, it clearly does not prevent the regeneration of large amounts of latex, even when a full spiral cut is employed. Translocation evidently continues through the uncut phloem elements within about 1 mm of the cambium, but Hebant *et al.* (1981) have reported that the older sieve tubes surrounding the latex vessels that contribute to latex flow are non-functional. These authors consider that the transport of water, ions and metabolites to these latex vessels takes place through the medullary rays. Recently, Premakumari *et al.* (1985) have observed the presence of intraxylary phloem as strands flanking the primary xylem vessels. This internal phloem resembles the external phloem but there are no laticifers associated with it. Similar internal phloem has been reported in a few other species. It occurs in tobacco, where girdling of the stem activates it to take on the function of the external phloem in transporting assimilates and endogenous auxins (Bonnemain 1980). It is possible that the internal phloem may be active in a similar way in *Hevea*.

Bark renewal

As a reaction to the paring away of the bark during tapping, the tree forms wound tissue. A newly formed phellogen differentiates cork cells on the outside and cork parenchyma on the inside. The vascular cambium is also activated and produces new phloem tissue, including rings of latex vessels, at an accelerated rate. The rate of bark regeneration is particularly fast during the first six months after tapping. Since the virgin bark above and below the tapping panel continues to increase in thickness as the tree grows older, the renewed bark will not reach the same thickness as virgin bark, but after about eight years the regeneration has progressed sufficiently for the renewed bark to be tapped. Under normal conditions of growth the number of latex vessel rings in the renewed bark is usually greater than that found in the original virgin bark.

Cytology of latex vessels

In the seed laticifers are present only in the cotyledons but on germination they develop rapidly and can be found in all parts of a four-day-old seedling except the primary root. They occur in all organs of a mature tree. The following account refers to the laticifers in the secondary phloem of the stem, where their most important concentration occurs. A full perception of the cytology of the laticifers was only obtained with the advent of electron microscopy. Dickenson published the first electron micrographs of *H. brasiliensis* latex vessels and described their initiation, development, structure and contents (Andrews & Dickenson 1961; Dickenson 1965, 1969). A review of this work and of more recent research by others has been given by Gomez and Moir (1979).

The latex vessels arise from longitudinally continuous rows of cells which initially closely resemble young phloem cells and possess cellulose walls, plasmalemma, nucleus and cytoplasm with associated organelles, including mitochondria, proplastids, Golgi bodies, ribosomes and endoplastic reticulum. As the cells develop, the cross-walls are wholly or partly dissolved to form tubes running more or less parallel to the main axis of the stem. Because the vessels are of the articulated anastomosing type their cross-sectional area is not uniform but is irregular and includes constrictions where the cross-walls have only been partially dissolved. (Fig. 2.3). Plasmodesmata occur in the walls between the vessels and adjacent phloem cells. In the maturing vessels the cytoplasm and its associated organelles gradually become restricted to the sides of the

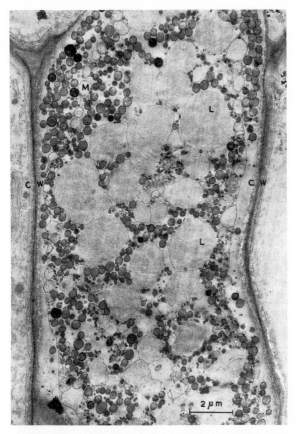

Fig. 2.3 Electron micrograph of young latex vessel in secondary phloem of green stem, showing cell wall (CW), lutoids (L), mitochondrion (M) and numerous rubber particles (×9000) (After Gomez & Moir 1979)

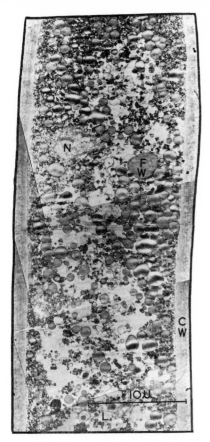

Fig. 2.4 Electron micrograph of latex vessel from mature bark close to cambium region showing cell wall (CW), Frey-Wyssling complex (FW), lutoids (L), nucleus (N) and numerous rubber particles (×4200) (After Gomez & Moir 1979)

vessel and at full maturity the lumen is wholly occupied by liquid latex containing very numerous rubber particles (Fig. 2.4). In the older vessels of the soft bark little cytoplasm remains and the vessels are packed with rubber particles together with fewer non-rubber particles (see below). Senescent vessels in the outer, hard bark contain mainly large rubber particles; other organelles are usually absent.

The rubber particles in young trees are spherical, but in mature trees the larger ones may be pear-shaped, the latter being frequent in some clones but rare in others. The size of the particles usually ranges from 5 nm to about 3 μm and each is surrounded by a surface film, approximately 10 nm thick, composed of proteins and lipids, including phospholipids, triglycerides, sterols and tocotrienols (Cockbain & Philpott 1963; Ho *et al.* 1976). This film carries a negative charge and is responsible for the stability of the rubber particles suspended on the latex serum. The latter, usually referred to as C serum, contains a number of proteins of low isoelectric point which are anionic at the normal pH (6.9) of the serum.

There are two main non-rubber particles in the latex. First, there are numerous vesicles or vacuoles, of between 0.5 and 3.0 μm diameter, bounded by a semi-permeable membrane about 8 nm thick consisting mainly of lipids. These were first reported by Homans *et al.* (1948) who called them lutoids because they appeared to be responsible for the yellow colour of a bottom fraction obtained by centrifuging latex. Although it was subsequently found that the colour is due to the Frey-Wyssling particles described below, the name lutoid has continued in use. The lutoids are fragile, have a liquid content (B serum) and are osmotically sensitive. The B serum has a pH of about 5.5, contains basic proteins of high isoelectric point and is capable of flocculating rubber particles if it is released from the lutoids. The lutoids therefore play an important part in the stability and flow of latex, which is described in Chapter 9.

Suspended in the fluid of lutoids from young tissues there are characteristic bundles of microfibrils, but these are absent from the lutoids of mature bark. Each microfibril is in the form of a helix coiled about a hollow core (Archer *et al.* 1963) and consists mainly of an acidic protein (Audley 1966). Microfibrils of a different kind have been observed in latex tapped from mature trees, specially in latex from trees treated with the yield stimulant, ethrel (Dickenson 1965). Minute spherical particles in Brownian motion have been observed within lutoids (Southorn 1961). These may possibly be rubber but, if so, they must have entered the lutoids either *in vivo* or during centrifuging, as the lutoids play no part in rubber biosynthesis. As lutoids contain hydrolytic enzymes enclosed within a semi-permeable membrane, Pujarniscle (1968) noted that they are comparable with lysosomes. Further evidence of the lutoids behaving as lysosomes was obtained by Tata *et al.* (1976, 1983), who showed that two basic proteins isolated from the bottom fraction of centrifuged latex (which consists largely of lutoids) had lysozyme activity in the standard assay using *Micrococcus lysodeikticus* as substrate.

The second non-rubber particle, much less numerous than the lutoid, is the Frey-Wyssling particle, which is mainly composed of lipid material and is yellow or orange in colour due to the presence of carotenoids. Frey-Wyssling (1929) observed that these particles commonly occurred in clusters of two or three, Southorn (1963) found that the clusters were enclosed in a membrane and Dickenson

(1969) demonstrated that the particles are part of a larger structure which he called the Frey-Wyssling complex. This comprises a spherical vesicle, 4–6 μm in diameter, easily deformed but stronger and less osmotically sensitive than the lutoids, bounded by a double membrane and containing several Frey-Wyssling particles together with other structures. The latter include a group of tubules embedded in a membrane-bound matrix and associated with 2–4 concentric lamellae formed from double unit membranes. There is also an elaborately folded invagination of part of the inner membrane of the double envelope surrounding the whole organelle. Adjacent to these folds there may be a membrane-enclosed group of particles, possibly, but not certainly, of rubber. The elaborate structure of the Frey-Wyssling complex suggests that it has important metabolic functions, but these are not understood. Dickenson (1969) has suggested that in mature vessels it may be a site for rubber biosynthesis supplementary to those in the cytoplasm of vessel primordia.

2.2.12 Tapped latex

Formerly, tapped latex was thought to be the sap contained in a large vacuole running the length of each vessel, but the work of Milanez (1948) and others, culminating in the electron microscope studies of Dickenson (1965), finally established its cytoplasmic nature beyond doubt. Tapping severs a number of latex vessel rings and the latex which flows out comprises the contents of vessels at different stages of development, together with small amounts of materials from other types of cell in the phloem. All the organelles mentioned above as occurring in the latex vessels can be found in tapped latex, but mitochondria and nuclei are rarely found, being mostly retained in the parietal cytoplasm remaining in the vessels. The commonest particles are the rubber particles, the lutoids, many of which are damaged, and the Frey-Wyssling particles, these last being much less numerous than the other two. The contents of tapped latex (rubber fraction and serum) are discussed in a practical context in Chapters 9 and 11.

2.2.13 Biosynthesis of rubber

The rubber hydrocarbon molecule, which has a molecular weight of 10^5–10^7, consists of long chains of the monomer isoprene, C_5H_8:

$$-CH_2C = CHCH_2 \ - \ CH_2C = CHCH_2 \ - \ CH_2C = CHCH_2 \ -$$
$$\qquad \underset{CH_3}{|} \qquad\qquad\quad \underset{CH_3}{|} \qquad\qquad\quad \underset{CH_3}{|}$$

It has generally been thought to be entirely *cis*-polyisoprene, but it is now considered that it may contain some *trans* double bonds at the start of the polymer chain. Much of the pathway of the biosynthesis of the monomer was elucidated during the 1950s. The research done during that period has been reviewed by Bonner (1961) and by Fournier and Tuong (1961), while more recent findings have been summarized by Archer *et al.* (1982).

The steps in the pathway are outlined in Fig. 2.5. Following glycolysis, acetyl CoA is converted by two condensations, first to acetoacetyl CoA and then to β-hydroxy-β-methylglutaryl CoA (HMGCoA). This is reduced in two steps to mevalonic acid (MVA), which is the first compound in the chain unique to isoprenoid synthesis. Two successive phosphorylations convert

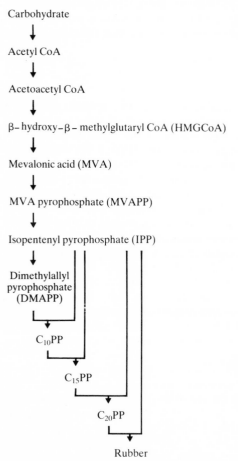

Carbohydrate

Acetyl CoA

Acetoacetyl CoA

β- hydroxy-β- methylglutaryl CoA (HMGCoA)

Mevalonic acid (MVA)

MVA pyrophosphate (MVAPP)

Isopentenyl pyrophosphate (IPP)

Dimethylallyl pyrophosphate (DMAPP)

$C_{10}PP$

$C_{15}PP$

$C_{20}PP$

Rubber

Fig. 2.5 Steps in rubber biosynthesis

MVA to MVA pyrophosphate. Decarboxylation and the simultaneous removal of water from MVAPP results in the formation of isopentenyl pyrophosphate (IPP). The step immediately preceding polymerization is the isomerization of IPP to dimethylallyl pyrophosphate (DMAPP) which acts as the initiator of polymerization. Polymerization then proceeds by repeated addition of IPP molecules to DMAPP with the elimination of pyrophosphate.

Although the above steps in the synthesis of the monomer have all been shown to take place in fresh latex withdrawn from the tree, until recently it seemed that the incorporation of IPP into the polymer took place only on the surface of existing rubber particles and was essentially a chain extension process. No chain initiation or *de novo* production of rubber particles had been observed *in vitro*, although it was evident that this must happen *in vivo* in view of the fact that rubber is continuously being formed in large quantities in tapped trees. Recently, evidence has been obtained of new rubber molecules being formed on the surface of rubber particles, but the mechanism by which new rubber particles are formed in the vessels after tapping is still unknown. It is very likely that they result from the division of existing particles and there is some evidence to support the suggestion that they are formed on small aggregates of protein suspended in the latex (Archer *et al.* 1982).

The later stages of rubber biosynthesis, and probably many of the earlier ones, take place in the latex vessels. Tupy and Resing (1968) showed that, apart from a small amount of raffinose, sucrose is the only sugar found in latex, provided that invertase activity is prevented during latex collection and analysis. D'Auzac and Jacob (1969) found that all the enzymes required to transform sucrose through the intermediate products of glycolysis to pyruvate are present in latex. Pyruvate is a very poor precursor for rubber formation in experiments with tapped latex, but this might well be due to the fact that most of the mitochondria needed for the conversion of pyruvate to acetyl CoA are retained in the vessels when the latex flows out.

Bealing (1969, 1976) has suggested that there may be alternative pathways of carbohydrate metabolism in latex *in vivo*, differing from, and in addition to, glycolysis and perhaps involving quebrachitol and L-inositol, which are always present in latex and might play a part in rubber biosynthesis. There is also the possibility that some steps in carbohydrate metabolism may not occur only, or at all, in the latex vessels and hence that rubber could arise from precursors originating in cells adjacent to the vessels. However, the view that rubber is formed *in vivo* from sucrose in the vessels is supported by evidence from recent experiments with radioactive tracers. These showed, first, that hexose sugars supplied to the bark were converted to sucrose during transport to the vessels; and, second, that labelled sucrose applied to the bark was extensively used in rubber biosynthesis (Tupy & Primot 1982; Tupy 1985).

As at least some of the enzymes active in the latex vessels must flow out with the latex on tapping, the vessels must possess the means of re-synthesizing the specific proteins involved, and the presence of DNA and of ribosomal-, transfer- and messenger-RNA within them has been demonstrated (McMullen 1959, 1962; Tupy 1969). Tupy noted that fresh, tapped latex could synthesize RNA and that the rate of synthesis in latex from high-yielding trees was faster than that of low-yielding trees of the same clone.

In untapped trees a spontaneous, random cross-linking process between the polyisoprene chains within the rubber particles eventually leads to fully cross-linked particles called microgel (Bloomfield 1951). Latex from untapped trees has a very high microgel content, but once tapping has started the content of the regenerated latex decreases markedly, although it remains high in those parts of the tree far removed from the tapping panel. Significant differences in microgel content between clonal latices have been demonstrated (Freeman 1954). The microgel content contributes to the degree of hardness of the rubber prepared from the latex.

2.2.14 The function of latex

There has been much speculation about the function of latex in the plant. The suggestion that it may serve for the transport and storage of food materials seems untenable. In the first place, the petiolar latex vessels cannot transport photosynthate from the leaves because they become blocked with callose at an early stage. Secondly, latex can hardly be regarded as a reserve food store for use under stress, since the rubber which forms a large part of it is not further metabolized. It has been shown that even in etiolated seedlings, after nearly 60 per cent of the carbohydrates have disappeared, the rubber content is not depleted (de Haan & van Aggelen-Bot 1948). As it coagulates on exposure to the air, the latex may have some protective action in the event of mechanical injury but there is no firm evidence for this, nor for the idea that it deters attack by animals, although the rubber tree has few insect pests. Fernando and Tambiah (1970b) have suggested that in view of the correlation of yield with rainfall and temperature the latex may function as a water-regulating system within the plant. Bealing (1965), noting that latex contains the enzymes for pentose phosphate metabolism and that there is a relation between the latter and rubber biosynthesis, has suggested that it is a function of latex to provide the surrounding tissue with a supply of pentose cycle derivatives. As none of the above hypotheses has been proven, the function of latex and rubber in the plant remains unknown.

Chapter 3

Rubber breeding

N. W. Simmonds

3.1 Historical

3.1.1 General

Rubber as a crop belongs entirely in this century. From the rubber 'boom' in South-east Asia (about 1910), the first great surge of planting, the crop developed remarkably, from a wild jungle tree to a major domesticate in about 50 years. This success was due to the systematic exploitation of agricultural research, the most important single component of which was surely breeding. The first steps in breeding were taken by Dutch workers in Java and Sumatra (the then Netherlands East Indies) in 1910–20. They showed that yields per tree were very variable, that clonal propagation by budding was possible and that individual clones differed. They went on to use mixed superior clones as parents in open-pollinating seed gardens in order to produce superior seedling progeny. The technique remains in use today (sections 3.4.1 and 3.4.3.6). From the 1920s onwards, the random open-pollinated seedlings upon which the industry had been founded were progressively supplanted by budded clones and superior seedling populations. Since the Second World War, clones have been dominant in all the most advanced sectors of the industry and rubber breeding has long been essentially a matter of selecting and testing clones from genetically variable seedling populations. Nevertheless, despite large genetic advances, so slow is the turnover of the crop that substantial areas of early clones survive today and even random seedling fields were not unknown as late as the 1970s.

3.1.2 Breeding programmes

A good many organizations, both commercial and governmental, have had a hand in rubber breeding at various times and in various places. Templeton (1978) gives a useful review. The pioneering Dutch work in the East Indies (started 1912–17) was succeeded by the present Indonesian Government Research Institute for Estate Crops in Java and Sumatra. Ridley's great work in the early days in Singapore led eventually to an industry-run institute, the Rubber Research Institute of Malaysia (RRIM, founded 1925), which remains prominent in rubber breeding today. In parallel with the RRIM in its early days, several plantation companies in Malaysia selected clones in their own seedling fields. Outstanding among them was Prang Besar Estate near Kuala Lumpur which selected several excellent early clones (e.g. PB 86) and went on to build up a major, and extremely successful, long-term breeding programme (section 3.4.3.6). Southern India and Sri Lanka (then Ceylon) were substantial early rubber producers and both have sustained breeding research programmes: those at the the the Rubber Research Institute of Ceylon at Agarawatta (RRIC) and the Rubber Research Institute of India in Kerala (RRII). In Thailand there is the Rubber Research Centre at Hat Yai. In West Africa, the Institut de Recherches sur le Caoutchouc (IRCA) is active in breeding at Abidjan, Ivory Coast and there is also the Rubber Research Institute of Nigeria. Several rubber-using companies have been active, at various times from the 1930s onwards, in trying to develop rubber plantations, mostly in Brazil. They include Ford, Pirelli and Firestone. The last, as the Firestone Plantation Company, also developed rubber breeding research at its estate at Harbel, Liberia, and at several sites in Guatemala. The Brazilian plantings all failed, due to South American Leaf Blight (SALB) (section 3.5) and this failure was, in fact, the stimulus for the Firestone breeding work in Liberia and tropical America. There is now a resurgence of interest in rubber research, including breeding, in Brazil; the leading agency is the national agricultural research system, EMBRAPA, which works on behalf of the national rubber authority SUDHEVEA. By far the strongest body of rubber breeding research has come from south-east Asia, especially Malaysia and Indonesia. With strong industrial interests in the products of breeding, international exchange of breeding materials has never been unrestricted, indeed has often been very narrowly confined. However, since the early 1950s the realization has been growing that at least relative freedom of movement of genetic material is to the general long-term good. Accordingly, extensive intra-Asian exchanges have been accomplished and an important Asian-American exchange took place in 1954 (section 3.6.2). One result of the latter has been that Brazilian clones (e.g. the IAN series) contain a large oriental genetic component, emphasizing the critical historical importance of Asian rubber breeding. The welcome move towards international collaboration, originally stimulated by fears of SALB but later appreciated to have a much wider basis, was formalized by the foundation of the International Rubber Research and Development Board (IRRDB), with headquarters now near Hertford, England, UK. A leading activity of the IRRDB has been, and will no doubt continue to be, the promotion of international exchange of genetic material (section 3.6.3). It is good to see the rubber world appreciating what workers on other crops have understood for decades, namely the merits of fluent exchange of information and materials through a dispersed international network of researchers.

Useful general references to rubber breeding are: Ferwerda (1969); Wycherley (1976b); Tan (1987). For more local treatments, see: *Malaysia* – Ang and Shepherd (1979), Baptist (1953, 1961), Ho (1979), Ross (1964), Ross and Brookson (1966), Sharp (1940, 1951), Shepherd (1969a), Simmonds (1986a), Subramaniam (1973), Wycherley (1969); *India* – Nair *et al.* (1976); *Sri Lanka* – Fernando (1966, 1969, 1977).

The abbreviations used in this chapter are listed below.

G,E,GE:	genotype, environment, genotype × environment interaction
GAM:	generation-wise assortative mating
GCA, SCA:	general combining ability, specific combining ability, the latter being interactions residual after GCA has been taken out
GLD, PLF, PML, SALB:	rubber diseases, respectively: *Gloeosporium* leaf disease; *Phytophthora* leaf fall; powdery mildew; South American Leaf Blight
GRC:	genetic resource conservation
HR, VR:	two major categories of disease resistance (Vanderplank's terminology): horizontal resistance (pathotype-non-specific, usually polygenic); vertical resistance (pathotype-specific, due to major genes)
MV, PI:	Mooney viscosity of rubber (a measure of molecular weight) and plugging index (a measure of latex flow)
LSCT, SSCT:	large-scale clone trial; small-scale clone trial
PBIG:	Prang Besar Isolation Gardens

3.2 Breeding system

3.2.1 Pollination

Floral morphology and flower behaviour are covered in Chapter 2. Here it is only necessary to recall that flowering generally occurs twice a year (e.g. in Malaysia), in two unequal bursts, that the flowers are small, numerous and insect-pollinated (mostly, it is thought, by midges and thrips). Earlier contentions notwith-standing, there is no evidence of self-incompatibility though there are reports of (rare) male-sterility (Majumder 1967; Wycherley 1971). Self-setting is certainly possible and is presumably usual in isolated trees.

No systematic study of the frequency of cross-pollination appears to have been undertaken. Since 'runts' (presumably mostly selfs) are fairly frequent, even in the products of good seed gardens, selfing is probably also frequent (see Simmonds 1986b). From Prang Besar data (Shepherd, personal communication), a rate of about 22 per cent selfing can be calculated for one clone, as follows. PB 5/51 is heterozygous for a recessive yellow gene, as evidenced by 25 per cent yellows in a selfed progeny. Crosses with other clones yield no yellows. Open-pollinated PB 5/51 seed from seed gardens gives about 4–7 per cent yellows. The range of selfing might thus be about 16–28 per cent.

3.2.2 Genetics

Hevea brasiliensis and its congeners are, so far as is known, uniformly diploid with $2n = 2x = 36$ (Majumder 1964a; Ong 1979). This is a common somatic chromosome number in the Euphorbi-aceae but several genera have basic $n = 7, 8, 9, 10, 11$. The 36-chromosome genera are therefore probably old tetraploids based on $x = 9$. A few experimental tetraploids ($2n = 72$) have been made (section 3.4.3.9) but, for practical purposes, all rubbers are diploid.

There has been, in effect, no formal Mendelian genetic analysis. The outbreeding habit of the crop would lead one to expect fairly frequent heterozygosity for more or less deleterious recessives (cf. the yellow mutant in PB 5/51 mentioned above). Several abnormal leaf forms have been picked up and propagated but not analysed. The R-gene-type hypersensitivities known in relation to SALB (section 3.5) and the isoenzymes now being used in analysing the variability of the species (Chevallier 1984) may all be presumed to be major-gene controlled. The few dwarfs that are known in the crop are as yet unanalysed genetically. In other crops, both domi-nant and recessive dwarfing genes are known and the matter is of some practical importance (section 3.4.3.8). Seed coat patterns are well known to be highly characteristic of clones (see Ch. 2) but there are no data on genetic control; analogy with other crops (e.g. many legumes) would suggest major-gene determination.

Economic characters in rubber (e.g. vigour, yield, most disease resistances) are polygenically controlled, i.e. quantitative in nature. They can only be treated by biometrical methods (section 3.4.2).

'Poly-' does not have to mean 'many', though it may do so. All that is implied is that the character in question is heritable, that an undefined number of genes is responsible and that there is an environmental component of expression.

Rubber is an outbreeder and it shows both inbreeding depression and heterosis ('hybrid vigour'); the one is the complement/converse of the other. To self (more generally, inbreed) a highly heterozygous plant incurs the segregation of homozygous recessive genes, whether major or minor in individual effect, hence inbreeding depression. To cross different heterozygous parents is likely to expose few deleterious recessives, hence little or no decline in average vigour. To cross widely distinct genotypes may, at the extreme, expose so few common recessives that F_1 vigour is markedly enhanced above the average; hence heterosis.

3.2.3 Practical implications

The practical implications for rubber breeding can be summarized in the following four points (see Simmonds 1979, for more general discussion). First, the rubber breeder has little or nothing to do with major genes; he is nearly always working with polygenic systems by biometrical methods. Second, he is, in general, seeking highly heterozygous clonal segregates in families derived from crosses between unrelated (or at least not-too-closely related) parents. Third, he may occasionally expect to find (and does find) heterotic combinations between very distantly related parents, that is, families of exceptional vigour (though not necessarily great worth). And fourth, given a finite gene pool, selection over generations incurs changing gene frequencies, narrowing of the genetic base, some degree of relatedness among survivors, hence some (more or less covert) inbreeding depression and a declining rate of progress.

Broadly, then, the rubber breeder seeks to raise frequencies of favourable alleles, over generations, in a shifting population of highly heterozygous clones. He must use biometrical methods to achieve this end, must minimize inbreeding, may be able to exploit interpopulation heterosis and must pay attention to the genetic base of his population.

3.3 Breeding objectives

3.3.1 Yield and vigour

Yield is always a substantial (usually the dominant) objective of any plant breeding programme. Rubber is no exception and, as we shall see, even what are usually called 'secondary characters' are mostly but aspects of yield. Yield is measured as weight of dry latex per unit area of land per unit of time, e.g. as t/ha/an. For practical purposes, diverse units such as t/ha over a panel or an arbitrary period of years may be used or, for experimental data, short-term yields may be stated on a per tree basis, such as grams per tree per tapping (g/t/t).

From the users' point of view, accumulated yield over years is what matters but with some extra emphasis on early yield. When discounted cash flow is considered, early opening for tapping and high early yields are obviously especially attractive (Lim *et al.* 1973).

Yield and vigour are hardly separable. Early opening and good early yield are only possible in a tree which grows very vigorously when young. After opening, photosynthate is partitioned between two competing sinks: latex offtake and tree growth. Accordingly, growth rate tends to decline under tapping and the breeder's task is to maximize latex yield in a tree which is still growing vigorously enough to sustain an upward yield trend for many years (Wycherley 1975, 1976a). Clearly, the breeder is trying to achieve both a maximal flow of photosynthate and an, in some sense, optimal partition between latex offtake and wood accumulation. Rather crudely, we may write (Simmonds 1982):

$$W = W_p - R/c \qquad [1]$$
$$R = kcW_p \qquad [2]$$

where W is wood increment under tapping

R is rubber yield

W_p is potential wood increment under a no-tapping regime (all on a dry matter basis)

k is the fraction of photosynthate converted to latex

c is a measure of the efficiency of conversion

Empirically, $k = 0.4$ to 0.5 and $c = 0.1$ to 0.2. The latter, c, may be taken to be biologically fixed so the breeder can manipulate W_p and k, maximizing potential growth (W_p) and optimizing partition (k). I shall argue later (section 3.4.2) that the outstanding success of rubber breeding may largely be attributed to the genetic responsiveness of k.

3.3.2 Secondary characters

These fall naturally into four categories, as follows:
(1) wind-fastness; (2) bark morphology; (3) latex vessel plugging; and (4) disease resistance. All four are, to some extent, simply aspects of yield.

Wind-fastness is an important objective because trees which are too susceptible tend to break (or, sometimes, uproot). The effect of tree losses is to diminish W_p and W on a per hectare basis. Because W declines as R increases (that is, growth is competitively diminished by yield – equations [1] and [2] above), the higher the yield the greater the risk of wind damage. Untapped trees are presumably at negligible risk. The breeder's objective is, therefore: first, to maximize W in competition with high R, that is, simply to maintain a high growth rate; second, high W is not the whole story because tree habit is also relevant (section 3.4.3.5). The objective therefore becomes fast-growing trees of appropriate habit.

Morphological characters of the bark include: smoothness (a rough, bumpy bark is difficult to tap efficiently); thickness (a thick bark is easier to tap and has a larger latex vessel system than a thin one); latex vessel ring numbers and diameters (related to the preceding). Of these, bark smoothness is an objective in its own right; the others are all essentially subsumed by yield.

Degree of latex vessel plugging (plugging index (PI) – see Ch. 9) is a feature of a clone but hardly provides an objective for the breeder. Management, by stimulating the trees or not, can be and is adapted to the plugging behaviour of the clone; a high-plugging clone responds to stimulation, a low-plugging one does not. The breeder needs to know about the responsiveness of his clones to stimulation but need not aim at any particular level of natural plugging. This character is therefore a matter of awareness rather than an objective *per se*. There is, however, a contrary argument (Wycherley 1975) to the effect that low plugging incurs long flow, therefore reduced growth and increased risk of dryness or breakage. On this argument, high-plugging clones should be preferred because their performance can be adjusted by judicious stimulation. In practice, selection for unstimulated early yield must imply unconscious selection for low plugging.

Disease resistance is essentially a matter of protecting yield against losses due to decrements of photosynthate. The diseases of which the rubber breeder must take account are, in effect, all leaf diseases (Ch. 10 and section 3.5 below). They diminish effective photosynthetic area and thus adversely affect both tree growth and rubber yield. The breeder's objective is thus to reduce the natural incidence of leaf diseases; immunity (i.e. no disease) would be good to have but is, in practice, a quite unrealistic objective (section 3.5). Reduction to a tolerable level (without being too precise as to what is 'tolerable') will be enough.

3.3.3. Quality

Quality is best defined as the aptness of the harvested plant product for a specified end-use. Most plant breeding programmes have to pay a great deal of attention to it. This is conspicuously true of food crops (for example, the quality of grains for milling, baking, malting or feeding, of tubers for cooking and processing, of fruits for storage, colour and flavour characters). Among industrial crops, sugar breeders tend to think of sugar contents of beet-roots or of cane-stalks as 'quality'. Rubber is exceptional in that the harvested product is, in effect, the end-product. The breeder's only concern with quality, so far, is that he has to discard a few clones with bad latex colour. The suggestion has sometimes been made that breeding might be directed towards low molecular weight (low viscosity) rubber for special purposes. Genetic variation in the character has indeed been detected but the objective does not seem worth while because the market is small and, in any case, crown budding (section 3.7.1) or post-harvest technology (Ch. 11) does what is required.

3.3.4 Conclusions

The rubber breeder's objective is high yield of rubber, with some emphasis on early opening and early yield. High yield is attained by high tree vigour, high partition of assimilates towards rubber, retention of vigour and growth under tapping, wind-fastness and resistance to leaf diseases. Quality factors are unimportant. While the 'perfect' rubber clone is conceivable, the attainment of it is not a realistic objective. All clones, even excellent ones, have defects. The 'balanced genotype', the one that will yield well under appropriate management and suffers from no serious defects is, as ever in plant breeding, the realistic objective.

3.4 Breeding plans

3.4.1 General pattern

The general pattern of rubber breeding is, as always in plant breeding, one of 'generation-wise assortative mating' (GAM), an animal breeders' term, but equally applicable to plant breeding. The process is cyclical, the best clones in one generation becoming the parents of the next cycle of seedlings among which successor-clones are identified, multiplied and used (Fig. 3.1). In each generation, the best are crossed with the best (hence 'assortative' in the phrase GAM), but with the minor qualification that useful parents are not necessarily successful agricultural clones. Figure 3.1 will serve as a reminder that random seedlings and monoclonal seedlings have long since been superseded, the latter because they suffer from inbreeding depression even if they come from a superior clone. Since yield in rubber has a large additive genetic element (see

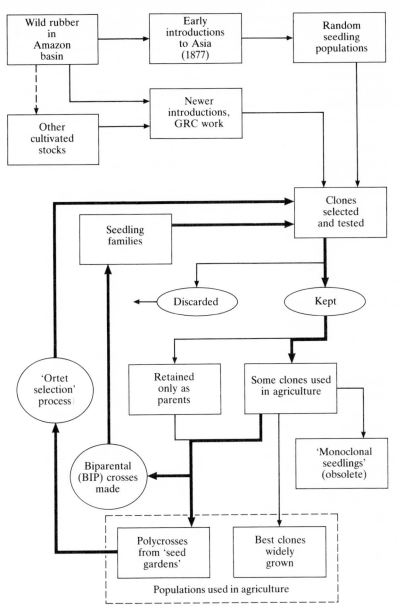

Fig. 3.1 General features of rubber breeding; the main (cyclical GAM) track is shown with bold arrows

section 3.4.2) 'polycross' or 'synthetic' seedling populations from mixed seed gardens of good clones are successfully used alongside the clones themselves. The breeder can select effectively in both biparental families (clone A × clone B) derived from hand-pollination and in polycross populations (clones A,B,C,D . . . randomly intercrossed). Seedlings from the latter are, in the rubber breeder's terminology, usually called 'illegitimate'. The 'ortet selection' shown in Fig. 3.1 selects among mature trees of very large (commercial) polycross populations in the confidence that such populations have both high means and genetic variances. An 'ortet' is a 'mother tree' that gives rise to a clone, as distinct from one that yields sexual progeny.

The top part of Fig. 3.1 brings out the fact that the effective breeding population in any one generation is not fixed by its previous history but is open to supplementation by introduction of new genetic materials. In rubber this process has not yet gone far but must, in the longer term, be of profound importance (section 3.6).

3.4.2. Biometrical features

A very brief sketch of the essential ideas is given below; for a fuller treatment and further references, see Simmonds (1979).

First, a 'heritability' (symbolized as h^2) is an estimate of the fraction of phenotypic variability that can be realized by selection. It is estimated, according to mating design and the nature of the data, either as a variance ratio or as an offspring-on-parent regression. The latter is usually more robust statistically. The variance ratio estimate may take several forms, for example: additive genetic variance divided by phenotypic variance (V_A/V_p); or total genetic variance divided by phenotypic variance (V_G/V_p). Since the denominator, V_p, contains an error component, V_E, any estimate of h^2 is always specific to an experiment and, in magnitude, h^2 is favoured by high genetic variance and low V_E. Given an estimate of h^2, genetic advance under selection is predicted by:

$$R = ih^2\sigma_p$$

where σ_p is the phenotypic standard deviation
i is a dimensionless statistical parameter defined by the intensity of selection

High R, measured as genetic deviation from the mean, is therefore favoured by intense selection, high genetic variability and high h^2 related to good control of error.

The idea of 'clonal repeatability' is closely related to heritability. It is the fraction of phenotypic variance among an array of clones that can be attributed to a genetic component. In effect, any clonal variety trial estimates (or can be made to estimate) such a ratio and the idea can be extended to a prediction of potential genetic advance given a large sample of clones under a specified intensity of selection and characterized by the same genetic parameters.

Combining ability ideas are statistical in character and embody no necessary genetic assumptions. Essentially, given an orderly set of crosses (or even, within limits, a disorderly one), statistically additive constants characteristic of parents may be fitted on the model:

$$Y_e = \bar{Y} + G_A + G_B + S_{AB} + E$$

Where Y_e is the expected parameter (e.g. yield)
$\bar{Y}$ is the general mean, the G are the additive parental constants
S is a specific (A–B) interaction
E is an error deviation
G represents general combining ability (GCA)
S represents specific combining ability (SCA)

Analysis of variance partitions sums of squares into G, S and E components (or G and S only if no E is available). The fraction of total parental sums of squares attributable to G is a measure of the predictability of performance of a cross. It is near to (but not identical with) a narrow-sense heritability (V_A/V_p). The GCA estimates are disturbed by inbreeding and heterosis effects (Gilbert *et al.* 1973) which appear as contributions to SCA, that is, to unpredictable residuals. Statistically, combining abilities are more robust than heritabilities and generally to be preferred. The idea, in the form of 'additive constants' is equally applicable to the analysis of stock/scion and trunk/crown combinations (see section 3.7.1).

Both heritabilities and combining abilities of economic characters have been investigated at the RRIM in the past decade or so. Studies are summarized by Tan (1987) and leading references are: Gilbert *et al.* (1973), Nga and Subramaniam (1974), Tan *et al.* (1975), Tan and Subramaniam (1976), Tan (1977, 1978a,b, 1981).

Examples are given in Tables 3.1 and 3.2 and Fig. 3.2. The striking feature is that both heritabilities and GCAs tend to be rather high, even very high. In other crops, yield generally has low h^2, even very low, usually less than 10 per cent. The rubber breeder's material therefore seems to be relatively highly predictable as to economic characters. This is probably a reflection both of the youth of the crop (it is at an early stage of selection and genetic variance is still largely additive) and of the (presumed) dependence of yield upon a genetically adjustable partition factor (roughly estimated by k in equation [2], section 3.3.1 above). Whatever the interpretation, the rubber breeder can have confidence that, so long as inbreeding is avoided, the results of crossing known

Table 3.1   *Proportions (%) of total sums of squares due to parental GCA in rubber crosses*

(1) Phase II data

	D.f. for GCA	D.f. for SCA	Yield (5 yr)	Yield (15 yr)	Girth (0)	Girth (9 yr)	Dry trees
A.	20	34	78	63	77	86	64
B.	15	26	81	68	81	86	67

(2) Phase III data

	D.f. for GCA	D.f. for SCA	Yield (1 yr)	Yield (5 yr)	Girth (0)
	13	15	80	73	82

(3) Nursery seedlings, 5-parent diallel

D.f. for GCA	SCA	Yield	Girth	Bark thickness	Latex vessel number	Latex vessel diameter	Plugging index
4	5	87	88	85	48	49	65

Sources: (1) and (2) from Gilbert *et al.* (1973); (3) from Tan and Subramaniam (1976)

Notes: Under (1) Phase II data, A refers to all parents, some relatedness/inbreeding disregarded; B refers to a reduced data set covering unrelated parents only; some improvement of B over A is expected and evident. Over (1) and (2) jointly a high GCA component is apparent, despite defects due to non-orthogonal, incomplete data. Set (3) shows that a high GCA component is revealed also in young seedlings.

Table 3.2   *Estimates of narrow-sense heritabilities for economic characters in rubber. Data for mature plants of various ages in three different mating patterns*

		A	B	C
Y	1 yr	—	21–34	—
Y	5 yr	56	11–14	27–34
Y	15 yr	48	—	—
G	0	47	27	17–24
G	5 yr	56	20–30	9–44
G	increment 0–5 yr	—	12–36	29–47
T	virgin	—	17–18	17–46
T	renewed	—	29–58	27–28

Sources: A, Nga and Subramaniam (1974); B, Tan *et al* (1975); C, Tan (1979a)
Notes: Y, yield of dry latex; G, girth; T, bark thickness.

parents are fairly predictable. As to parents unassessed for breeding performance, their worth will be at least roughly related to their worth as clones but always with the possibility that wide crosses will reveal unpredictable SCA in the form of heterosis.

The deleterious effects of inbreeding are shown in Table 3.3 (and

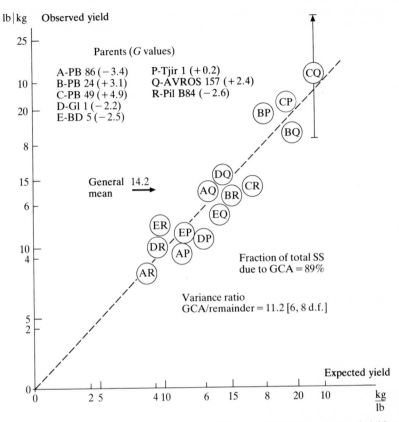

Fig. 3.2 Combining ability for yield in a 5 × 3 set of rubber crosses; observed yields plotted against expected on basis of GCA. Means per tree per year for 15 years. At the top right, a plausible genetic range for family CQ is plotted as $\pm 2\sigma_G$ assuming that $\sigma_G = 0.1\ X$ (Data of Ross and Brookson 1966 reanalysed by Simmonds 1979)

see Simmonds 1986b). It is impossible to get accurate estimates of the effect of selfing because family means are biased upwards by infertility, inviability and the inevitable discard of runt seedlings. The attempt made in Table 3.3(2) to correct for the bias only takes account of runts. No correction is possible for embryo inviability (which was obviously severe). Of the four selfed families, Tjir 1 gave the best yield *after culling* but suffered as severely as the others overall. Clearly, 'monoclone seedlings' (Fig. 3.1) are a very bad choice of planting material but, historically, they were used and, not surprisingly. 'Tjir 1 monoclone' was about the best of a bad lot. The matter is also relevant to choice of rootstocks (section 3.7).

Innumerable (probably many hundred) correlation coefficients have been calculated from rubber-breeding data but, with few

Table 3.3   *The effect of inbreeding in rubber*

(1)

	5 families, parents unrelated	7 half-sib families	2 full-sib families
Girth:			
Observed	22.2	19.9	17.2
Expected	20.8	20.5	20.0
	+1.4	−0.6	−2.8
Yield:			
Observed	17.6	12.0	10.2
Expected	15.2	14.0	15.1
	+2.4	−2.0	−4.9

(2)

	9 crosses	1 backcross (half-sib)	5(4) selfs
Pollination success (%)	3.7 (1.0–7.6)	2.8	0.5 (0–1.2)
Runts discarded (%)	7.4 (3.3–17.0)	24.5	46.8 (42.8–57.1)
Girth (cm):			
Survivors	5.8 ± 0.22	4.4	5.4 ± 0.38
Weighted	5.5 ± 0.24	3.6	3.5 ± 0.35
Yield (g/t/t):			
Survivors	6.7 ± 0.43	4.7	3.1 ± 0.86
Weighted	6.3 ± 0.46	3.9	2.1 ± 0.67

Sources: calculated from (1) the data of Gilbert *et al.* (1973) and (2) the data of Tan and Subramaniam (1976)

Notes: (1) Based on a GCA analysis of yield (lb/tree/yr for 5 years) and girth at opening. GCA estimates are slightly biased by inbreeding so observed greater than expected for crosses but the reverse is evident for inbred families. All these data are biased by the early discard of runt seedlings (see following).
(2) Seedling nursery data from a 5 × 5 diallel cross. Inbreeding reduces fertility, vigour and yield and increases the runt frequency. Of the five selfs, one was sterile so seedling data refer to four families. The weighted estimates of girth and yield were calculated on the assumption that the runts would have averaged one quarter of the survivors for each character. Ranges rather than standard errors are given for the first two entries. Yields and girths are fairly highly correlated. Several points regarding fertilities and the effects of inbreeding are further explored by Simmonds (1986b).

exceptions, have not proved to be very useful. Historically, however, they did contribute to our present understanding that vigour and yield are both fairly highly heritable and related to each other. So correlations for both characters between successive stages and between yield and girth are generally positive, though widely variable in actual magnitude. By contrast, but as would be expected from simple physiological considerations (section 3.3.1, above), yield and girth increment under tapping tend to be negatively correlated. Nor does the fact that a thick bark and a large latex vessel system are often correlated with yield come as any great surprise. The breeder is offered no short cuts because he still selects on yield and vigour even if he notes the vessel system and bark characters as he does so.

Substantial bodies of correlation data are given by: Narayanan and Ho (1970, 1973), Ho *et al.* (1973b), Narayanan *et al.* (1973), Narayanan *et al.* (1973), Ho (1976), Hamzah and Gomez (1982).

There are, however, four correlations which are important and have potential for helping to frame economical breeding plans. These are: (1) between test tapping of young budded plants and performance in subsequent clone trials ($r = 0.73$, rising to 0.85, Ho 1976); (2) between test tapping of young nursery seedlings and yield in small-scale clone trials (SSCT) (about 0.3, Tan 1987); (3) between yields in SSCT and in subsequent large-scale clone trials (LSCT) ($r = 0.5$–0.79 for yield and 0.63–0.72 for girth in 700 Series trials, Ong 1981); and (4) between GCA estimates for seedlings and for the same genotypes as clones after five years of tapping (Tan 1978a). I return later (sections 3.4.3.4, 3.4.3.5) to how these results are or could be used. Meanwhile, I conclude that, though correlation studies contributed historically to appreciation of heritability/repeatability, they leave the rubber breeder where he always was, selecting empirically for vigour, yield, bark characters and wind resistance.

3.4.3 Practical procedures

3.4.3.1 General

The general routine of breeding at the RRIM (and, with minor modifications, at other places, too) is shown in Fig. 3.3. Crosses are made between chosen parents, seedlings are test tapped in the nursery, selected and the survivors go into SSCT. The survivors of SSCT go to the LSCT and the best of these emerge to recommendation. Overall survival is of the order of 10^{-4} and the process is slow. Even preliminary (Class III) recommendation takes about 20 years and large-scale (Class I) recommendation about 30.

There are two supplementary sources of seedlings among which selection is practised, namely: mature commercial trees from poly-cross seed from commercial seed gardens, which give rise to the so-called 'ortet selection'; and polycross seedlings from specially constructed polycross seed gardens. Numerically, these sources vastly enhance the relatively tiny population of hand-pollinated seedlings.

Efforts to shorten the otherwise very prolonged breeding cycle have been developed in recent years in the form of 'promotion

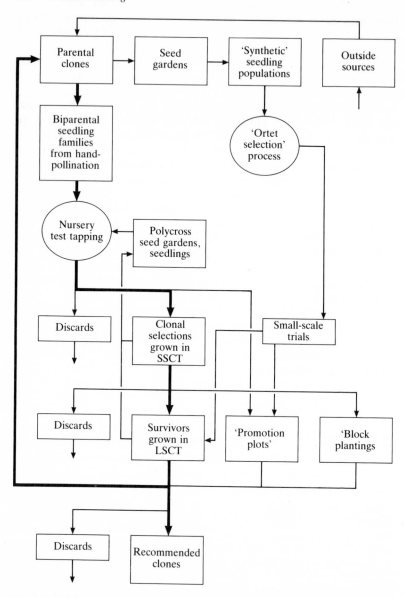

Fig. 3.3 Rubber breeding procedures. SSCT is small-scale clone trial (RRIM, equivalent to Prang Besar preliminary proof trial); LSCT is large-scale clone trial (RRIM, equivalent to Prang Besar further proof trial)

plots' and 'block plantings'. Both are essentially speculative choices of promising clones multiplied and tested quickly on a bigger scale than they would be under the standard routine.

3.4.3.2 Choice of parents

Most crosses are made with parents which are either of proven excellence as rubber-producing clones or are known to be reasonably good clones with some peculiar extra merits, such as wind or disease resistance. Some crosses are therefore made with the object of combining good characters from different sources. The biometrical evidence summarized above (section 3.4.2) gives some assurance that families will be, on the whole, of high average performance, provided always that inbreeding is avoided. If estimates of parental combining abilities are available from previous data, the crosses will be made with extra confidence. GCA estimates are especially valuable in focusing attention on good combinations that ought to be tried but have not been (Gilbert *et al.* 1973; Tan 1978a). Biometrical information is not yet available for disease resistances; it is reasonable to assume provisionally that it will reveal a predominantly additive pattern (section 3.5).

To confine the crossing to proven clones would, however, incur long delays in exploiting good parents. Biometrical theory shows that, given a large additive element in the genetic determination of economic characters, it is favourable to advance the generations rapidly by exploiting promising parents as early as possible. Thus it would be reasonable to use seemingly outstanding clones at, say, the SSCT stage, as parents, on the grounds that SSCT predict LSCT performance rather well (section 3.4.2). The process is not without risk, however, because information about wind and disease susceptibility may be very deficient. Pedigree data may allow some plausible guesses but the risk remains of producing advanced-generation material suffering from unforeseen defects. Nevertheless, some resources must be put into early, somewhat speculative crossing. Tan's (1978a) finding that seedling-based GCAs predict clonal GCAs supports the idea.

There is another category of cross that must also be considered, namely wide crosses between well-established parental clones and foreign materials of effectively unknown worth, either as clones or as parents. These may be expected occasionally to give large SCA effects apparent as heterosis (Ho & Ong 1981). If the foreign materials are ill adapted locally, however, as they probably will be, such families, however heterotic, may be unproductive of useful selections.

To conclude, therefore, much of the crossing will be devoted to proven parents of known good performance and breeding value; but some effort will also be devoted to somewhat speculative early crosses among promising, locally adapted but imperfectly known new clones; and some also to wide crosses with foreign materials, of prospectively poor intrinsic worth, in the hope of useful heterosis.

3.4.3.3 Pollination

Hand-pollination of protected female flowers is done after removal of male flowers from the inflorescences. The success rate varies widely with season and clone, from <1 to 8 per cent or so by season and from 0 to 10–12 per cent (occasionally more) for a fertile clone in a good season (Wycherley 1971). Disease, especially *Oidium*, tends to depress setting and sulphur dusting is sometimes worth while. *Hevea* fruits with less than the full complement of three seeds rarely survive and Ghandimathi and Yeang (1984) found that pollination may be limiting when the standard hand-pollination technique is used. They propose a somewhat different technique whereby, in two experiments on RRIM 600 as female, fruit set was approximately doubled. But the technique is demanding and it is not yet clear that such large gains are in practical reach.

Hevea seed is short-lived ('recalcitrant', an unfortunate term), cannot be dried off and must be sown quickly. Given care, though, viability of well-formed seeds is high. Husin *et al.* (1981) found that the best germination was obtained by harvesting fruits shortly before dehiscence, when natural seed moisture content was least.

Most pollinations are done on well grown trees, an expensive and laborious process. Efforts to induce early flowering on buddings in pollination nurseries have been in hand for many years (Ong 1972; review in Tan 1987) but have not reached practical use. The most promising techniques appear to involve ring-barking or girdling, with or without a coumarin spray. The matter is of considerable practical importance because an effective nursery technique might be economical and it would permit the early pollination of advanced selections (cf. section 3.4.3.2).

Natural fruit setting (usually 1–2 per cent) is even less than that following hand-pollination. Since clones differ genetically in fertility, a mixed stand (as in a seed orchard) is unlikely to yield polycross progeny equally representative of parent clones. Wycherley suggests that an exceptionally good seed orchard might yield about 160,000 seeds per hectare (780 kg) but guessed that 10,000 seeds per hectare might be a realistic average for ordinary plantings. The variable seed fertility of clones has implications for the breeder seeking to use polycrosses to advance the generations (section 3.4.3.6).

Wycherley's (1971) data on self- and cross-fertility by clones suggested that selfing was nearly always less productive than crossing, with average success rates of 4.8 per cent for crossing and 2.7 per cent for selfing. These data exclude RRIM 501 as a parent, which seemed to be exceptional in being about as fertile selfed as crossed. However, Tan and Subramaniam (1976) did not find that RRIM 501 was especially self-fertile and they found a much greater contrast between selfing and crossing in respect of rate of success (3.7 as against 0.5 per cent – Table 3.3). It is tempting (and probably correct) to regard this generally poor set on selfing as an early expression of inbreeding depression, the excess of failed fruits representing fruits with one or more lethally homozygous embryos, very early expressed runts, one might say (see Simmonds 1986b).

The relation between seed fertility and genetic advancement would be worth investigation. Analogy with other crops suggests that infertility would tend to be selected, over generations, as a correlated response to selection for the vegetative characters, growth and latex yield (Simmonds 1979, 1985). This has not yet been investigated but there is certainly plenty of room for such a response. The rubber breeder of the future may have to struggle to get crosses of some outstanding parents.

3.4.3.4 Nursery

In most crops, selection at the nursery (single-plant) level for yield is virtually useless because of low heritability/repeatability of the character. This is not so in rubber because yield is moderately heritable, i.e. correlated as between nursery and SSCT performance (section 3.4.2). Accordingly, nursery seedlings are grown at close spacing and test tapped by the Hamaker–Morris–Mann (HMM) method at 2–3 years of age (Tan & Subramaniam 1976; Tan 1987). Because the correlation with SSCT is low (*c.* 0.3) selection is, or should be, generous (at least 20 per cent). Other characters are also watched and selection is intuitively adjusted for girth, habit (light lateral branching favoured) and diseases (obviously susceptible plants are discarded).

There is an argument for growing the nursery at field spacing so that: (a) a second, later, round of selection can be made on better yield data; and (b) the families can be made to give genetic information on diverse characters. However, to do so takes more space and retains a lot of rubbish; also the needs of genetic studies can generally be met by budding random samples of families into separate plots. Another possibility is to bud all the seedlings and select on the results of test tapping, as before. The correlation with SSCT rises (section 3.4.2): selection becomes more efficient and can, in principle, be further improved by selecting against plugging (i.e. for low PI) (Ho 1976). However, a great deal of budding work and some delay is incurred.

No individual 'best' practice is apparent and the matter might usefully be the subject of an optimization study. At all events a fair share (we do not know exactly what proportion) of the best genotypes are carried forward to SSCT. The selection rate mentioned above, of at least 20 per cent, is on a 'whole population' basis. Figure 3.2 shows that families will be very unequally represented. Nursery test tapping may not identify superior plants with high reliability but will give quite good estimates of family means. Selection is therefore automatically concentrated on the best families where the prospect of genetic advance is greatest (Fig. 3.2, family CQ, top right).

3.4.3.5 Clone trials

The two cycles of clone trials, small-scale (SSCT) and large-scale (LSCT), are intended progressively to identify the best clones that emerge from nursery selection. SSCT are essentially fairly rough 'sorting trials', with very numerous (several hundred) entries, small row-plots and little replication. Balanced lattices with two replicates are usual and reasonably efficient (Narayanan *et al*, 1974b); but the modern 'generalized lattice' designs and rectangular plots might well be more efficient, in view of the numerous entries and large areas involved. Tree girths are measured intermittently as a measure of vigour and yields are recorded on a (monthly) sampling basis from regularly tapped, unstimulated trees. Traditionally, decision was based on yield over five years (i.e. the first panel), qualified by observations of tree habit, wind-fastness (if any wind damage occurred), incidence of dryness and diseases. Control clones are, of course, included and selections are required to exceed the middling–good control (currently RRIM 600) and at least closely to approach a precociously high-yielding standard. Given the increasing emphasis on early yield (section 3.3.1) and the need for speed, the present tendency is to select first for LSCT after only two years of tapping in the knowledge that non-precocious high-yielders can be picked up later.

LSCT have rather wider functions than SSCT. They are aimed, not only at identifying the highest yielders, but also at producing information upon which to base recommendation. That is, they must include all candidate clones, whether from local or foreign sources, and the trials must be widely enough distributed ecologically to give at least some idea of possible local adaptation. Lattices, again, are used, normally with three replicates of 8 × 8 tree plots of 20–30 clones. These are big trials (each about 10–14 ha, say 200 ha/an). The RRIM aims at 10 sites for each trial, well spread over environments diverse as to climates, diseases and soils. The normal girth and yield measurements are taken, together with observations of wind damage, dryness, diseases, bark characters, plugging and viscosity. The first two panels are yield-recorded, again on a sampling basis, without stimulation and under fairly low intensity of tapping d/3 (see Appendix 9.1). At least two years of yields on renewed bark are then recorded before the trial is closed or given over to secondary experimentation (e.g. on stimulation response). Early yield on renewed bark is regarded as an important indicator of the capacity of the clone to go on to complete a full economic life.

On any reckoning, these LSCT are large and prolonged experiments and they could not be otherwise. They take about 5 + 10 + 2 = 17 years before reasonably well-founded information on early yield, yield on renewed bark, vigour under tapping, wind-fastness and disease responses can be accumulated.

The accuracy of trials deserves some comment. SSCT have plot errors (CV) of about 25 per cent, which are rather high. The use of unguarded row-plots probably contributes to this (by competition) and the numerous entries (implying large area and soil heterogeneity) must also be relevant. Rectangular plots and generalized lattice designs would almost certainly help. LSCT have lower errors (plot CV 10–20 per cent) but, again, generalized lattices would probably be favourable. In all trials, there is obvious tree-to-tree variation within plots, due presumably to a mixture of rootstock (section 3.7.1) and random effects. Clonal rootstocks, if and when available, would probably be favourable but how favourable, in terms of lowering the CV, is unpredictable.

On the evidence of CVs, the LSCT are reasonably accurate experiments; given data from 10 sites, they would be expected to predict commercial yields fairly well and, in fact, they do so (Simmonds 1982). Commercial yield data for 24 clones that had reached recommendation were compared with LSCT yields. Means were: trials 1.40 t/ha, commercial 1.33 t/ha. The correlation was +0.8572 and the regression was:

$$Y_c = 0.403 + 0.6646 \, Y_t$$

That trials tend to overpredict commercial yields is probably due to the presence of greater error variance in the trials data and to 'attenuation', only the top end of the trials distribution having been selected for recommendation. There are some disagreements as to rank order but, broadly, the agreement is quite good. Curiously, the predictive ability of variety trials has been little investigated in other crops despite the importance of the fundamental assumption that underlies all such tests. What few data there are (in sugar cane, potatoes and cereals in various countries) suggest that trials are by no means always good predictors of commercial performance.

Any set of trials replicated over sites and seasons can be made to yield information on the relative constancies of varietal performance, that is on genotype × environment (GE) interactions (Jayasekara *et al.* 1977; Tan 1987). In a perennial crop such as rubber, varieties × years interactions may well occur but are inaccessible to study. Interactions with sites are to be expected and can indeed be detected by joint analyses of variance of sets of trials (review in Simmonds 1979). In principle, regression analysis of variety on variety and/or of variety on trials means could be informative in identifying genotypes which were more (high b) or less (low b) responsive to environmental 'quality'. In practice, a set of only 10 trials is insufficient to give more than hints of such differing reactions; there are signs that rubber clones do indeed differ in responsiveness but the data are insufficient for critical analysis (Tan 1987 and personal communication; see also section 3.8).

Decisions as to whether or not to recommend new clones for planting are based largely on the data accruing from LSCT. The subject is treated in Chapter 4 and we need only note here that yield dominates the decision-making, qualified by reference to other characters (section 3.3, above). In practice, a sort of 'intuitive selection index' (Simmonds 1979) is used, weighting characters according to some judgements of their economic values (Sultan 1973). Given that areas diverse as to climate, diseases and soils are to be served, the intuitive index may have to be adjusted in the light of information from LSCT as to potential for *local* adaptation (the Enviromax principle – Ho *et al.* 1974; Rubber Research Institute of Malaysia 1975b, 1983c).

3.4.3.6 Polycrosses

Polycrosses find two uses in breeding clones, namely: 'ortet selection' and those derived from specially constructed seed gardens (sections 3.4.1, 3.4.3.1 and Fig. 3.3). The latter have not yet been exploited to any significant extent but surely offer an attractive means of using somewhat speculative, advanced but unproven, parents relatively quickly and cheaply. Polycross progeny would be selected in the nursery (Fig. 3.3) and the breeder would have to be prepared to forgo information as to male parents. The procedure will no doubt be especially attractive for the exploitation of new genetic resources (section 3.6.3, Fig. 3.4).

'Ortet selection' is well established and early trials results look very attractive (Ho *et al.* 1980; Khoo *et al.* 1982). The approach depends upon the existence in Malaysia of large areas of superior seedlings, 'synthetic' populations (Fig. 3.1, 3.3). These derive from the Prang Besar Isolation Gardens (PBIG) of which a whole series of progressively more advanced ones have been developed over the past 50 years (Shepherd 1969a). Starting in 1972, the first phase of the work surveyed some 1.5 million trees in commercial cultivation; with much help from estates and tappers, some thousands of potentially superior trees were identified and a proportion was short-listed for yield on test tapping. To date, about 2000 have been cloned into small-plot trials and early yields have been recorded from the first two plantings. On yet scanty data, the results look spectacular: about 33 per cent yield as much as or more than RRIM 600 and 14 per cent as much as or more than the precocious control clone PR 255. Since early yield on young buddings is rather highly correlated with yield in mature trials (section 3.4.2), some truly superior clones must be present (despite 'attenuation' of the early yield data). The best 57 were deemed worthy of inclusion in promotion plots (section 3.4.3.7). Genetically, the PBIG seedlings are equivalent to the biparental families that rubber breeders have been looking at for several decades; but mean genetic values should rise the later the seed garden source, so subsequent cycles of selection should show sequential improvement. The strength of the procedure lies in the very high intensity of selection that can be applied: that is, i in the equation $R = ih^2\sigma$ can be far higher than normal, even if h^2 is rather low. A proper biometrical analysis of the trials will be of extraordinary interest.

An early example of 'ortet selection', in Indonesia about 1930, is described by Ferwerda (1969). About 6 million trees were assayed and intensely selected on their own performance under tapping. Only 260 were cloned into trials and only 3 emerged ultimately as good clones. In retrospect, with benefit of biometrical hindsight, it is evident that selection was grossly over-optimistic; relatively weak selection among promising trees, followed by rough preliminary trials of numerous clones (as in the current RRIM programme) would have been far more effective. This is, I think, a valid generalization for all clonal breeding.

There remains the question of the place of polycross/synthetic seedling populations in commercial rubber production. The trend has been downwards, to replace them with clones, but they persist in large areas and are still subject to Class I recommendation in Malaysia (Table 3.5). They are usually held to be robust and reliable, though not so high-yielding as the best clones. Their future is uncertain. If it were held that, on socio-economic grounds, they should continue to be made available to rubber growers, the genetic principles of seed garden composition are fairly clear from Fig. 3.2 (and see Simmonds 1986b). From an array of potential parent clones having varying GCA values, a selection of those having the highest values is made; from Fig. 3.2 it is easily calculated that a random polycross of clones B,C,P,Q would have yielded a mean of about 8.85 kg/tree or 137 per cent of 6.45 kg/tree, the mean of the whole population of crosses. In principle, more (say 10) rather than fewer parents should enter a seed garden to keep inbreeding down and choice would best be based on GCA estimates coupled with knowledge of parentages. In practice, not many good but unrelated clones are available, so parents are usually fewer than 10 and they are chosen on performance *per se*. A general theory of polycross design is given by Simmonds (1986b).

3.4.3.7 Shortening the cycle

The idea of 'promotion plots' was adopted by the RRIM in the early 1970s as a means of shortening the breeding cycle, from about 25–30 years to 15–20 years (Ong *et al*. 1984). The essential principle is to rely, once again, on the predictive power of nursery yield data and take outstanding clones straight from the nursery to a kind of LSCT, replicated over sites. The first eight clones were established in 1972–74 and yield data are available for the first panel (Ong *et al*. 1984). Many more selections have since been made, including 57 clones from the 'ortet selection' programme (section 3.4.3.6).

Yields of the first 8 clones were, as percentages of RRIM 600: 130, 145, 112, 60, 94, 90, 112, 173. Thus 1 looks poor, 4 look middling and 3 look good to outstanding, a salutary reminder that nursery selection is only moderately efficient (as is also regularly revealed, of course, by the results of SSCT). Of the 3 really promising yielders, 2 show signs of weakness against wind, for which no prior selection could have been practised. One looks forward with interest to critical analysis of more extensive data on more numerous selections. Meanwhile, it seems as though some high-yielders can be picked and even a 10 per cent overall success rate would be very well worth while.

3.4.3.8 Dwarfness

Dwarfing mutants have been identified in many crops and have been of great economic importance in several (e.g. wheats, barleys, rices, sorghums, bananas, coffee, coconuts, several temperate fruits). Some are dominant, some recessive and, in general, they act by way of disturbance of gibberellic acid metabolism. Typically (though not always), internodes are shortened but the plant retains its ordinary capacity for biomass accumulation; agriculturally, they must normally be grown at higher plant density. The effect of lessened stem height growth is to enhance potential for partition towards the economic product, with secondary (and usually favourable) effects on wind resistance and ease of harvest.

In rubber, the search has been going on for some 20 years but without success. Several putative dwarfs or semi-dwarfs have been found (e.g. at RRIM and Prang Besar) and several possible new ones have been identified among recent collections from South America (section 3.6.3). Some were simply stunted plants, others 'grew out' of the habit; none, so far, has been the vigorous, fast-growing plant with thick stems and short internodes that is desired.

In rubber, any useful mutant would have to be dominant or semi-dominant (as in coconuts, coffee and oil-palm) and it could only be used by systematically crossing it into high-yielding genetic backgrounds; at first occurrence it would almost certainly be in a poor to indifferent background. How useful such a mutant would be cannot be predicted, but there must be at least fair prospects of wind-fastness, enhanced partition and high yield (probably under closer field spacing). Closer spacing would have unwelcome implications for tapping costs, of course, but might still be economically attractive.

The prospects are interesting, and the search must continue, but practical achievements seem far off.

3.4.3.9 Polyploidy and mutation induction

This subject has been reviewed by Ong *et al.* (1984) and by Tan (1987). Briefly, nothing useful has emerged. Putative polyploids (from colchicine-treated buds) have been examined in several countries over a period of 20 years (e.g. Shepherd 1969b). The products have been plagued by chimeral problems and, though there have been some indications of enchanced yields, these have not been critically tested. Some materials, at least, have been characterized by reduced wintering, fragile/drooping branches and bumpy bark. As to mutation induction, rubber is a clonal outbreeder and, in practical effect, only dominant mutants could be detected and used. Since dominant dwarfing genes are known in other plants (sections 3.4.3.8 and 3.6.3) these might also be sought in rubber and indeed would be, in my opinion, the only sensible objective for a mutation induction programme in the crop. The basic requirements for any such programme (Simmonds 1979) are that the mutants sought should be genetically 'reasonable' and easily detectable.

3.5 Disease resistance

3.5.1 General

Diseases are covered thoroughly in Chapter 10 and all that is required here is a brief outline of plant breeding implications, which have also been treated by Ho (1986).

It should be noted that despite a long total list of pests and diseases, rubber (so far) suffers from only one really devastating pathogen, South American Leaf Blight (SALB, section 3.5.1.1. below). Of the rest, several moderately damaging leaf diseases have alone been judged worthy of attention from rubber breeders in Asia. A curious feature of rubber diseases is the lack of viral pathogens; most other major clonal crops are attacked by viruses, and often very damaging ones at that.

It is generally agreed that root, bark and stem diseases are well enough controlled by a combination of husbandry and, when necessary, chemical methods (Wastie 1975). The Malaysian rubber breeder can, and does, ignore them (though panel diseases need attention in Sri Lanka – Fernando & Liyanage 1976). As to the leaf diseases, three are of some consequence: *Gloeosporium* leaf disease (GLD – *Colletotrichum gloeosporioides*, perfect stage *Glomerella cingulata*); powdery mildew (PML – *Oidium heveae*; and *Phytophthora* leaf fall (PLF – *Phytophthora* spp.). On present (uncritical) evidence, genetic control of response to all three is quantitative; that is, there are no signs of vertical resistance (VR – see section 3.5.1.1.) In general, adequate horizontal resistance (HR) has been built up in relation to *locally* significant pathogens by a policy of what might be called 'watchful neglect'. Since the diseases differ ecologically (e.g. GLD is worse in wetter areas, PLF in drier, more seasonal climates), arrays of local clones show different spectra of resistances (Wycherley 1969). This is well illustrated by Wastie's (1973a) data on GLD, reanalysèd by Simmonds (1982, 1983). Selection in Malaysia and Indonesia produced (on a resistance scale 0–9) a middling-high level of resistance (5.6 ± 0.13); clones from the more seasonal environments of Sri Lanka and Indo-China were less resistant (3.8 ± 0.68) and South American clones were more resistant (7.4 ± 0.43), all reflecting local adaptation rather than conscious breeding. A policy of 'watchful neglect' merely implies that too-susceptible segregates are thrown away in the ordinary course of selection and trials; survivors are at least moderately resistant and, HR being fairly highly heritable (on general experience), the next generation of parents produces slightly more resistant offspring. In practice, most diseases are controlled in most crops this way but the fact only becomes apparent when varieties are tested in different environments or a 'new' disease arrives. One notes, also, that even if an otherwise excellent clone is rather susceptible to one disease or another (and they nearly all are!), then Enviromax planting recommendations are perfectly capable of recognizing the fact and adjusting planting decisions accordingly (Rubber Research Institute of Malaysia 1975b).

A neat example of local adaptation to disease is provided by *Corynespora cassiicola*, long known in Malaysia as a minor leaf spot. That the disease is potentially very damaging and that the Malaysian Wickham population carries an effective HR became apparent only when three 'foreign' clones proved to be very susceptible. One comes from Thailand, one from Sri Lanka and one is an American-Asian hybrid. The only actions that Malaysian breeders need take are to avoid those clones as parents and maintain the usual watchful neglect (see also section 3.6.3).

3.5.1.1 South American Leaf Blight (SALB)

There is now a large, scattered and diffuse literature on SALB, *Microcyclus ulei* (reviewed by Chee & Wastie 1980), which can, however, be rather simply summarized from the breeding viewpoint (Simmonds 1982, 1983; Ho 1986). Since the 1930s, immunities have been known, some derived from wild relatives, *H. benthamiana, H. pauciflora* and *H. spruceana.* Sporadic attempts to use these immunities by crossing to *H. brasiliensis* have produced numerous clones with transient, non-durable resistance. If any such clones still survive uninfected, they must be few and they must be expected to fail soon. This is a case of 'vertical resistance' (VR – Vanderplank's term). VR genes are (normally dominant) hypersensitivities which fail in response to newly evolved pathotypes of the fungus; they are *pathotype-specific.* Experience of other crops gives no support to the idea that multiplication of VR genes ('pyramiding') can do any more than delay the appearance of the new pathotype a little. By the use of 'differential host' series of cultivars, arbitrary 'races' of the pathogen can be distinguished; for practical purposes, it matters not at all whether 10 or 20 or 100 'races' can be identified, because the number is, in effect, set only by the patience of the investigator.

In annual crops supported by vigorous plant breeding, by rapid turnover of varieties and by excellent seed-multiplication-delivery systems, VR has often been quite useful, occasional disasters notwithstanding. In rubber, a long-lived perennial growing in a nearly non-seasonal environment, the effective use of VR is inconceivable and it must be judged to have totally failed. Furthermore, the derivation of VR genes from wild species means that the breeder would have to start from a very low genetic level in respect of other characters; several backcrosses and many decades of effort would be implied. I should add that no proper genetical study of VR genes in rubber has ever been done but that the inferences stated above seem clear enough for practical purposes.

The only alternative recourse is to 'horizontal resistance' (HR) which is genetically polygenic/quantitative and pathotype-non-specific. The 'minor' leaf diseases mentioned above (section 3.5.1) are minor because they are, like a multitude of similar diseases in diverse crops, controlled by HR developed to a level sufficient for the local environment.

Unlike resistance to these minor leaf diseases, obtained after several cycles of selection, there is no history of selection for HR to SALB. That HR could be built up by several cycles of selection *in the presence of the disease* is not, I think, in doubt but a large and prolonged effort in the American tropics is implied. SALB could, and some day very probably will, be reduced to the present status of GLD in Malaysia.

The core of any such work must lie in the field, not in the nursery, laboratory or glasshouse. Only the behaviour of substantial plots of mature plants can reveal the functional level of HR in a genotype. When, but only when, a good range of standards with known levels of HR have been assembled, then spot tests in nursery or laboratory may prove useful for preliminary screening or diagnosis – but spot tests cannot be invented *a priori* (although they all too often are).

The polygenic nature of HR determines breeding methods. Assuming (plausibly, on other experience) that inheritance is largely additive and h^2 fairly high, the obvious process of crossing resisters and selecting in the progeny would be appropriate. Polycrosses of resisters would also be attractive, followed by intense selection in large progenies.

At the mature-plant, field level there is yet, as far as I know, no clear evidence of HR. On nursery tests in Trinidad, Chee (1976a) discerned differences between clones, with a distinct hint that RRIM 600 and PR 107 had a little resistance. However, it has recently been noted (Tan, personal communication) that RRIM 600 under severe infection in Brazil is very badly attacked. The need for field experience is plain and one recalls one of the lessons of GLD: that varying levels of resistance are appropriate to different environments. It may be that, for drier areas, quite modest HR to SALB would suffice. This is merely universal experience for diverse airborne fungal pathogens; enough resistance is enough. Indeed, in Brazil, current plans call for an extension of rubber planting into 'escape' areas, where a marked dry season checks proliferation of the disease, and where tree growth and yield can be satisfactory (see Ch. 4, pp. 137–8).

While the long-term objective of breeding for HR to SALB would be, no doubt, the production of sufficiently resistant clones, there is a potentially very valuable intermediate step, namely crown budding (section 3.7). Early resisters, even if they were indifferent clones *per se*, could be used as crowns with very fair prospects of success. A sensible starting strategy would therefore be to concentrate on vigour and resistance, thus permitting relatively intense selection unimpeded by consideration of yield (Simmonds 1986a). As to materials, the new collections (section 3.6.2 below) should be screened in conditions of both moderate and severe natural infection; it would be surprising if some resistance did not show up in many thousands of quasi-random genotypes. A further recourse that might be productive would be to search old, abandoned seedling fields in tropical America in the hope of finding trees that had survived under local natural selection.

The prospect of trying to build up HR to a major disease in a long-lived tree crop might seem daunting; but, given crown budding and intense selection in large populations, reasonably rapid progress is far from inconceivable.

3.6 The genetic base

3.6.1 General

The early history of the introduction of *Hevea* to Asia is reviewed in Chapter 1. Essentially, a substantial collection of 70,000 seeds made by Sir Henry Wickham in 1876 in the Tapajós River area of Brazil suffered considerable losses in transit. Robert MacKenzie Cross's small Lower Amazon collection is widely thought to have failed but this view may not be entirely correct (see Ch. 1). In the outcome, only about 2000 plants reached Asia, mostly Ceylon (Sri Lanka). Singapore and Indonesia received a few dozen apiece and only a very small minority of the total contributed to the Asian genetic base. Whether or not any Cross plants survived, the great bulk of rubber production rests on the very narrow Malaysian-Indonesian source customarily referred to as the 'Wickham Base'. Over the years there have been sporadic exchanges of materials, within Asia and between Asia and America. One outcome was the series of 'exchange clone trials' initiated in the 1950s (Brookson 1956; Subramaniam 1969, 1970) but these, though informative, can hardly be held to have had much practical impact. Materials introduced from Brazil to Asia were selected for SALB resistance and all turned out to be poor producers from inferior, unselected *H. brasiliensis* sources or crosses with inferior wild species such as *H. benthamiana*; further, the best of the American materials were themselves crosses with Asian Wickham clones. Collectively, then, these introductions have done nothing for the genetic base.

Some awareness that the base needs to be broadened goes back a good many years (e.g. Baptist 1953; Brookson 1956) but acute awareness is more recent, maybe 20 years old (e.g. Simmonds 1969; Subramaniam & Ong 1974; Wycherley 1968a,b, 1969). The arguments in favour of widening the base are twofold. First, there is the evidence of pedigrees that shows that current populations go back to few 'primary clones' (about seven in Malaysia – Tan 1987), coupled with the consequence, that some degree of inbreeding is certainly occurring, probably, indeed, at a higher rate than is implied by pedigrees. Second, there are signs of diminished response to selection (Tan 1987; see also section 3.8.2 and Table 3.4). Some diminution of response must be expected, on physiological-genetic grounds, whatever the genetic base, but there is yet no compelling reason to suppose that rubber approaches a limit (though what that limit might be cannot yet be convincingly calculated).

Table 3.4 *Progress of yield improvement in the RRIM breeding programme; yields (kg/ha/an.) of successive phases*

Period of yield	Stage, advancement index, and material					
	Early	1920s	Phase I 1928–31	Phase II 1937–41	Phase III 1947–58	Phase IV 1959–65
	(0) Unselected Seedlings	(1) Pil B 84	(2) RRIM 501	(2) RRIM 600	(2.5) RRIM 712	(3) RRIM 803
1–5 yr	450	1000	1300	1550	1750	1650
6–10 yr	650	1350	1550	2450	2500	2350

Source: Tan (1987)

Notes: Based on diverse trials, with some extrapolation. RRIM 600 is one of the great rubber clones. The best Prang Besar (PB) clones have generally exceeded the last two entries (see section 3.8.1) so the overall decline in response is not so great as it might seem. For advancement index, see Table 3.5.

In general, the arguments in favour of a large, systematic programme of base-broadening are very strong indeed; it is a pity that serious action has been so long deferred.

3.6.2 Recent collecting

A major, internationally supported collecting expedition was undertaken in Brazil in 1981 and the background to it and preliminary results have been described by Ong *et al.* (1983). Plans were first mooted at an IRRDB meeting held in Colombo in 1976 and a plant breeders' workshop followed, in Kuala Lumpur, in 1977. A preliminary mission to South America in 1978 laid the groundwork and the expedition proper started in January 1981. It was staffed by eight scientists from Malaysia, Thailand, Indonesia, China, Ivory Coast and Nigeria and it concentrated on three chosen areas of the Amazon basin where there was expectation of good *H. brasiliensis* material: the States of Acre, Mato Grosso and Rondonia. The collectors pursued the genetically recommended strategy of taking moderately sized seed collections from numerous, widely dispersed and quasi-random seed sources. Besides about 65,000 seeds, they also collected about 1500 m of budwood from 194 selected trees thought to be potentially high-yielders. In accord with established plant collecting convention, half the seeds were left in Brazil, the country of origin; the remainder were distributed 75 per cent to the Rubber Research Institute of Malaysia, Malaysia and 25 per cent to the Institut de Recherches sur le Caoutchouc en Afrique, Ivory Coast. Ong *et al.* (1983) describe the very careful phytosanitary precautions, including an intermediate repacking and chemical treatment in England, that were adopted to minimize the risk of SALB transfer.

In the outcome, the RRIM has established about 14,000 seedlings and 100 clonal buddings (rising to 150). In comparison with the original Wickham base, numbered in tens of genotypes, this is a major achievement.

I note in passing that shoot-tip culture techniques (section 6.4.8) could be of great value in easing the problems of international transfer of vegetative material. One hopes that they will become available for the use of future collecting expeditions.

3.6.3 Exploitation

Preliminary observations of the Brazilian seedlings shows that, as expected, there is a great deal of variability in vigour, foliage characters and nursery disease reactions (Ong *et al.* 1983). In time, analysis of observations will no doubt lead to a greatly improved understanding of geographical variation in the species. More immediately, there is the question of how to frame a reasonable strategy for practical exploitation. On present evidence, there is no reason to favour any particular provenance; for a start, anyway, useful materials had better be assumed to be randomly distributed in the population.

The underlying genetic arguments on exploitation are as follows. Recalling an appropriate form of the response equation (section 3.4.2), namely:

$$X_N = \bar{X} + i \sqrt{h^2}. \, \sigma_G$$

we see that high future performance (X_N) will depend on a high starting mean ($\bar{X}$) and high genetic variability (σ_G). Since poorly yielding parents cannot be expected to give a high $\bar{X}$ it is clear at the outset that only fair–good yielders from the new populations can be expected to make really useful new parents; further, σ_G must be sustained in selecting or constructing those parents, so narrowly based selection would be self-defeating.

The elements of a strategy are therefore as follows (Fig. 3.4). First, there may be fairly high yielders that could be attractive, either as such or as parents, in the population as introduced. They will be revealed by the usual procedures of test tapping, selection for yield and other characters and small-scale trial against standards. They are, on *a priori* grounds, unlikely to be frequent because the source population is essentially unselected. However, some clones with the order of performance of Malaysian primary clones might reasonably be expected to occur.

Second, there is some, not very critical, evidence of interpopulation heterosis for vigour in rubber (Ho & Ong 1981). So any fair

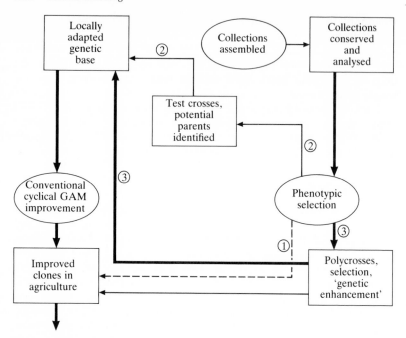

Fig. 3.4 Widening the genetic base, 1. Three main tracks (1, 2, 3) to utilize the new materials can be identified (see text). The most important routes are shown by bold arrows

yielders might, if crossed with standard Malaysian parents, give outstandingly vigorous families. These would be more likely to yield good crowns than good clones *per se* but, certainly, test crosses of any promising selections should be pursued with this in mind.

Third, and in the long term by far the most important, will be the genetic upgrading of a large population to a level at which X is high enough and σ_G has been well sustained. This implies generous (not intense) selection over two or three cycles. In effect, the programme must seek to re-enact the evolution of the crop in Asia but do it quickly and on a wide genetic base. The use of polycrosses is, of course, indicated, but otherwise, the normal rubber breeding procedures of nursery test tapping and visual selection will be used. Such a programme would, nowadays, often be called 'genetic enhancement'. Similar programmes are being carried on in diverse crops; they vary in detail but all appeal to the same underlying principle, of relaxed mass selection over cycles of improvement. No genetically sensible alternative is available.

At best, therefore, and with a good deal of luck, a few new clones might be entering routine rubber breeding in a decade or so.

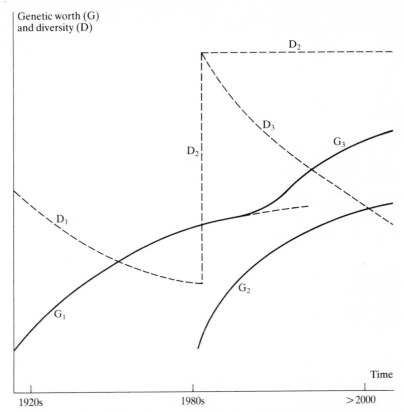

Fig. 3.5 Widening the genetic base, 2. As initial diversity (D_1) and response to selection (G_1) decline, introduction of new diversity (D_2), even in the form of initially inferior material (G_2), will generate a new (but lagged) response to selection (G_3). New diversity (D_2) is maintained in the collection but working diversity (D_3) declines as selection proceeds

More likely, any substantial impact will take several decades (Fig. 3.5).

So far we have been considering the general strategy of exploitation. Some details are worth noting. First, several putative dwarfs or semi-dwarfs have been identified (Ong *et al.* 1983 and see section 3.4.3.8 above). Their behaviour on propagation and crossing will be very interesting and of great potential value. Second, among the leaf diseases noted in the nursery in Malaysia, *Corynespora* has been rather prominent and has reached high frequencies in some provenances. Recalling the status of the disease in Malaysia (section 3.5.1. above), it looks as though some discard of too-susceptible materials will be necessary.

3.7 The compound tree

3.7.1 Genetic aspects of rootstocks and crown budding

One simple statistical fact dominates the relationships of stock with scion and of trunk with crown: effects are overwhelmingly additive and variance is therefore well accounted for by additive constants (equivalent to the plant breeder's general combining abilities – section 3.4.2). Thus, Ng *et al.* (1982), in an experiment with 6 stocks and 6 clones, found a narrow range of stock means (32.6 to 39.9 g/tree/tapping over 10 years) and a much wider range of scion means (16.9 to 47.0) but no evidence of interaction. Fitting additive constants accounts for 97 per cent of the variation (my calculation). There is room to choose improved stocks but not much; probably, the best stocks will prove simply to be vigorous, outbred seedlings, themselves potentially good rubber trees. (Curiously, this does not seem to have been properly tested.) If clonal rootstocks were ever to become available (see below) it is reasonable provisionally to assume that the same principle of additivity would apply. For the breeder, the conclusion is that he can use any stock he likes so long as any one experiment is homogeneous in this respect; choice of stock will not distort clonal comparisons.

With certain qualifications, similar results apply to crown budding (Tan 1979b). Again, additive effects predominate so that a good crown is such whatever trunk it is budded on but it need not itself be a good rubber producer. However, recent studies (Simmonds, in preparation) suggest that there is a larger interaction component than has hitherto been appreciated. Indeed, a multiplicative model based on analysis of logarithms serves very well. This finding is physiologically intelligible if we assume that a vigorous, wind-fast, disease-resistant crown need only provide a good flow of photosynthate to a trunk capable of high partition (Simmonds 1982). The partition effect suggests a biological basis for the fact that a multiplicative model works well.

For the breeder, the implications are that experiments testing trunks do not need a multiplicity of crowns and, similarly, that potential crowns need not be tested on numerous trunks. The choice of statistical model, additive or multiplicative, does not alter judgements as to worth of crown or trunk clones.

The prospect of clonal rootstocks would be attractive, not only for commercial planting (because the best would presumably be better than any seedling population) but also for rubber breeding (because any workable clonal rootstock should, in principle, reduce between-tree-within-plot variability and so enhance accuracy). So far, as is well known (see Ch. 6), rubber cuttings can be mist-propagated but lack tap-roots and tend to be unstable. Recent work (Leong & Yoon 1985) hardly encourages the idea that complex horticultural manipulations will be practically feasible but there are, nevertheless, interesting prospects that tap-rooted cuttings will ultimately become available. Some *in vitro* manipulations are also relevant to this idea (Ch. 6).

Finally, the additivity principle also seems to apply to latex viscosity (Mooney viscosity, MV, a measure of molecular weight – Ch. 11). This is indicated by the work of Leong and Yoon (1978) and of Leong *et al.*(1986). In general it looks as though the MV of trunk latex can be well described by a relation of the form:

$$V = a + bV_C + cV_T$$

where V_C and V_T are MV of the crown and trunk grown clonally. This is not, it should be noted, a relation based upon additive constants: it is a regression in which the coefficients b and c may be nearly equal to each other and to 0.5. Thus the signs are that trunk MV is, in principle, predictable. In particular, a middling MV can be expected from a high trunk on a low crown and vice versa. This essential intermediacy of MV of a compound tree deserves some biological explanation, which is not yet available. How does the crown influence the MV of latex in the trunk?

3.7.2 Breeding for the compound tree

So far, the breeder has sought good rubber clones and the results summarised above show that it does not matter what stock he tests them on; clonal stocks, if/when available, would probably enhance experimental precision, however. As to crown budding, it has not yet been widely adopted in practice and the breeder has had to do no more than be aware of the possibilities of occasionally picking out potential crowns on the one hand and high yielding but weakly trunk clones on the other. The logical conclusion of this dichotomy would be a dual breeding programme, aimed at crowns and trunks as separate objectives. Such a division would be genetically attractive because it would permit a great (at least tenfold) increase in selection pressure at no great increase in cost of testing (because of essential additivity) (Simmonds 1986a). All this, however, is for the future. If the industry were prepared to adopt crown budding as a general strategy, the breeder's task would be lightened and progress enhanced. The first clear objective for crown breeding *per se* has, of course, already become apparent in connection with SALB (section 3.5.1.1).

3.8 Achievements and prospects

3.8.1 Historical

Rubber breeding in south-east Asia is one of the outstanding success stories of plant breeding and has clearly been shown to be so in formal economic terms (Pee & Khoo 1976; Pee 1977). With qualifications (see below) it has been responsible for something like a four- or fivefold increase of yield in about 70 years. If we call the raw Wickham population generation 0, then the first cycle of selection ('primary clones') can be assigned an advancement index of 1, the second cycle of clones (from 1 × 1 crosses) scores 2, backcrosses 1.5 and so on. An unselected seedling population scores the same as its parents. The index is thus a measure of distance from the Wickham base. Tan (1987) gives pedigree data from which the index can be calculated. From Table 3.4 we see that the step from unselected seedlings (0) to RRIM 600 (2) and RRIM 712 (2.5) procured a roughly fourfold increase in yield. The last two clones are the highest yielders under current Class I recommendation (Table 3.5). Class II recommendation includes (mostly) promising but as yet incompletely evaluated clones for which only small- to moderate-scale planting is advised. The general average of Class II clones, it will be seen, is unremarkable in comparison with Class I (Table 3.5). However, a breakdown by source shows that, though the RRIM components (despite an increase in index) seem unpromising, the three Prang Besar (PB) clones for which 10-year yield data are available show signs of maintaining an overall upward yield trend with rising index. Thus it looks as though the potential for

Table 3.5   *Planting recommendations in Malaysia*

	Material	Advancement index	Yield (t/ha)		
			1–5 yr	6–10 yr	1–10 yr
Class I:	6 clones	2.1	7.5	11.4	18.9
	PBIG seedlings	[2]	5.9	7.4	13.3
Class II:	14 clones	2.5	8.2	10.8	18.9
	3 PB clones	2.8	9.6	12.7	22.3
	5 RRIM clones	2.5	7.5	9.7	17.2

Source: based on data from Rubber Research Institute of Malaysia (1983c)
Notes: The advancement index is the mean number of generations away from the Wickham base (= 0); necessary pedigree data are given by Tan (1987). Yield data for 6–10 yr (second panel) are incomplete for the 14 Class II clones but the last two lines contain only clones for which 6–10 yr data are given. The bracketed [2] given for the index of PBIG seedlings indicates that the figure is necessarily a rough average.

well over 2 t/ha/an. for the first 10 years of tapping is there. Good comparative data for older trees are wanting but plenty of yields over 3 t/ha/an. have been recorded. Nevertheless, however successful the best of the Class II clones turn out to be, it is fairly clear that the rate of increase per generation is slowing down.

Three other points about Table 3.5 are worth making. First, the average index of over 2 for clones conceals the fact that two outstanding primary clones (1) survive: the Indonesian GT 1 in Class I, regarded as an exceptionally reliable and hardy clone, though of somewhat less than average yield; and PB 28/59 in Class II. Second, the three outstanding PB clones in Class II (235, 255, 260) are all in the index range 2.5 to 3. And, third, PBIG seedlings are far superior to the Wickham base but well short of the clones in yield; 'ortet selection' among them (section 3.4.3.6) will yield clones of index about 3 but only experience will show whether the yields of the best will transgress those of the outstanding PB clones now in Class II.

The data of Tables 3.4 and 3.5 come from experiments carried out at a good level of estate management and we have seen above (section 3.4.3.5) that experiments are fairly predictive of commercial estate yields. Such data tell us nothing about genotype-environment (GE) interactions and, indeed, no really good data exist. General experience, however, rather strongly suggests that positive GE interactions, in the form clones × husbandry, have been an important feature of the historical improvement of rubber yields. Agricultural research on rubber has generated both improved genotypes and improved husbandry. Say that husbandry has risen from a low level (E_1) to a high (E_2) and that genotypes have improved from the Wickham base (G_1) to good modern clones (G_2), then we have a two-by-two table of the form given in Table 3.6. The figures in this table are frankly 'guesstimates' but they are based on much discussion with well-informed rubber researchers and represent, I believe, a reasonable synoptic view. However the table is read, the G effect is much larger than the E and there is a large GE component; really high yields are attained only with good clones and good husbandry. If, somewhat arbitrarily, we assign the GE component pro rata to the G and the E, we conclude that breeding (G) has contributed 70–80 per cent of total rubber yield improvement, husbandry (E) some 20–30 per cent.

In temperate annual crop agriculture, at high inputs, there has been a general experience of doubling of yields during the past 40 odd years and probably rather more than doubling on the 70-year time-scale with which we are concerned in rubber. It is commonly held that husbandry and breeding have borne roughly equal shares of the gain but there have been no critical studies and the GE component has normally been ignored even though, on scrappy

Table 3.6 *Genotype, Environment and GE interaction effects in rubber from a historical viewpoint (yields, kg/ha/an.)*

	E_1	E_2	
G_1	300	500	$\triangle (E) = +200$
G_2	1000	2500	—
$\triangle(G) =$	700	—	$\triangle(G + E + GE) = 2200$

	kg/ha	%	Analysis of variance	Sums of squares
$\triangle(G)$	700	32	G	61
$\triangle(E)$	200	9	E	24
$\triangle(GE)$	1300	59	GE	14
G effect		78	G effect	72
E effect		22	E effect	28

Notes: See text; the lower part of the table gives two interpretations, the first based on marginal differences, the second on proportions of sums of squares; an analysis of variance is based on an additive model which automatically minimizes the interaction component; in both interpretations, allocation of GE pro rata to G and E gives $G \gg E$.

evidence, it must have been substantial. In rubber, the overall gain in yield has been roughly eightfold (Table 3.6) and it is reasonable to enquire why it should have been so much greater than in temperate annuals and why the G share should have been so large.

I think there are two main reasons. First, the crop started from a wild tree in which, given reasonable vigour, yield is dominated by what turned out to be a rather highly heritable partition; so vigorous breeding and clonal propagation permitted exceptionally rapid genetic advance. And, second, the crop was domesticated, far away from its place of origin, in the absence of the critical disease SALB (section 3.5.1.1), the presence of which in Asia would, at best, have slowed the rate of advance and might even have prevented domestication.

3.8.2 The future

What, then, are the prospects for the crop? A crucial, but yet unsolved, problem is the biological cost of rubber to the tree. Templeton (1969b) put the 'yield summit' at about 7–11 t/ha, assuming a low cost; the much higher costs estimated empirically imply a much lower 'summit' (Simmonds 1982). It is not, therefore, known whether the evidence of declining response to selection for yield (section 3.8.1) is to be interpreted as an approach to a physiological limit or as a consequence of neglect of the genetic base (or perhaps a bit of each). I incline to the 'bit or each' idea but we really do not know.

If rubber yields are to continue to rise in the future, three things must probably happen (Simmonds 1986a). First, the knowledge of husbandry already available must be more generally used in practice in order to exploit fully the favourable *GE* interaction component discussed above. Many rubber fields, after all, are less than excellently managed. Second, husbandry (*E*) research has had, historically, a lesser practical effect than breeding/genetic (*G*) research and its share of future progress can only be less still. In a sense, husbandry research has been 'done'; no more than marginal gains can be foreseen for the future and its function is likely to become, I believe, essentially a matter of 'fine-tuning' field practice to the favoured genotype, to exploiting *GE* effects to the maximum, in fact. Third, it follows that breeding, always a weighty component of rubber research, will have to become the dominant component because nearly all the expectation of progress lies there and future genetic gains will be harder to make than the early ones were; this is simply a universal plant breeding principle: progress becomes more difficult the nearer the limit, whether genetic or physiological. The idea is explored a little more fully below.

If these arguments seem radical in nature, we should recall that they apply to all intensive, technology-based agricultures. In the sense that we now nearly always know how to grow clean, well-fertilized, weed-free crops at high yield, husbandry research has generally been 'done' and future research in that area must mostly lie with adapting the *E* bit to exploit the *GE* effect. As in rubber, any future gains must largely lie with breeding, in crops in which, as we are now beginning to understand, physiological and genetic limits are already being pushed.

Putting all this together, I think we can distinguish four main elements in future rubber breeding. First, there is the necessary widening of the genetic base (section 3.6) in the hope of replenishment of the additive genetic component of performance; whether or not this hope is realized, it is likely that the non-additive fraction of genetic variance will increase which will demand increasing biometrical sophistication from the breeders but is no great disadvantage in a clonal crop. Second, sheer numbers (of families, seedlings, clones) will have to increase, not only to accommodate the base-broadening operation, but also to enhance selection pressure as progress becomes harder. Third, rubber breeding (and, for practical purposes *only* breeding) has the potential to take the crop into environments which are now difficult, even impossible. I am thinking of monsoonal climates with severe seasonal drought and areas beset by SALB (section 3.5.1.1); more generally there will be endless opportunities for the exploitation of *GE* effects by the fine adaptation of clones to local environments. Fourth, the crop offers opportunities (unparalleled, I think) for the breeder to exploit the compound tree (section 3.7.2). To be able to breed the three components (the clonal rootstock, the trunk and the crown) separately, but with reasonable confidence of additivity of performance, would have immense attractions, the greatest of which would be the release of trunk breeding from the restraints of crown characters, with all their negative implications for selection pressures (Simmonds 1986a).

Rubber breeding has been very successful; future progress will probably be less spectacular and harder won but the opportunities are great and there is still much to do.

Chapter 4

Climate and soil

G. A. Watson

Hevea brasiliensis is indigenous to the forests of the Amazon basin, occurring generally within 5° latitude of the equator. Here, the climate is dominantly of the wet equatorial type (Strahler 1969), with no marked dry season – Af, Am under Köppens classification (Köppen 1923).

Commercial development of the crop first occurred in areas of similar latitude and climate, particularly in South-east Asia, but has since extended away from the equator to latitudes as far north as 29° N in India, Burma and China, down to 23° S in São Paulo State, Brazil. This extension has mainly occurred in the wetter areas of tropical wet-dry climate (Aw), including some areas where there is a marked dry season (Fig. 4.1). It is a tribute to the robust nature of *Hevea* that it can be successfully grown under the widely varying conditions that are encountered, but to obtain good tree performance a great deal of adaptive research has been required. This has involved a close study of the main environmental factors that affect the growth, yield and disease pattern of rubber – soil, temperature and rainfall.

4.1 Effect of rainfall on *Hevea*

Hevea has been thought generally to perform best in climates of the tropical lowland, evergreen rainforest regions, with an annual rainfall of 2000–4000 mm/an., evenly spread through the year and with not more than one dry month. The tree will 'winter' during this dry period and acquire a healthy new flush of leaves without undue damage by disease. Ideally, the number of rainy days should range from 100–150, for with any great excess above this number, conditions can lead to interference with the tapping process, and favour the development of leaf and stem disease.

The great bulk of commercial rubber, in South-east Asia, is indeed grown under conditions closely related to the above, and to appreciate the effect of rainfall on *Hevea*, its performance there must be compared with that in areas where rainfall is marginal.

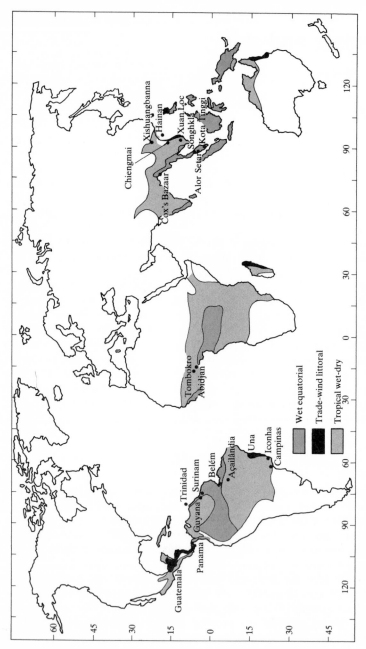

Fig. 4.1 Climatic regions associated with *Hevea* cultivation (After Strahler 1969)

4.1.1 Rubber cultivation under marginal rainfall conditions

With extension away from the equator into regions of tropical wet-dry climate, large areas of rubber are now grown in localities with a dry season of up to 5 months with total rainfall down to 1500 mm/an. The effect of this marginal rainfall on tree performance is confounded with that of temperature and other environmental factors. In the equatorial regions of rubber cultivation, mean annual temperature is 28 °C $\pm$ 2 °C, and diurnal variations about 7 °C (Barry & Chorley 1976). Annual mean and minimum temperatures decrease with distance from the equator, but because of increased day length and reduced cloudiness the amount of insolation increases. As a result, diurnal temperature differences are greater and the highest mean maximum temperatures are found around 5° N and S of the equator. Because of the higher total annual input of radiation energy, there is a greater potential for dry matter production from photosynthesis away from the equator (Oldeman & Frère 1982). In the absence of other limiting factors, this should have beneficial effects on tree performance.

Establishment of rubber under marginal rainfall conditions

In southern Malaysia, with around 2500 mm rainfall per annum and no severe dry period (Fig. 4.2a), conventional planting techniques give satisfactory establishment of rubber. In the north of the country, however, there can be periods of severe moisture deficit. Experience at Alor Setar (6° N, 1770 mm rainfall with moisture deficits for 4 months of the year – Fig. 4.2b) has shown that budded stumps of RRIM 600 and GT 1 planted in July may start well but be very susceptible to drought and failure during the subsequent dry period over December–March (Pushparajah 1983b). Buddings grown in polybags to the two-whorl stage proved much better suited to withstand the drought.

Further north still, at Cox's Bazaar (23° N) in Bangladesh and Xuan Loc (12° N) in Vietnam (Figs 4.2c, 4.2d), severe winter dry seasons will also demand appropriate planting practices, with particular attention paid to the use of vigorous planting material, early planting in the wet season, and subsequent protection against drought, wind damage and sun scorch. The situation is exemplified by data from Thailand. In that country, rubber traditionally has been cultivated in the peninsular south of Bangkok, Phuket Province, latitude 6°–12° N with rainfall of around 2000 mm. Preliminary trials, however, show that cultivation may be feasible much further north, at latitudes of around 18° N, and with rainfall of 1200–1500 mm falling over 120 days (Figs 4.3a, 4.3b). In these areas there is a marked dry season of 6 months, with a severe moisture deficit, and temperatures may range between 14 and 38 °C, with

128   *Climate and soil*

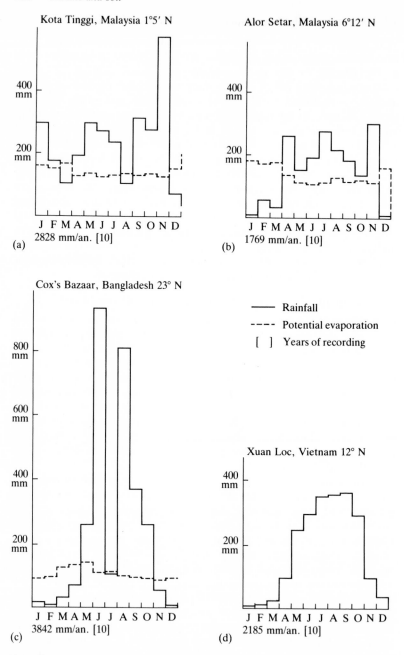

Fig. 4.2 Rainfall characteristics of rubber-growing localities in Malaysia, Bangladesh and Vietnam (*Source:* Pushparajah 1983b)

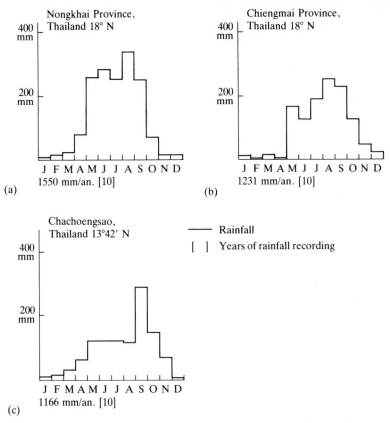

Fig. 4.3 Rainfall characteristics of marginal rubber-growing localities in North Thailand (*Sources:* (a),(b) Saengruksowong *et al.* 1983; (c) Prawit *et al.* 1983)

lows of around 5 °C for a few days. Results (Table 4.1) show that growth will be slower than in south Thailand, with trees taking at least 6 months longer to reach tappable size. Under the circumstances, however, this must be considered satisfactory, and early latex flow tests have been promising (Saengruksowong *et al.* 1983).

Greater difficulty has been experienced in attempts to establish rubber at Khao Hin-Suon in Chachoengsao Province, latitude 13°42′ N and 120 km east of Bangkok. This locality is typical of a large tract of land where deforestation, repeated cassava cultivation, and a fall in mean annual rainfall over the decades 1951–60, 1961–70 and 1971–80, from 1430 mm to 1349 mm and 1166 mm respectively, have presented particularly adverse conditions (Prawit *et al.* 1983). Annual rainfall and rainfall days for 1981 and 1982 were 1209 mm and 101 days and 1239 mm and 114 days

Source data for Figure 4.2   *Rainfall (R) and Potential Evaporation (E) Data (mm)*

Figure	Site		J	F	M	A	M	J	J	A	S	O	N	D	Total
4.2a	Kota Tinggi	R	284	174	101	194	249	262	231	101	314	277	565	76	2828
		E	154	152	164	127	137	120	128	136	135	137	123	156	1668
4.2b	Alor Star	R	1	52	37	255	145	184	268	212	180	138	295	3	1769
		E	174	166	168	131	108	102	109	122	112	119	109	156	1574
4.2c	Cox's Bazar	R	17	13	33	65	250	940	102	817	361	258	57	10	3842
		E	92	98	125	135	140	107	107	97	95	91	82	93	1262
4.2d	Xuan Loc	R	10	15	22	96	241	249	355	362	376	286	95	36	2186

Source: Pushparajah 1983b
R = rainfall
E = potential evaporation

Source data for Figure 4.3   *Rainfall data*

Figure	Site	Rainfall (mm)												Total		Years
		J	F	M	A	M	J	J	A	S	O	N	D			
4.3a	Nongkhai	10	14	20	79	254	280	247	337	242	61	13	16	1550	121	[10]
4.3b	Chiengmai	17	4	19	5	165	126	190	250	229	124	45	25	1242	115	[10]
4.3c	Chachoengsao	14	15	34	62	118	118	119	179	291	148	64	6	1166	—	[10]

Source: 4.3a & 4.3b: Saengruksowong *et al.* 1983
             4.3c: Prawit *et al.* 1983

Table 4.1 *Growth of rubber in marginal rainfall areas of north Thailand compared with that in south Thailand*

Site	Latitude	Annual rainfall (mm) (rainfall days)	Planting material*	Time from planting (years)	Mean girth (cm)
Phon Pisai, Nongkhai Province (north-eastern region)	18° N	1550 (121)	GT 1, PB 28/59, Tjir 1, RRIM 500, PB 5/51	4.5	33.6
			Seed-at-stake budded at 11 months: GT 1, PB 28/59, Tjir 1, RRIM 600, PB 5/51	4.5	28.7
Chiengmai Province (northern region)	18° N	1242 (115)	GT 1, RRIM 600	2.5	12.6
Phuket Province (southern region)	8° N	2163	GT 1, RRIM 600	2.5	21.2
			GT 1, RRIM 600, PB 5/51	4.5	39.5

Source: Saengruksowong *et al.* (1983)
* Except where otherwise stated, bare-root green-budded stumps were used.

respectively, and moisture deficits must be expected for 6 months of the year (Fig. 4.3c).

A pioneer planting of RRIM 600 in this area, at 14 years after planting, is said to yield around 2000 kg/ha with 8 months of tapping out of 12, but new plantings have been less successful. Budded stumps of RRIM 600 and GT 1, planted in 1980, suffered heavy casualties over the first two years after planting, despite early watering, mulching and whitewashing of the stems. Drought, sun scorch and wind damage were responsible for the losses, and future attempts at establishment are to include: soil improvement through deep cultivation, and use of legume covers; use of polybagged planting material and high-level budded stumps; early planting in the wet season; mulching; better care of the young trees using stem sleeves, pruning and branch induction; and use of windbreaks (Thainugul & Sinthurahas 1982). The use of drought-resistant root-stocks and clones (Combe & Gener 1977b; Pushparajah 1983b) would be of additional value.

Yield potential under marginal rainfall conditions

Away from the equator, the period of wintering and low yield in mature rubber coincides with that of low temperatures so that any adverse effects of the latter are minimized. On the other hand, during the summer wet season, with no constraints due to drought or low temperature, potential for latex production should be high. Provided that the trees are not put under excessive tapping stress during the dry season, overall productivity may even be enhanced compared with that experienced at the equator. Pushparajah (1983b) reports that at Cox's Bazaar in Bangladesh (23° N, Fig. 4.2c) with a monthly mean temperature of about 18 °C in December, January and February, it might take 7 years or more to bring a young planting into tapping compared with about 6 years in Malaysia. However, a stand of RRIM 600 in the first 18 months of tapping gave a yield of 980 kg/ha, comparable to that obtained nearer the equator.

Trials at Tombokro in Ivory Coast, at latitude 6°54′ N, with annual rainfall of 1200–1500 mm, a 5-month dry period (Fig. 4.4a), and an annual rainfall deficit of 250–400 mm, showed that growth of GT 1 was reduced in comparison with that experienced in the

Source data for Figure 4.4 *Rainfall data (mm)*

Figure	Site	J	F	M	A	M	J	J
4.4a	Tombokro	23	54	99	119	187	180	101
4.4b	IRCA-Abidjan	39	71	101	145	232	561	239

Source: Omont 1982

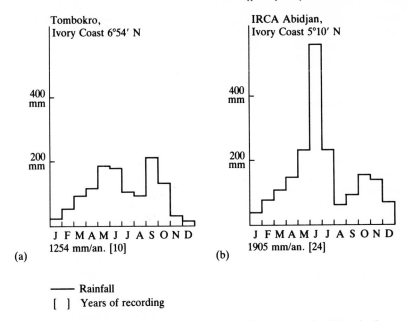

Tombokro,
Ivory Coast 6°54′ N

(a)   1254 mm/an. [10]

IRCA Abidjan,
Ivory Coast 5°10′ N

(b)   1905 mm/an. [24]

——— Rainfall
[   ] Years of recording

Fig. 4.4 Rainfall characteristics of marginal rubber-growing localities in Ivory Coast (*Source:* Omont 1982)

wetter Abidjan area (Fig. 4.4b). However, differences decreased with increasing age of the trees (Omont 1982). Irrigated trees in these trials were opened for tapping at $5\frac{1}{2}$ years after planting, and unirrigated trees at 7 years. By the fifth year of tapping yields were, respectively, 1641 and 1614 kg/ha/an., compared with levels of 1904 kg/ha expected of the clone at Abidjan. Malaysian work shows that irrigation after dry periods can reduce plugging and increase yields of GT 1, RRIM 612, and RRIM 703 (Haridas 1984) but this practice is rarely practicable or economic in the field.

It seems probable, however, that use of drought-resistant rootstocks and clones, combined with good husbandry, which involves timely establishment of polybagged plants, mulching and protection against wind and sun, could ensure the establishment of productive

A	S	O	N	D	Total	Year
95	214	130	35	16	1254	[10]
60	94	154	137	73	1905	[24]

rubber in all but the very worst circumstances. Once into maturity, any significant area of rubber will exert a beneficial effect on the local microclimate, the trees providing mutual protection against wind damage and sun scorch. In exposed areas such as those at Chachoengsao, shelter belts of drought-resistant trees such as *Leucaena* and *Casuarina* spp. would act as suitable 'nurses' to any new *Hevea* plantings.

4.1.2 Diurnal pattern of rainfall and interference with tapping

Heavy tropical rain can interfere badly with the tapping process, either by direct discouragement of the tappers, or more generally by canopy and branch runoff trickling down the trunk and interfering with the flow of latex into the collection cup. Interference with tapping is related to the time of rainfall and is classified as follows (Wycherley 1967):

1. 'Late tapping': when rain falls before tapping, the trees are wet and water trickles down the trunk so that tapping must be postponed.
2. 'Early collection': when rain falls during tapping, but there is time to carry out an early collection of latex before any interference. If early collection is not possible, and heavy rain washes the latex out of the cups, the situation is termed a 'wash-out'.
3. Very heavy rain before or during the normal tapping may prevent trees being tapped at all and results in a 'tapping day lost due to rain'.
4. Rain after tapping may result in loss of the 'late drip'.

A survey of 70 estates in Malaysia over a 5-year period gave a mean of only 5.3 per cent of tapping days lost due to rain, and 15.8 per cent in which rain interfered at all. These surprisingly low figures are due to the fact that rainfall in Malaysia follows a diurnal pattern. Over 17 recording stations, with mean annual rainfall of 2430 mm, 21 per cent of this rain fell between midnight and 0600 hours, 16 per cent in the normal tapping period of 0600–1200 hours, 35 per cent from 1200–1800 hours, and 28 per cent during the remainder of the day. Thus, the least rain fell during the tapping period itself, while most fell during the afternoon allowing the maximum period for the trees to dry off before tapping started the next day. Rainfall patterns for Alor Setar and Kluang in Malaysia are given in Fig. 4.5; obviously rainfall interference with tapping is more likely at the former, and was indeed confirmed in the actual survey (Wycherley 1963a).

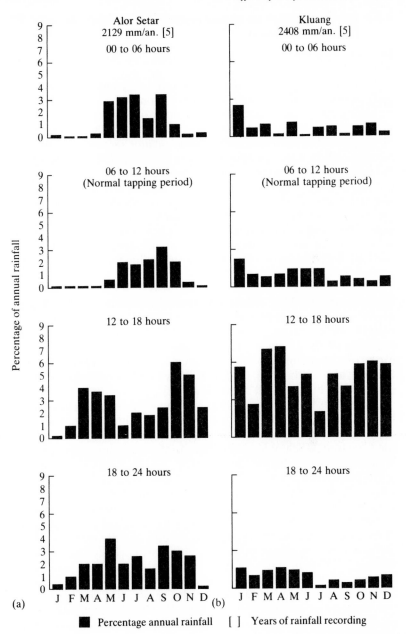

Fig. 4.5 Diurnal distribution of rainfall at two localities in Malaysia (*Source:* Wycherley 1967)

4.1.3 **Effect of rainfall on disease pattern** (see also Ch. 10, Diseases and pests)

In areas characterized by a well-defined dry season, *Hevea* will winter during that period and acquire its new canopy of leaves without undue damage by disease. In other areas where there is no such period or where there are heavy monsoon rains, and where wintering must take place during periods of intermittent rain and high humidity, the sustained wet conditions favour the development of *Oidium, Phytophthora* and *Colletotrichum* diseases, as well as *Microcyclus* in Latin America (Wastie 1972, 1973b; Peries 1979). All can severely damage the tender young leaf flushes that develop during the wintering period. These may absciss and fall, stimulating successive leaf flushes, any further loss of which may lead to invasion by secondary pathogens, stem dieback and general unthriftiness. The most severe expression of such a condition occurs in the wetter coastal areas of Brazil where South American Leaf Blight (SALB) due to *Microcyclus ulei* is a major limitation on the crop.

Rainfall and the problem of *Microcyclus ulei* in Latin America

South American Leaf Blight was first observed in 1900 on leaves of *H. brasiliensis* collected from the upper Amazon valley and eastern Peru, areas of wet-equatorial Af climate (Rogers & Peterson 1976). Damage caused by the disease was so severe as to force the abandonment of early plantings in Surinam, Guyana, Panama and Costa Rica, and at Belterra and Fordlandia in Brazil – all similar high-rainfall areas (Fig. 4.1).

The spores of *M. ulei* require a period of at least 10 consecutive hours of relative humidity above 95 per cent, with optimum average daily temperatures of 24–26 °C, to permit germination and infection (de Camargo & Schmidt 1976). Periods of intermittent rain are most favourable to the disease; in Trinidad, SALB is less severe in the north-east of the country (2500 mm rainfall) than in the north-west, where there is less rainfall (1300–2500 mm) but no distinct dry season (Chee, undated; Rao 1973). Similarly in Guatemala, disease is more severe at Navajos in the east, with rainfall of 2500 mm, than at Los Clavellinas in the west, with rainfall of 3500 mm but where there is a severe dry season. The situation has been described by Holliday (1970) in terms of high, intermediate and low incidence. Incidence will be high where annual rainfall is 2500 mm or more, fairly well distributed with no long dry season, i.e. not more than two consecutive months with less than 70–80 mm per month. Intermediate incidence can be expected where annual rainfall is less than 2000 mm and well distributed with no long dry season. Incidence will be low in areas with a variable annual rainfall of 1300–2500 mm but with a long dry season of at least four consecutive months with a rainfall of not more than 70–80 mm per month.

Fig. 4.6 Defoliation of *Hevea brasiliensis* due to *Microcyclus ulei*, Bahia, Brazil

In Brazil the disease causes little damage in plantings located along the banks of large rivers, where higher net radiation and air turbulence minimize germination of the fungal spores and subsequent infection (Bastos & Diniz 1980). Elsewhere SALB is severe along the coastal strip exending southwards from Belém, the main rubber growing area. Rogers and Peterson (1976) report severe disease at Marathon Estate, 120 km east of Belém, where the rainfall pattern is very like that of Fig. 4.7a, with no overall moisture deficit and rain continuing through the refoliation period of July and August. Disease incidence continues to be high down as far as Una (15°30′ S) in Bahia State, and beyond (Fig. 4.7b), but at Vitoria (20°20′ S) and Iconha (21° S) in Espirito Santo (Fig. 4.7c,d), where marked dry seasons occur over June–September, disease incidence is much less (Chee undated).

Incidence of SALB is also low at inland sites in Brazil as far apart as Açailândia (4°56′ S) in Maranhão State and Pindamonhangaba (23° S) in São Paulo, both experiencing a marked dry season. At Açailândia (Fig. 4.7e), with a moisture deficit of 335 mm, no symptoms of SALB were apparent in one clone trial planted in 1967, and yields of the order of 2000 kg/ha, have been obtained (Pinheiro *et al.* 1981). At Pindamonhangaba, and Campinas immediately to the west, rubber has been planted on well-drained upland sites to avoid frost damage (see Fig. 1.2); SALB is present in the area, but has only been serious in low-lying areas with cold and saturated night air (Chee undated).

Attempts to control SALB include the use of resistant varieties, top budding with such varieties, and application of chemical fungicides. It would appear more promising to shift rubber cultivation to those areas favoured by a distinct dry period, and in Brazil increased plantings in such 'escape' areas are planned (de Barros *et al.* 1983).

Effects of flooding on *Hevea*

Hevea brasiliensis roots largely in the upper soil layers, but can be quite sensitive to high water tables. With a permanently high water table, when the water level may commonly be within 60 cm of the surface, root development is restricted, and the tap-root has been known to grow back on itself to form a bulbous mass of corky tissues (Rubber Research Institute of Malaya 1959a). Under such conditions anchorage is poor and severe leaning and wind-throw can occur.

In exceptional circumstances, when land is inundated following very heavy rainfall, severe tree damage and death can occur. One Malaysian nine-year-old planting of Tjir 1 clonal seedlings on coastal alluvium suffered severe tree damage following inundation under 0.7–1 m of flood water for 25 days. Stem bleeding and some defoliation took place, but after wound dressing and resting recovery seemed probable (Rubber Research Institute of Malaya 1964b). Less fortunate were mature plantings of mixed clones and seedlings that were under 1–2 m of flood water for 40 days: 75 per cent of the trees were killed, the only survivors being small trees

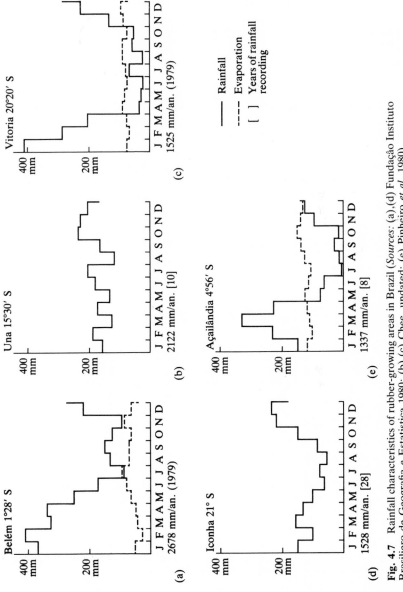

Fig. 4.7 Rainfall characteristics of rubber-growing areas in Brazil (*Sources:* (a),(d) Fundação Instituto Brasiliero de Geografia e Estatistica 1980; (b),(c) Chee, undated; (e) Pinheiro *et al.* 1980)

Source data for Figure 4.7   *Rainfall (R) and Potential Evaporation (E) (mm)*

Figure	Site		J	F	M	A	M	J	J	A	S	O	N	D	Total	Years
4.6a	Belem	R	367	407	330	342	256	167	81	136	150	127	88	226	2678	1979
		E	52	34	44	49	62	81	83	78	72	71	80	62	768	1979
4.6b	Una	R	159	187	130	179	137	178	201	115	161	236	234	205	2122	[10]
4.6c	Vitoria	R	411	284	201	35	29	18	61	23	49	47	134	228	1525	1979
		E	64	62	74	81	80	69	74	84	84	73	82	87	913	1979
4.6d	Iconha	R	150	104	159	130	101	64	74	59	81	155	220	231	1528	[28]
4.6e	Açailândia	R	145	236	324	229	77	61	6	14	26	14	94	111	1337	[8]
		E	118	105	115	120	122	110	104	124	135	142	133	127	1445	

Source: 4.6a & 4.6d: Fundação Institute Brasiliero de Geografie e Estatistica 1980
4.6b & 4.6c: Chee
4.6e: Pinheiro *et al.* 1980

and those untapped because of brown bast or pollarding for control of wind damage.

Clones may differ in their susceptibility to flood damage. In one catastrophic flooding in Malaysia in 1967, following rainfall of 2850 mm over four months, certain plantings were under 4.6 m of water for 6–7 days: clones RRIM 601 and RRIM 615 were defoliated, while RRIM 600 and RRIM 650 retained their leaves (Newall 1967). Obviously the planter or land development planner cannot make any provision against such events when they are unpredictable, other than normal maintenance of drainage outlets, but the probability is that rubber will recover quite well from a short period of shallow inundation. Constantly high water tables can, however, affect the overall performance of *Hevea*, and this is considered further on pages 179–82.

4.2 Effect of temperature and altitude on *Hevea*

The mean annual temperature of the lowland humid tropics is about 28 °C, and decreases at about 0.6 °C per 100 m in height. In such conditions, *Hevea* will grow most rapidly below 200 m, and trees require approximately 3–6 months longer to reach tappable size with each rise of 200 m above sea level, so that plantings above 600 m are not normally advisable. Nevertheless, in Java many estates laid out in a belt between 600 and 700 m above sea level have proved successful because of good siting with favourable exposure, soils and rainfall. Some clones, including LCB 510 (PR 107) and Tjir 16 have performed only slightly less well at 515 m.a.s.l. than at 250 m (Dijkman 1951).

The effects of altitude and temperature on *Hevea* have been most extensively studied in China (Huang & Zheng 1983; Pan 1983). Faced with a need to increase the area under cultivation, expansion has taken place mainly on the island of Hainan in Guangdong Province, and in Xishuangbanna Prefecture of Yunnan Province (Fig. 4.1). Of a total of 185 000 ha producing 140,000 t/an., some 39,000 ha are grown in the latter area in moist valleys running up to 800–1000 m.a.s.l. and produce 12,000 t/an. dry rubber (Anon 1983).

4.2.1 Cold damage to *Hevea*

The rubber-growing areas in China lie between 18° and 24° N and are affected by typhoons, cold climate and marginal rainfall (Table 4.2). The cold weather originates in Siberia and reaches the rubber-

Table 4.2 *Meteorological record in China's rubber-growing areas compared with that of other rubber-growing countries*

Element	Manaus, Brazil	Kuala Lumpur, Malaysia	Songhkla, Thailand	Location Danxian County, Hainan Island, China	Yaxian County, Hainan Island, China	Jinghong County Yunnan Province, China
Temperature (°C)						
Annual mean	26.9	27.1	27.4	23.4	25.4	21.7
Annual range	1.7	1.1	7.7	10.7	7.7	9.9
Yearly accumulated temperature of daily means 10°C	9,819	9,892	10,001	8,451	9,271.0	7,921
Coldest month mean	26.2	26.6	26.5	17.0	20.8	15.6
Extreme minimum	17.6	17.1	19.1	1.5	5.1	2.7
Annual rainfall (mm)	1,996	2,499	2,163	1,766	1,247	1,209
Rainfall days	171	195	159	162	114	n.a.
Number of days with wind of force 8	n.a.	3	n.a.	7	7	4
Maximum wind speed (m/s)	n.a.	n.a.	39	28	74	20
Sunshine hours	2,125	2,200	n.a.	2,177	2,498	2,153
Total solar radiation (kcal/cm²/an.)	n.a.	166	n.a.	115	132	n.a.
Latitude	3° 08' S	3° 07' N	7° 12' N	19° 30' N	18° 14' N	21° 52' N

Source: Huang and Zheng (1983)

growing areas in a series of waves. Resultant damage to the rubber can be of two types, radiative and advective.

The radiative type of cold damage can occur when night temperature falls sharply to less than 5 °C, with below 0 °C on the leaf surface and day temperature rising sharply to 15–20 °C. Rubber trees encountering extremely cold and hot conditions within one day will probably suffer cold damage. In milder cases, leaf margins shrivel, spots appear on leaf lamina, shoots die back and burst, and there is latex exudation. In severe cases the whole leaf becomes discoloured and withered, and shoots may die off.

The advective type of cold damage in south China is ascribed to temperature fall along with high wind in the early stage of a cold wave invasion, followed by an unbroken spell of windy and cold weather. It occurs particularly on flat terrain with no natural shelter from the wind, and is most likely to occur when the daily mean temperature remains below 8–10 °C, with a daily minimum of around 5 °C, and a wind velocity of over 3–4 m/sec for 3 or more consecutive days. The lower the temperature, the higher the wind speed, and the longer the duration of this kind of weather, the more severe the damage will be. Symptoms are that spots appear first on foliage and twigs, enlarging progressively before withering; the spots may spread to branches and trunks in some instances, and in severe cases the whole plant is affected and even the roots may die.

Different clones or individuals vary greatly in cold-hardiness. For some common clones, winter injury symptoms may be observed at 4–5 °C. Long-term tests have shown that both GT 1 and Haiken 1 (a China-bred wind-fast clone) can endure a low temperature of 0 °C for a short time, but the best cold-endurant clone bred in China, 93–114, can stand an even lower temperature of −1 °C, and has given yields of 750–900 kg/ha in the northernmost fringe of China's rubber-growing region with an absolute minimum temperature of 0 °C (Huang & Zheng 1983).

The latter clone, with hardy seedlings, is recommended as a main cultivar for severely cold-ridden areas, with GT 1 as a secondary choice. GT 1 is the preferred clone for moderately cold areas, while the less cold-tolerant clones PR 107, Haiken 1, PB 86 and RRIM 600 should be restricted to milder areas. Under such conditions RRIM 600 is quoted as yielding 1800–2250 kg/ha/an. (Huang & Zheng 1983).

In these cold areas of China, cold-air stagnation is avoided by increasing inter-row spacing to 12 m, by cutting undergrowth and by trimming of any windbreak. Lower tapping frequency with shallow tapping is adopted in winter, and tapping panels are dressed or sealed for a seasonal break in tapping.

4.3 Wind damage

4.3.1 Tree losses directly due to wind damage

In general, *Hevea* is not grown in areas badly affected by high winds. In the main south-east Asian producing territories, for example, wind speeds are generally in the region of 1–3 m/sec, with stronger winds in coastal areas, at times up to 4 m/sec (Rubber Research Institute of Malaya 1959b; Oldeman & Frère 1982). Storms do occur, however, and strong winds may cause tree loss with damage taking several forms. Trees may be uprooted altogether on soils with impeded drainage or concretionary pans where rooting has been limited to the surface soil layers (Rubber Research Institute of Malaya 1959b). Young plantings with heavy leaf canopies may show stem bending, and require corrective pruning and roping.

In Sri Lanka (Silva & de Silva 1971), uprooting is the most common form of wind damage, occurring principally near gaps in the stand and in hilly areas. Branch breakage and trunk snap are next in order with Tjir 1 particularly susceptible to the former. There is a marked increase in susceptibility to both uprooting and trunk snap as the trees reach mid-age, and a sharp decrease thereafter. Trunk snap is thought to be related to commencement of exploitation and reduced girthing.

In the typhoon-susceptible rubber-growing region of China, mean wind velocities of below 1 m/sec have no adverse effect on *Hevea* (Huang & Zheng 1983). At 1–1.9 m/sec there is some hindrance to growth, and above 2 m/sec windbreaks or shelter belts are required to prevent adverse effects on growth and yield. The adverse effects of these moderate wind speeds must be associated with a 'chill' factor unknown in the lower latitudes. With winds of Beaufort force 5–6 (8–13.8 m/sec) young leaves become crinkled or lacerated, and damage is aggravated by any accompanying cold waves. At winds of Beaufort force 8 (more than 17.2 m/sec) wind-susceptible clones are subject to branch break and trunk snap, and above Beaufort force 10 most rubber trees will suffer branch break, trunk snap and uprooting.

Use of shelter belts to prevent wind damage

In China, shelter belts are now widely used to protect *Hevea* in the more wind-affected areas, while the rubber is planted in square blocks of 1 ha in highly wind-prone areas, and rectangular blocks of 2–3 ha in less windy regions. The shelter belts consist mainly of *Eucalyptus* spp., with *Acacia confusa*, *Homalium hainanesis*, *Michelia macclurei* and *Schima superba*, together with *Camellia oleifera*.

In these areas, clones PR 107 and Haiken 1 are the dominant cultivars (Pan 1983), but RRIM 600 is also used, because of its good crown renewal capacity after wind damage. Tree densities are increased from 375 to 630 trees/ha in order to provide mutual shelter and thinner, less wind-susceptible canopies.

4.3.2 Tree losses in the early years of tapping

In Indonesia, serious tree losses due to wind damage were first observed with the clone AV 36 in the early 1930s (Dijkman 1951). Other wind-susceptible clones were BD 10 and Tjir 1, and Tjir 1 progeny, while AV 50 showed some resistance. Susceptibility to wind damage was greatest at the time of maximum girthing and canopy development, and was at that time related to branching structure, trees with narrow crotches being particularly prone to breakage. Concurrent work demonstrated a reduction of wind losses in high-density plantings, even with the wind-susceptible Tjir 1 (Table 4.3), and use of the higher planting densities permitted the planting of large areas of this clone on the wind-affected east coast of Sumatra. In north Sumatra, wind resistance is still regarded as the most important secondary characteristic of a clone (Harris & Siemonsma 1975).

In Malaysia, wind-damage problems only became serious in the 1950s, with the large-scale planting of clone RRIM 501. In the original trials of this high-yielding clone, the trees were opened for tapping at a larger than standard girth of 64 cm and seemed perfectly wind-fast. In subsequent commercial replantings, no wind damage occurred during the immature phase, but extensive trunk snap occurred at 2–3 years after opening for tapping at the standard 51 cm girth (Rubber Research Institute of Malaya 1959b; Wycherley *et al.* 1962). One of the parents of RRIM 501, and of its sibling RRIM 509, was LUN N, itself susceptible to trunk snap. Their susceptibility, and that of other high-yielding and trunk-snap-prone clones such as RRIM 613 and RRIM 614 (both Tjir 1 × RRIM 509) was not discovered until all had been used extensively in breeding programmes.

4.3.3 Effect of tree structure and wood strength on wind damage

Early investigation into the problem centred on studies of tree structure, wood strength, and the effect of fertilizer and cover plant use on wind-damage incidence (Rubber Research Institute of Malaya 1959b, 1961; Rosenquist 1961). Faults in the crown development of susceptible clones (Rubber Research Institute of Malaya 1967a were listed as:

Table 4.3 Effect of planting density on wind damage in five clones: Indonesia

Planting density (trees/ha)	Number of wind-damaged trees*					Total	As % of total number of trees present (including supplied trees)
	Tjir 1	Tjir 16	BD 5	Bd 10	Av 50		
250	80	43	36	43	7	209	9.8
400	63	44	37	34	16	194	5.6
550	68	54	50	36	16	224	4.7
700	52	78	39	40	20	229	3.9
Total damaged	263	219	162	153	59	856	

Source: Dijkman 1951
* Data are from experiment 1, planted in 1931/32, by Heubel (1939): with plots of approximately 2 ha in size.

Fig. 4.8 Typical trunk snap of clone RRIM 501 (Rubber Research Institute of Malaysia)

1. A very tall canopy.
2. Development of one or more secondary branches.
3. Heavy lateral branching especially with a narrow angle of union with the main stem.
4. Loss of stem dominance especially in the low to mid-regions of the crown, followed by development of a whorl of secondary leaders.
5. V-forks at the top of the trunk.

The above are certainly factors that predispose a tree to wind damage, and are countered by corrective pruning, as described in Chapter 6, section 6.6.2.

In studies on wood strength, workers in Ivory Coast found no relation between torsion couple (resistance to twist) and wood density (Institut de Recherches sur le Caoutchouc en Afrique 1976). Wood density was, however, a clonal factor and related to the age of the tree and to wind damage (Table 4.4).

In Malaysia differences in the modulus of rupture of green timber could not be related to wind damage (Rubber Research Institute of Malaya 1959b). Rapid growth during immaturity following use of leguminous covers and fertilizer applications did not result in weak wood, and detailed studies of the physiology of the tree during the early years of tapping indicated that other factors may be

Table 4.4   *Trunk wood density and wind breakage*: Ivory Coast*

Clones	RRIM 623	RRIM 605	IR 7	IR 22	GT 1
Trunk wood density	4778	4867	5153	5151	5186
Duncans 5%†					

Source: Institut de Recherches sur le Caoutchouc en Afrique (1976)
* Clones in italic are susceptible to wind breakage.
† Densities joined by the same line are not significantly different at the 5% level.

involved. After opening for tapping these combine to reduce the growth rate of all clones, and in particular to depress the rate of girthing more than growth of the crown. With some high-yielding clones this results in an imbalance, with the trees becoming top-heavy and prone to trunk snap.

4.4 Soils

Hevea has sometimes been described as a weed that will grow on most soils, and thrive where other tree crops might fail. This is an exaggeration. The crop will certainly grow on the vast majority of the acid soils of the humid tropics, but its performance and economic viability can be restricted severely where, for instance: there is deep, very acid peat; concretionary or rocky parent material is present and sufficiently massive to restrict root growth; drainage is excessive or impeded; or soil pH values are in excess of 6.5, as on young limestone or coralline-derived soils. Nutrient deficiencies themselves do not present a major limitation, for any shortfall can generally be made good by application of appropriate fertilizers.

4.4.1 Soil and land classification

In considering the effect of soil characteristics on the growth of *Hevea*, it is salutary to remember that the vast majority of early smallholder plantings in Africa and Asia were sited on areas chosen solely on grounds of availability and convenience. Commercial planters were guided by the condition of existing vegetation, and looked for freely draining soils of open texture to permit good rooting, on level or gently rolling terrain for easy access and absence of soil erosion. In general, they were very successful, although at times forced to accept marginal areas of peat swamp, ravines and steep slopes within concession areas. As plantation centres developed in Brazil, West Africa and Asia so also did planters' appreci-

AV 2037	PR 107	*Tjir 1*	PB 86	Y 427/3
5289	5594	5617	5683	5819

ation of their local environment. Soil surveys identified local associations and classified soils under series names within traditional nomenclature. National systems of soil classification were developed, to which the bulk of past agronomic work refers (Owen 1951; Joachim 1955). However, in order to facilitate the extrapolation of experience and experimental work between regions, attempts have been made to fit the rubber-growing soils into international systems of nomenclature (Silva 1969; Hardjono & Angkapradipta 1976; Panichapong 1984). Four main systems, reviewed by Beinroth (1975), Sanchez (1976), Young (1976) and Landon (1984), are particularly relevant to rubber growing: one developed in Brazil for the local Red-Yellow Podzolics and Latosols (da Costa 1968); that developed by the Commission for Technical Co-operation in Africa (CCTA) in producing a soil map of Africa based on French and Belgian traditions (D'Hoore 1964); that used by FAO in producing its Soil Map of the World (FAO-UNESCO 1974); and the system known as the 'seventh approximation' developed in the United States (United States Department of Agriculture 1975).

The latter system has received wide attention, but presents a problem in that while it may facilitate pedological comparisons between different climatic zones and cropping areas, its very detail makes it unsuitable for comparison of fertility and production potential of rubber-growing soils within the relevant and restricted zone of the humid tropics. With a perennial tree crop such as rubber, the effects of management husbandry on soil characteristics on the one hand, and adaptation of the tree to local soil and climatic features on the other, combine over the years to blur the finer differences between soil types in respect of their suitability for rubber growing. In consequence, tree performance may be similar on taxonomically different soils. There is in addition a limitation to the extent to which field management can be modified to take into account the different taxonomic phases that may be identified within any one planting unit. To meet this situation supplementary land use capability systems are required, grouping soils generally on the basis of crop performance rather than on taxonomic considerations.

A suggested land capability classification based on the FAO system

For rubber, Sys (1975) has put forward a land capability classification based on FAO principles, that defines orders, classes and sub-classes, and units of land suitability according to the number and severity of limiting factors that are present. These factors are defined in Table 4.5 and include those that can be corrected under good management (d, n, t), and those that cannot be corrected (s, c, l).

Soils with not more than 'moderate' limitations are grouped in Order I, as suitable for rubber. Soils with severe or very severe limitations that can be corrected are considered potentially suitable and are placed in Order II. Those with severe or very severe limitations that cannot be corrected are grouped in Order III, unsuitable for rubber. The orders are then subdivided into classes according to the number and degree of limiting factors (Table 4.6) with sub-classes eventually indicated by lower-case letters according to the nature of the limitations (Table 4.5), and into units determined by the overall nature and degree of the limitations. Thus the unit gives all information on land suitability. For example I.3d2 would indicate.

I   Order I, suitable.
I.3   Class, moderately suitable.
I.3d   Sub-class, moderately suitable, drainage limitations.
I.3d2   Unit, drainage limitations moderate: a soil suitable for rubber cultivation, but with moderate drainage limitations.

Classification of Malaysian soils for *Hevea* cultivation

The process of classifying rubber-growing soils under an international system and yet also developing a local capability classification, has been taken further in Malaysia than in any other country. In view of the comparative wealth of data available from that country its experience is detailed below, in the hope that useful analogies may be drawn for use elsewhere.

In Malaysia commercial and experimental data on *Hevea* performance and nutrient status have been related to: soil texture and slope; soil properties as influenced by ground covers; soil nutrient status; and other characteristics such as structure, aggregation, water-holding capacity, bulk density, presence of peat and acid sulphate conditions (Chan & Pushparajah 1972; Chan *et al.* 1972). Since all rubber is grown in the lowlands, altitude is of relatively minor importance. Criteria for grading soil suitability for rubber are quoted in Table 4.7, and have permitted the grouping of soils into five capability classifications (Table 4.8). Highest yields are obtained on deep, freely draining Oxisols and Ultisols, with

Table 4.5 Range in degree of soil limitations for *Hevea* cultivation; used to calculate orders, classes sub-classes and units of land suitability

Limiting factors*	None 0	Slight 1	Degree of limitation Moderate 2	Severe 3	Very severe 4
s Permanent soil limitations:					
Depth of soil (cm)	200	150–200	100–150	50–100	50
Texture†	Sandy clay Clay loam Silty clay loam	Fine sandy clay loam Loam Clay Silty clay	Coarse sandy clay loam Sandy loam	Loamy sand	Sand
Gravel and stones (%)	15	15–50	50–90	90	90
Reaction of subsoil (pH)	5–6	4.5–5	4–4.5 6–6.5	6.5–7	7.0
Drainage due to ground-water level	Well drained Water table 2 m	Well drained Water table 1.5–2 m	Moderately well drained	Imperfectly drained	Poorly and very poorly drained
c Months of dry season	1	1–2	2–4	4	4
l Altitude (m)	200	200–500	500–600	600–800**	800
d Drainage due to pseudo-gley	Well drained	Well drained	Moderately and imperfectly drained	Poorly drained	Very poorly drained
n Nutrition status	High to medium	Low to very low	Low to very low	Low to very low	Low to very low
t Slope (%)	3	3–8	6–20	20–35	>35

Source: Sys (1975)

* Factors d, n and t should be susceptible to correction under good management.
† Terms for texture are as used by USDA/FAO; see Landon (1984).
§ Experience suggests that, given appropriate fertilizer dressing, *Hevea* suffers no limitation to growth over pH range 4.0–6.0. On some basalt-derived soils, excess uptake of Mn may present a limitation if pH falls below 4.5. See also Table 4.7.
** But see experience in China and Java quoted in section 4.2.

Table 4.6 *A suggested system of land suitability classification for Hevea, using orders (derived from Table 4.5) and classes*

Number of limiting factors	Degree of limitation				
	None 0	Slight 1	Moderate 2	Severe 3	Very severe 4
0	I1 Very suitable	—	—	—	—
1	—	I2 Suitable	—	II1 Potentially suitable	II2 Potentially suitable
2	—	I2 Suitable	I2 Moderately suitable	II2 Potentially suitable	II3 Potentially suitable
3	—	I2 Suitable	I3 Moderately suitable	II2 Potentially suitable	II3 Potentially suitable
4	—	I3 Moderately suitable	I4 Marginally suitable	III Unsuitable – one or more very serious limitations that cannot be corrected by management (s,c,l in Table 4.5)	
5	—	I3 Moderately suitable	I4 Marginally suitable		

Source: Sys (1975)
Note: Limitations on Order I and II soils are those that can be corrected by good management.

performance falling off on soils that are subject to drought, impeded drainage, lateritic concretions and other adverse characteristics.

Soil capability and clonal effects

Several factors combine to complicate the simple picture of soil capability; their importance cannot be overestimated, for it is these factors that determine the ultimate degree of success of a planting. There is, for instance, a strong clonal factor that needs to be taken into account. Clones with a heavy canopy such as RRIM 600 tend to suffer severe losses due to uprooting, this being more severe on soils with an impenetrable concretionary layer, for example the Malacca series soil (Chan & Pushparajah 1972). However, clones like PB 51, GT 1 and PB 86 have generally smaller canopies and are better able to withstand the force of winds even though root depths may not be great. This feature is complicated by the fact that in any area of Malacca series soil the concretionary layer may be at varying depths from the surface, so causing variation in tree growth and wind-fastness, and perhaps requiring the use of a number of clones to achieve optimum performance overall.

In Malaysia it is also felt that clones may show differential adaptation to soil type. Data from a study of commercial plantings over some 1800 sites (Chan & Pushparajah 1972), showed that on Munchong and Holyrood soils, for example, RRIM 600 gave highest yields, but that on Rengam and Selangor soils this clone was inferior to GT 1 (Table 4.9). Rootstock selection could have an important part to play here, for strong rootstocks may boost growth in the early stages (Buttery 1961; Ng *et al.* 1982; Ng 1983) and help to minimize differences due to soil type.

Effect of management on soil/clone interaction

Experience from Malaysia also confirms that while the more productive classes of soils will give better growth and yield of rubber, they are also much more responsive to management inputs. Given satisfactory basic soil characteristics, the full benefits of good management practices, such as use of leguminous ground covers and appropriate fertilizer programmes, can be realized. The above survey (Chan & Pushparajah 1972) has shown, for instance, that on Munchong series, Class I soils, clone RRIM 600 gave mean yields of 1214 kg/ha/an. under conditions of minimum input, but that on the better holdings with full inputs, yields rose to a maximum of 2106 kg/ha/an. (Table 4.10). On poorly drained, Selangor series, Class V soils, however, yields were much lower and also much less responsive to high inputs. Under the same conditions, a more adaptable clone, GT 1, showed better response to management inputs, although yields still did not reach the levels attained on better soils.

The data quoted in Tables 4.9 and 4.10 are of the greatest

Table 4.7 *Principal criteria in grading of soil suitability classes for rubber (Malaysia)*

Soil properties	Desirable range	Degree of severity in limitations		
		Minor limitation	Serious limitation	Very serious limitation
I. Physical properties				
Rock outcrop (%)	Absence	50	50–75	75
Effective depth* (cm)	100	60–100	25–60	25
Texture	Almost equal amounts of sand and silt + clay	(i) Sandy loam (50–70% sand) (ii) Clayey (50–70% clay) (iii) Silty clayey (50–70% silt + clay)	(i) Very sandy (70–90% sand) (ii) Very clayey (70–90% clay) (iii) Very silty clayey (70–90% silt + clay)	(i) Extremely sandy (90% sand) (ii) Extremely clayey (90% clay) (iii) Extremely silty clay (90% silt + clay)
Consistency dry/moist	Soft/friable	Slightly hard/firm	Hard/very firm, loose	Very hard/extremely firm, very loose
Consistency wet (i) stickiness (ii) plasticity	Slightly sticky Slightly plastic	Sticky Plastic	Very sticky Very plastic	Extremely sticky Extremely plastic
Internal drainage†	Class D – well drained	Class C – moderately well drained	Class B – imperfectly drained Class E – somewhat excessively drained	Class A – poorly drained Class F – excessively drained
Peaty characteristic	Absence	Acid peat 50 cm from surface; 25 cm thick peat	Acid peat 25–50 cm from surface; 25–50 cm thick peat	Acid peat layer 25 cm from surface; 50 cm thick peat
Acid sulphate characteristic	Absence	Acid sulphate layer 50 cm from surface; 25 cm thick	Acid sulphate layer 25–50 cm from surface; 25–50 cm thick	Acid sulphate layer 25 cm from surface; 50 cm thick

	150 cm	100–150 cm	50–100 cm	50 cm
Moisture retention (based on available water in cm/m)	150 cm	100–150 cm	50–100 cm	50 cm
Permeability†	Moderate	Moderately slow or moderately rapid	Slow or rapid	Very slow or very rapid
Erodibility†	Class 1 (slightly eroded)	Class 2 (moderately eroded)	Class 3 (severely eroded)	Clas 4 (very severely eroded)
II. Chemical properties				
pH	Mean: 4.5 Range: 4.3–4.6	4.6–5.0	5.0–6.0	6.0
III. Physiographic features				
Slope	0–8°	8–15°	15°–33°	33°
Susceptibility to flooding	No flooding	Floods after very heavy downpours	Floods after heavy downpours	Floods after light downpours
Stagnation of water at surface	No stagnation	Water stagnates for a few hours	Water stagnates for 3 days	Water stagnates for 3 days

Source: Rubber Research Institute of Malaysia (1977c).
* Considers depth to hard pan (limestone, laterite layer, quartz vein, compact parent material, etc.) or to permanent water table.
† Based on USDA definitions.

Table 4.8 *Soil productivity ratings of some common Malaysian rubber-growing soils*

Yield categories	Soil suitability classes	Series name	Derivation	Taxonomic classification Order	Sub-group	FAO classification	Physical limitations
Above average 1250 kg/ha	I(a)	Munchong	argillaceous shale	Oxisol	Tropeptic Haplorthox	Orthic Ferralsol	None
		Kuantan	basalt	Oxisol	Haplic Acrorthox	Acric Ferralsol	None
	I(b)	Rengam	granite	Ultisol	Typic Paleudult	Orthic Acrisol	None
Average 1000–1250 kg/ha	II(a)	Harimau	older alluvium	Ultisol	Typic Paleudult	Orthic Acrisol	One or more minor limitations – moderate drainage, weak structure, susceptibility to flooding
	II(b)	Serdang	sandstone	Ultisol	Typic Paleudult	Orthic Acrisol	One or more minor limitations – susceptibility to soil erosion
	III	Holyrood	sub-recent riverine alluvium	Inceptisol	Oxic Dystropept	Dystric Cambisol	Minor limitations – weak structure and sandy within 90 cm. One serious limitation – susceptibility to moisture stress

	IV(a)	Batu Anam	argillaceous shale	Inceptisol	Aquoxic Dystropept	Plinthic Acrisol	More than one serious limitation – poor structure, strong compaction, poor permeability and infiltration
	IV(b)	Malacca	argillaceous shale	Oxisol	Petroplinthic Haplorthox	Plinthic Ferralsol	More than one serious limitation – laterite pans near surface and rock outcrops
Below average 1000 kg/ha	V(a)	Selangor	marine alluvium	Inceptisol	Sulfic Tropaquept	Gleysol	At least one very serious limitation – heavy clay, high permanent water table, and potential acid sulphate conditions
	V(b)	Sungei Buloh	recent riverine alluvium	Entisol	Orthoxic Quartzipsamment	Ferralic arenosol	At least one very serious limitation – very sandy, structureless

Table 4.8 (continued)

Yield categories	Soil suitability classes	Series name	Derivation	Taxonomic classification Order	Sub-group	FAO classification	Physical limitations
	V(c)		peat	Histosol	Tropohemist	Dystric Histosol	Several very serious limitations – high permanent water table, acid peat, no mineral component and unavailability of trace elements

Sources: Chan and Pushparajah (1972); Chan (1977); Noordin (1981)

Table 4.9 *Ranking of clonal yield performance (kg/ha/(an.) on different soil types in Malaysia (rank in parentheses)*

Clone	Munchong I		Rengam I		Holyrood III		Selangor V	
					Soil type and productivity rating			
RRIM 600	1736	(1)	1248	(2)	1290	(1)	897	(2)
RRIM 605	1522	(2)	1234	(3/4)	1118	(3)	690	(4)
GT 1	1452	(3)	1349	(1)	1102	(2)	984	(1)
PB 5/51	1270	(4)	1234	(3/4)	1125	(2)	872	(3)
Means	1495		1266		1159		861	
(Range)	(1736–1270)		(1349–1234)		(1290–1102)		(984–690)	

Source: after Chan and Pushparajah (1972)

Note: Yields are of panel A, over 2–4 years of tapping

Table 4.10 *Effect of soil type and management inputs on rubber yield: Malaysia*

Clone	Level of management inputs	Years of recording*	Mean yield (kg/ha/an.)			
			Class I Productivity rated soils		Class IV and V productivity-rated soils	
RRIM 600	max.	2–4	Munchong	2106	Selangor	985
	min.			1214		760
GT 1	max.	2–4	Rengam	1830	Selangor	1215
	min.			610		787
PB 5/51	max.	2–4	Munchong	1891	Selangor	1044
	min.			883		795
RRIM 623	max.	2–5	Munchong	1777	Selangor	1229
	min.			768		890
Tjir 1	max.	2–5	Rengam	1420	Selangor	925
	min.			471		690
PR 107	max.	2–8	Munchong	2000	Selangor	1344
	min.			978		842
PB 86	max.	3–10	Rengam	2122	Batu Anam	1216
	min.			837		1035
Means (Range)	max.			1878		1137
				(2106–1420)		(1344–925)
	min.			833		828
				(1214–471)		(1035–690)

Source: after Chan and Pushparajah (1972)
* Data are inclusive over years after opening for tapping.

importance. They indicate that with minimum management inputs only poor performance can be expected even on the better types of soil. With good management, however, involving the use of leguminous covers, good weed control, and particularly use of appropriate fertilizer regimes, major increases in productivity can be achieved. These increases may amount to more than a doubling of the yields that might otherwise be obtained, and are greater than can be achieved by the use of high-yielding clones alone. They can mean the difference between success and failure of a planting.

4.5 Selection of planting material

In Malaysia clonal and seedling material for conventional planting is classified into three main classes, depending on the amount of experimental and commercial experience that has been acquired:

Class I: material recommended for planting on a wide scale. Performance of material in this class, as recorded in experimental trials, has generally been confirmed by records from commercial areas.

Class II: promising clones that are not as fully tested as Class I material.

Class III: experimental material, divided into two classes, Class IIIA for promising clones to be planted in blocks of up to 10 ha per clone, and Class IIIB for mainly new selections for very small-scale planting.

When required, a secondary classification of material for crown budding is also available (Rubber Research Institute of Malaysia 1983c).

From the foregoing sections, however, it can be seen that the environment can have a major influence on *Hevea* performance. Most rubber-growing territories acknowledge this fact by basing recommendations for planting material on the results of local clone trials, but often the link between research and extension remains weak, in so far as environmental effects are taken into account in clone selection.

4.5.1 The Malaysian 'Enviromax' system for selection of planting material

In peninsular Malaysia, clonal recommendations are carried out according to an 'Enviromax' system (Rubber Research Institute of Malaysia 1983c). In this system, the rubber-growing areas have been divided into environs according to the factors (e.g. wind damage and fungal diseases) that act as constraints in the selection of clones.

Each environ is identified by a boundary and a distinctive colour or combination of colours and alphabetical codes indicating local wind characteristics and major disease incidence (Fig. 4.9). Soils are grouped, as already explained, into five broad, suitability classes (Table 4.8); these are colour-coded, and the overall environ is then characterized by a colour or combination of colours and codes which

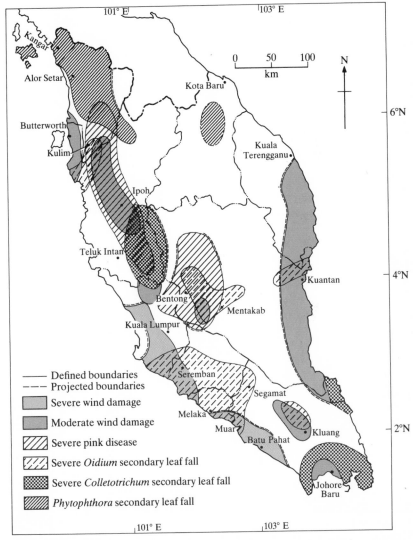

Fig. 4.9 Enviromax system, Malaysia; delineation of rubber-growing areas according to occurrence of wind damage and high incidence of disease (*Source:* Rubber Research Institute of Malaysia 1983c)

Table 4.11 *'Enviromax' system, Malaysia: characterization of recommended clones*

| Clone | Susceptibility to wind damage | | Severe susceptibility to | | | | Suitability for steep terrain (16°) | Suitability for shallow soil | Suitability for close planting (3 m) |
	Moderate	Severe	Pink	Oidium	Colletotrichum	Phytophthora			
Class I									
RRIM 600			0			0	acceptable	unsuitable	unsuitable
RRIM 712			0			0	acceptable	acceptable	unsuitable
PB 217							suitable	unsuitable	acceptable
PR 255				0			unsuitable	acceptable	unsuitable
PR 261					0		unsuitable	acceptable	unsuitable
GT 1					0		suitable	suitable	acceptable
Class II									
RRIM 527			0		0		acceptable	acceptable	acceptable
RRIM 623		0	0				acceptable	unsuitable	unsuitable
RRIM 628	0		0				unsuitable	unsuitable	unsuitable
RRIM 701	0		0				unsuitable	unsuitable	unsuitable
RRIM 703		0	0				unsuitable	unsuitable	unsuitable
RRIM 728	0				0		acceptable	acceptable	acceptable
RRIM 729			0				acceptable	acceptable	acceptable
PB 230							acceptable	acceptable	acceptable
PB 235		0					unsuitable	acceptable	acceptable
PB 255			0				acceptable	unsuitable	unsuitable
PB 260	0						suitable	acceptable	acceptable
PB 280		0	0				unsuitable	acceptable	suitable
PB 28/59	0				0		unsuitable	unsuitable	unsuitable
AVROS 2037						0	unsuitable	unsuitable	unsuitable

Source: Rubber Research Institute Malaysia (1983d).
Susceptibility to a malady is indicated by 0. Each defect is given a colour code which is the same as that in Fig. 4.9. Only the significant defects are shown against the recommended clones. Class I clones are recommended for both estate and smallholder sectors of the industry. Discrimination is used in planting Class II clones.

Shallow soils are soils where:
(i) Parent rock or an impenetrable layer is reached at 75 cm or less from the soil surface.
(ii) An impenetrable layer may be a layer of compact laterite, stones, pebbles or a cemented horizon.
(iii) Water table is present at 75 cm or less from the surface; rock outcrop forms 50% of the soil surface.

Table 4.12 *'Enviromax' system, Malaysia:*
Ranking of clones according to yield potential on the various soil classes

Soil class	Class I clones	Class II clones
I, II & III	RRIM 712, RRIM 600, PR 255, PR 261, PB 217, GT 1	PB 235, PB 255; RRIM 703, PB 260, PB 28/59, PB 280, RRIM 729, RRIM 628, RRIM 728, PB 230, RRIM 527, RRIM 623, RRIM 701, AVROS 2037
IV	RRIM 712, PR 255, PR 261, GT 1	PB 255, PB 260, RRIM 527.

Source: Rubber Research Institute of Malaysia (1983d)

Notes:
(a) Information on the performance of PR 255, PR 261, RRIM 712 and PB 217 on various soil types is limited. Clones PR 255 and PR 261 resemble RRIM 600 in yield level and habit. They are, therefore, provisionally ranked after it.
(b) Class II clones are ranked on their yield performance in clonal trials. On Class IV soil, however, only the more wind-fast Class II clones with relatively lighter crowns and shorter habit are recommended.
(c) Class IV includes Durian, Malacca, Batu Anam, Sogomana, Sitiawan, Gajah Mati, Apek and Marang soil series.
(d) Class V soil is not recommended for rubber.

reflect both the important environmental constraints and soil type. Clones are characterized according to yield and secondary characteristics, and major defects are colour-coded so as to facilitate the mechanics of clone selection for a particular planting environ (Table 4.11).

In selecting material for planting in a particular locality, within constraints imposed by the overall classification noted above, the following system is used:

1. Eliminate from the list (Table 4.11) all clones which are susceptible to the maladies (i.e. wind damage and fungal diseases) which occur in the area (Fig. 4.9).
2. Ascertain the suitability class (Table 4.8) of the soil to be planted, and choose from the clones remaining those that yield best on the soil class concerned, as indicated in Table 4.12.
3. Refer again to Table 4.11 to determine which of the clones is most suitable for the topography and soil depth in different parts of the locality and for the system of planting intended. Clones described as suitable for close planting, for example, have relatively small canopies and could be recommended for hedge-row plantings, or for high-density plantings in wind damage susceptible areas. On smallholdings, a simplified letter-coded system is available (Rubber Research Institute of Malaysia 1983c).

Chapter 5

Preparation of land for planting and replanting

C. C. Webster

5.1 Clearing forested land

New rubber plantations are mostly established in areas of primary or secondary lowland tropical forest. The virgin forest consists of three or four storeys of evergreen trees, the top storey being scattered, very large, buttressed trees, over 30 m high, not forming a continuous cover but emerging here and there from the canopy formed by the next storey below. The lower tree storeys occur at about 25 and 15 m. Numerous climbing, woody lianas are a conspicuous feature. Below the mature tree storeys there is a layer of young sapling trees and shrubs and below this some under-shrubs and herbaceous plants, but in virgin forest this undergrowth is not dense because of the low light intensity at this level.

Secondary forest is the regeneration which occurs after the land has been cleared from virgin jungle, cultivated for a time and then abandoned. Usually the initial clearing has been partial, the larger trees being left standing, and the cultivation has only been of a few years' duration. The nature of the secondary growth depends on the time it has had to regenerate but most rubber is planted after high secondary forest rather than after young regeneration. High secondary forest has fewer large trees and denser undergrowth than primary forest but the methods used for clearing the two types differ little.

Most of the soils on which rubber is planted are not highly fertile and the growth of luxuriant forests on them is maintained by a nicely balanced, virtually closed, cycle of nutrients between the vegetation and the soil. The forest annually contributes a large amount of organic matter to the surface soil in the form of leaf and stem fall and dead roots, which is rapidly broken down by the soil flora and fauna. Nutrients released by the decomposition of the organic matter are speedily reabsorbed by the mass of mainly shallow roots with the result that the loss of nutrients by leaching is slight. Almost all the nutrients are held in the vegetation and the surface soil, the subsoil usually being poor in most nutrients. The good structure of the humic topsoil is protected from the impact of high-intensity rainfall by the closed forest canopy and by the deposited litter so that rainfall readily infiltrates into the soil and there is little surface runoff except on steep slopes during heavy storms.

When the land is cleared, and especially if it is cultivated, the closed nutrient cycle is broken, the water regime is changed and the productive potential of the site can deteriorate rapidly unless it is carefully managed. Exposure of the bare soil to the sun and to the impact of rainfall results in accelerated decomposition of the organic matter, leaching of nutrients, breakdown of the aggregate structure of the surface soil, diminished infiltration of the rainfall and, except on flat land, much runoff of water and erosion of the nutrient-rich surface soil. To avoid these ill effects as far as possible, exposure of the bare soil should be minimized and the clearing and cultivation methods employed should disturb or compact the soil as little as possible and thus do the least damage to its structure. At the earliest opportunity after clearing, terraces or other soil conservation works should be constructed on sloping land and either a ground cover of controlled naturally regenerating vegetation established or leguminous cover plants sown between the rubber tree rows (see Ch. 7).

The importance of employing appropriate clearing methods and other well-tried measures designed to conserve soil and water is nowadays generally appreciated, but, partly because of the cost, the required procedures are not always fully implemented. However, avoidance of the capital loss represented by the erosion of valuable, and irreplaceable, surface soil is normally well worth the initial expenditure on effective soil conservation.

Before starting clearing, it is necessary to make a rough survey of the area to be dealt with. This is done by clearing a path around the boundary and making a grid of inspection traces which are cut on compass bearings at intervals of about 200 m. By walking these traces and noting the occurrence of various features at intervals, a sketch map can be prepared which shows the lie of the land, variations in soil, presence of ravines, rock outcrops, swamps, etc. It can then be decided whether or not certain areas should be cleared and planted. Undrainable swampy areas must clearly be excluded. Steep-sided, narrow ravines and steep hilltops, especially those with exposed rocks, will often best be left under forest. It is difficult and expensive to put in effective soil conservation works on steep, rocky slopes and if even modest erosion starts at a hilltop it can cause great damage lower down.

Another essential preliminary is to prepare a schedule of the time, labour and equipment required for clearing and planting the area. Strict adherence to a time schedule is essential for the efficient conduct of operations in field and nursery. In most places the various operations have to be timed to fit in with seasonal variations in the rainfall. For example, in parts of Malaysia it may be necessary to fell the forest in October–December, leaving two months for the felled vegetation to dry before burning in the drier weather of January–March and then to carry out subsequent operations before planting at a time of reliable rainfall in August–September. Similar considerations will apply in other countries with variations depending on the incidence of drier seasons. This kind of schedule tends to create intermittent demands for much labour and/or machinery over relatively short periods. Because of this there is an increasing tendency for largely mechanized clearing and land preparation to be done by contractors who are able to make use of skilled labour and expensive machinery for most of the year by moving to different parts of a country in accordance with differences in the rainfall regime.

5.1.1 Outline of clearing procedures

The main operations in clearing and preparing land for planting might be carried out in the following sequence.

1. Clearance of undergrowth.
2. Felling of trees, perhaps accompanied by extraction of stumps if clearing is mechanized.
3. Burning, after allowing about two months for drying of felled timber.
4. Second burning, after cutting and stacking, or restacking, unburnt timber.
5. Constructing roads and soil conservation works.
6. Establishment of ground cover plants.
7. Lining.
8. Holing.

Some of these operations will overlap in time. The precise sequence and degree of overlapping will depend on the rainfall regime, the topography, the clearing methods used and the size of the area to be cleared. The important operation of drainage has not been listed above as it may be carried out before or after clearing, but it is discussed in section 5.4.

5.1.2 Manual clearing

Nowadays much clearing is mechanized but manual methods are still employed in some places where they are cheaper, or on sites unsuitable for heavy machinery, such as steep slopes, areas where there are frequent rock outcrops, or where drainage is poor. They are also used for small areas, both on estates and smallholdings, where clearing with large machines is uneconomic or impracticable. Often a combination of the two methods is used, especially where there are many large trees; underbrushing and felling may be done manually but machines brought in for subsequent operations.

The first step is to cut all undergrowth, lianas, creepers and small saplings as close to the ground as possible with cutlasses or brush hooks. This underbrushing facilitates the movement of the tree fellers who follow and also provides a lot of vegetation which dries out to give readily combustible material useful in ensuring the success of the burn which comes later.

The trees are felled by skilled men using axes or chain saws. Chain saws, which are powered by petrol engines and are available in various sizes from about 0.3–1.2 m, are quicker and more efficient than axes but require careful daily maintenance. Smaller trees are cut close to the ground, larger ones 0.5–1.5 m above ground and very large, buttressed trees are cut from platforms 2 m or more above ground at the height where the buttresses arise from the trunk. Felling the trees in one direction, parallel to the intended rubber rows on flat land or on the contours on sloping land, will lessen the cost of subsequent clearing of the planting rows.

Although it involves certain disadvantages, which are mentioned in section 5.1.5, every effort is normally made to obtain a thorough burn after felling. This has practical benefits in destroying most of the debris and leaving the land weed-free for a time. It allows ready access for workers and supervisors and reduces expenditure both on further clearing and on the establishment and weeding of leguminous covers. The branches of the felled trees are lopped and, together with the slashed undergrowth, piled around and over the larger timbers. The material is then left to dry for a period which depends on its bulk and on the weather but is usually 6–10 weeks. Burning is done during the drier season of the year and preferably after at least a week without rain.

The actual burning operation must be carefully organized to ensure that it is effective and to avoid the danger of uncontrolled fire. It is best done on a day when there is a light wind and started after the sun has had a few hours to dry off superficial moisture from the vegetation. A common procedure is for a line of men, placed about 20 m apart, to start together, on a given signal, from one side of the clearing facing into the wind. Each man carries a torch and a bottle of kerosene and they advance in line setting fire to dry material at about every 20 m and using the kerosene where necessary. Close supervision is necessary to make sure that no one lags behind to be caught by the fire and a roll-call is taken when the line reaches the other side of the clearing. Various additional measures may be taken to ensure safety and a thorough burn. For example, narrow traces may be cleared through the vegetation to facilitate uniform and safe progress of the men. Combustibles, such as oil-soaked sacking and small dry branches, may be placed in advance at intervals to aid in starting the burn. It is commonly necessary to return to the area as soon as it is cool enough after the initial burn to cut any unburned, smaller timber (up to about 20 cm diameter) into manageable lengths for reburning. Stacking and reburning is done around the larger tree stumps so as to destroy these as far as possible. Apart from this, stumping is usually limited to the removal of larger ones that are found to be in the vicinity of planting points after lining.

5.1.3 Mechanical clearing

All the operations involved in full clearing and stumping can now be done mechanically, most of them more quickly than manually. On estates and on the large areas developed for tree crops by government-sponsored settlement schemes, machines have largely replaced men for most operations although felling, and to a lesser extent underbrushing, are still often done manually. The heavy machinery required is expensive in initial cost, in maintenance and to move from place to place. It has to be operated by skilled men who can demand relatively high wages. To protect equipment and operators from the hazards of falling trees and branches, tractors must be fitted with heavy duty cabs and screening together with guards for hydraulic cylinders and lines, cable controls and radiators. As the use of this expensive, specialized equipment is only profitable if it is used efficiently and kept fairly fully employed, mechanized clearing is mostly done by contractors.

Clearance of undergrowth

Where the trees are sufficiently widely spaced to allow the passage of the equipment, the undergrowth and small saplings can be cleared before felling, using a bulldozer or one of the shearing blades described below.

Felling and stumping

Felling is often done with chain saws and this is the only practicable method with the biggest trees, but quite large trees can be cut at ground level with shearing blades fore-mounted on powerful tracked tractors. With this equipment the full power and weight of the tractor is applied to a cutting edge which can be periodically resharpened by means of a small, portable grinder. The blade is controlled either hydraulically or with a cable and is provided with a flat sole to prevent it from digging into the ground, as tends to happen with bulldozer blades. The V-type blade (Fig. 5.1) has angled, serrated cutting edges surmounted by guide bars to push the vegetation off both sides of the tractor and a central, heavy duty

Fig. 5.1 'Fleco' V-type shearing blade (Balderson Inc.)

'splitter', or 'stinger', projecting in front of the V. The 'stinger' enables large trees to be split by one or more thrusts before they are felled with the cutting edge but with smaller trees the equipment can move continuously forward. The other type of shearing blade is the angled blade, such as the Rome K/G blade (Fig. 5.2). This is fore-mounted on the tractor at an angle of 30° and has the leading end drawn out into a strong point or 'stinger'. The cutting edge and the 'stinger' are replaceable and can be resharpened. The blade can be raised, lowered or tilted by hydraulic controls and is surmounted by a guide bar, or push frame, to control the direction in which the trees fall. With the use of the 'stinger' to split the base of the trunk quite large trees can be felled with this blade in several passes.

The tree stumps are usually left in the ground, only those near planting points being removed after lining. However, if mechanized cultivation is intended after clearing, which may be the case if intercropping the rubber is proposed, the stumps and all the roots of a size capable of impeding or damaging tillage implements must be removed. Large tracked tractors with bulldozer blades have commonly been used to uproot trees, thus felling and stumping in one operation. Such equipment works fairly well in secondary forest with small to medium sized trees but is not effective with large trees. Also, saplings and bushes tend to bend and allow the blade to ride over them. Another disadvantage is that the bulldozer blade tends to dig into the ground, thus disturbing and shifting the topsoil. Better work can be done with fore-mounted tree dozers (Fig. 5.3)

Fig. 5.2   Rome K/G shearing blade mounted on Caterpillar D8K tractor (Rome Industries)

Fig. 5.3   'Fleco' tree dozer (Balderson Inc.)

having a hydraulically or cable-operated boom or probe which is raised to contact the tree 2 or 3 m above the ground and push it away from the tractor while immediately afterwards a lower blade is driven into the base of the tree and subsequently raised to uproot it. Uprooting large trees requires time and manoeuvring with either the tree dozer or the bulldozer and after uprooting there is a mass of roots and soil which is difficult to burn and a large hole which has to be filled in before planting. If removal of stumps is required after felling has been done manually or with a shearing blade, it can be done with a tractor-operated winch or with the point of a Rome blade.

Stacking and windrowing for burning

After being left where it has fallen to dry for some weeks, the timber is usually burnt *in situ*, although sometimes it may be stacked in windrows for further drying and then burning. In either event, after the first burn there will be a residue of unburnt material which requires stacking in windrows for reburning. On flat land the windrows will usually be parallel to the intended rubber rows, but on sloping land they should be on the contour. The work can be done with a bulldozer but this is not entirely satisfactory as the blade scrapes up surface soil and deposits some of it in the rows of piled debris. In wet areas this may increase the difficulty of getting

Fig. 5.4 'Fleco' clearing rake (Balderson Inc.)

a good burn. It is also undesirable to have topsoil removed from the greater part of the area and concentrated in mounds which may subsequently require levelling. The Rome K/G blade is better for windrowing than the bulldozer but a heavy rake fore-mounted on a tractor does a faster and cleaner job and moves much less soil (Fig. 5.4). The soil can pass between the tines and the operator is able to loosen some of the soil clinging to roots by shaking the rake as he moves a load, or by dropping the load and picking it up again.

Root eradication

If subsequent mechanized cultivation is intended, roots remaining in the soil will need to be removed after burning. Root eradication involves root cutting, or ripping, and raking. Root cutting can be done with an implement having several heavy tines, with bases shaped like a V on its side, towed by a medium powered crawler tractor. The tines can work to a depth of 45 cm and after the larger root pieces have been cut they ride up over the tines to the surface. Specially designed root ploughs which can be rear-mounted on heavy crawler tractors are also available (Fig. 5.5). These consist of a horizontal blade carried on two vertical cutting standards equipped with replaceable shins. The blade is pulled through the

Fig. 5.5 Rome/Holt root plough mounted on Caterpillar D7 tractor (Rome Industries)

soil at a depth of 20–45 cm controlled by a depth-setting mechanism. Fins welded to the top of the horizontal blade force the larger roots to the surface. After they have been cut and brought to the surface the roots can be gathered for burning by means of a fore-mounted rake of the kind already mentioned.

Limitations of mechanized clearing

Clearing with heavy machinery has some disadvantages. There is bound to be appreciable disturbance of the surface soil and destruction of its structure, especially if the bulldozer is used, when topsoil may be concentrated in some places and subsoil exposed in others. On sloping land loss of structure enhances runoff and erosion, which may be accentuated by rill erosion developing in vehicle tracks. Inevitably there is some compaction of the soil, although it should be possible to keep this within reasonable limits by minimizing manoeuvring and not working when the soil is too wet. Heavy equipment cannot be used efficiently on wet land with a high water table, on deep peats or on areas with many rock outcrops.

5.1.4 Poisoning forest trees

Poisoning jungle trees and allowing them to fall when dead has been tried as an alternative to felling. Marty (1966) reported that the method was successful and reliable in the New Hebrides and that costs were only half those of felling. An arsenical weed killer was applied to frill girdles made with an axe around the tree trunks at 15 cm above ground level. By 15 months later 90 per cent of the trees had fallen, mostly breaking off at the frill girdle and leaving stumps which could be removed with a bulldozer. The remainder were brought down by burning around their bases.

Poisoning has not generally been found satisfactory and has been little used for forest clearing, although sometimes employed to kill old rubber before replanting. The main problem is that during the long period after poisoning and before the trees die and fall it is unsafe to work in the area and a massive growth of undergrowth develops as the forest canopy thins out. In trials in Malaysia lining was done through standing secondary jungle, 1.2 m strips cleared along the planting lines and rubber planted immediately after poisoning the trees. In light secondary jungle the fall of dead trees did relatively little damage to the young rubber but the greater fall of timber in thicker forest did considerable harm. Moreover, although initial clearing costs were relatively low, there was little advantage in the long run compared with conventional felling and burning because the growth of the young rubber was somewhat retarded and weed control was more expensive during the longer period of immaturity.

5.1.5 The effects of burning

Burning destroys most of the organic matter in the felled vegetation and in the leaf litter accumulated on the ground, but does not usually result in appreciable loss of the humified organic matter in the surface soil, except where timber is piled for burning. Assuming that the land is to be left untilled after clearing, as it usually is under rubber, the destruction of the leaf litter is a physical, as well as a nutritional, disadvantage, as if it remained it would afford protection to the topsoil against the impact of rainfall for some time after clearing, thus increasing infiltration of rainfall and reducing runoff. Burning results in most of the carbon, nitrogen and sulphur in the vegetation and litter being lost by volatilization, but the other nutrient elements remain in the ash which enriches the topsoil in phosphate and exchangeable cations and raises its pH. However, on sloping land part of the ash is liable to be washed away by rain falling on the bare ground soon after burning. There will also be some loss of nutrients by leaching, although this should be reduced as far as possible by the early establishment of ground cover plants.

The direct effects of heat on the soil when felled forest is burned are probably confined to limited areas where much timber has been piled and burnt. Heating may improve the physical condition of some soils, especially clays, rendering them more friable, easier to cultivate and more permeable to rainfall. However, the improvement is unlikely to be significant for rubber growing as it probably lasts no more than a year. Of greater significance is the temporary increase in the rate of mineralization of nitrogen that has been observed after burning. This effect is presumably partly due to the fact that drying or heating a soil results in a temporary reduction in the microbial population and decomposition of some of the humic substances in the soil. On cooling and rewetting there is an increase in the number of micro-organisms, especially bacteria, and a flush of nitrate production from the decomposition of the solubilized humic substances and of the microbial cells killed by the heating and drying. In addition the rate of nitrification is enhanced by the increase in available cations and the rise in pH consequent upon the deposition of ash.

As burning results in the loss of the organic matter in the vegetation and litter, and probably in the leaching of part of the nutrients contributed by the ash, it would seem to have an adverse effect on soil fertility compared with leaving the felled vegetation to rot. But under some conditions this effect may be slight and of little or no practical importance. Thus Sly and Tinker (1962) reported that in three trials comparing burning with no burning before planting oil-palms in Nigeria the effects on soil nutrient status were quite small at 9, 10, and 20 years after planting. The only significant yield differences found were in the first four years of one experiment where increased yields were obtained from the burnt plots. There is little evidence from other investigations but on some soils burning may have a more adverse effect on fertility. For example, Gunn (1960) reported that in an experiment on a poor soil in Nigeria the initial yields of oil-palms were less on burnt than on unburnt plots.

These disadvantages of burning have sometimes been avoided by simply piling the felled vegetation between the planting rows and leaving it to rot. Alternatively, the loss of organic matter may be lessened by having a light burn to destroy only the smaller timber and debris while leaving the larger timber to rot. Both these methods have the advantage of reducing the risk of erosion but they have the great disadvantage of leaving the land encumbered with a mass of timber and debris between the planting rows. This greatly impedes the movement and supervision of labour and can make terracing difficult on slopes. The debris harbours vermin, especially rats which can damage young plants, and presents a risk of subsequent accidental fire in areas subject to a spell of dry weather. In addition, regeneration of ground vegetation is quicker than it is after burning and weeding costs are higher. Consequently, although burning has the disadvantages mentioned above, the almost universal practice is to burn as thoroughly as possible in order to minimize subsequent expenditure on further clearing and on the maintenance of the plantation in its early years.

5.2 Clearing non-forested land

In some places, especially where an annual drier season occurs, land that has been cleared of forest and cultivated for a period has reverted to a cover of shrubs and grasses which has been maintained as a sub-climax vegetation by the frequent passage of fire. In parts of south-east Asia there are considerable areas of such land where the dominant species is the grass *Imperata cylindrica*, commonly known as lalang. This grass spreads by strong rhizomes to form a dense cover and is a most persistent weed since it can regenerate from quite small pieces of rhizome left in the soil after weeding. If the period during which the land was under cultivation was long, there may also have been considerable erosion and loss of fertility so that extra fertilizer will be needed during the immature period of the rubber. Depending on the lapse of time since cultivation was abandoned and, especially, on the frequency of fires, there may be some regeneration of woody shrubs and trees.

The main problem in preparing such land for rubber is to get rid of the lalang. This can be done manually by first slashing and/or burning the grass and then digging to about 20 cm deep to turn up clods and expose them to the sun. Several weeks later a second digging is required to break the clods and, after a further period of exposure, several rounds of fork hoeing are likely to be needed to dig out any regenerating plants and to collect the rhizomes for burning. Lalang can also be eradicated by disc ploughing, usually to a depth of 30 cm, followed by disc harrowing. It is an advantage if the ploughing is followed by a dry spell but in most rubber-growing areas this cannot be reliably expected. About five rounds of harrowing are usually needed and after that some regenerating plants will have to be dealt with by hoeing or by 'wiping', i.e. drawing a rag soaked in herbicide over each plant from the base upwards.

Nowadays, lalang, whether on land to be cleared or as a weed in the rubber plantation, is normally killed with herbicides. Spraying is most effective when the grass is in active growth and in the absence of a mass of dead or senescent stems. Hence, on land to be cleared it may be best to burn off the grass during a dry spell and spray the regeneration when it is about 50 cm high. The commonest herbicides now used are dalapon, which is applied at 16–22 kg of active ingredient per hectare, and glyphosate, which is used at 2.2–3.3 kg a.i./ha. Both are translocated herbicides which are usually applied in about 700–1000 litres of water per hectare. Dalapon is rendered more effective by a wetting agent, which is incorporated in some proprietary formulations. To be effective, glyphosate needs six hours without rain after spraying. Of the two, dalapon acts more quickly; in trials in Malaysia it gave a maximum kill of 80–90 per cent in four weeks whereas glyphosate took 2 months to give a 90–98 per cent kill. A second spray is usually needed to reduce the grass to a level at which it can subsequently be dealt with by hoeing or wiping. It was found that with dalapon this second spray is best given about 5 weeks after the first, while with glyphosate there should be an interval of 9 weeks between sprays (Yeoh & Pushparajah 1976).

5.3 Soil conservation measures

On sloping land, soil conservation works need to be constructed as soon as possible after clearing. Where the slope exceeds 8 per cent, the common procedure is to construct a terrace on the contour for each planting row, the distance from centre to centre of adjacent terraces being equal to the normal inter-row spacing of the rubber, usually from 6 to 9 m. A base line is marked out at right angles to the contours down an average slope of the clearing, pegs being put in at intervals equivalent to the inter-row planting distance. From each of these pegs contour planting lines are marked out using a dumpy level or a road tracer.

Unless the slope is uniform, which it rarely is, the distance between the contour lines will vary, as they will come closer together on steeper land and farther apart where the slope is gentler. If the distance between contour lines does not vary greatly, it will be possible to maintain a fairly uniform planting density by varying the spacing between the planting points in the rows. Where this is not practicable, it will be necessary to discontinue terraces that come too close together or, if they become too far apart, to introduce short additional terraces. A rule of thumb is that where the distance between terraces is reduced to two-thirds of the normal distance, alternate terraces are stopped, and where it exceeds $1\frac{1}{3}$ of the normal, short additional terraces are interposed.

Terraces can be cut with a wheeled tractor with a grader blade on the gentler slopes or by a crawler tractor and an angle dozer on steeper land. The blade cuts 1.0–1.8 m into the solid, upper side of the slope and moves the earth to the lower side where it is consolidated to build the outer half of the 2.0–3.6 m wide terrace. The terrace is sloped back into the hill at an angle of 10°–15° to check runoff. Although they are made as precisely on the contour

Fig. 5.6 Rubber planted on contour terraces

as is practicable, terraces are seldom strictly level and it is usual to leave stops, or pillars, of uncut earth protruding from the back wall at intervals of 10–20 m to slow down the lateral flow of excess water. The planting holes are dug in the middle of the terrace at the original soil level (Fig. 5.6). Contour terracing is initially expensive but is essential on relatively steep land because without it erosion is inevitable. The terraces make all operations conducted along the rows, such as weeding, pruning, tapping and collection, easier and cheaper. To save money initially, planting is sometimes done on small platforms, 90–120 cm square, cut out of the slope at appropriate intervals on the contour and each providing a planting place for one tree. These so-called 'individual bench terraces' are less effective against erosion than full terracing.

On gentler slopes with permeable soil, adequate protection against erosion may be provided by constructing ridges of earth, from 1.2–3.6 m wide at the base and 30–60 cm high, along the contour with a channel on the upper side. But these contour ridges tend to retain too much water in heavy rainfall areas and are not much used in rubber plantations. Silt pits, which are discontinuous short lengths of ditch with the spoil thrown below them, are also sometimes used on moderate slopes. Whatever kind of soil conservation works are constructed, the early establishment of ground cover plants is essential if erosion is to be minimized.

5.4 Drainage

An integrated plan for the drainage and road systems needs to be prepared as soon as the land has been cleared or, if practicable, at an earlier stage once a survey has been done. This should aim at maximizing the land available for rubber and facilitating the movement of workers and equipment. As it depends on the topography, the drainage system has to be planned first but it must also fit in with the predetermined planting distance and pattern. Field drains, roads and paths should all, as far as possible, run parallel to the tree rows so as to cause least interference with the movement of workers and machinery along the rows and to minimize the number of bridges required.

It is mostly on flat or gently sloping land having a permanent, or seasonal, high water table that drains are needed. For good growth of rubber the water table should be not less than 1.0–1.2 m below the surface. The drainage system should be designed to keep the water table at or below this depth over the whole of the area it serves and virtually at all times of the year. In some places where the rainfall is unevenly distributed this may mean that a system designed to prevent the water table rising above a depth of 1 m during the wettest season will render the soil less moist than desirable during a drier season, with the possibility of some degree of water stress developing in the trees. However, such drainage will improve soil aeration and enable the trees to develop a good root system which can explore the soil for water and nutrients to a depth of at least 1 m. On the other hand, a lesser degree of drainage may allow the water table to rise high enough in the wetter part of the year to restrict the development of the root system to shallow depth so that the trees have little ability to withdraw water from the deeper soil layers in a drier time.

5.4.1 Construction of drains

Open surface drains are invariably used because sub-surface drains of tile or plastic are much more expensive and are liable to be blocked or disrupted by roots. When draining flat or gently sloping land, the first step is to find and, if necessary, deepen a good outlet or outlets. The main drain is then cut away from the outlet right through the area to be drained, keeping it as straight as possible. Next, the collection drains are dug at intervals of 200–400 m along the main drain to link up with the field drains, which run parallel to the tree rows at intervals depending on the intensity of drainage required, usually of the order of 25–50 m. The sizes of the different types of drain depend on the rainfall and the soil, but an indication of the likely dimensions is given in Table 5.1. To allow for an adequate fall in the drains and to accommodate the increasing volume of water as it flows towards the outlet, the smaller dimensions indicated are used towards the head of the drain and the larger towards the outlet. To maintain the water table fairly uniformly at a depth of approximately 1.2 m throughout the area, the collection drains need to be at least 1.5 m deep at their outlets and the main drain at least 2.0 m deep. Field drains need not be cut to full depth at the time of planting; they can be gradually deepened over 1–2 years ahead of the developing rubber roots until the water table is lowered to 1.2 m. The slope of the sides of the drains depends on

Table 5.1 *Dimension of drains (in metres)*

| Type of drain | Width | | Depth |
	At top	At bottom	
Field	1.0–1.5	0.3–0.5	1.2
Collection	2.0–2.5	0.6	1.2–1.7
Main	2.5–5.0	1.0	2.0–2.5

the soil; on stiff clay the sides can be nearer vertical than in loam or sandy soil.

The gradient of the drains must be sufficient to prevent rapid silting up but not so great as to cause scouring. Provided that the drains are well maintained, a gradient of 1 in 5000 to 1 in 3000 is usually adequate. The frequency of field drains depends partly on the rainfall but mainly on the soil permeability. Drains need to be closer together on heavy clay soils than on lighter soils through which the water moves more rapidly. It is usual initially to put them in at the lowest frequency which seems practicable and then, if necessary, to put in additional intermediate drains later. A need for more drains is indicated if the growth of trees adjacent to existing drains is better than that of trees midway between drains. With some tractor-powered operations, such as fertilizer application or tree spraying, it is usual to treat two rows at the same time, so it is desirable that the interval between drains should, if practicable, be a multiple of four rows in order that the tractor can travel up and down without any unproductive runs and with the minimum number of drains (bridges) to cross.

Field drains and collection drains can be made, and also cleaned, with ditchers or excavators operated from light tractors with hydraulic controls. There are many types of such equipment; with some the tractor straddles the ditch, while with others it works from the banks; some have buckets or scoops and some have rotary trenchers. Those commonly used have buckets 45–60 cm wide and are capable of digging to a depth of 3 m. Large self-propelled ditchers, employing a drag line, are used for digging and cleaning main drains.

Peat soils are generally unsuitable for rubber but the crop is sometimes planted on them. The water table is usually high, sometimes within 30 cm of the surface. Some drainage is necessary before any vehicles or clearing equipment can be brought on to the land, but deep or frequent drains cause the peat to shrink and bring about changes in its composition which restrict water movement. It is advisable to put in outlet and sub-main drains not more than 1 m deep about a year before felling. After felling, collection and field drains can be put in, the latter rather sparsely and not deeper than 0.5 m. Subsequently, the frequency and depth of the field drains can be gradually increased as necessary until they are 1.2–1.5 m deep when the rubber reaches maturity.

On hilly land without a high water table the natural drainage is usually satisfactory provided that the soil is permeable to rooting depth. Some valley bottoms or wet pockets may require drains but, provided that outlets can be found, the methods for this work will be the same as already described for flat land, although on a smaller scale. In some situations, however, small valley bottoms or depressions may receive a considerable amount of water as seepage from surrounding higher land, in which case it is advisable to construct a foothill drain on the contour at the bottom of the slope to collect the seepage and convey it to a main drain. Artificial drainage is not usually needed on slopes but if it is required it must be constructed as part of the soil conservation works so that water is conveyed across the slope on a slight gradient and does not cause erosion.

Drainage may be poor in some areas which do not have a high water table but where the physical condition of the soil impedes the percolation of rainfall. This can occur on heavy clay soils due to their texture and/or poor structure. Digging shallow drains can sometimes be of limited help in removing water standing on the surface but it is more effective to grow deep-rooted cover plants which will open up the soil, improve its structure and protect its surface from the impact of rainfall. Any necessary tillage should be done only in dry weather. Impeded drainage may also be due to the presence of a sub-surface layer of low permeability, such as a layer of clay accumulation or of ironstone. Here again it will not usually be practicable to do more than plant deep-rooted cover plants. However, if, as occasionally happens, there is a thin, hard layer at shallow depth and the profile below it is free-draining, it may be possible to effect an improvement by digging fairly frequent field drains or by deep ploughing, or subsoiling, to break up the impermeable layer. The latter will only be practicable on land newly cleared from forest if the stumps and roots of the trees are first removed.

5.4.2 Maintenance of drains

It is advisable to inspect main and collection drains during periods of heavy rain and to deal promptly with any blockages caused by the collapse of banks or the piling up of debris behind obstructions such as fallen branches. This enables the drain to be cleared more quickly and cheaply than if silt is allowed to collect behind the blockage. The depth and efficiency of drains is decreased by the growth of water weeds and the gradual build-up of silt and debris, so it is important that they are cleaned out at regular intervals. Main drains require cleaning out once every 1–3 years, depending on conditions, using a drag line or other kind of large bucket excavator. Smaller drains usually need cleaning once or twice a year, either manually or using tractor-mounted bucket scoops. Vegetation growing along the drain edges is kept in check with herbicides, usually every 3–4 months, as part of the field weed control programme (see pp. 279–90).

5.5 Preparation of land for replanting

Fields are usually replanted on estates when the trees are 25–30 years old. As a rule the old trees can be replaced by higher-yielding clones, or clonal seedlings, which have become available as a result of progress made by the plant breeders. Smallholders tend to delay replanting until the rubber is over 30 years old, principally because they can ill afford the loss of revenue over several years before the new rubber is ready for tapping. Measures taken in various countries to encourage, organize and subsidize replanting by smallholders are described in Chapter 12.

On estates it is necessary, as with clearing land from jungle, to prepare in advance a schedule of the time, labour and equipment requirements for the various operations in the field and nursery. Adherence to a predetermined timetable will allow operations such as burning and planting to fit in with seasonal variations in rainfall and will also minimize the period during which the land is not carrying tapped rubber. Furthermore, the plan should aim to minimize fluctuations in labour requirements, especially for tapping, and in the total output of rubber from the estate. Labour requirements are heaviest during clearing, planting and the two years thereafter, so it is advantageous if a proportion of the old rubber can be replanted every third year. Fluctuations in total crop are to some extent reduced because untapped, replanted areas are offset by fields nearly due for replanting where high yields are obtained from intensive ('slaughter') tapping.

5.5.1 Improvement of field conditions prior to felling

It is important that replanting be on clean land. There is an advantage in eradicating weeds as far as possible before felling while their growth is limited by the shade of the old rubber trees. At this time less herbicide is required and it is easier and cheaper to get rid of weeds than it is when they have been stimulated to make vigorous growth by exposure to full sunlight. At least a year before felling, woody shrubs and self-sown rubber seedlings should be eradicated (either manually or with appropriate herbicides such as those containing 2,4-D with 2,4,5-T or picloram) and herbaceous weeds, especially lalang and other competitive grasses, sprayed out. This will aid in the early establishment of leguminous covers. When clearing and burning has been done there is usually a rapid growth of legume cover plants resulting from the germination of previously dormant seeds shed by the cover plants grown during the immature period of the old stand. Although it will usually be necessary to sow fresh cover plant seeds, if these self-sown plants can develop with little or no weed competition they will contribute considerably to the establishment of a pure legume cover and will lower costs.

It is advisable to examine the drainage system and to clean out and improve it, where necessary. Some of the drains provided when the original stand was planted after clearing from forest may have become neglected because they ceased to flow (except briefly during heavy storms) once large amounts of water were transpired by the mature rubber trees. After felling has stopped water uptake by the old stand, some areas that were previously dry may become wet, thus necessitating the opening up of the old, neglected drains. Where the replanting is contiguous with a mature stand, it is usually worth while digging a boundary drain between the two areas to reduce root competition from the older trees and to check invasion of weeds from the mature area.

5.5.2 Manual felling and clearing

Felling the old stand with axes or chain saws is usual where the area to be replanted is small and also wherever there is a market for the trees as timber or firewood. In the latter case, felling followed by chopping off the branches and removal of the usable wood may be done free of charge by a contractor who wants the timber. Where the wood is not saleable, it may be necesssry to cut the tree trunks and, perhaps, some of the larger branches into sections to facilitate stacking and burning. The stumps remaining after felling are usually poisoned to hasten their decay and to aid in the control of root diseases. The usual procedure is to paint the bark with a 5 per cent (acid equivalent) solution of 2,4,5-T tree-killer in diesoline or 5 per cent triclopyr in diesolene or kerosene. To prevent infection of healthy stumps by spores of root-disease fungi it is also recommended that the cut surfaces be painted with a wood preservative, such as creosote.

5.5.3 Tree poisoning

If the timber is not required, the old rubber trees can be poisoned and left to die and fall. Formerly, this was done by applying a sodium arsenite solution to a frill girdle made around the tree with an axe but on account of its high mammalian toxicity the use of this poison is now uncommon and has been legally banned in some countries. The materials now most commonly used are 2,4,5-T (although this is banned in some countries) and triclopyr, formulated as described in section 5.5.2, and painted or sprayed on to the basal 40–50 cm of the tree without any need for ring-barking or frilling. Other materials which can be used include a mixture of 2,4-D and 2,4,5-T ('Finopal DT'), ammonium sulphamate and picloram

('Tordon'), the last requiring injection into the tree.

Poisoning is usually done at, or shortly before, planting the new stand, and after lining and holing has been done under the old trees. Branch fall starts 6–8 months after poisoning and continues for some months. Trunks may remain standing for 2–3 years but those remaining after about 18 months are usually felled. Falling timber is something of a hazard to workers and inevitably does some damage to the young stand. The extent of this damage is variable but it is usually slight. Poisoning long enough before replanting to avoid the hazards of falling timber is impracticable as it would extend the period during which the land is unproductive by 1–3 years. Poisoning the trees has been found more effective than felling and then poisoning the stumps for reducing the incidence of root disease in the replant (Newsam 1967).

5.5.4 Mechanized felling and clearing

Rubber trees can be uprooted by winches or stumping jacks operated by hand or by light tractors. Prior cutting of the roots is not usually necessary and the trees can be pulled out with most of the tap-root and part of the laterals. However, the method is slow and requires a good deal of labour for fixing the pulling wires, which need to be hitched around the trunks or branches about 5 m above ground. The equipment more usually employed comprises medium or heavy crawler tractors with bulldozers, Rome K/G blades, tree dozers or special stumping and grubbing blades, the use of which has already been described for clearing forest. The tree dozer is probably the most efficient implement for uprooting rubber trees as it can push them out fairly cleanly with the tap-root and laterals attached and without great disturbance of the soil. The work is usually done by the tractor retreating along the row from each tree after felling it so that movement is not impeded by trees already felled.

It is convenient to fell the trees in 3–5-row units, first felling the centre row straight down the line of the old trees and then bringing down the outer rows so that the crowns fall inwards towards the centre row, leaving only the lower trunks to be pushed inwards to complete windrowing before burning. On sloping land both the old and the new planting rows should be on the contour and it is desirable to fell and form windrows on the contour. On terraced land this will normally be done along each terrace. If there are no terraces, the lowest of three rows can be felled first and the two rows above then felled and pushed down to it to form the windrow. As mentioned in the section on clearing forest, windrowing is best done with a fore-mounted rake.

Several rows of rubber trees can be simultaneously uprooted by two heavy tractors moving forward on a parallel course and linked by a heavy chain. A heavy steel ball of 1–2 m diameter may be attached to the middle of the chain to prevent it riding up too high on the trees. As the tractors move forward, the loop of the chain pulls the trees out of the ground with most of their roots, after which they can be pushed into windrows by tractors equipped with fore-mounted rakes. The method can bring down the trees speedily but has the disadvantages that uprooting may not be complete and that the trees are left lying untidily so that subsequent windrowing is rather slow.

Root and stump removal, as described in section 5.1.3 is sometimes done before replanting but is only essential if mechanized cultivation is intended. Full mechanical clearing of old rubber with removal of stumps and roots can be done speedily and leaves the site completely clear and weed-free after burning, which is an advantage for the establishment of cover plants. It has also been shown that the removal of roots significantly reduces losses from root disease in the replant. However, the lower loss of trees and the savings in disease control measures after planting may not offset the higher costs of mechanized stump and root removal compared with those of tree poisoning and felling, or stump poisoning after felling (Newsam 1967; Ng 1967). Root removal is, therefore, probably not worth while for the purpose of reducing root disease.

5.6 Planting density and pattern

In the early days of the industry, trees on estates were usually planted wider apart than they are nowadays; for example, in Malaysia spacing at 6.1 m square (20 ft) was common, giving an initial stand of 270 trees/ha. It was soon realized that, while this gave strong growth and high yields per tree, larger yields per hectare could be obtained by closer spacing, but there was much diversity of opinion on the number of trees per hectare which should be planted for maximum profit. There was also some difference of opinion about the best planting pattern, particularly regarding the value of wide rectangular, or 'hedge' planting, which is discussed in section 5.6.4. Initially, it was difficult to reach sound conclusions because little information was available on the effect of different planting densities on a number of factors affecting the performance and profitability of a plantation over the whole of its productive life. Apart from direct effects on growth and yield, the density of the stand can affect the duration of the immature period, the percentage of trees reaching tappable size at different ages, the thickness of the bark and the rate of its renewal after tapping, the incidence of panel diseases, the yield per tapper and the cost of tapping and collection.

5.6.1. Effect of initial planting density on growth and yield

To obtain information on the matters just mentioned, a number of experiments comparing different planting densities were started in Indonesia in the 1920s and in Malaysia from 1930. For example, in a trial which was started in Malaysia in 1930 and which ran for 28 years, the clone AVROS 50 was square-planted at distances ranging from 3.05 to 9.14 m, giving initial densities of 1074, 746, 548, 309, 267 and 119 trees/ha (Buttery & Westgarth 1965). Fertilizers were applied at a uniform rate per unit area so that the amount per tree varied with planting density. No thinning of the initial stands was done apart from the removal of diseased or damaged trees. This, and other, experiments gave results, outlined below, which were in general agreement.

For a year or two after planting there is no competition between the young trees, but as the crowns develop a stage is reached, at an age dependent on density, when they begin to compete for light, nutrients and water. From then on, growth, as indicated by girth increment rate, progressively declines with increasing density, with consequent delay in reaching tappable size.[1] From about the sixth year after planting the percentage of tappable trees diminishes with increasing density but the number of tappable trees per hectare increases. At the highest densities some trees may never reach tappable size. For example, in the experiment mentioned above 90 per cent of the trees planted at 119/ha were tappable in the third year of tapping but at 1074/ha 31 per cent remained too small when the trees were 19 years old.

It was early observed that with increasing density the height of the lowest branch, or crotch, of the mature trees increased, due to the heavier shade causing more branch shedding. This has been confirmed with modern clones in more recent trials, which also showed that, as might be expected, the restricted space per tree at higher densities resulted in smaller crowns (Napitupulu 1977; Leong & Yoon 1982b).

The thickness of virgin and renewed bark decreases with increasing density and the rate of bark renewal is slower. In some experiments the renewed bark at the closest spacings was not thick enough for tapping by the time that the tapping of the virgin panels was complete. As a rule, the incidence of panel diseases was greater at the higher densities.

With increasing density, yield per tree decreases but mean yield per hectare per annum, and cumulative yield per hectare over a period of years, increases up to an optimum density at which the increase in yield per hectare due to the greater number of tappable trees is balanced by the decline in yield per tree due to reduced girth and bark thickness. This optimum density for yield per hectare has been found to vary considerably with clone. For example, Eschbach (1974) reported that maximum yield in the fourth and fifth years of tapping was obtained from planting densities of 650/ha for clone PB 86 and 550–600/ha for PR 107. In a recent Malaysian trial the cumulative yield data over 11 years' tapping showed that the best yield of RRIM 600 was obtained where the density *at planting* was 741 trees/ha whereas it was at 399/ha for the more vigorous RRIM 701 (Rubber Research Institute of Malaysia 1983a). However, in this experiment, and in some other trials, it is noticeable that yield per hectare has varied relatively little over quite a wide range of planting densities.

At higher densities the number of trees per tapping task can be increased because the tappers have less distance to walk between trees. But this is more than offset by the decrease in yield per tree so that the yield per tapper per task declines with increasing density while the cost of tapping and collection per kg of rubber increases. At closer spacings the dry rubber content of the latex is slightly decreased and the proportion of cup lump and scrap rubber brought in by tappers is slightly increased, but the magnitude of both effects is quite small.

5.6.2 Initial planting density and thinning

The number of trees initially planted per hectare must be greater than the number of mature, tappable trees finally desired in order to allow for losses from root disease, wind damage and other causes. Furthermore, from the results reported above it appears that as the trees become more competitive with age the number per hectare required for maximum yield per hectare must decrease. Hence, it would seem desirable to have a relatively large number of trees per hectare at the time of opening and then progressively to thin the stand as the trees grow older in order to maintain the maximum production per hectare. However, as yields per hectare do not vary greatly over a fairly wide range of densities, and as gross profits per hectare vary even less (see section 5.6.3), the value of thinning after opening is limited.

On estates it is normal to remove the proportion of poorly grown plants, or runts, which becomes apparent in any plantation up to the third year. Such plants will always remain small and backward because of competition from more vigorous neighbours and should be removed as early as possible along with damaged and diseased trees. If such plants are removed before there is strong competition from their neighbours, they may be replaced with advanced planting material (see p. 220) from the nursery. Beyond this there is no standard procedure for the degree and timing of thinning. As there is usually considerable tree-to-tree variation in growth and yield, even in buddings, due to environmental factors and influence of the stock, some selective thinning based on performance may be done. During immaturity, selection based on test tapping is sometimes done with clonal seedlings, but this may result in the selection of some precocious yielders which fail to maintain their early promise. Girth is the more usual criterion for selection and with buddings it is as effective as test tapping. Such selective thinning of the poorer trees may slightly reduce the immature period of the plantation and improve the growth and yield of the remaining trees by giving them more light and nutrients. In the early years of tapping, some low-yielders may be removed. These are usually selected by visual observation of the contents of the cup, or by rough measurement with a dipstick in the cup, rather than by any more accurate yield recording. The more uniform the planting material the less is the need for thinning. With advanced planting material that has been subjected to selection in the nursery, little or no thinning should be required beyond removing diseased and damaged trees.

5.6.3 Effect of planting density on profit

While experiments can show the relation between planting density and yield, they cannot fully achieve the practical objective of esti-mating the number of tappable trees per hectare that will maximize profit over the productive life of the plantation. Assuming that overhead charges are not much affected by planting density, the number of trees per hectare that will give *maximum gross profit* (gross income less tapping and other costs related to weight of rubber and size of tapping task) will increase when rubber selling prices and yields are higher and tapping costs lower but will decrease when prices and yields are lower and tapping costs higher. As the estimate of optimum density is required at the time of planting it involves the difficult task of attempting to predict the future levels of these three variables.

Barlow and Lim (1967) examined the effect of density of tapp-able trees on gross profit per hectare using the same experimental data as Buttery and Westgarth (1965) but adjusting yield levels to those expected of more modern clones and employing regression equations to predict yields at densities other than those used in the field experiment. For the estate situation they estimated that with RSS 1 (ribbed smoke sheet grade 1 – see Ch. 11 and Appendix 11.1 (p. 545)) prices in the range of $M 1.10–1.35 and medium task size, gross profit would be maximized with a stand of about 312 tappable trees per hectare. But their figures showed that estimated gross profit varied little over the range of 247–494 trees per hectare.

Paardekooper and Newall (1977) estimated the effect of mature tappable density on gross profit per hectare by means of a mathematical model embodying relations between density and yield and density and task size derived from experimental data. Their estimates indicated that gross profit per hectare is almost constant over the range 200–600 tappable trees per hectare. They noted three factors that militate against the use of high densities. First, higher densities cost more for planting material and manuring. Second, for higher tappable densities progressively higher initial planting densities are needed and the higher the latter the greater the proportion of trees that fail to reach tappable size. Third, at higher densities slower growth means a longer immaturity period before tapping can begin. They concluded that it will rarely pay to aim at more than 400 tappable trees per hectare which, assuming uniform planting material, good transplanting, upkeep and disease control, should be achieved by planting an extra 10 per cent, i.e. 440/ha. With more variable planting material and greater expected losses it might be necessary to plant up to 500 trees/ha, thinning to 400/ha by the fifth year after planting.

Owing to uncertainty about the future levels of the several variables involved and because of the probable influence in the future of other unpredictable factors, it is impracticable at the time of planting to decide with much precision what planting density will maximize future profits. But in view of the evidence that gross profit does not vary greatly over a considerable range of densities it would appear satisfactory for estates to select a density near the middle of the range which is convenient for local conditions and tapping procedures. In Malaysia an initial planting density of 400 trees/ha is currently recommended for estates, assuming good and uniform planting material.

Higher densities may be more profitable on smallholdings if they are worked with family labour or if tapping is done on a share system, which usually means that about 60 per cent of the latex goes to the owner and the remainder to the tapper. Under such conditions Barlow and Lim (1967) found that profit increased up to the densest stand considered – 617 trees/ha. But again, variation in profit was small over the range 247–617 trees/ha. Paardekooper and Newall (1977) considered that the optimum for individual smallholdings would be the density which would maximize yield per hectare over a long tapping life of up to 35–40 years. As indicated above, this would vary with the clone, but they thought that the necessity for good bark renewal might limit the maximum tappable density to about 500 trees/ha. A more recent analysis of data from 185 independent smallholdings and 149 on settlement schemes in Malaysia indicated that the highest family income was obtained from 400–500 trees/ha in tapping on independent holdings and from 350–400/ha on the settlement schemes, where planting and maintenance in immaturity had been supervised. It was suggested that appropriate initial planting densities would be 600–750 trees/ha for independent holdings and 500–600/ha for settlement schemes (Sepien 1980).

5.6.4 Planting pattern

Except on sloping land, or where intercropping is intended, planting is most commonly done on the equilateral triangle system. This allows each tree the same amount of space in all directions and is generally considered to give the most uniform growth and the shortest immature period. Spacings of 5.1–4.9 m are suitable, giving initial stands of 444–481 trees/ha. Assuming 15 per cent for losses and thinning, these spacings will leave a final stand at 10–12 years after planting of 378–409 trees/ha.

On sloping land the trees will normally be planted on the contour and usually on terraces. The planting system usually approximates to rectangular with the distance between contour rows greater than that between the trees in the rows. The wider the inter-row spacing and the closer the trees are planted in the row the shorter is the total length of row required for a given number of trees per hectare and the lower are the costs of constructing and maintaining terraces. Small savings also result from reduction in the distance that labourers have to walk in doing such operations as tapping and carrying fertilizer. However, it has been shown that planting closer than 2.0 m in the row reduces growth and yield and delays maturity. In Malaysia the commonest spacing is probably 30 × 8 ft (9.15 × 2.44 m) but 24 × 10 ft (7.32 × 3.05 m) and 20 × 12 ft (6.09 × 3.66 m) are also used and all three spacings give 448 trees/ha.

A spacing of 30 × 8 ft (9.15 × 2.44 m) is also common on smallholdings in Malaysia, even on flat land, as the wide inter-rows are advantageous for the intercropping which is normally practised for the first three years after planting. From what has been said above about optimum densities on smallholdings it can be argued that better returns would be obtained in the long run by planting a higher density on the equilateral triangle system, which would allow only one year's intercropping but would give higher yields of rubber. However, to the majority of smallholders the income, or food, from intercropping is essential during the first few years after planting, or replanting, when there is no return from the rubber trees. On terraced land intercropping is never recommended as it would usually result in erosion.

Hedge planting

An extreme form of rectangular planting, known as hedge, or avenue, planting, was originally used for mixed planting of rubber and coffee in Java. The rubber was either spaced 2–3 m apart in rows 8–9 m apart, which enabled the coffee planted between the rubber rows to be kept for about 12 years before it was shaded out, or at 1 m apart in rows 16–20 m apart, to allow the coffee to be permanently retained. About 30 per cent of the rubber trees planted were gradually removed by selective thinning. Compared with conventional square or rectangular planting, growth was slower and the period of immaturity was lengthened but it was claimed that yields were considerably higher. This superiority was attributed to the high degree of selective thinning and the relatively large, but asymmetric, development of crowns and root systems permitted by the space and light available in the wide inter-rows (Heubel 1939; Schweizer 1940; Dijkman 1951).

Experiments with hedge planting were carried out in Malaysia and it was also tried out on estates. Typical results were obtained from one experiment which compared conventional spacings of 30 × 8 ft (9.15 × 2.44 m) and 20 × 12 ft (6.09 × 3.66 m) with three within-row spacings, each in rows 50, 60 and 80 ft (15.2, 18.3 and 24.4 m) apart, using four clones. Cumulative yields over eight years of tapping were highest for the conventional spacings and progressively decreased with widening of the inter-rows except that one avenue system (15.2 × 1.1 m) gave yields that did not differ significantly from those of the conventional spacings. The hedge plantings had a longer immaturity period, suffered greater losses from wind damage and root disease, and the cost of cover plant maintenance was increased by the wide inter-rows. A small reduction in tapping and collection costs, due to the shorter walking distance in the avenues, did little to offset the disadvantages (Barlow *et al.* 1966). Experience on estates was in line with these findings and hedge planting was abandoned in Malaysia some years ago.

5.7 Lining and holing

The preparation of the land for planting is completed by marking out the alignment of the trees (lining) and digging holes for planting. The production and planting of the young trees is described in Chapter 6.

5.7.1 Lining

For contour planting the procedure already described in section 5.3 is followed to demarcate the terraces and the planting points are then pegged out along them at appropriate intervals. On flat or gently sloping land, wires, marked with coloured tags at intervals according to the spacing required, are used to mark out the planting points. The first step is to fix a base line, which may be done, for example, on the longest straight boundary of the clearing running north and south by extending the wire from a corner peg on a due north compass bearing and placing a second corner peg at its end. Other pegs are then put in on the base line at each planting point marked on the wire. For square or rectangular planting the wire is then successively extended due east or west from the north and south corner pegs and planting points pegged along the guide lines, thus completing three sides of a square or rectangle. Starting from the base line, the wire is then held at its north and south ends by two men who advance east or west, stopping at each peg on the guide lines while other men place pegs at the planting points along the wire. When one square or rectangle has been completed, the process is repeated. For quincuncial or equilateral-triangle planting the procedure is similar except that the guide lines are put in at an angle of 60° to the base line.

5.7.2 Holing

The size of the planting hole varies somewhat with the kind of planting material to be used and with the nature of the soil. Manually dug pits are usually in the range 45–75 cm 'cubes', but the pits taper with depth. Holing is often done mechanically with tractor-mounted augers of diameter 22–45 cm digging to 45 cm depth. Seedling or budded stumps, or stumped buddings (see pp. 213–15), need the larger pits while smaller ones suffice for normal-sized polybag plants. Larger pits may have an advantage on stiff, compact or stoney soils but unless there is some definite impediment to root development the cost of digging very large holes is not justified. The main requirement is that the pit shall be big enough to accommodate the root system of the transplant and to permit consolidation of the soil around the roots of stumps.

When digging the holes the topsoil is placed on one side and the subsoil on the other. If stumps are to be planted, the pits will not be filled until the time of planting, but with other types of planting material they should be refilled as soon as convenient after checking in order to allow the soil to settle. After removing any water or debris that may have collected in the pit, it is filled with topsoil in two stages, the bottom half being trampled down before soil is put into the top half. It is generally a good practice to mix phosphate fertilizer with all the soil put into the hole since surface applications of phosphate after planting only penetrate to depth very slowly.

Notes

1. Later work (Mainstone 1970) showed that in some circumstances the initial period during which girth is unaffected by density may be followed by a transitory phase of about six months when girth increments are greatest at the *highest* densities.

This is because the earlier canopy closure at closer spacings suppresses the leguminous cover, thus reducing its competition for water and causing the release of nutrients by its decomposition. There follows a further short period of up to six months when girthing is not affected by density because the effect of suppression of the cover plants is balanced by the effect of inter-tree competition.

Chapter 6

Propagation, planting and pruning

C. C. Webster

6.1 Introduction

In the early years of the industry all the rubber in south-east Asia was grown from unselected seed. When these plantations came into tapping it was found that the trees were very variable in vigour and yield and that a small proportion of them gave much above average yields. For example, one study showed that 28 per cent of the total crop was produced by only 9.6 per cent of the trees (Whitby 1919). Planters in Java and Sumatra therefore began to use open-pollinated seed taken from the best-yielding trees. Certain estates went into the business of collecting and selling this so-called 'mother tree seed'. Appreciable yield increases over unselected seedling plantations were claimed and it was estimated that between 1917 and 1932 about 190,000 ha were planted with this material (Dijkman 1951).

The vegetative propagation of selected high-yielding trees by budding on to seedling rootstocks was initially developed by Van Helten, who planted the first clone trials in Sumatra in 1917 (Dijkman 1951). Following upon the results of the clone trials, budded rubber began to be planted on estates in the early 1920s. At first the use of the material necessarily developed slowly because of the time needed to test clones and multiply bud-wood, but in due course budding on seedling stocks became the main method of propagating rubber on estates and has remained so ever since. However, for many years the majority of smallholdings continued to be planted with unselected seed.

The selected seedling from which a clone is multiplied has commonly been called the 'mother tree' or 'clone mother' but, as this could imply parentage by sexual reproduction instead of by vegetative propagation, the more recently used term 'ortet' (from the Latin *ortus*, origin or source) seems preferable. Initially, primary clones were made from high-yielding trees selected on estates. Later, as detailed in Chapter 3, secondary clones were propagated from the best individuals in the legitimate seedling families produced by the plant breeders by crossing primary clones.

Over the years a large number of clones have been produced, each of which is identified by a number preceded by letters indicating the name of the research station or estate where it originated e.g. RRIM 600 (Rubber Research Institute of Malaysia), Tjir 1 (Tjir-andji Estate). A list of the abbreviations used to denote the more important sources of clones is given in Appendix 2.1 (p. 541).

Although plantations of buddings give better yields and are more uniform than seedling plantations, it has long been known that the high yield and other desirable characteristics of a selected ortet are not usually fully reproduced in its clone and that there is appreciable variation between individuals within a clone. There are several reasons for this. First, if the selected ortet is a mature tree in the field, it may owe its good performance more to a favourable environment – such as a good patch of soil or limited competition from weaker neighbours – than to its genetic constitution. Second, although the scions of a clone are genetically identical, their performance is influenced by the variable seedling rootstocks on which they are grafted. Stock-scion relations are dealt with later in this chapter (see section 6.4.5). Third, the buddings used commercially are 'mature type' buddings grown from buds taken from high on a seedling or after passage through several bud generations. As already mentioned (p. 71) the scion of a mature-type budding is not identical with its seedling ortet because it does not possess the juvenile features of the latter. That is to say, that the trunk formed from the scion is almost cylindrical and the thickness of the bark and the number of latex vessel rings therein varies little with height, in contrast to the tapering juvenile trunk in which bark thickness and number of latex vessel rings decrease with height.

For the above reasons, the correlations for vigour and yield between mature-type buddings and their ortets are not necessarily high. In consequence, the selection of outstanding ortets in commercial plantations, or from the plant breeder's legitimate families, must be followed by trials of their mature-type buddings before clones can be selected and recommended.

Early work by plant breeders showed that some of their crosses between high-yielding mother trees or clones resulted in legitimate seedling families which gave high average yields per tree. This led to the idea of obtaining superior seed from natural cross-pollination, in isolated seed gardens, between budded trees of clones selected for their ability to produce good seedling families. The first of such seed gardens was planted by Heusser (1919) who recommended establishing them in forest or other tall trees, such as oil-palms, where they could be separated from other rubber by at least 1 km so that pollen of other clones, or seedlings, could not be brought in by insects. Later, it was found satisfactory to isolate a seed garden on a rubber estate by surrounding it with a 100 m belt of one of the clones in the garden. The production of superior seed in this way expanded as more information became available on the performance of seedling families derived from crossing between the various clones used as parents. In Malaysia, the first of the well-known Prang Besar isolated seed gardens was planted by Gough in 1928. The seed and seedlings obtained from such gardens have long been known as 'clonal seed' and 'clonal seedlings', although it is the parents which are clonal, not the seed. A distinction can be made between polyclonal seed obtained from a garden containing several clones and monoclonal seed resulting from self-pollination in an isolated garden containing only one clone. Polyclonal seed from reputable seed gardens has long been planted on part of both the estate and smallholder area, although the proportion of the total area of both sectors so planted has declined in recent years. A comparison of buddings and clonal seedlings is given later in this chapter (section 6.4.4) but it may be noted here that although the yield level of the seedlings is lower they have certain advantages, especially in earlier maturity and lower tapping costs.

For many years two methods were used in the establishment of the majority of estates and smallholdings. One was to sow seed in the field and bud the resulting seedlings there about a year later. Usually, three germinated seeds were sown at each point and the most vigorous one budded. As budding was done on a strong seed-ling stock with a well-developed root system, a full stand was usually obtained with little difficulty. This method is now little used as it has the great disadvantage that a year's field maintenance costs are incurred before budding, although this period can now be shortened by several months by green budding (see p. 208). In the other method seedlings were raised in the nursery, budded at about a year old and then transplanted to the field as budded stumps with bare roots, after cutting back the stock stem just above the bud-patch and pruning the tap-root and laterals. Alternatively, if clonal seed-lings were used, they were raised in the nursery and transplanted after a year as bare-root stumps with the stem cut back to 40–60 cm and the roots pruned. Compared with budding in the field, planting stumps has the advantage that the time taken from planting in the field for the trees to reach tappable size is shortened by about nine months, or rather more in the case of clonal seedling stumps, as explained in section 6.4.3. Although the successful establishment of budded stumps is very dependent on the weather following trans-planting, they are still widely used and various means of countering the ill effects of dry spells are mentioned later (p. 212).

In the 1960s two developments occurred which have made it easier to establish a full, vigorous stand and have also helped to shorten the unproductive period of immaturity in the field. The first is the use of plants raised and transplanted in polythene bags with their root systems intact so that there is little or no check from transplanting. The second is the production of more advanced plants in the nursery, such as buddings that have grown one or two metres of scion and which can then be transplanted either as intact plants in polythene bags or as stumped buddings (see p. 213).

6.2 Propagation from seed

At present all rubber planting involves propagation from seed whether the seedlings are used as rootstocks for budding or grown to maturity as clonal seedling trees. As will be detailed later (section 6.4.5) there appears to be an advantage in using certain clonal seedlings as rootstocks, so it seems likely that an increasing proportion of the seed sown will be clonal seed from seed gardens rather than unselected seed.

6.2.1 Seed gardens for production of clonal seed

Selection of parent clones

The earlier seed gardens were planted with one or more primary clones, among which Tjir 1 was popular. Trials in Malaya (Burkill 1959) indicated that Tjir 1 monoclonal seedlings were among the most vigorous and high-yielding of those tested and these seedlings were widely used both as clonal seedlings and as stocks for budding. However, it later became evident that selfed progenies were generally less vigorous and lower-yielding than those resulting from crossing between selected clones. For example, Malaysian trials showed that seedlings from the Prang Besar isolated polyclone gardens yielded better than monoclonal seedlings of a number of high-yielding secondary clones (Rubber Research Institute of Malaya 1973). The degree to which selfing reduced vigour and yield was found to vary somewhat with the clone. Selfed seedlings of a few clones, e.g. RRIM 605 and PB 5/51, approached the performance of comparable cross-pollinatd seedlings but the general inferiority of monoclonal seedlings was clearly established.

In the polyclonal seed gardens now grown, natural pollination results in some seed being derived from crossing between clones and some from self-pollination since there is no evidence of self-incompatibility in the species. A low level of selfing is desirable, not only on account of the generally inferior performance of selfed progenies but also because most clones set more seed when cross-pollinated than when selfed, although a few are equally fertile whether crossed or selfed (Wycherley 1971). Obviously, it is possible to obtain cross-pollinated seed by planting only two clones in a garden but in order to minimize selfing it is advisable to plant at least five clones. To maximize the production of cross-pollinated seed it is desirable that all the clones in a garden should flower at the same time and that all should be of similar vigour so that no clone is suppressed by competition from stronger neighbours.

A prolonged search has been made for male-sterile clones. If male-sterile clones of high potential could be found, each could be planted as a female parent along with a high-yielding male-fertile clone in a two-clone seed garden to produce improved, cross-pollinated seed of known parentage. But so far all the clones found to be largely or completely male-sterile have also proved to have a very low female fertility.

Clones vary greatly in fertility, some being almost sterile and others very prolific. Obviously, reasonably fertile clones must be chosen in order to get good yields from a seed garden. Reciprocal cross experiments have shown that it does not matter which of a pair of clones is the female parent and which provides the pollen, except that seed yields are liable to be reduced if clones susceptible to the leaf diseases caused by *Colletotrichum gloeosporoides* and *Oidium heveae* are used as mother trees (Wycherley 1971). It is inadvisable to include with clones chosen for their ability to produce high-yielding progenies others chosen because they are highly fertile, as the latter may well produce much selfed seed which will give rise to inferior plants.

The genetic principles underlying the selection of clones for seed gardens have already been stated in Chapter 3. As a rule they will be proven high-yielding clones already used as parents in plant breeding programmes and selection will be based on the performance of their legitimate (hand-pollinated) seedling progenies in the replicated trials conducted by the breeders. The need is for several clones capable of producing high-yielding progenies when crossed with one another in all combinations. Hence, selection should ideally be based on the results of diallel-type trials, i.e. trials in which there are progenies derived from crossing all parent clones in all possible combinations. However, such trials have been few, so selection has usually been based on the results of less comprehensive trials.

In the absence of any such information from plant breeding programmes, an alternative approach is to plant several clones that have performed well in large-scale plantings. In due course, seed can then be harvested separately from each of the clones in the garden and planted in a trial to compare the clonal seedling progenies. (This can be done even if the seed has fallen because the seed of each mother clone can be recognized by its shape and the pattern of markings on the testa.) When this progeny trial comes into tapping, any clones which, as female parents, have produced poor progenies can be cut out of the seed garden. It will require 6 years for the garden to produce seed for the progeny trial which will then need to run for about 10 years before the yields can indicate which, if any, of the clones should be eliminated from the garden. Thus, it will be about 16 years before the production of reliably good clonal seed can be achieved, but this is no longer than is needed for the alternative procedure mentioned above, if allowance is made for the time required for the breeder's progeny trials.

Plant spacing and arrangement

Heavier seed production is generally observed along roads and field boundaries adjacent to open ground than within a plantation. Hence, it is usual to plant seed gardens on a rectangular pattern at a rather lower density than the normal commercial one in order to get good crown development and greater exposure of the canopy to light. In the planting density trial with AVROS 50, mentioned on p. 187, it was found that with increasing density there was a significant decrease in the number of seeds produced per tree and a steady, but non-significant, trend towards fewer seeds per hectare. Records of seed production in the trial mentioned on p. 192, in which four clones were grown on the avenue system, showed the seed yield per hectare progressively declined with increase in avenue width; thus contradicting an earlier view held by some people that the corrugated canopy of avenue planting would enhance seed production because it exposed more foliage to the sun and also gave rise to more flowering top branches. Bearing in mind that the yield of rubber from a seed garden is also of some significance, it seems suitable to plant initially at 9.4×3.0 m, giving a stand of 360 trees/ha which may be reduced by subsequent thinning to 250–300/ha.

The insects which pollinate rubber, mainly thrips and midges, are weak fliers and do not travel far. Hence, in order to minimize selfing and to give an equal opportunity for all possible crosses to take place, it would seem best to plant single trees of each clone in such a way that each is surrounded by trees of all the other clones (Fig. 6.1a). However, it is rare for all the clones to be of equal vigour and, if planted singly, the weaker clones are likely to be suppressed by the stronger ones. So it is usual to plant groups of 3–5 trees of each clone in succession along the rows with different clones opposite each other across the inter-rows (Fig. 6.1b). Thinning of the more vigorous trees where they begin to suppress the weaker ones then permits the survival of at least part of each group. A simpler design, easier to lay out but rather less effective in preventing competition and maximizing crossing, comprises complete rows of the same clone (Fig. 6.1c).

To prevent contamination of the seed garden by pollen other than that of the clones therein there must be a surrounding boundary belt, at least 100 m wide, which is either planted to one

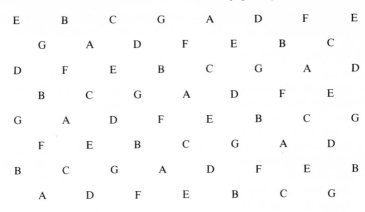

```
E     B     C     G     A     D     F     E
   G     A     D     F     E     B     C
D     F     E     B     C     G     A     D
   B     C     G     A     D     F     E
G     A     D     F     E     B     C     G
   F     E     B     C     G     A     D
B     C     G     A     D     F     E     B
   A     D     F     E     B     C     G
```

(a)

DDDDDBBBBBCCCCCBBBBBAAAAAEEEEEDDDDDEEEEEAAAAACCCCC

BBBBBDDDDDAAAAAEEEEECCCCCAAAAABBBBBCCCCCDDDDDEEEEE

CCCCCEEEEEDDDDDCCCCCBBBBBDDDDDEEEEEAAAAABBBBBAAAAA

DDDDDBBBBBCCCCCBBBBBAAAAAEEEEEDDDDDEEEEEAAAAACCCCC

(b)

```
A ――――――――――――――― A
B ――――――――――――――― B
C ――――――――――――――― C
D ――――――――――――――― D
E ――――――――――――――― E
```

(c)

Fig. 6.1 Arrangement of trees of different clones in seed gardens: (a) 7 clones planted on the equilateral triangle system so that each clone is surrounded by trees of all the other clones; (b) 5 clones planted on the rectangular system with groups of 5 trees of each clone in succession along the rows; (c) 5 clones in complete rows. (After Wycherley 1964)

of the clones in the garden or else consists of forest or another tall tree crop. If it is planted with rubber, seed is not collected from this belt for planting as clonal seedlings but may be used to raise stocks for budding. A square garden is preferable to a markedly rectangular one as the latter will increase the area required for the boundary belt.

Manuring and disease control

Seed yields are usually improved by applying fertilizer, especially nitrogen which also improves germination rate. Fertilizers do not appear to affect average seed weight. Details of fertilizers for seed gardens are given in Chapter 8.

In areas subject to the occurrence of the disease due to the fungus *Oidium heveae*, which can destroy flowers, dusting of seed gardens with sulphur at 10-day intervals just before and throughout the flowering season is recommended. Where pod rot due to *Phytophthora* spp. is a problem, control with a copper-based fungicide is necessary. Further details on the control of these maladies is given in Chapter 10.

Seed collection

About five months after flowering the ripe fruits dehisce explosively, scattering the seeds over the ground. In monoclonal areas most of the seed falls within a period of one or two weeks but in a polyclone garden it may extend over one to two months. Shortly before seed fall begins, the ground must be clean-weeded so that the seed is not hidden by the undergrowth, and any old seed should be collected and destroyed. To avoid the risk of erosion following clean-weeding, seed gardens are best sited on flat or gently sloping land. As the seeds lose viability rapidly, especially on exposure to the sun and/or drying, they need to be collected at least every two or three days, and preferably daily during peak seed fall.

Yields

Yields of seed gardens show great variation depending on a number of factors, especially the weather, the incidence of disease and the clones planted. In Malaysia it is reported that an average of 150 seeds per tree, or 37,500 seeds/ha at 250 trees/ha, may be expected from a well-maintained seed garden, provided that sulphur dusting is carried out where necessary, but yields of up to 100,000 seeds/ha have been known (Rubber Research Institute of Malaya 1965a). Determination of the mean seed weight for 17 clones showed a range of 3.59–5.81 g with a mean of 4.878 g, equivalent to 205 seed/kg. Thus, a yield of 37,500 seeds/ha is equivalent to 183 kg/ha. It is stated that in India an average of 250 kg/ha is reasonable (Pillay 1980).

6.2.2 Seed storage and packing

Seed is best germinated as soon as possible after collection as it loses moisture and viability fairly rapidly at ambient temperatures, which will usually be in the range 23–35 °C, unless measures are taken to prevent this. According to Chin *et al.* (1981) fresh seed has a moisture content of about 36 per cent and loss of viability roughly parallels loss of moisture down to a content of 15–20 per cent, below which seeds are not viable. Seed can be stored at ambient temperature in perforated polythene bags or sleeves with an equal volume of damp sawdust (10 per cent moisture content) or powdered charcoal (with 20 per cent moisture) for two months without loss of viability, provided that the bags are spread out in a thin layer so that there are air spaces between them. Under these conditions loss of viability is usually only slight after three months. Storage without air spaces between the small bags, or storing in large bags, results in an appreciable heat build-up and loss of viability. If the moisture content of the sawdust or charcoal is much above that indicated above, some seeds are likely to germinate during storage. Loss of viability is lessened by storing in a cool place. If it is possible to maintain the temperature at 4 °C, so as to restrict the rate of respiration, good germination can be obtained after four months (Rubber Research Institute of Malaysia 1974a).

Seeds which are to be sown within a few days of collection can be transported in sacks but if there is to be a longer delay before sowing and they have to be transported over some distance, they are usually packed in boxes or cartons with damp sawdust or powdered charcoal of the moisture contents mentioned above. Transport of germinated seeds is only satisfactory if the period between despatch and planting is no longer than three days. They should be packed within two days of the first sign of germination in boxes or cartons with about an equal volume of sawdust of 85 per cent moisture content by weight, and there should be a good layer (say 2 cm) between the seeds and the side of the container.

6.2.3 Germination

Before being planted in nursery beds in polythene bags or in the field, seeds are usually germinated. Germination beds about 1 m wide are sited on level ground in the nursery area. They may be made by raising soil beds about 15 cm high and placing a 5 cm layer of coarse river sand or aged (not fresh) sawdust on top of the soil. A shade of palm fronds, grass or hessian should be provided at a height of about 75 cm above the beds. The seeds are pressed into the sand or sawdust closely adjacent to one another with their flatter sides down, leaving part of the curved side just visible. A square metre of bed should take about 1200 seeds. Watering is usually done twice daily, morning and evening. The seeds may be covered with hessian or coir mats which may reduce the amount of watering needed but, provided that the overhead shade is good, it is probably better to leave the seeds uncovered so that they are more easily inspected.

When germination begins about seven days after sowing, the beds should be inspected daily and germinated seeds removed for planting in the nursery before the radicle reaches a length of 2 cm. It has long been recognized that seeds which are slow to germinate tend to produce weak plants and it is advisable to discard seed which has not germinated after 21 days. Studies in Sri Lanka showed that plants grown from seed germinating early (after 12–24 days) had faster growth rates, greater stem girth and larger leaf area up to the age of 18 months than those grown from seed which germinated after 24 days (Jayasekera & Senanayake 1971; Senanayake *et al.* 1975) It may well be that seed size is the important factor since Amma and Nair (1977) found that heavier seeds germinated more rapidly than lighter ones and grew into plants which had greater height and girth at 10 months after planting. The viability of seed is liable to be rather variable, possibly being affected by, among other factors, the physiological condition of the tree at the time of collection. In Malaysia fresh seed collected during the main seed fall season in August–September gives an average of about 90 per cent germination whereas that collected during the secondary seed fall in February–March gives only about 60 per cent. It is usual to germinate twice as many seeds as the number of plants required for field planting.

6.3. Nurseries

Nurseries are of two kinds; those in which plants are raised in polythene bags and those in which they are grown in the ground for the production of seedling and budded stumps, stumped buddings, three-part trees, or 'source bushes' for the multiplication of budwood.

6.3.1 Ground nurseries

The site for a ground nursery should be near a reliable, year-round water supply on level, or gently sloping, land with deep, well-structured and well-drained soil, preferably of loam or clay loam texture. It should be sheltered from high winds but should not be subject to shading or root competition from adjacent trees. In clearing the site from forest or old rubber it is necessary to extract all stumps and roots and to remove or burn them together with the above-ground timber. This will usually be done mechanically as described in Chapter 5, after which the land should be ploughed to a depth of 30–45 cm and harrowed to produce a tilth suitable for sowing germinated seeds. Initial manuring of the nursery will depend on the soil and is dealt with in Chapter 8. A nursery site can be used more than once, provided weeds are controlled and soil fertility is maintained.

The dimensions of any beds and paths and the spacings at which germinated seeds are sown in the nursery depend on the purpose for which the seedlings are to be grown. They will usually be used as stocks for budding and details of the spacings adopted for the production of several sizes of budded plants are given in section 6.4.2. To a limited extent clonal seedlings are still grown in ground nurseries for transplanting to the field as stumps. For small stumps, germinated seeds are planted at a spacing of 60 × 20–30 cm and allowed to grow for 6–8 months. They are then extracted with a lever jack, the stems cut back to 30–45 cm of brown wood, the tap-roots severed at 30–40 cm and the lateral roots pruned to about 10 cm. These small seedling stumps were formerly widely planted on smallholdings as they could be transported considerable distances from a central nursery without suffering damage, but they have now largely been superseded by buddings. Much larger clonal seedling stumps ('maxi stumps') are raised by planting germinated seeds at a spacing of 75 × 75 cm and leaving the plants to grow for 12–14 months before pollarding them in brown wood at 2.0–2.5 m with a sloping cut made just below a whorl of buds. The subsequent treatment, transport and planting of these large stumps is as described for stumped buddings on p. 215.

Thorough weed control is essential in the nursery. While the plants are young and tender this is usually done by hand with hoes but, after they have developed brown bark, herbicides can be used, as described on p. 288.

Watering, or irrigation, has been found beneficial even under the well-distributed rainfall of Malaysia where a small amount of water applied during dry spells significantly improved growth. Where a distinct dry season is experienced, growth will not be satisfactory without watering; for example, in Kerala, India, nurseries must be regularly watered during the dry season from December to March. As might be expected, mulching is beneficial; apart from conserving soil moisture it also lowers soil temperature, checks weed growth and adds organic matter to the soil. In experiments in Thailand mulches of sawdust, rice husks or dry grass increased the girthing rate of plants even in months with adequate rainfall and mulched plants could be budded earlier than unmulched, with consequent savings in weeding and fertilizer costs (Greenwood & Promdej 1971). No significant difference was found between the effects of the different mulches, which was also the case when different mulching materials were compared in India (Potty *et al.* 1968).

6.3.2 Polybag nurseries

The site requirements for polybag nurseries are similar to those for a ground nursery but, since the ground is not to be planted, complete extraction of stumps and roots and overall cultivation are not necessary. Nevertheless, there must be a good supply of surface soil for filling the bags either on site or close at hand. A reliable water supply is essential as polybag plants will usually need watering frequently.

The plants may be grown in the bags to any size from 2–3 leaf whorls to 6–7 whorls before being transplanted to the field. The use of large polybag plants, which have virtually intact and undisturbed root systems, has the great advantage of appreciably shortening the period of immaturity in the field before the trees are ready for tapping. The bags, which are normally black and are provided with perforations for drainage, are filled with good, friable surface soil which is freed of roots, stones and clods before use. The soil should be of clay loam, or similar, texture so that it remains bound to the roots when the bag is stripped away at transplanting. It is usual to incorporate 50 g of ground rock phosphate with the soil for each bag. Plants grown in polybags may be either clonal seedlings or buddings. The procedure for growing and transplanting them is the same except for the bud grafting. Details for plants of various types and sizes are given in section 6.4.2.

6.4 Vegetative propagation

6.4.1 Bud grafting

Brown budding

Until the early 1960s all budding was done with axillary buds taken from the matured parts of shoots where the leaves had fallen and the bark had turned from green to brown. The shoots ('bud-sticks') which normally had a diameter of 2.0–2.5 cm and were about a year old, could be cut in limited quantity from mature trees but were usually obtained from source bushes in multiplication nurseries (see p. 210). Budding was generally done on seedling stocks of similar age and size to the bud-sticks, either in the field, or in the nursery for the production of budded stumps. Brown budding of stocks smaller than about 2 cm in diameter tends to result in poor scion growth but larger stocks, up to 8 cm diameter, can be successfully budded and scion growth rate usually increases somewhat with stock size up to at least a diameter of about 4 cm.

For brown budding the stock is prepared by making two vertical cuts 7.5–10.0 cm long and 2 cm apart and joining them at the bottom with a horizontal cut at about 2.5 cm above ground level.

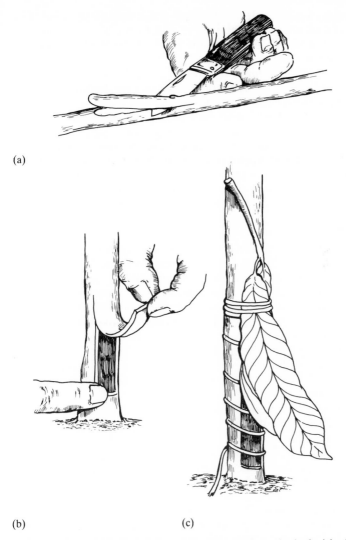

(a)

(b)   (c)

Fig. 6.2 Brown budding: (a) cutting the bud-slip from the bud-stick; (b) inserting the bud-patch into the panel prepared on the stock; (c) binding and shading

A bud-patch is obtained by cutting from the bud-stick a strip of bark and wood, about 5.0–7.5 cm long by 2.0–2.5 cm wide, containing a bud and a leaf scar, removing the sliver of wood by bending it away from the bark (Fig. 6.2a) and trimming the patch to a size to fit the panel prepared on the stock. The flap of bark cut on the stock is then peeled upwards, the patch inserted into the panel beneath with the leaf scar below the bud (Fig. 6.2b) and the flap of bark is pulled down over the patch and bound to the stock with twine. Two or three leaves are tied in above the bud to shade it from the sun (Fig. 6.2c). During the operation the exposed cambium layers of the stock and scion must be kept clean and untouched. At 18–21 days after budding, the binding is removed, the flap of stock bark cut off at the top and the bud-patch scratched to see if it is alive and hence successfully grafted. The shade leaves are than replaced for 10–14 days before cutting back the stock 7.5–10.0 cm above the top of the bud-patch. The scion bud usually starts growth one or two weeks after cutting back. Shoots developing from the stock are regularly removed. With large stocks it is advisable to cut back 30–40 cm above the patch, to leave this snag of stock stem until the scion has grown 45–60 cm of shoot with brown bark and then to remove the snag with a cut sloping away from the base of the scion.

Green budding

Green budding of rubber was developed in North Borneo in 1958–60 (Hurov 1961) and, after some modifications to the original technique, it largely replaced brown budding. Initially it was thought that only the buds in the axils of the scales on the non-leafy parts of terminal flushes could be used, but it was eventually shown that buds from the leaf axils of the terminal, and the two preceding, whorls of leaves of a shoot, as well as the scale buds, could all be used without any significant difference in budding success or in rate of emergence and growth of scions (Gener 1966; Templeton & Shepherd 1967; Plessis 1968; Samaranayake & Gunaratne 1977; Leong & Yoon 1979).

It is now usual to green bud seedling stocks in the field or nursery when they are 5–6 months old, with the bark at the base of the stem turning from green to brown, and with a diameter of 1.2–1.5 cm just above ground. The stock is prepared in the same way as for brown budding except that the width of the panel is less (usually about one third of the circumference of the stock) and that after peeling up the flap of bark it is cut off at the top, leaving only a small projecting piece. The bud-patches from the young green terminal shoots are also prepared in the same way as for brown budding, but where the buds are in the axils of growing leaves the petioles must be cut off at their bases, either several days before the bud-wood is used or at the time of budding. The trimmed patch is placed on the panel with its upper end slipped under the tongue of stock bark remaining from the flap (Fig. 6.3a) and secured in place by binding from the top downwards with transparent poly-thene tape which is tied at the bottom (Fig. 6.3b). The bud-patch is visible through the tape and no shade is needed; in fact, shade has been proved to be deleterious.

Three weeks after budding, provided that the patch is still green,

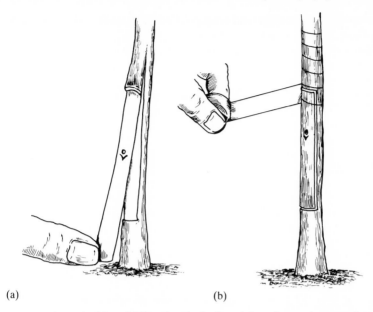

(a)   (b)

Fig. 6.3 Green budding: (a) inserting the bud-patch into the panel prepared on the stock; (b) binding with transparent tape

the stock is cut back 5.0–7.5 cm above it and the tape removed. No ill effects result from delaying cut-back for up to three months should this prove desirable; for example, on account of dry weather. The buds normally begin to grow 7–10 days after cut-back but a small proportion may take longer. Any that remain dormant as long as a month are unlikely subsequently to grow out.

Considerably larger stocks than those mentioned above can be green budded provided that good contact can be made between the cambiums of stock and scion by inserting extra thicknesses of polythene over the bud-patch before binding. Budding seedlings in the field younger than 4 or 5 months old is inadvisable as this often results in dieback of scions and death of an appreciable proportion of the plants. But in the nursery polybag seedlings no more than 6–8 weeks old can be successfully budded, even though the width of the grafting panel may be almost half the circumference of the stock. With these so-called 'young buddings' it is best to cut back about 20 cm above the bud-patch, subsequently pruning off any developing stock shoots in order to hasten sprouting of the scion bud. Growth of the young buddings is improved by spraying with a foliar feed twice a week until the scion shoots are hardened. Using these methods, about 90 per cent success can be obtained (Ooi *et al.* 1976).

Multiplication, storage and transport of bud-wood

Bud-wood is normally obtained from multiplication nurseries in which seedling stocks have been budded with the required clones. For brown bud-wood, plants may be spaced 90 × 60 cm and the scions initially allowed to grow for a year to a height of about 2.5 m before harvesting bud-wood by sawing off at 15–30 cm above the union. Two scion shoots can then be allowed to grow for 12–18 months to provide a second harvest after which four shoots are left for each subsequent harvest.

For production of green bud-wood in Malaysia, plants spaced at 120 × 120 cm are pollarded at 75–90 cm after 8–9 months' scion growth and 4–5 side branches allowed to grow. When these have grown for 6–9 weeks they are pruned off near their bases (but above several dormant buds) to provide bud-wood and then allowed to regenerate 3–4 shoots each. By repetition of the procedure a 'source bush', shaped rather like a tea bush, is formed from which green bud-wood can be regularly obtained by pruning branches just above a whorl of dormant buds 6–8 weeks before the bud-sticks are required for use. Established bushes can be pruned overall four times a year to provide bud-wood over about eight years before replanting.

In Thailand, where very large quantities of green bud-wood are produced for issue to smallholders, plants are spaced at 2 × 1 m in the nursery and the procedure is, initially, the same as in Malaysia, scions being tip pruned to produce 4–6 lateral branches which are harvested as bud-wood. After this, the main scion stem is pruned again above the next whorl of dormant buds lower down to produce another 4–6 laterals for harvest and this can be repeated for a third round of bud-wood harvesting. At the end of the season all the scions are cut back hard and in the second year the whole nursery is allowed to regenerate two scion shoots per plant, which are subjected to pruning and harvesting as in the previous year. In the third and subsequent years, four scion shoots per plant are allowed to grow after cutting back. The nursery is divided into three (or six) sections and each section pruned at fortnightly (or weekly) intervals so that there is regular harvesting of laterals after six weeks' growth; if the harvesting interval is longer, the bark will harden. A mature nursery yields about 200,000 bud-sticks per hectare annually. In areas with a marked dry season, irrigation greatly extends the season during which bud-wood may be harvested.

Brown bud-wood is best cut early in the morning. For use on the same day it should be carried to the field or nursery in polythene bags or moist sacking and kept in the shade until used. If it is to be used later or at a considerable distance, it is cut into convenient lengths and the ends immediately dipped into molten wax to prevent drying out. For transport it is packed in boxes with layers of bud-wood alternating with layers of damp sawdust. The more tender green bud-wood should be cut early in the morning, the leaves removed and, if it is to be used the same day, carried to the work place in polythene bags and kept as cool as possible. For transport to a distance, it is treated as for brown bud-wood. So treated, green bud-wood can be stored in a cool place for up to six days without appreciable loss of viability or buddability.

6.4.2 Production and transplanting of various types of budded plants

This section describes the production of four types of budded plant in the nursery and their transplanting to the field. Of these, budded stumps and stumped buddings are transplanted with bare roots while buddings in soil cores or polybags have more or less intact root systems. Budded stumps, which are transplanted with a dormant bud-patch, have been planted for many years and continue to be widely used. They are cheap to produce and can be easily handled and transported with little risk of damage, which is particularly advantageous for their distribution to scattered smallholdings. But the percentage of stumps successfully established in the field is liable to be poor if dry weather follows transplanting. The main advantages of using plants with scions already grown out in the nursery are greater reliability in achieving a full, uniform stand and shortening of the unproductive immature period in the field. The production and transplanting of these more advanced plants requires more skill and costs more than budded stumps and they are not yet in general use. After describing the techniques for their production, a comparison of the costs of the different types of plant and of their effect on the length of the immature period is given.

Budded stumps

For the production of budded stumps, germinated seeds are sown in the nursery at a spacing of about 60 × 25 cm and the plants either green budded when they are 5–6 months old or brown budded at a year old. Three weeks after green budding the tops of the successfully budded stocks are bent over and broken 30 cm or more above the bud and 4–5 days later the plants are extracted with a lever jack, or pulled by hand after digging a trench along one side of each row. The stems are then cut off at 5 cm above the bud-patch, the tap-roots severed at 30–40 cm and the lateral roots pruned back to 10 cm (Fig. 6.4). The polythene tape can be kept intact to protect the bud-patch until after the stump has been planted in the field.

Brown buddings have the binding removed 10–14 days after budding. They are cut back 10 cm above the bud-patch 7–10 days

Fig. 6.4 Budded stump

before it is desired to extract them so that most of the buds have begun to swell when they are transplanted. Extraction is done as with green budded stumps.

The successful establishment of budded stumps in the field is very dependent on the weather following transplanting; if there is a dry spell, there may be dieback or death of a number of the young scion shoots. This is partly because after transplanting the stumps are rather slow to initiate new roots, which may take 6–8 weeks. To try to overcome this problem, experiments were conducted with a number of growth regulators (Pakianathan & Wain 1976; Pakianathan *et al.* 1978) and it was found that root initiation could be considerably accelerated, and growth of roots and stems improved, by dipping the pruned roots of the stumps before transplanting into a liquid formulation of kaolin and ethanol containing 2000 ppm of indolebutyric acid, 1.0 per cent potassium nitrate and 5 per cent Captan 50.

To prevent damage to, and dessication of, green-budded stumps during transport from central nurseries to smallholdings, the stumps, with the polythene tape retained around the bud-patch, may first have the cut stems dipped in wax and all the lateral roots pruned off. They are then dipped in the formulation mentioned above and packed into bundles of 50 with moist sawdust wrapped around with several layers of newspaper and an outer cover of polythene sheet. It is reported that this procedure can check dessication for 20–30 days but it is preferable to plant after not more than 10–15 days, when an establishment in the field of 95 per cent should be obtained with fairly uniform, rapid bud break (Pakianathan & Tharmalingam 1982).

Another problem with budded stumps is that with some clones the emergence of a proportion of the scion buds may be considerably delayed with consequent slow growth and lack of uniformity. Recent experiments (Jaafar 1984) have shown that bud break can be accelerated by treating the bud-patches with a formulation containing 2500 ppm Atrinal® (dikegulac-sodium), 5 per cent dimethyl sulfoxide and 10 per cent Triton X100 in 50 per cent ethanol, which was sprayed or painted on six times on alternate days. This increased the bud break of four clones 30 days after transplanting by 10–29 per cent, but the longer-term effects of the treatment have not yet been reported.

If mature-type buddings are deeply planted so as to bury part of the scion stem, the latter does not produce any roots, even if notching or ring-barking is done to try to induce rooting. On the other hand, if seedlings are budded high and planted deeply with the bud-patch just above ground level, the buried part of the stock stem develops a taper like a tap-root and produces lateral roots. It was suggested that such deep planting of budded stumps might improve establishment success under dry weather conditions, or on soils with a low water-holding capacity, by enabling the roots to exploit moisture reserves at greater depth and from a larger volume of soil than could be tapped by normal planting. On certain soils, the incidence of wind damage might also be reduced. A further possibility envisaged was that the decline in yield commonly experienced as the tapping cut approaches the union of stock and scion might be lessened by increasing the distance between the bud-patch and the roots (Yoon & Ooi 1976). Whether these possibilities are realizable remains to be investigated, but clearly they could only be advantageous if deep planting has no adverse effect on growth and yield. Some evidence on the latter point is available from an experiment in which seedlings were budded with clone PB 260 at 30, 15 and 5 cm above ground and planted as bare-root stumps buried up to the position of the bud-patch. The girth and percentage of trees tappable at opening five years after planting and the yields over the first six months of tapping have all been slightly better from the deeper-planted trees than from the controls (Rubber Research Institute of Malaysia 1985a).

Stumped buddings

To produce large stumped buddings, or 'maxi-stumps', germinated seeds are sown in the nursery at a spacing of 90 × 90 cm and budded as described above for budded stumps. After cutting back they are left in the nursery for a further 12–18 months to produce a scion with 2.25–2.50 m of brown bark. Six to eight weeks before it is desired to transplant, a trench is dug along one side of each row of plants to expose the roots so that the tap-root can be cut off at 45–60 cm and the exposed laterals pruned to 10–15 cm. This procedure is sometimes called 'tailing'. The trench is then loosely refilled, which may result in some new roots being initiated from the pruned laterals. Ten to fourteen days before transplanting the plants are topped at 2.0–2.5 m with a sloping cut in wood with brown bark just below a whorl of dormant buds. The cut is treated with a protective dressing and the whole stem is white-washed with hydrated lime to prevent sun scorch (Fig. 6.5). By the time the

Fig. 6.5   Stumped budding (Rubber Research Institute of Malaysia)

plants are extracted from the nursery the buds near the top of the cut stem will have started to shoot (although they should not have grown more than about 0.5 cm) and care is needed in transporting to the field in order to avoid damage to these shoots and to newly developed roots. Single layers of stumps (with bare roots) on lorries may be separated by layers of grass, or the vehicles may be provided with racks to carry single layers of plants. Before transplanting the roots may be treated with the indolebutyric acid formulation to promote rapid root development, as described for budded stumps, but using a higher concentration of 3000 ppm of IBA (Pakianathan *et al.* 1980).

Stumped buddings are normally transplanted in the field into holes 60 cm square (or diameter) and 40–50 cm deep with a small hole cut deeper into the centre of the main hole into which the tip of the tap-root is inserted and packed with soil. This should hold the plant upright while the large hole is filled in.

Production and transplanting of smaller stumped buddings, known as 'mini-stumps', is similar to the above. Germinated seeds are planted in the nursery at 90 × 30 cm, scions are topped to give 60 cm of stem with brown bark after seven months' growth from budding, and no 'tailing' is done – only root pruning at the time of extraction.

Soil core buddings

The procedure for producing soil core buddings (Sergeant 1967) is initially the same as for budded stumps, but after cutting back the plants are left in the nursery until the scions have grown 2–5 whorls of hardened leaves before they are extracted with an intact cylinder of soil around the roots. The extraction is done with special equipment (Fig. 6.6). The metal cylinder (a) is placed over the plant and driven down with the double-headed 'hammer' (c) until the arms of the cylinder reach soil level (d). The cylinder and plant are then levered out of the ground with a crowbar, or a lifting fork, inserted under the arms of the cylinder. That part of the tap-root protruding from the base of the soil core is pruned off and the cylinder containing the plant is placed on top of the ejector piston (b). Pressing down on the arms of the cylinder pushes out the plant with its soil core (e) which is then wrapped in newspaper or polythene and tied with string (f). Transplanting to the field can follow immediately if the weather is suitable but, if not, the plants can be kept under light shade and watered periodically until they can be planted out. After the plant is placed in the field planting hole, the wrapping is removed and the soil filled in and compacted around it without treading on top of the soil core itself.

Cylinders measuring 40 × 18 cm diameter have been used for plants with 2 whorls of leaves and those of 48 × 18 cm for plants

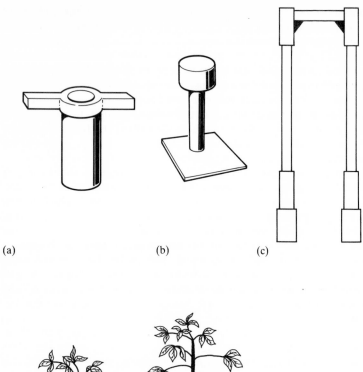

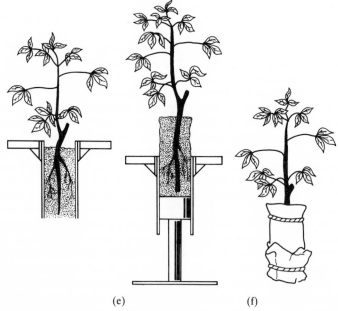

Fig. 6.6 Soil core budding; for explanation see text (Rubber Research Institute of Malaysia)

with 3–5 whorls. The soil type in the nursery is critical for the success of the method as it must contain sufficient clay to form a cohesive core which will not break up during extraction and transport. The method has advantages in that it involves little root disturbance and it enables transplanting to the field to be done at almost any time of the year in most rubber-growing areas, but the extraction procedure is slow. Soil core buddings are not much used for large-scale planting but could be useful for supplying vacancies.

Budded plants in polybags

For the production of polybag buddings the seedling stocks may be grown and budded in the bags or, alternatively, the stocks may be grown, budded and cut back in the nursery beds and the small budded stumps then transferred to bags and allowed to grow therein until they are ready for transplanting to the field. The second method gives more opportunity for selection of the most vigorous stocks and avoids wastage of bags containing poor seedlings or budding failures. If budding is to be done in the bags, it is best to sow two germinated seeds per bag and to remove the less vigorous one after a month, leaving the other to grow on for green budding at 5–6 months old.

After budding and cutting back, the scions are allowed to produce anything from 2–7 whorls of leaves before transplanting. The smaller plants are raised in bags measuring 40 × 25 cm when laid flat and holding about 9 kg of soil. Plants with 6–7 whorls of leaves are best grown in 65 × 40 cm bags which weigh about 23 kg when filled and are rather difficult to handle when they contain the large plants. As mentioned on p. 209, stocks only 6–8 weeks old can be budded to produce 1–2 whorl plants in small bags. These are cheaper to produce and transplant than the larger plants but their comparative performance has not yet been reported.

It is desirable to arrange the bags in the nursery so that as far as possible self-shading and competition for light between plants is minimized. This is especially important during the later stages of growth of 6–7 whorl plants which have to remain in the nursery for 12–14 months. One way of doing this with large bags is to place them in double rows with 20 cm between bags in the row, 30 cm between the rows and 75 cm between pairs of rows. In order to prevent heating up and loss of moisture on exposure of the bags to the sun, it is advisable to put them in trenches 30 cm deep, using the soil dug from the trenches to mound up around the bags, leaving only the upper 5–10 cm exposed (Rubber Research Institute of Malaysia 1975a). The bags are not usually shaded and watering is done immediately after sowing the seeds, after each monthly fertilizer application (see p. 343) and during dry weather.

Plants are transplanted to the field when they have reached the desired stage of growth, preferably when the top whorl of leaves has hardened. Unless there has been very recent rain they are watered before removal from the nursery. Any lateral roots protruding from the bag are cut off and if the tap-root has grown through the bottom of the bag it is severed at this point. Precautions must be taken to prevent damage to the root system of the plants during transport, as this can easily occur if the bags are roughly handled, bent or allowed to topple over. When the larger polybag plants were first used, such damage led to poor field establishment in a number of cases. The smaller, 2–3 whorl plants can be easily carried and loaded on to a vehicle and packed close together to prevent much movement during transport. The 6–7 whorl plants are not easily handled by one man and on some estates they are moved in specially made boxes, with handles, which can hold three plants and are carried by two men. Some means of support may also be needed to prevent damage to the stems during transport on lorries or trailers. On arrival at the planting point in the field, the bag is stripped off the soil core which is placed carefully in the planting hole without disturbing the root system. The hole is then filled in with topsoil which is pressed down sufficiently firmly to anchor the plant.

6.4.3 Reducing the period of immaturity by propagation and planting methods

For all tree crops, efficient methods of propagation, transplanting and maintenance during the early years in the field are important in order to obtain a full, uniform stand which will grow vigorously without check and reach a productive stage as soon as possible. With rubber it is particularly important to minimize the relatively long period which must elapse after planting in the field before the trees have attained sufficient girth for tapping to begin. On efficient estates the immature period is about $6\frac{1}{2}$ years if budding is done at stake in the field, 6 years if budded stumps are planted or 5 years where clonal seedlings are grown. It is generally longer on small-holdings, whether they are on supervised settlement schemes or individually held, provided that the recommended girth standard for opening is observed. A survey in Malaysia in 1970 showed that the immature period on settlement schemes for smallholders ranged from 6.9 to 8.1 years with a mean of 7.5 years (Lim & Chong 1973). During the immature period an estate has no return from its invest-ment in the land and a smallholder has no income from his rubber. On a settlement scheme, the longer the period of immaturity the larger will be the amount of the loan which the smallholder has to repay to the sponsoring authority.

The economics of shortening the immature period

The economics of shortening the unproductive phase can be examined by estimating costs and returns over the whole life of the plantation for different immature periods and calculating discounted cash flow measures, such as the internal rate of return (IRR), the net present revenue (NPR) and the benefit : cost ratio (B/C). This has been done by several workers for example, Barlow and Ng (1966), Lim *et al.* (1973), Rubber Research Institute of Malaysia (1974b), Chong and Pee (1976). The figures reported vary according to the level of yield, selling price and interest rate assumed but, as might be expected, and as indicated in Table 6.1, they all show considerable financial advantage from shortening the immature period.

Table 6.1 *Relative merits of different immature periods for clone RRIM 600*

Period of immaturity (years)	Internal rate of return (%)	Benefit cost ratio	Net present revenue ($M/ha)
4	21.7	1.5	4662
5	18.9	1.4	3854
6	16.9	1.4	3143
7	16.0	1.3	2720

Source: Rubber Research Institute of Malaysia (1974b)

As the NPR measures the absolute amount of present revenue, the difference between the values for any two immature periods represents; (a) the gain in revenue derivable from achieving the shorter period; and (b) the maximum additional expenditure which may be incurred to achieve the shorter period without reducing the NPR below that obtainable with the longer one. Thus, from Table 6.1 the additional expenditure permissible to achieve 4, 5, or 6-year periods instead of 7 years are, respectively, $M1942, $M1134 and $M423. These maximum amounts represent break-even points in each case and are considerably in excess of the costs of the procedures normally used to shorten immaturity.

Means of shortening the immature period

An obvious, but not always fully implemented, means of shortening immaturity, as well as improving growth and yield, is to carry out rigorous selection at all stages in the nursery – of germinating seeds, of seedling stocks and of plants after budding – to ensure that only really good, vigorous plants are taken to the field. A second means is to raise plants that have reached as advanced a stage as practicable in the nursery. Thirdly, the check to growth in transplanting should be minimized, which is greatly aided by the use of polybags. There are a number of other means of shortening immaturity: they include the branch induction and pruning procedures mentioned later in sections 6.5 and 6.6 the use of the somewhat more precocious clones now becoming available, together with manuring, mulching, maintenance of legume covers and weed control, which are covered in Chapters 7 and 8.

Effect of type of planting material on costs and on the duration of the immature period in the field

A number of studies have been made to compare the effects of using different planting materials on costs and on the length of the immature period in the field. Based on several investigations with the clone RRIM 600, Sivanadyan *et al.* (1973) assessed the costs during immaturity of five kinds of plants, assuming normal commercial practice in respect of establishment of legume covers, weed control and manuring (Table 6.2). Compared with the old, standard practice of planting seed at stake and budding in the field, large polybag buddings (with 6–7 whorls of leaves) and stumped buddings were estimated to shorten immaturity by 18 and 30 months respectively. The higher costs of producing and transplanting these more

Table 6.2 *Comparative costs of various planting techniques during immaturity ($M/ha)*

Nursery and field operations	Budding in the field	Bare-root budded stumps	Budded stumps in polybags	Large polybag buddings	Stumped buddings
Immature period in field (months)	78	69	69	60	48
Holing	26	36	36	32	32
Budding	31	22	20	20	19
Nursery operations	—	27	98	157	77
Planting or transplanting	20	35	56	96	136
Total, nursery and transplanting	77	120	210	305	264
Weed control	1129	1063	1063	813	642
Manuring	455	424	406	384	294
Total	1661	1607	1679	1502	1200

Source: Sivanadyan *et al.* (1973)

advanced plants were more than offset by the lower costs of weeding and manuring during a shorter immature period. Consequently, the total costs per hectare for the large polybag plants and the stumped buddings were respectively $M159 and $M461 less than those for seed at stake planting. Thus, there were actual savings in costs to maturity, quite apart from the increase in net present revenue to be expected from shortening the immature period, as indicated in Table 6.1.

Shepherd (1967) and Shepherd *et al.* (1974) reported the results up to the fifth year of tapping of a randomized block experiment with 6 replications which compared 6 types of plant of clone PB 5/51 (Table 6.3). The best results in terms of shortening immaturity, yield over 5 years of tapping and discounted revenue over the first 9 years from planting were obtained by the use of stumped buddings, with the large polybag buddings coming second best.

Table 6.3 *Relative performance of different kinds of planting material of clone PB 5/51 in a six-replicate experiment*

Type of plant	Girth at 114 months (cm)	Immature period (months)	5-year cumulative yield (% of control)	Discounted* revenue over 9 years ($M/ha)
Stumped buddings	68.6	46	151	1061
Soil core buddings, 4-whorl	66.0	55	121	399
Polybag budded stumps, 2-whorl	65.8	57	126	602
Polybag buddings 2-whorl	63.5	59	106	141
Green budding in field (control)	64.3	61	100	91
Brown budding in field	63.2	67	87	−247

Source: Shepherd *et al.* (1974)
* Assuming RSS 1 price $M1.60/kg and discount rate 10%.

Other studies have also shown that the greatest reduction of the immature period in the field and, usually, the greatest financial advantage, is obtained by planting stumped buddings ('maxi-stumps'). Large polybag buddings with up to 6–7 whorls of leaves are probably the next best material and in some cases have not differed from maxi-stumps in time to maturity (Sivanadyan *et al.* 1976; Pushparajah & Haridas 1977). Soil core buddings with 2–4 whorls of leaves have done as well as stumped buddings in some trials (Nor *et al*. 1982) but have given much poorer results in others, the variation possibly being related to soil type. Although these advanced planting materials have been shown to shorten the immature period compared with field budding or budded stumps, their use is still rather limited, possibly because their production and reliable establishment in the field require more care, or on account of uncertainty regarding the costs and benefits involved.

Replanting in the shade of an old stand

When replanting, it is practicable to shorten the unproductive period between ceasing to tap the old stand and beginning to tap the new by planting the new trees before removing the old ones. Davidson (1962) described a procedure used successfully in replanting large areas over several years. After removing all undesirable ground vegetation, budded stumps with dormant bud-patches and about 1 m of stock stem were planted between the rows of the old stand. About 10 months later the old trees were poisoned and then left for 6 weeks before felling. Shortly before felling, the stock stems of the young plants, which had grown numerous shoots, were cut back just above the bud-patches. The falling timber did negligible damage to the budded stumps, which made rapid scion growth because of their already established root systems. The new stand was ready for tapping 4½ years after planting, which was claimed to shorten the unproductive period by at least 2 years compared with clearing the old stand before planting seed at stake for budding.

This method may have advantages when replanting on sloping land where the new rubber can be planted on existing terraces between the old trees, which can subsequently be precision-felled to fall between the terraces. On such land this can be better than complete clearing before replanting in that it involves less soil disturbance and risk of erosion. Obviously, the procedure can be modified in various ways. For example, faster scion growth can be obtained if budded stumps are planted 2 years before felling and then treated as described above, or if seedlings are planted 2 years before felling, budded a year later and cut back at felling time.

6.4.4 Comparison of buddings and clonal seedlings

As clonal seed production must be preceded by the selection and testing of the parent clones used in the seed gardens, it is evident that the yield level of the best available clonal seedlings is likely to lag behind that of the most advanced clones. The seedlings also yield less than buddings because they have to be tapped less intensively in order to avoid a proportion of the trees suffering from dryness or brown bast (see section 9.4.4). Commonly, buddings have been tapped on a half spiral cut alternate daily while with seedlings the same length of cut has been tapped every third day, resulting in lower yields but also lower tapping costs. Clonal seedlings are more variable in growth vigour, bark thickness and yield than clones. This is usually partly countered by planting them at a higher initial density than buddings and later thinning the stand by removing the less vigorous and lower-yielding trees but this adds somewhat to the cost of establishment.

Although seedlings do not match buddings for yield, they do have some advantages. They are generally regarded as being more robust and easier to establish. Under tapping they produce thicker renewed bark and show better girthing than buddings; with less wind damage, but these features are a consequence of their lower yield level. Their greatest advantage is that their immature period in the field is a year less than buddings. This is mainly because tapping can begin when the conical trunk of a seedling reaches a girth of 50 cm at 50 cm above ground level whereas the cylindrical trunk of a budding takes longer to attain the required girth 45–50 cm at 125–150 cm above the union (see section 9.2.1).

Despite their lower yield level, clonal seedlings have long been planted on a proportion of the estate area in Malaysia because of their earlier maturity and lower tapping costs, although the proportion planted with seedlings has been declining in recent years. Formerly, clonal seedlings were widely planted on smallholdings, perhaps because their transport and establishment as seed at stake or stumps was easier and more reliable than for buddings, but in Malaysia they are no longer recommended for smallholdings as they cannot achieve the objective of maximizing yields per hectare.

6.4.5 Stock-scion relations

Early observations that the average yield of a clone budded on seedling rootstocks was usually somewhat inferior to that of its mother tree led to investigations to assess stock effects and to look into the possibility of selecting superior stocks which would improve scion performance. Evidence of stock influence on scion was obtained from some experiments started in the 1930s in which twinned seedlings were established by splitting young plants longitudinally between the cotyledon petioles and one of each pair budded while the other was grown as a seedling. In some of these trials it was found that there was a significant correlation between the budded and unbudded twins in growth and yield and in one trial buddings on the six most productive seedlings yielded considerably more than those on the six poorest (Dijkman 1951). In others, for example in two trials started in Malaya in 1932 and 1934, significant correlations between budded and unbudded twins were only established for a few combinations of stock and scion (Buttery 1961).

Later, a number of experiments compared illegitimate seedlings of different clones as stocks (Schweizer 1938; Schmöle 1940; Paardekooper 1954; Campaignole & Bouthillou 1955; Buttery 1961; Combe & Gener 1977a; Abbas & Ginting 1981). Most of these trials demonstrated marked stock effects. For example, differences between the yield of a scion clone on different stocks of up to 20 per cent were reported by Schmöle (1940) and of up to 18 per cent by Paardekooper (1954). On the other hand, in five Malaysian experiments started between 1931 and 1941, which included a number of scion clones and illegitimate seedling families as stocks, clear evidence of significant stock influence on girth and yield of scions was obtained in only two of the trials (Buttery 1961). In all the trials in Indonesia and Malaysia the effects of the stocks on yield were much smaller than those of the scion clones.

It is not surprising that these experiments failed to give entirely consistent results since in some of them there were only quite small differences in vigour between the seedling families used as stocks. There was much variation within the seedling families, presumably partly due to trees of the mother clone being fertilized with pollen from several neighbouring clones in the plantation. On account of this latter circumstance the samples of the seedling families used in the experiments could never be precisely reproduced again. This limited the chances of improved stocks being brought into use as a result of these trials.

More recent experiments have used reproducible and less variable material as rootstocks in the form of monoclonal, or selfed, seedlings obtained from monoclone blocks of buddings isolated from foreign pollen. For example, in one long-term trial 6 clones were budded on 5 monoclonal seedling stocks and on unselected seedlings (Ng *et al.* 1982). The results up to the 10th year of tapping showed that there were significant differences between stocks in their effects on girth, percentage tappability and yield of the scions (Table 6.4). The best girthing and tappability were given by the PB 5/51 stocks followed by unselected seedlings and RRIM 623. The PB 5/51 stocks also gave the highest mean yield per tree over 10 years tapping (producing 23 per cent more than unselected seedlings and 20 per cent more than Tjir 1) while RRIM 623 was second best and the unselected seedling stocks gave the poorest yield. However, when yield per hectare is considered, the unselected stocks took third place because they gave greater girth and higher percentage tappability than the other three stocks. Monoclonal seedlings of PB 5/51 and RRIM 623 were also found to be significantly superior to other stocks in another trial (Ng 1983).

It is clear that the scion has the major influence on yield and can

225 Vegetative propagation

Table 6.4 *Mean girth, percentage tappability and yields of six scion clones on different monoclonal seedling stocks and on illegitimate seedling stocks*

| Rootstock seedling family | Scion girth (cm) | | Tappability at opening (%) | Mean scion yield averaged over 10 years | | | |
| | At opening | After 11 years tapping | | per tree | | per hectare | |
				g/tapping	% control	kg	% control
PB 5/51	53.2	78.0	67.8	39.9	122.6	1878	116.1
RRIM 623	51.3	74.9	57.4	36.9	113.2	1688	104.4
RRIM 501	49.1	70.5	48.4	34.9	107.2	1563	96.7
RRIM 600	49.2	70.6	46.2	34.5	106.1	1459	90.2
Tjir 1	50.8	72.1	56.6	33.2	102.2	1611	99.6
Unselected	51.9	75.1	63.4	32.5	100.0	1617	100.0

Source: Ng *et al.* (1982)

also affect the growth of the stock, a vigorous scion increasing the growth of a weaker stock, but the stock can also have a marked influence on yield. The latter is partly due to stock effects on scion growth during immaturity, but by about 10 years after planting differences in scion girth due to stock are usually no longer significant (Paardekooper 1954; Combe & Gener 1977a). Other mechanisms, related to the genetic yield capacity of the stock, are evidently involved in its influence on yield. Paardekooper (1954) noted that the effect of stock on yield was greatest when the tapping cut was close to the union, which seems likely to be due to latex being withdrawn from the stem and root of the seedling stock. But it has also been observed that the stock can affect the thickness of the scion bark and the number of latex vessel rings therein (Ng & Yoon 1982) and by this means influence the yield of the whole of the tapping panel.

Despite the long period during which trials have been conducted, only tentative recommendations regarding stocks can be made to growers are present. In Sri Lanka open-pollinated seedlings of RRIC clones 36, 45 and 52, and of PB 86 and GT 1 have been found preferable to Tjir 1 seedlings (Samaranayake 1975). In Malaysia, monoclonal seedlings of PB 5/51 and RRIM 623 are tentatively recommended on the basis of their performance in the experiments mentioned above. In areas where moisture deficits are experienced, monoclonal seedlings of clone GT 1 may be useful as this stock has been found to improve the yield of some scions under these conditions in the Ivory Coast (Combe & Gener 1977b). The absence of stock-scion interactions in the experiments suggests that the ability of a good stock to improve scion performance is likely to extend to all, or most, clones rather than that improved stocks will be found for specific clones.

With few exceptions, selfed seedlings are less vigorous than those resulting from cross-pollination. Thus, seedlings raised from seed that results from crossing between high-yielding clones in the polyclone seed gardens described on p. 198, are likely to make better stocks than monoclonal seedlings; the former should be more vigorous and might also possess higher yield potential than the latter. This possibility has not yet been properly tested but the use of polyclonal seedling stocks has been tentatively recommended in Malaysia.

Seedling stocks of *H. spruceana* and *H. pauciflora* were found to give poorer growth and yield of scion clones than did a number of illegitimate families of *H. brasiliensis* (Paardekooper 1954; Buttery 1961; Yahampath 1968). The use of inter-specific hybrid seedlings, resulting from crossing *H. brasiliensis* with other *Hevea* spp., does not seem to have been adequately investigated, although Schmöle (1941) claimed that hybrid *H. spruceana* × *H. brasiliensis* stocks gave better growth and yields of three scion clones than did stocks of *H. brasiliensis*.

More precise studies of stock-scion relations might be made and, perhaps, improved performance achieved, by the use of genetically uniform, vegetatively propagated, clonal stocks. No adequate experiments have yet been done with such materials owing to difficulties in obtaining wind-fast plants from cuttings (see section 6.4.7).

6.4.6 Other methods of grafting

Techniques for several types of grafting have been described by Teoh (1972) and by Ooi *et al.* (1976). Terminal shoots of 0.7–1.0 cm diameter, growing on source bushes, were approach-grafted to 5–6-month-old seedlings of similar diameter raised in polybags and placed adjacent to the scion shoots (Fig. 6.7). About 80 per cent success was achieved. For cleft grafting, the scion was prepared by taking the apical 8–10 cm of semi-hardened shoot from a source bush, cutting off the terminal bud and the leaves and making the lower end into a wedge which was then inserted into a short vertical cut in the decapitated stem of a 7–8-week-old polybag seedling (Fig. 6.8). About 60–70 per cent of the grafts were successful.

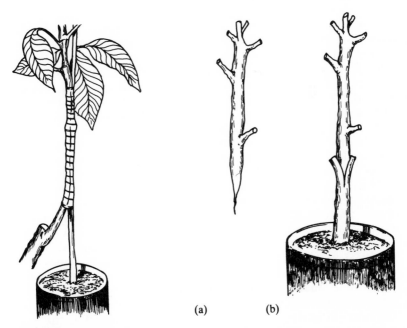

(a)   (b)

Fig. 6.7 Approach grafting   **Fig. 6.8** Cleft grafting

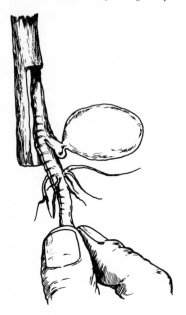

Fig. 6.9   Root grafting

For root grafting, scions were 35 cm long apical shoots, with dormant terminal buds, cut from source bushes and prepared by removing the lower leaves and peeling back a strip of bark 0.5 cm wide by 2–3 cm long at the base. The stocks were one-week-old seedlings. The epicotyl of the seedling was cut obliquely, the cut face slipped under the raised strip of bark at the base of the scion and binding done with clear tape (Fig. 6.9). Grafting success varied from 50–85 per cent with various clones.

Grafting by the above procedures could be done on stocks appreciably younger than the 5–6-month-old seedlings used for normal green budding. It was, therefore, thought that such grafting might produce plants with a lower root/shoot ratio than normal budding, thus enabling smaller bags to be used, and also that it might result in more advanced scion growth at the time of field planting and hence a shorter period of immaturity. This was investigated in an experiment comparing various types of plant (Ooi *et al*. 1978) in which root and cleft grafts and young buddings grew rather better, and had higher mean girth and percentage tappability at opening, than approach grafts and normal buddings (Rubber Research Institute of Malaysia 1984b). However, no significant reduction in immaturity seems to have been achieved as the trees were not opened for tapping until $5\frac{1}{2}$ years after field planting and it remains doubtful whether any of these other grafting procedures has a significant advantage over green budding.

6.4.7 Cuttings

The rooting of stem cuttings on a large scale could be expected to provide a simple means of propagating clones on their own roots or on genetically uniform clonal rootstocks. Cuttings from the basal juvenile part of young seedlings were rooted at Kew and in Singapore in the late 1870s and subsequently by other workers in several countries. Wiersum (1955) obtained up to 40 rooted plants from one young seedling by cutting it back several times and rooting the induced shoots. Many attempts to root mature-type stem cuttings failed but Stahel (1947), using a modification of the technique developed for cocoa propagation by Pyke (1932), obtained a low rate of success with cuttings from the tops of two-year-old seedlings. Eventually Levandowski (1959), using mist propagation, was able to report the rooting of leafy, mature-type cuttings taken from commercial clones of buddings. By modifying Levandowski's technique, Tinley and Garner (1960) developed a method by which they could root cuttings of most clones in quantity, although rooting success varied considerably with clone, being 60–80 per cent with most clones but less than 10 per cent for a few of those tested.

Fig. 6.10 Mist spray propagation of cuttings (Rubber Research Institute of Malaysia)

The essentials of their procedure, with minor subsequent modifications (Leong *et al.* 1976) are briefly summarized here. Terminal shoots with fully expanded leaves and dormant apical buds are cut from source bushes and carried to the mist propagation nursery in polybags. There they are trimmed to 30–35 cm with a slanting cut just below a node, the lower leaves removed and the larger leaflets trimmed off the upper leaves to reduce transpiration. They are then placed in a bucket of water for a few minutes after which the coagulated latex is peeled off the basal cut and the bottom 5 cm of the cutting dipped in 'Fermate' powder (ferric dimethylthiocarbamate) which not only has a fungicidal effect but has also been shown to stimulate rooting (Tinley 1961). The cuttings are set about 10 cm apart in rows 25 cm apart in sand beds which are provided with shade at about 1.2 m above ground. Except when it rains, misting is continuous from 07.30 to 18.30 hours and at nightfall sufficient nutrient solution is supplied through the nozzles to wet the leaves thoroughly, with the aim of replacing the nutrients lost by leaching. Rooting occurs in 6–9 weeks, after which the plants are transferred to polybags and hardened off under gradually reduced shade and decreasing frequency of misting (Fig. 6.10).

Experiments were started in 1960 to compare clones grown as cuttings on their own roots and as buddings on rootstocks raised from cuttings or from seedlings. But these trials came to nought because the cuttings only developed a shallow, spreading root system without a tap-root; consequently, they were not wind-fast and were mostly uprooted. To try to overcome this problem, pseudo-tap-roots were induced by removing from a cutting all the roots except the one which grew most vertically downwards from the base and planting it in a polybag with the union of root and stem about 2 cm above soil level (Fig. 6.11). About 90 per cent of the cuttings so treated formed a pseudo-tap-root similar to a seedling tap-root (Yoon & Leong 1976). Trees grown on their own roots as cuttings with pseudo-tap-roots were included in the experiment mentioned on p. 228. It is reported that they did not suffer wind damage and that they have achieved girthing and percentage tappability similar to those of young buddings (Rubber Research Institute of Malaysia 1984a). However, this technique has so far only been used experimentally.

Another possibility for obtaining clonal rootstocks with tap-roots is suggested by the indication obtained by Muzik and Cruzado (1958) that juvenile *Hevea* tissues may transmit a substance, or substances, to mature tissues which induces the later to assume juvenile characteristics. They grafted buds from the branches of 8–10-year-old trees on to the base of 12-month-old seedlings. When the scions had grown out 1.0–1.5 m, they made cuttings from part of them for planting under mist spray while at the same time buds

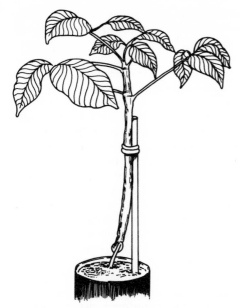

Fig. 6.11   Induction of a pseudo-tap-root on a cutting

from them were also grafted on to other seedlings. This sequence of budding and planting cuttings was repeated four times. It was found that 30 per cent of the cuttings from the fourth and fifth graftings formed tap-roots while all those from previous graftings failed to do so. A similar procedure is now under investigation in Malaysia.

Some difficulty has been experienced in budding on to young plants raised from cuttings because dieback frequently occurs after cutting back the stock. The best results have been obtained by delaying budding until the cutting has made at least two new whorls of leaves, cutting back just below the terminal bud and making an inverted U isolation cut around the bud-patch at the same time. Most of the stock leaves are then retained until the scion has made a good start (Yoon & Leong 1976). Because of the problems relating to lack of a tap-root and difficulty in budding, cuttings have not yet been used commercially.

6.4.8 Organ, tissue and cell culture *in vitro*

Several centres have been working on the application to *Hevea* of *in vitro* techniques for the culture of embryos, meristems, callus, cells and protoplasts and for the regeneration of complete plants from these cultures.

Plantlets have been obtained from the culture of immature or mature embryos and successfully established in soil (Paranjothy & Ghandimathi 1976; Rubber Research Institute of Malaysia 1983b). With some other species embryo cultures have been known for decades as a useful technique for rescuing embryos that might otherwise die and it might yet prove to have uses for rubber.

Shoot apices 2–3 mm long taken from 2–3-week-old seedlings have been grown into rooted plantlets in both liquid and solid media but shoot tips from clonal source bushes only partially expanded their leaves and did not root. Attempts to obtain plantlets from axillary buds, either as nodal cuttings or with attached strips of bark, have so far failed in Malaysia (Paranjothy & Ghandimathi 1976; Rubber Research Institute of Malaysia 1977a, 1983b). However, Enjalric and Carron (1982) have reported the *in vitro* propagation of 'micro-cuttings', consisting of 3–4 cm portions of stem, each bearing one or more axillary buds, which were obtained by cutting up shoots arising from the cut-back stumps of clonal seedlings of GT 1, PB 5/51 and RRIM 623. After soaking in a solution of benzyl adenine for one hour, the micro-cuttings were placed on a nutrient agar containing sucrose and growth regulators, where they developed shoots 2–4 cm long. The latter were then detached, dipped in an auxin solution and transferred to a fresh medium where they developed roots. The value of the resulting plants as rootstocks has yet to be established and the propagation technique, while useful experimentally, may not be commercially viable.

Formation of unorganized callus has been induced on media containing auxin and cytokinin from 1 mm thick slices of plumules of 3–4-day-old seedlings (Chua 1966) and from explants of the somatic tissue of anther walls (Satchuthananthavale & Irugalbandara 1972), of roots (Paranjothy & Ghandimathi 1976) and of leaves (Carron and Enjalric 1982). The development of embryoids and subsequent root formation has been reported for callus grown from stem tissue (Wilson & Street 1975; Toruan & Suryatmana 1977) and anther walls (Paranjothy & Ghandimathi 1976). The frequency of somatic embryogenesis from anther wall cultures has varied considerably with clone. With two clones, RRIM 600 and GT 1, a few of the embryoids formed by such cultures have grown *in vitro* into plantlets with tap-roots and lateral roots, but none have yet been successfully transferred to soil (Rubber Research Institute of Malaysia 1984c).

If a technique for sterile shoot-tip culture of clonal material could be perfected, it would be extremely useful for transporting planting material internationally without risk of transferring pathogens. A reliable technique for *in vitro* embryogenesis and plant regeneration from callus cultures might serve the same purpose and might also allow the propagation of clones having tap-roots which could be planted as such or used as rootstocks. But this would only be possible if the products were free of the damaging mutations (so-called somaclonal variation) which commonly occur in callus and cell cultures.

In vitro embryos or adventive shoots developed from pollen grains (microspores) are a potential source of haploid plants in which chromosome doubling could then be induced to give homozygous diploids. Several attempts to produce such plants in rubber have yet to yield decisive results (Tan 1987); on *a priori* grounds, since rubber suffers severely from inbreeding depression, the products would be expected to be inviable or, at best, weakly and would not be helped by the (probably inevitable) somaclonal variation. Even in intrinsically favourable genetic materials (such as barley) microspore-derived haploids must be judged to have failed. And there are formidable technical problems, such as deciding whether a regenerant from an explanted anther derives from a microspore or from maternal tissue.

Protoplasts have been released by enzymatic treatment of leaf mesophyll, of the pith of young green shoots and of cell suspensions prepared from callus derived from anther walls. In culture the protoplasts remained viable for only a few days; some regenerated cell walls, but no cell division occurred (Othman & Paranjothy 1980b). Presumably, this work was undertaken with a view to the possible use of protoplast fusion but it is difficult to see how this could serve a useful purpose in the foreseeable future especially as, in addition to the technical difficulties, it is already possible to cross all *Hevea* spp. using the normal method of artificial pollination.

It will be seen that *in vitro* organ and tissue culture is still very much in the experimental stage and the value of these techniques, if perfected for rubber, remains to be seen. To generalize, it would seem that shoot-tip culture would almost certainly be useful, somatic embryogenesis and embryo culture might be, but all else is of doubtful or, at best, very remote impact.

6.5 Branch induction

Some clonal seedlings and certain clones (e.g. RRIM 600, GT 1) tend to grow very tall before they branch and then to form crowns which are initially relatively small. Because of their smaller leaf area, such trees increase in girth more slowly and take longer to reach tappable size than those that branch earlier and lower. To stimulate lower branching it was formerly common practice to cut off, or pinch out, the terminal bud of the young plant. This was not really satisfactory as it often caused three or four branches with weak crotches to arise close together on the trunk, with consequent risk of wind damage. After investigation of several methods of branch induction (Yoon 1973; Rubber Research Institute of Malaysia 1976a) two are currently recommended in Malaysia.

The first is to make a double-ring cut around the main stem with a device comprising two V-cut steel blades held parallel and 20 cm apart by two metal separators (Fig. 6.2). The blades are pressed into the stem at 1.8–2.1 m above ground so as to reach the wood without penetrating it and then rotated around the stem. It is preferable to make the cuts within a length of stem where there are dormant axillary buds remaining from a flush of leaves rather than in an internode; the former produces suitably spaced shoots both between the two cuts and below the lower one whereas the latter results in too many shoots close together below the bottom cut. A number of shoots emerge 2–3 weeks after ringing and, where necessary, it is advisable to thin to 4–5 shoots which are clearly separated vertically and evenly spaced around the trunk.

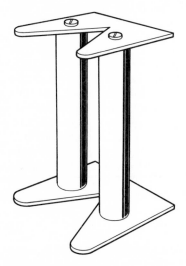

Fig. 6.12 Implement for double-ringing stems to induce branching (Rubber Research Institute of Malaysia)

The double-ringing method is only satisfactory on stems with greenish brown or brown bark. To get the greatest benefit it is desirable to induce branching as early as possible; hence a technique has been devised for use on young stems with green bark. It consists in enclosing the apical bud in a cap of leaves and resembles the method of Mulder (1941) which used a cap of dark, waxed paper. If the terminal whorl of leaves is well developed or fully hardened, the upper leaves can be folded down to enclose the apical bud and secured with a rubber band but, if this is impracticable, the cap can be made with two or three detached leaflets.

Branch induction usually causes a slight initial check to the growth of the tree but subsequently the rate of growth is faster than that of untreated trees of high-branching clones and the period of immaturity before tapping can begin is somewhat shortened.

6.6 Pruning

6.6.1 Pruning to increase leaf area and growth rate

In order to produce a suitable length of clean trunk for tapping it has been customary for all branches developing on young plants up to a height of about 2.4 m to be regularly pruned off at intervals of 2–3 months. As tapping is normally started no higher than 1.5 m above the union, it would seem that leaf area could be increased, and the growth rate of the tree consequently enhanced, by allowing branches to develop as low as, say, 1.7 m above ground. Furthermore, provided that the pruning wounds would heal so as not to interfere later on with tapping, leaf area and growth might be further augmented by allowing branches below 1.7 m to grow for a time instead of removing them by frequent pruning rounds. This was tested in Malaysia in a number of experiments with clones RRIM 600 and GT 1 (Yoon *et al.* 1976; Leong & Yoon 1982a, 1984).

It was shown that, compared with normal estate pruning to a height of 2.4 m at intervals of 2–3 months, there were advantages in leaving all branches produced above 1.7 m and only removing those produced below that height when they had reached a basal diameter about half that of the trunk. This treatment resulted in greater leaf area, larger crowns and earlier closure of the canopy, with the incidental benefit that less weeding was needed. It increased girthing and in some of the trials gave a higher percentage tappability at opening. In the most favourable case the effect was equivalent to shortening the immature period by six months. Yield per tree was not affected but in most trials there was a small increase in yield per hectare over the first 2–4 years due to the improved percentage tappability. All pruning wounds were fully healed before the trees were opened for tapping and there was no effect on the number of latex vessels or stone cells in the bark. Accompanying this pruning with branch induction, where it was needed, had a small beneficial effect. The current recommendation is to induce branching at 1.8–2.0 m where necessary and to prune all the branches developing below 1.7 m only after two whorls of branches have grown out above them, or, in the absence of the latter, when they have produced four flushes of leaves.

The benefit obtainable from this kind of pruning and branch induction will not be fully realized if the occurrence of leaf disease later results in a thin canopy. It should also be noted that the effects of the treatment in producing lower branching and larger crowns become less pronounced as the trees grow older, owing to competition between crowns and the normal shedding of some of the lower branches.

6.6.2 Corrective pruning to minimize wind damage

Although rubber trees are not very resistant to wind, serious damage from wind is restricted to certain susceptible clones and/or to areas which are liable to high winds or have soil conditions that prevent the root system from providing firm anchorage. Damage may be leaning, bending, branch break, trunk snap or uprooting. Mature trees of some high-yielding clones are prone to trunk snap because tappping reduces girthing of the trunk more than the growth of the crown, with the result that the trees become top-heavy. The different forms of wind damage have already been discussed in Chapter 4, to which the reader should turn for further details. This section is concerned with pruning to prevent, or minimize, wind damage but, as already mentioned on p. 145, in areas subject to high winds other measures, such as shelter belts and the use of wind-resistant clones, will be needed.

A strong branching system, resistant to wind damage, is ideally one in which the main stem persists into the upper part of the tree giving rise at short vertical intervals to many relatively light branches evenly spaced around the trunk and at wide angles to it (Fig. 6.13). Pruning can correct some undesirable features which commonly occur. If a main stem divides to form a V fork, one arm of the V should be cut off flush with trunk below. Where a heavy branch develops on one side of the tree unbalanced by a similar branch, or branches, on the other side, it should be removed. Sometimes a number of branches emerge very close together vertically and the leader, or terminal shoot, dies out or is retarded. It is then advisable to leave the branch most likely to develop into a leader and to prune off the others. In areas with high risk of wind damage it may be advisable to try to induce the formation of low crowns by topping back leading branches two or three times or, alternatively, to thin crowns by pruning off shoots in the middle so as to give more light to lower branches and thus to minimize their shedding.

Any corrective pruning needed should be done as soon as practicable after planting. With some clones it should be possible to identify and correct most of the undesirable branching by several rounds of pruning during the first year after planting. Where this

Fig. 6.13   Desirable branching system (Rubber Research Institute of Malaysia)

is not practicable, inspection and pruning at intervals of 4–6 months should enable the work to be completed within $2\frac{1}{2}$ years of planting. If early pruning is neglected, it will be more difficult and expensive to deal with bad branching in trees 3–6 years old. Nevertheless, if there is a risk of wind damage, it becomes advisable to carry out pruning at this late stage. If undesirable heavy branches or V forks occur in these older trees, it may be better to shorten the branches to an appropriate length rather to remove them entirely as the latter will depress girthing and yield.

.All branches removed should be cut off flush with the main stem; leaving stumps of branches is liable to result in their producing numerous shoots which may lead to the development of a heavy and badly shaped crown. To avoid tearing and splitting when sawing off branches an undercut should be made partly through the limb before making the main cut through it from above. When short-ening branches it is best to make cuts just below a whorl of buds as this results is fewer and better spaced shoots growing out than if the cut is made immediately above. Cuts should always be sloped so that the rain water runs off them and should be painted with a fungicidal wound dressing. Whenever the removal of a large portion of the canopy results in exposure of previously shaded trunks or branches to full sunlight, it is necessary to white-wash the exposed parts in order to prevent damage from sun scorch.

6.6.3 Remedial pruning

When wind damage occurs, any practicable remedial measures should be taken promptly. This is particularly necessary when the canopy has closed, because a gap in it caused by a damaged tree, or trees, allows the wind to enter the stand lower down instead of flowing smoothly over the top and this results in damage to neigh-bouring trees and widening of the gap. Where trees are leaning on boundaries, or into gaps in the stand, it is advisable to shorten the branches on the side to which they are leaning in order to lessen the weight on that side of the crown. Broken branches have to be sawn off at appropriate points and the wounds protected. Uprooted trees can sometimes be saved by pollarding them at mid-crown level or lower, pulling them upright, covering exposed roots with soil and stay-roping them to pegs in the ground.

When trunk snap begins to occur in a mature stand, the nature of the practicable remedial measures will depend on the suscepti-bility of the clone to further damage. Where a few trees have snapped in an unusual storm and there is little risk of further damage, it may be sufficent to prevent enlargement of the gaps in the canopy by pruning the surrounding trees to make them wind-fast. It is best to try to arrange that the canopy rises gradually from the gap to the normal level by pruning several circles of trees around the gap with decreasing severity. Where a susceptible clone has developed an unstable structure so that further breakages are likely, it is necessary to prune all the trees without delay. The severity of the pruning required will depend on the condition of the trees and the degree of wind hazard. It may vary from partial pruning of the crown at a height of about 7.5 m through shortening of all branches down to mid-crown level to full pollarding at as low as 3.5–4.5 m. However, experience in Malaysia was that, when dealing with susceptible clones in areas liable to high winds, pruning was not a satisfactory solution to trunk snap as it reduced yields for five years or more and damage usually recurred when the crowns of the pruned trees regenerated. Fortunately, trunk snap of mature trees in tapping is now far less common that it was in the 1950s and 1960s because the susceptible clones are no longer planted.

6.7 Crown budding, or double working

The well-known horticultural technique of double working seems to have been first tried out with rubber in Java in 1926 at the suggestion of J. S. Cramer, who envisaged the possibility of forming a tree with improved rootstock, trunk and crown components (De Vries 1926). Since then many experiments have been made with three-part trees formed by budding a seedling rootstock with a high-yielding clone which is subsequently budded at a height of 2.0–2.5 m with a second clone selected to form a crown which will have a beneficial effect on the performance of the tree. Crown clones have been selected with the aim of improving performance in one or more of the following ways:

1. Giving resistance to wind damage.
2. Conferring on a susceptible clone resistance to leaf diseases such as *Microcyclus ulei* (South American Leaf Blight), *Oidium heveae, Colletotrichum gloeosporioides, Phytophthora* spp., or to pink disease (*Corticium salmonicolor*).
3. Lessening the yield depression which occurs with some clones during refoliation after wintering.
4. Increasing the rate of girthing of some trunk clones of high-yield potential but deficient in vigour.
5. Modifying rubber properties, especially Mooney viscosity.

6.7.1 Propagation of three-part trees

Top working was originally done by brown budding, but this resulted in a marked setback to growth and some losses from dieback of, or wind damage to, the emerging crown shoots. It was found to be better to top work plants at an earlier stage by green budding, so that the amount of the trunk scion removed at cut-back was minimized and a number of functioning leaves could be left below the upper bud-patch. This reduced losses from dieback and lessened the setback to growth (Yoon 1972).

It is now usual to raise stumped three-part trees in the nursery. The procedure is similar to that described for stumped buddings on p. 214. After base budding, the scion is allowed to grow 3.0–3.5 m before it is green budded at about 2.4 m above ground level. Three or four weeks later the successful buddings are cut back and all the trunk shoots pruned off. After the bud shoot has emerged and grown 2–3 cm, two or three stock shoots about 30 cm below it are allowed to grow for a time. This reduces losses due to dieback and encourages growth by providing more leaf area. When the crown scion is well established these trunk shoots are removed. The procedure for extraction and transplanting is as described for stumped buddings.

Interrelation of crown and trunk clones

Some characteristics of three-part trees can be illustrated by reference to the first crown budding trial in Malaysia, which was started in 1948 and aimed at obtaining information of the interrelation of crown and trunk by using clones having a range of vigour, yield potential and susceptibility to leaf diseases (Rubber Research Institute of Malaya 1960; Yoon 1967, 1972; Tan 1979b). The clones were Tjir 1 (vigorous, fairly good yield), LCB 870 (vigorous, poor yield), BD 5 (poor vigour, susceptible to leaf diseases but a good yielder in suitable areas) and a clone of *H. spruceana* (vigorous but very low-yielding). The trial was of the diallel type, each clone being used as trunk and crown.

From Table 6.5 it can be seen that there was little difference between the trunk clone means for girth but a significant effect for crowns (the poor canopy of BD 5 having a large depressing effect) indicating that the crown can have a marked effect on the growth of the trunk. For yield (Table 6.6) there was a large difference between trunk means but lesser, although appreciable, differences between crown means, indicating that the trunk has the major influence on yield but that the crown may exert a considerable modifying effect, perhaps chiefly due to its effect on the girthing of the trunk. Thus, the yield of the *H. spruceana* trunk was low because it has relatively few latex vessels, and it was little improved by even the most vigorous crown. On the other hand, the production of BD 5, which has a high yield potential but lacks vigour owing to its thin canopy and susceptibility to leaf disease, was more than trebled by a vigorous Tjir 1 crown.

It should be mentioned that similar indications to those just described had been obtained in an earlier trial started in Java in 1932 (Ostendorf 1948). This included the use of the species *H. guianensis*, *H. spruceana* and *H. pauciflora*, all of which were lower-yielding and less vigorous than *H. brasiliensis*. As crowns for *brasiliensis* clones all the other species depressed girth and yield compared with *brasiliensis* crowns. The use of vigorous *brasiliensis* clones as crowns did not significantly improve the yield of the trunks of the other species, presumably because the latter possessed relatively poor latex vessel systems.

Evidence was soon obtained that the crown could influence yield by mechanisms other than its effect on trunk girthing. It was found that bark thickness and the number of latex vessel rings might be increased or decreased depending on the crown-trunk combination. For example, in one experiment the crown RRIM 612 decreased bark thickness and latex vessel numbers in trunks RRIM 613 and RRIM 605 but effected increases in Nab 17 trunks (Leong & Yoon 1976). It was also observed that the plugging index might be influ-

Table 6.5 *Girth at opening (a) and after six years' tapping (b) of four trunk clones in a diallel experiment (cm)*

Crown clone	Trunk clone								Crown clone means	
	Tjir 1		LCB 870		H. spruceana		BD 5			
	(a)	(b)	(a)	(b)	(a)	(b)	(a)	(b)	(a)	(b)
Tjir 1	52.8	77.0	56.9	89.4	57.7	98.0	57.4	88.9	56.2	88.3
LSB 870	53.8	83.3	58.7	93.0	54.1	99.8	55.6	88.6	55.6	91.2
H. spruceana	49.3	72.1	48.8	80.8	54.9	93.2	48.5	82.8	50.4	82.2
BD 5	35.8	52.8	40.1	65.0	38.6	60.7	36.8	64.3	37.8	60.7
Trunk means	47.9	71.1	51.1	82.0	51.1	87.9	49.6	81.1	50.0	80.6

Source: Tan (1979b)

Table 6.6 *Average yield of four trunk clones over five years' tapping on panel B in a diallel experiment (g/tree/tapping)*

Crown clone	Trunk clone				Crown clone means
	Tjir 1	LCB 870	H. spruceana	BD 5	
Tjir 1	37.59	13.49	3.80	55.54	27.61
LCB 870	29.61	9.23	3.69	41.82	21.09
H. spruceana	17.25	5.59	2.72	29.50	13.77
BD 5	9.16	6.48	2.54	17.82	9.00
Trunk means	23.40	8.70	3.19	36.17	17.86

Source: Yoon (1972)

enced by the crown. Thus, in one trial the plugging index of PB 28/59 trunks was increased by crowning with PR 107, RRIM 612 and Ch 30 but decreased by a GT 1 crown (Leong & Yoon 1976). Furthermore, the crown may influence the response of a trunk to application of a yield stimulant. For example, in the first Malaysian experiment described above the mean yields for two years averaged over all four trunks showed that ethepon stimulation of trees with BD 5 and Tijr 1 crowns gave 126 and 115 per cent increases respectively over unstimulated trees while with LCB 870 and *H. spruceana* crowns the responses were only 54 and 28 per cent respectively (Rubber Research Institute of Malaysia 1976b).

Observations in a number of experiments have shown that crown clones can influence the dry rubber content and the concentrations of nitrogen, phosphate, potassium and magnesium in field latex, the mechanical stability of concentrated latex and the Mooney viscosity of the rubber (Leong & Yoon 1976; Tan & Leong 1976). The Mooney viscosities of rubber from three-part trees are not typical of those of either of the component clones but intermediate between the two.

Using a vigorous clone to top work a non-vigorous trunk clone of high yield potential can substantially increase the yield of the latter (as indicated in the case of BD 5 in Table 6.6) but this is of limited value in practice because most of the popular high-yielding clones have adequate vigour. In almost all the numerous experiments with commercial clones it has been found that, in the absence of damage from wind or disease, top working has given yields which have either been less than, or not significantly different from, those of the uncrowned trunk clone. One of the very few exceptions is that in three trials the yield of the clone PB 28/59 has been substantially increased by top working with RRIM 612 or GT 1. Clone PB 28/59 is susceptible to *Oidium* and *Colletotrichum* but there is evidence that the yield increases from top working cannot be wholly accounted for by the crowns conferring resistance to leaf disease (Rubber Research Institute of Malaysia 1981). If this is so, it suggests that certain especially favourable, specific trunk-crown combinations may exist which could give higher yields than either of the component clones as a result of some mechanism unrelated to wind or disease resistance. As suggested by Tan (1979b) this possibility could be investigated by experiments of the diallel type, using a number of clones differing markedly in such characters as yield, growth vigour, plugging index and dry rubber content of the latex.

6.7.2 Practical application of crown budding

The major practical value of crown budding has been in enabling some high-yielding clones which are susceptible to damage from wind or disease to be grown in areas where these maladies are liable to occur. An example of how crown budding can reduce wind damage in a susceptible clone and increase yield per hectare is given in Table 6.7. Top working for disease resistance appears to have been first attempted by Mass (1934) who used the *Oidium*-resistant clone LCB 870 but this was not successful either in Java (Radjino 1969) or in Sri Lanka (Chandrasekera 1980). The crown budding of high-yielding Asian clones with allegedly *Microcyclus*-resistant clones started in South America in 1936 (Sorenson 1942), but has not succeeded fully because material with adequate resistance to the various races of the fungus has not become available. In Malaysia, clones resistant to *Oidium* and *Colletotrichum*, such as GT 1, AVROS 2037 and PB 260, which also improve resistance to wind damage, have been used with some success to protect against these maladies. Crowns of GT 1 have also been used to protect susceptible clones from pink disease (*Corticium salmonicolor*) but have only achieved a moderate reduction in the disease where its incidence has been high.

Table 6.7 Wind damage (%) and yield (kg/ha) in a crown budding trial

Crown-budded trees *and controls*	*Year of tapping*				
	1	*2*	*3*	*4*	*5*
	Wind damage (%)				
Control RRIM 613	10.7	40.0	46.3	48.6	54.3
RRIM 613 with RRIM 612 crown	1.0	1.9	4.8	4.8	5.7
RRIM 613 with PR 107 crown	—	1.3	1.9	3.2	5.7
	Yield (kg/ha)				
Control RRIM 613	1100	900	1000	870	830
RRIM 613 with RRIM 612 crown	870	1380	2100	2860	3100
RRIM 613 with PR 107 crown	910	1300	1880	2470	2610

Source: Yoon 1972

It has also been suggested in Malaysia that crown budding may be useful in localities where steep slopes, shallow soils or high water table limit the range of clones that can be used and where certain crown-trunk combinations perform better than the few clones normally recommended. However, three-part trees are not now widely planted, partly because their benefits have not always justified their extra cost but largely on account of the increasing availability of high-yielding clones with improved wind and disease resistance.

Chapter 7

Field maintenance

G. A. Watson

Rubber is grown under a wide range of management conditions. Where wild trees are exploited, in the Amazonas region of South America, the only field input is to maintain access paths to the trees scattered through the forest. Another minimal input system is found in some traditional Indonesian smallholdings, where unselected rubber seedlings are planted into forest clearings as conditions deteriorate after 2–3 years of food cropping. This serves to stake ownership in the land, with the hope that the trees will withstand up to 7–9 years of competition from weeds and forest regrowth, but eventually grow to reach tappable size. Such plantings may then be tapped for perhaps 20–30 years until bark reserves are exhausted. Management inputs are minimal, and with self-seeding and weed invasion the planting will eventually become a virtual forest, unless favourable price conditions or government assistance encourages replanting.

A more general practice is for smallholders to plant rubber immediately into land cleared from jungle, but to intercrop the trees with rice, maize, cassava, bananas and other crops for the first 2–3 years, or until further cropping is inhibited by the developing tree shade. With some income from the food crops management tends to be better, but inputs are still kept at a minimum. Weeds are controlled by hand slashing, and since there is little if any fertilizer use, tree performance is limited.

Intercropping may still play a part in some government-financed smallholder planting and replanting programmes, but in large government development schemes and private sector estates, *Hevea* is generally planted as a monoculture, using high management inputs to produce rubber as economically as possible, by shortening the period of immaturity and maximizing yields during tapping. To ensure profitability, flexibility is also retained in the cropping cycle, so that existing *Hevea* clones may be replaced by others as higher-yielding material becomes available, or even replaced entirely by other, more profitable crops as market conditions change, for example cocoa and oil-palm (Watson 1962; Pushparajah & Chan 1973).

These various systems may conveniently be considered in terms of the immature period from planting to opening for tapping, a period of 5–7 years, and the mature period or tapping phase that may extend for up to 25 years on estates and perhaps 30 years on the less intensive smallholdings, before replanting.

7.1 Management during immaturity

When land is cleared from forest or old rubber, generally by felling and burning, the ground is bare, subject to drought and erosion, and the planter's first task is to establish conditions that will protect the soil, favour crop growth and ease maintenance. Where necessary, drainage and soil conservation systems must be installed, the planting lines must be prepared and ground covers established. Indonesian experience in these respects has been discussed by Dijkman (1951) and Malaysian by Edgar (1958), and is reviewed in Chapter 5, Preparation of land for planting and replanting.

7.1.1 Use of ground cover plants

The use of ground covers in the rubber plantation has been a subject of much controversy and investigation, with early work reviewed by Dijkman (1951), Watson (1957a) and Wycherley (1963b).

When first introduced into the Far East, *Hevea* was felt to grow best under competition-free, clean-weeded conditions with the soil kept bare of all vegetation and soil erosion kept in check by silt-pitting. In the depression years of the 1920s, this method became too expensive to maintain and two main practices subsequently developed, one making use of spontaneous plant growth as a ground cover, the other involving the establishment and maintenance of a cover of leguminous creepers or shrubs.

Natural covers

The first took its most extreme form in the 'forestry' system of planting advocated during the early 1930s (Rubber Research Institute of Malaya 1932). This system attempted to replicate the natural conditions under which rubber is found, young self-sown rubber seedlings being allowed to establish and grow to tapping size within the existing rubber stand, and other plants being selectively eliminated or slashed to provide a general ground cover. However, the great length of time required to bring trees to tappable size, together with other difficulties, prevented the system being adopted on any scale. The value of maintaining a shade-tolerant ground cover of mixed vegetation for erosion control was, however, demonstrated, and such covers are now a common and tolerated feature in many plantations during the mature phase.

Creeping legumes as cover plants

The second main practice, now commonly used over the great bulk of new rubber plantings, is the use of creeping legumes as cover plants. Immediately after clearing from forest or old rubber, a mixture of leguminous creepers, typically of *Pueraria phaseoloides, Centrosema pubescens* and *Calopogonium mucunoides*, is sown between the rows of young rubber. These covers are kept free of weeds and fertilized with rock phosphate in order that they may rapidly produce a complete ground cover and build up a vigorous mat of vegetation. In the process they effectively eliminate soil erosion, but compete with the rubber, accumulating large amounts of nutrients. After reaching a peak of vigour at 2–3 years after planting, these covers eventually die out under the shade of the developing trees. The consequent return of nutrient and organic matter to the soil produces major beneficial effects that have been the subject of detailed investigation, particularly in Malaysia, with trials covering a range of cover plant species and conditions. The role of covers in the nutrient cycle of the rubber plantation is discussed in detail in Chapter 8, Nutrition.

Fig. 7.1 A young clone trial with *Pueraria phaseoloides* ground cover, Porto Velho, Brazil

Effect of legume covers on tree performance

One of the earliest trials studying the effect of legume covers on tree growth was laid down in 1951 on Dunlop Estates, Malaysia (Mainstone 1963c). Covers of *Pueraria phaseoloides* mixed with *Centrosema pubescens* were compared with non-leguminous naturally occurring covers (naturals), in the presence of high and low N fertilizer regimes during immaturity. Tree growth was significantly better with the legume covers, even with the high N regime, enabling the first trees to be opened for tapping at 67 months after budding. Others were opened for tapping at 4-monthly intervals over the succeeding 2 years, until the last, in the naturals/low N treatments, were opened at 91 months from budding. At this stage the legume treatments showed a mean advantage in tree girth of 8.2 cm over the naturals.

Due to the earlier opening, over the first 4 years of tapping the legume treatment plots outyielded the naturals, by 74 per cent in the low N treatments, and by 31 per cent in the high N. High N had no effect on yield of the legume plots, but increased that of the naturals by 34 per cent (Table 7.1). Over the first 10 years of tapping, a mean advantage of 20 per cent in cumulative yield was recorded, but in the 10th year the mean advantage had fallen to only 5 per cent, at an overall yield level of 2253 kg/ha (an equivalent yield, obtained by calculation from plot yield, and so perhaps rather flattering). Girth differences persisted during this period, but became less marked by the tenth year of tapping (Mainstone 1969).

In a review of this trial and others, Broughton (1977) calculated by regression equation that the yield benefit due to legumes would expire after 18.6 years of exploitation, extending to 20.9 years if only the data from the low N regime were considered. As trees in the legume/low N plots had been no more expensive to bring to tappable size than those in the naturals/high N plots, the earlier and higher yields of the former brought very significant advantages in profit.

The above effects are confirmed by data from other trials in Malaysia (Warriar 1969), and in particular one series of long-term trials comparing the effect of leguminous creeping covers on soil characteristics and tree performance with those caused by other, commonly encountered covers: mixed grasses dominated by *Axonopus compressus* and *Paspalum conjugatum; Mikania micrantha*, a non-leguminous creeper; and natural covers of mixed shrubs and ferns (Watson *et al.* 1964b; Ti *et al.* 1972). In the first two years after planting in these experiments, as the covers established themselves and competed for soil moisture and nutrients, differences in tree growth were slight. On two sites of reasonably good nutrient status, one cleared from jungle on a granite-derived soil (Typic Paleudult), and the other on coastal alluvium (Sulfic Tropaquept),

Table 7.1 *Effect of legume covers and nitrogen fertilizers during immaturity on yield of dry rubber*

Period	Recording	Low nitrogen		High nitrogen		s.e. ±
		Naturals	Legumes	Naturals	Legumes	
First 4 years of tapping (67–116 months after budding)	Mean number of months of tapping	28.5	43.0	35.5	45.0	1.2
	Number of trees tapped per hectare over period of tapping	156	259	203	257	7.4
	Total yield during 4-year period (kg/ha)	2,335	4,068	3,140	4,115	46
	Number of trees tapped per hectare at end of 4-year period	301	306	311	309	4.9
10th year of tapping	Yield (kg/ha)	2,104	2,297	2,251	2,361	64
First 10 years of tapping	Cumulative yield (kg/ha)	12,450	16,009	14,413	16,166	m.s.d. 1,107 $P = 0.05$

Source: Mainstone (1963c, 1969)
Notes: Experiment sited on replanting with PB 86 on Batu Anam/Malacca series shale-derived soil.

few long-term effects developed. However, on two replanting sites, on shale/schist-derived soils of poor to moderate fertility, clear differences became apparent: trees grown together with leguminous creepers developed faster and reached tappable size up to 18 months before those grown with non-legumes. Indications are that on fertile sites, and particularly those newly cleared from forest, where organic matter and N levels are high, legume covers may show little advantage over natural covers. On poor to moderately fertile replanting sites, however, earlier and higher yields can be achieved by use of legume covers. These will more than offset their costs of establishment, and give substantially higher cumulative discounted returns (Table 7.2).

Table 7.2   *Effects of cover plants on time to tapping, early yields and discounted returns*

Experiment Cover treatment	Time to tapping[†]		Cumulative yields (kg/ha)		Cumulative discounted returns ($M/ha)	
	B	S	B	S	B	S
Legumes	61	56	9,096	10,619	2,014	2,799
Grasses	68	59	7,371	9,369	1,421	2,279
Mikania	80	63	6,365	9,615	1,003	2,288
Naturals	68	67	7,895	9,028	1,601	2,038

Source: Ti *et al.* (1972)
* Experiment B: sited on Malacca series soil (Petroplinthic Haplorthox), clone RRIM 623; experiment S: sited on Munchong series soil (Tropeptic Haplorthox), GGI seedlings
† Time to first tapping in months from planting; cumulative yields were recorded from that time onward, in experiment B over the first 7 years of tapping, and in experiment S over the first $8\frac{1}{2}$ years

In further recordings in experiment S, a replanting on a moderately fertile shale-derived soil, over a period of $14\frac{1}{2}$ years of tapping, *Hevea* grown together with leguminous creepers during immaturity gave a total of 19,668 kg/ha dry rubber. This was 2895, 1471 and 2088 kg/ha more than the cumulative yields produced under conditions of grass, *Mikania* and natural covers respectively (Pushparajah & Wahab 1978). The latter required total additional quantities of N fertilizer equivalent to 800, 195 and 900 kg N per hectare respectively to give similar yields to those obtained by use of leguminous covers. Such additional application of N fertilizer would not, however, be economic except in those conditions where it is difficult to establish legumes. These include steep, erosion-susceptible areas where clearing to establish legumes might be dangerous, and ravine areas difficult of access; under these circumstances the use of controlled, naturally regenerating vegetation with heavy dressings of nitrogenous fertilizer to the rubber may be the most appropriate technique to follow.

(a)

(b)

Fig. 7.2 A comparison of young clonal rubber grown in association with *Mikania micrantha* (a) and with mixed leguminous creepers (b) – experiment B in Table 7.2. Tree vigour is related to degree of canopy shade

Nitrogen contribution of legume covers

The data quoted above, and in Chapter 8 on N fertilizer/cover type interactions, show that legume covers mobilize considerable amounts of N for subsequent release to the rubber. Some of this must be due to N-fixation by root-nodule bacteria, but to what extent is uncertain.

In lysimeter trials, *Centrosema pubescens* has been shown to accumulate an advantage equivalent to 241 kg/ha of N over a period of 5 months' growth, compared with the non-legume *Mikania micrantha* (Watson 1957b). In pot culture, N fixed by *Pueraria phaseoloides* over a period of 3 months' growth has been estimated as equivalent to 65 kg/ha (Joseph 1970). Measurement of total nutrient content of the aerial parts and litter of legume covers in one field trial has demonstrated a net fall in N content between the second and fourth years of planting, of 112–178 kg/ha, with 110–152 kg/ha still remaining in the green material and litter of the cover for later release (Watson *et al.* 1963). In another trial (Watson *et al.* 1964a), the N content of legume covers fell by 240 kg/ha N between the third and fifth year of planting with 99 kg still held in the covers at the fifth year. Over the periods concerned, legume covers displayed higher N contents than did non-legumes, and were associated with a marked proliferation of rubber roots in the surface soil layers, increased rubber leaf N contents, increased tree girth and weight of rubber leaf litter returned to the ground (Table 7.3).

The build-up of organic matter under legume covers, their high N content and the relative ease with which this and other nutrients can become available to the rubber, are the principal reasons for their beneficial effects on tree performance. The dead litter under legume covers has a C/N ratio of approximately 15, while that of *Mikania micrantha* is between 20 and 25, and that of grass approaches 40 (Watson 1961). In the prevailing moist surface conditions, decomposition of the legume litter is rapid and nutrient uptake is enhanced with consequent benefits to the tree.

Effect of covers on soil characteristics

Research workers have generally taken the role of ground covers in minimizing soil erosion as self-evident, but limited data on legume covers during immaturity (Haines 1932), and on shade-tolerant natural covers during the mature phase (Pushparajah *et al.* 1977) confirm that they can play a valuable role in reducing erosion losses.

Effects of covers on soil characteristics follow a classical pattern. Experiments have shown that after clearing in preparation for planting, the establishment of covers will keep soil temperatures down and so prevent the very high rates of mineralisation observed under bare soil conditions (Table 8.2). This, together with the

Table 7.3 *Changes in total nitrogen content of aerial parts of cover plants with time, and effect on Hevea nutrient status and root growth*

	2	2–3 Net change	3	3–4 Net change	4	4–5 Net change	5	s.e. ±
Total N content (kg/ha)								
Legumes	253.5	+85.6	339.1	−152.5	186.6	−87.7	98.9	
Grass	68.1	+ 3.4	71.5	− 6.9	64.6	−15.5	49.1	
Mikania	114.2	−31.4	82.8	− 36.1	46.7	+ 3.1	49.8	
Total N return over years 3–5 (kg/ha)	Legumes 240.2		Grass 22.4		Mikania 33.0			s.e. ±
N content of covers (%)								
Green material	2.60		1.22		1.17			0.047
Litter	2.83		1.18		1.44			0.081
N content of rubber leaf (%)	3.46		3.27		3.12			0.026
Rubber leaf litter fall (g/tree/an.)	6710		3181		2502			293
N content of leaf litter (g/tree)	99.3		46.7		32.9			
Rubber roots, surface and 0–8 cm (kg/ha)	7633		3605		3977			

Source: Watson *et al.* (1964a)

Notes: Data on N content of covers and of rubber leaf are means of 4 annual samplings; leaf litter was analysed over 4th year after budding, rubber roots weighed during 3rd–4th year after budding.

retention of water and nitrogen by the covers, reduces leaching of nitrate N and cations. Higher levels of soil N have been recorded under legume covers than under non-legumes and have been associated with higher levels of nitrate N and lower levels of exchangeable K (Watson *et al.* 1964a). These effects have been reflected in high N and Mg contents in the rubber leaf, and lower K contents, but whether the effect is due to leaching of the K, to some interaction at the root/soil exchange zone, or to a direct dilution effect within the tree is not known (Watson *et al.* 1964a, b).

There is little detailed information on the effect of covers on soil physical characteristics. However, at 16 years after planting, Soong and Yap (1976) found significant residual effects due to cover regimes during immaturity in the experiment S already mentioned. Legume and natural covers left the soil in much better condition than did grass or *Mikania* covers, with lower bulk densities and higher pore space resulting in better water infiltration (Table 7.4). They concluded that within limitations imposed by the inherent physical characteristics of the soil, the effect of cover plants in improving soil structure depends particularly on the quantity of decomposable organic matter which the covers add to the soil.

Effect of legume covers on root disease incidence

One additional benefit due to legume covers recorded in several experiments, has been a reduction in tree losses due to root disease (*Rigidoporus lignosus*). This is probably largely due to the enhanced rate of decay of woody residues in the soil caused by the moist conditions and the high N status of the cover and its litter (Watson *et al.* 1964b; Wycherley & Chandapillai 1969). In Indonesia, Soepadmo (1981) has shown that a two-year fallow period under legume covers between clearing of old rubber and replanting can significantly reduce root disease incidence in the new planting, but such a policy could not be recommended as an economic proposition. The use of legume covers within the replanting, combined with other measures in an integrated approach to the problem, should be sufficient answer to this particular problem (Fox 1977). This reduction in root disease incidence is important, for when combined with other effects directly benefiting tree growth, it permits a higher proportion of the stand to be exploited when opening for tapping, with consequently higher and more profitable early yields.

Alternative leguminous and non-leguminous covers

While the use of creeping legumes as ground covers has been commonly accepted among rubber growers, they do have certain shortcomings. They are susceptible to invasion by grass weeds, and if not controlled will cover the planting strips and climb over young rubber. Good management is required for satisfactory establishment

Table 7.4 *Effect of ground covers on soil physical properties (0–15 cm)*

Experiment, soil type, cover*		Organic C (%)	Bulk density (g/ml)	Permeability (cm/h)	Aggregation (%)	Mean weight diameter (mm)
SE S	Legume	1.34	1.04	90.0	93.9	3.77
Munchong series	Grass	1.35	1.11	65.0	91.1	2.67
	Mikania	1.21	1.21	52.5	88.3	2.99
	Naturals	1.53	1.00	105.0	90.0	3.22
	s.e. ±	n.a.	0.028	n.a.	1.30	0.085
SE 81/1	Grass	1.42	1.07	146	88.5	2.07
Serdang/Munchong	*Nephrolepis* fern	1.23	1.09	114	84.5	1.76
association	Clean cultivation	1.19	1.21	43	90.3	1.80
	s.e. ±	n.a.	0.030	n.a.	0.64	0.087

Sources: Soong and Yap (1976); Pushparajah *et al.* (1977c)
* In SE S, cover treatments were applied during the immature phase, with soil sampling at 16 years after planting; in SE 81/1, cover treatments were applied 6 years after planting, in the mature phase, and soil sampling 7 years later

and maintenance, and after dying out under the developing shade of the rubber trees they often leave bare soil conditions with attendant risk of erosion. A number of alternative cover plants and associations of cover plants have been studied, in order to identify those that may be quicker to establish and easier to maintain, more resistant to weed and pest invasion, and more shade-tolerant, at the same time with a relatively low C/N ratio, so that litter returned to the soil may be easily decomposable and its nutrient content become available for uptake by rubber roots.

Alternative legume covers

Results from a series of six Malaysian trials confirmed that legumes promoted better growth than did non-legumes, and also reduced root disease incidence in the rubber, but that light grasses or heavily controlled natural shrubby covers were equally satisfactory in some cases (Wycherley & Chandapillai 1969). The leguminous creepers *Pueraria phaseoloides*, *Centrosema pubescens* and *Calopogonium caeruleum* showed up well. The shrub legume *Desmodium ovalifolium* was regarded as a useful component of a legume mix, although this species may not persist strongly due to its susceptibility to *Meloidogyne* nematodes (Rao 1964a). *Moghania macrophylla*, a vigorous bushy legume, had a variable effect on tree growth but was also favoured, as it showed greater shade tolerance than the creepers. *Stylosanthes gracilis* sometimes depressed tree growth, possibly because of its competitive vigour allied with the fibrous nature of its litter (C/N ratio of 19.9). The unnodulated *Cassia cobanensis* proved positively detrimental to the rubber, and, together with other members of the sub-genus *Senna* and those genera in the Caesalpinioideae which lack nodules, is now regarded as a harmful weed. *Mimosa invisa* var. *inermis* has provoked interest elsewhere. It can develop an attractive mass of leaf and stem but during any dry season will desiccate badly and present a fire hazard. Additionally, invasion by the thorned *Mimosa* can take place undetected to present a difficult weed problem.

Among other legumes, *Mucuna cochinchinensis* has since been identified as a deep-rooting creeper that will establish faster and more vigorously than the conventional legumes. It will survive for only 8–10 months, however, and so may only be worth planting in erosion-susceptible or weedy areas where quick establishment of a cover is essential, and then only in mixture with other creepers. Its large seeds are commonly eaten by rural people, and the foliage can be used as an animal feed (Rubber Research Institute of Malaysia 1977b; Teoh *et al.* 1979). *Psophocarpus palustris* has been recommended as an ideal ground cover for rubber and oil-palm in Indonesia (Tan *et al.* 1961). *Psophocarpus tetragonolobus*, the winged bean, is another useful creeper, producing edible seeds and tuberous roots as additional benefits (Purseglove 1968).

Calopogonium caeruleum, a legume creeper indigenous to Central America, has proved very suitable for cultivation in association with rubber. Observations in Malaysia have shown that this creeper is more vigorous and shade-tolerant than the conventional legumes, and is also more tolerant to drought and less susceptible to insect damage (K. H. Tan *et al.* 1976). *Calopogonium caeruleum* is said to seed satisfactorily in north Thailand, but in Malaysia it is not a prolific seeder, and may need to be propagated by cuttings (Teoh *et al.* 1979).

Alternative non-leguminous covers

Among the non-legumes, grasses have been closely investigated as the most commonly encountered weed in replantings. In the trials quoted in Tables 7.2 and 7.3 the grass covers consisted principally of *Axonopus compressus* and *Paspalum conjugatum*, and generally depressed tree growth. The ubiquitous *Imperata cylindrica* (lalang, cogun, illuk illuk) is even more depressive and is regarded as a noxious weed in rubber (Rubber Research Institute of Malaya 1963), but other grasses may be more acceptable. For instance, in the work reported by Wycherley and Chandapillai (1969), *Brachiaria mutica* depressed tree growth but the less vigorous *Ottochloa nodosa* was relatively beneficial. However, apart from their occasional use to counter erosion on steep hillsides, grasses are never planted as covers; they are tolerated only as weeds following poor establishment of legume covers, or after cessation of intercropping with food crops.

After clearing from jungle, the vegetation that would normally regenerate in a young rubber planting can include a great variety of non-leguminous shrubs, ferns and creepers. Of the shrubs, *Chromolaena odorata* (*Eupatorium odoratum* – Siam Weed), *Melastoma malabathricum* (Straits rhododendron), *Cordia cylindristachya*, *Ficus*, *Macaranga* and *Mallotus* spp., have been shown to be particularly aggressive competitors (Wycherley and Chandapillai 1969). These shrubs need to be controlled by regular slashing, and with good management this can be done so as to permit the development of a shade-tolerant ground cover association that can effectively prevent soil erosion and persist through to replanting. In Malaysia, the fern *Nephrolepis biserrata* and wild ginger, *Hornstedtia* spp., are prominent members of mixed cover associations, and have also been found to establish well and provide good soil protection when introduced into mature rubber (Dale-Rudwick & Hastie 1970).

Of the non-leguminous creepers, *Passiflora foetida* is aggressive and will depress tree growth. *Tetracera scandens* and *Merremia* spp. have also presented weed problems at times, but *Mikania micrantha* is the most noxious. This is a quick-growing plant, favoured by moist conditions, that has been found to depress rubber tree growth to a disproportionately great extent in relation to its competition for water and nutrients (Watson *et al.* 1964a, b). In field trials this creeper has depressed tree growth, NO_3 N levels in the soil, and rubber leaf N content, and the effects have been linked to a depression of N mineralization effects in pot-incubation trials, suppression of rubber seedling rooting in pot culture, and growth inhibition of root disease fungus *Rigidoporus lignosus* in bio-assay tests (Wong 1964). These toxic effects are thought to be due to essential oil and phenolic constituents in the plant and require that it be kept under close control. Where the creeper dominates, additional quantities of N fertilizer will be required to promote tree growth.

Classification of natural covers

The Rubber Research Institute of Malaysia has attempted a classi-fication of the naturally occurring covers: those that may be encour-aged (Class A); those of some value but requiring care in control (Class B); and those (Class C) that are generally undesirable (Rubber Research Institute of Malaysia 1972b). As a general guide, it can be taken that covers showing soft, lush growth, preferably legumes, are likely to be beneficial during the immature phase; those of aggressive growth with vigorous rooting will be useful in checking erosion but will need to be kept under control by regular slashing; and those of modest growth, with some shade tolerance, will be of particular value as a persistent cover under mature rubber. Cover plant practice in Indonesia during pre-war years reflected similar thinking, but occasionally went to extremes in planting leguminous trees such as *Albizzia falcata* and *Leucaena glauca* (Dijkman 1951). These would not now be tolerated because of their competitive power, but find a place in agro-forestry systems in association with food crops (Spears 1980).

Establishment of legume covers

Where legume covers are to be established, in either a jungle clearing or a replanting, good husbandry is required for the oper-ation to be successful. Conditions of seeding are often harsh, since the soil surface is subject to drought and erosion, and aggressive weed invasion a constant threat. Under these conditions, good seed, vigorous early growth and prevention of weed growth, are all required in order to establish and maintain the cover at an accept-able cost.

Seed quality and treatment prior to sowing

Available legume cover seed is often of low quality and its condition should be checked by petri dish germination tests or by biochemical tests using 2,3,5 triphenyl-tetrazolium chloride as an indicator (Chee *et al.* 1982a).

A proportion of all legume seeds are 'hard seeds', with an impermeable seed coat that prevents absorption of water and so reduces early germination. The presence of this hard seed coat is of some benefit in nature as it can help to spread germination and so avoid complete seeding failure during drought, but it is a disadvantage to the planter. To break seed dormancy, some suppliers may scarify the seed by tumbling it in a drum lined with sandpaper, but in the absence of such treatment other methods must be used. The most satisfactory is to soak the seed in a glass or ceramic basin for ten minutes in concentrated sulphuric acid and then wash in running water for one hour. For best results the treated seeds should be sown immediately. Where acid treatment is not possible, seed may be soaked for two hours in hot water, initially at a temperature of 65 °C, prior to sowing. In tests with *Pueraria phaseoloides, Calopogium mucunoides, C. caeruleum* and *Centrosema pubescens*, which gave an average of 30 per cent germination when untreated, mean increases in percentage seed germination of 124, 149 and 44 were obtained, respectively, with scarification, sulphuric acid and hot water treatment (Chee & Tan 1982).

Rhizobium inoculation

The N contribution of legume covers to the nutrient cycle of the rubber plantation is dependent on their symbiotic association with rhizobia, root-nodule bacteria, of the 'cowpea miscellany group'.

These rhizobia absorb N from the soil air, which is then converted into ammonia and, via the nodule cytoplasm, into amino acids which may be taken up directly by the *Hevea* roots, or become available to the plant on death and decay of the nodules. Such rhizobia are present in most tropical soils, but differ in their ability to infect legumes and fix N. Some may form many nodules and fix much N. Others will be ineffective, forming few or many nodules and fixing little or no N. Their effectiveness may also be affected by low soil pH and nutritional levels. The beneficial effects of lime, phosphate and molybdate application on cover plant performance, reported in Tables 8.22 and 8.23 of Chapter 8, may be related in part to this fact.

Legumes, themselves, differ in their rhizobia requirements. *Calopogonium, Pueraria* and *Mucuna* spp. are non-specific, nodulating freely and establishing effective symbioses (Graham & Hubbell 1975) even with native rhizobia; as a result inoculation will often fail to improve yields. *Desmodium, Centrosema* and *Stylosanthes* spp. nodulate with many rhizobia, but often ineffectively. *Glycine* (soya) and *Leucaena* spp. are highly specific and require inoculation with proven rhizobia. Even if a legume is inoculated with a selected, effective strain of rhizobia, there can be a problem of competition for inoculation sites with the native rhizobia in the soil.

This is a difficult area for field experimentation and there are no data that show clear benefit from inoculation at sowing of legumes used as cover plants in rubber plantations. Where a legume cover plant is indigenous, or has been established in the locality for many years, a satisfactory strain of rhizobia will nearly always be present in the soil. As a precautionary measure, however, inoculation with effective strains is recommended generally, and extension centres stock appropriate cultures. In Malaysia, rhizobia inoculates are prepared on a medium of soil and coir dust, treated with calcium carbonate. This is mixed as a wet slurry with the seed in the presence of a sticking agent, and the moist seeds are then rolled with rock phosphate until evenly coated and readily handled (Faizah 1983a, b).

Seeding rates and methods

Methods of cover establishment may vary significantly depending on local experience and conditions. One typical system is to use a mixture of *Pueraria phaseoloides, Calopogonium caeruleum* and *Centrosema pubescens* at 3.5, 0.5–2.0 and 2.0 kg/ha respectively, sown in one or more drills along the centre of inter-row areas or, on steep terrain, along terrace edges. *Mucuna cochinchinensis* may, additionally, be sown at about 1 kg/ha, with seedling points about 1.5 m apart along the centre of the avenue, in order to obtain a ground cover rapidly without suppressing the other legumes (Teoh *et al.* 1979).

Formerly, all weeding of covers was conducted manually, but with increase in labour costs the use of herbicides is becoming more common, a practice which is often facilitated by sowing in cultivated bands. One system recommended (Chee *et al.* 1982b) is to broadcast a 1:1:1 mixture of *P. phaseoloides, C. pubescens* and *Calopogonium mucunoides* at 2.2–5.7 kg/ha of treated seed on to cultivated strips 60 cm wide, situated 1 m away from the trees on both sides of the planting strip. The seeds are mixed at planting with an equal weight of rock phosphate, and 500 kg/ha of rock phosphate is incorporated in the soil at the time of cultivation or dusted on to the covers within six months in two or three split applications. Weeds in and between the sown strips are controlled by chemical herbicides.

Weed control in legume covers

Before covers are planted all unwanted vegetation is removed, either by slashing, burning, cultivating or by using herbicides as appropriate. Immediately after cover seeds are sown in the cultivated strips prepared for them, the surface is sprayed with pre-emergent herbicide. Alachlor ('Lasso') at 4 litres/ha, metalachlor ('Dual') at 1.5 litres/ha, and oxyfluorfen ('Goal') at 1.75 litres/ha, applied in 500 litres per sprayed ha, have been found effective in suppressing the majority of weeds encountered (Ibrahim 1979; Tan & Pillai 1979; Chee *et al.* 1982b). Between the cover strips, weeds are controlled by appropriate contact and translocated herbicides, including paraquat ('Grammoxon'), diuron ('Karmex'), glyphosate ('Roundup'), 2,4-dichlorophenoxyacetic acid (2,4-D) and disodium methylarsonate (DSMA).

As the covers develop, every effort must be made to maintain dominance of the creeping legumes, in order to obtain maximum benefit from their presence. This is not easy, for some grasses and broad-leaved weeds, notably *Mikania micrantha*, find a congenial environment in the moist, fertile conditions under the covers and can develop strongly if not controlled. Hand weeding can be very effective but is expensive, and trials have demonstrated that 4-(2,4-dichlorophenoxy) butyric acid (2,4-DB), 4-(4-chloro-2-methyl-phenoxy) butyric acid (MCPB) and dinoseb amine are selectively effective against the broad-leaved weeds, particularly *M. micrantha*, and that grasses can be kept under control by two or three rounds of glyphosate per annum at a rate of 0.8–1.0 litres ha, albeit at the risk of some check to the legumes. Alternatively, application of the selective fluazifop-butyl ('Fusilade') at 1–1.5 litres/ha can offer effective grass control.

Good management and expertise in herbicide techniques is required in order to maintain legume purity using chemical herbicides, but there are possible cost advantages in the more intensive use of herbicides. One semi-commercial scale trial (Teoh *et al.* 1979) has indicated that the total upkeep cost of an integrated system, using manual weeding together with pre-emergence and post-emergence herbicides during the establishment and maintenance phases, was at least 40 per cent lower than that of estate practice involving manual weeding and spot spraying with a paraquat/diuron mix during the first five months, after which maintenance was entirely manual (Table 7.5).

As legumes will compete with young trees, the area within a 1 m radius of the tree collar should be clean-weeded during the first 18 months, by hand or by use of herbicides such as paraquat, glyphosate, diuron or DSMA, taking care not to contact the young rubber. Subsequently, the legumes may be allowed to grow over the strip. They must be checked periodically, however, by pulling back the runners or by spraying with a contact herbicide to prevent the creepers from climbing the trunks. The latter approach is preferred as it leaves a layer of mulch around the tree. A secondary hazard associated with unchecked growth of legumes round a tree is that mycelia of the root disease fungus, *Rigidoporus lignosus*, may be encouraged by the moist conditions to grow across the soil surface.

Table 7.5 *Legume cover establishment and maintenance costs over first two years after planting: comparison of two weed-control systems*

No. of drill rows of legume seed	Method of upkeep*	Herbicide spraying (No. rounds)	($M/ha)	Manual weeding (No. rounds)	($M/ha)	Total weeding costs ($M/ha)	Planting cost ($M/ha)	Total cost ($M/ha)
3	Chemical/manual	7	92.96	13	160.59	253.55	50.04	303.59
5	Chemical/manual	6	70.23	13	157.95	228.18	53.74	281.92
5 (1½ normal seed rate)†	Chemical/manual	8	84.41	12	133.66	218.07	66.10	284.17
3	Estate practice	5	31.78	19	396.05	427.83	50.04	477.87
5	Estate practice	3	18.36	18	389.43	407.79	53.74	461.53
5	Estate practice	2	14.95	18	396.18	411.13	66.10	477.23

Source: Teoh *et al.* (1979)

* Chemical/manual policy: oxyfluorfen sprayed over legume drills immediately after sowing, followed by paraquat/diuron mix and glyphosate spot spraying. Estate practice: manual weeding/spot spraying with paraquat/diuron mix during first 5 months after sowing, thereafter only manual weeding.
† Normal seeding rate: 3.4 kg *Pueraria* + 0.6 kg *Calopogonium caeruleum* per ha of rubber.

Unrestricted by the microbial competition found normally in the soil, these mycelia will quickly reach the tree, ring the stem, penetrate the bark and kill the tree.

Fertilizing of covers

In addition to the rock phosphate used at sowing and immediately afterwards, the young covers will benefit from an application of 50 kg/ha of NPKMg fertilizer broadcast along the seedling rows at two or three weeks after germination. Further dressings will depend upon local conditions, but application of 100–250 kg/ha of rock phosphate in the second year after planting can have a significant beneficial effect on the vigour of the covers and the ultimate extent of nutrient return to the soil (Watson *et al.* 1963) (Table 7.6). Under certain conditions, leguminous covers may also respond to lime and molybdate fertilizer application (see Ch. 8, Tables 8.22 and 8.23).

The pests and diseases affecting legume covers are considered in Chapter 10, but from the foregoing it can be appreciated that the establishment and maintenance of legume covers is a management-intensive operation. On the majority of rubber-growing sites, low in fertility and subject to erosion, their use will be justified by tree performance. On areas that are difficult of access, or when the soils are particularly fertile, the softer, naturally regenerating grasses (not *Imperata cylindrica*), shrubs and ferns present a cheaper and acceptable alternative. In all cases, the rubber will require fertilizer dressings appropriate to the local circumstances.

Mulching – a supplementary technique for marginal areas

In areas of marginal rainfall, or on steeply sloping terrain suscep-tible to erosion, mulching can be helpful to young rubber at planting by protecting the soil, maintaining cool and moist surface condi-tions, and by suppressing weeds immediately round the planting points (Rubber Research Institute of Malaya 1956). For the same reasons, mulching is sometimes practised in nursery plantings on drought-susceptible sites.

Bulky grasses such as *Pennisetum purpureum, Panicum maximum* and *Tripsacum laxum* may be grown specifically to provide mulching material (Baptist & Jeevaratnam 1952), but more generally roadside or inter-row vegetation will be used for the purpose. Lalang and the leguminous creepers are most generally available and are cut and laid round the young plants to provide a mulch that should persist for three months or more. Actual contact of the mulch with the plants should be avoided, for under the warm and moist conditions that pertain some bark scorch may occur.

Mulching is a laborious and costly practice, only used under particular circumstances, and so has received little detailed study. There has been some evidence that release of K from some K-rich

Table 7.6 *Effect of rock phosphate application on nutrient content of legume covers and net return of nutrients to the soil (kg/ha)*

Treatment*	Dry weight	N	P	K	Ca	Mg
Year after planting						
2						
p_0	10,328	222	9.2	129	44	25.6
p_1	12,525	305	17.9	143	77	29.2
p_2	13,040	330	29.2	140	103	31.1
4						
p_0	4,575	110	4.0	17	33	10.4
p_1	5,137	127	8.4	14	46	8.6
p_2	6,161	152	13.8	15	58	8.1
Total net return to the soil over 2–4 years						
p_0	5,753	112	5.2	112	11	15.2
p_1	7,388	178	9.5	129	31	20.6
p_2	6,879	179	15.4	125	45	23.0
		Ammonium sulphate (21%N)	Rock phosphate (38%P_2O_5)	Muriate of potash (60%K_2O)	Limestone (56%CaO)	Kieserite (26%MgO)
Fertilizer equivalent of nutrient return						
p_0	—	533	31	224	29	100
p_1	—	848	57	259	78	132
p_2	—	848	93	252	111	148

Source: Watson *et al.* (1963). Data from experiment S1 on Seremban/Munchong series schist-derived soil (Tropeptic Haplorthox), with NH_4F extractable P of 2.9 ppm)

*p_0: no fertilizer

p_1: approximately 250 kg/ha rock phosphate in first year after planting, and 110 kg/ha in the second

p_2: approximately 500 kg/ha rock phosphate in first year after planting, and 250 kg/ha in the second

mulches may induce Mg deficiency in the rubber, but this can be compensated for by light dressings of magnesium sulphate fertilizer. In studies on methods of reducing the period of immaturity (Sivanadyan *et al.* 1973), mulching with lalang during the first two years after planting at a cost of $M120–140/ha, increased the density of feeder roots in the upper soil layers, increased the rate of girthing, and gave an estimated three months' reduction in the time of immaturity (Table 7.12).

However, good organization would be required for mulching on any significant scale to be effective and economic, and could only be justified in those situations where the practice has a particular value in countering adverse physical and climatic circumstances.

7.1.2 Multicropping with *Hevea*

In general, rubber will grow best in monoculture, with inter-row areas protected by leguminous ground covers during immaturity and by shade-tolerant covers thereafter. With such a system, management (whether smallholding or estate) can be concentrated on optimizing performance of the rubber. However, during the first 2–3 years after planting, before the tree shade closes over, it is possible to cultivate a variety of crops in the inter-row areas. Estates have done this during times of emergency, and it is commonly practised by smallholders in an effort to raise some income while waiting for the rubber to come into tapping. In organized development schemes, intercropping can reduce project costs by generating income during the period of rubber establishment, so lessening the need for subsistence subsidies. However, successful intercropping will be dependent on availability of a market outlet for surplus crops, and must be set against the extra management problems involved, and loss of the long-term beneficial effects of the legume covers that would otherwise be used.

Intercropping with annual food crops

The method and degree of success of intercropping will depend on several factors. Plantings on poor soil and in areas of marginal rainfall cannot be expected to support the demands of an intercrop without detriment to the rubber. On the better soils and with adequate rainfall, however, a review of general experience shows that intercropping with food crops for the first two years after planting can be successful in the short term at least (Tan *et al.* 1969; Watson 1980, 1983). In Asia, traditional cash crops have been pineapple, bananas, ginger, cucurbits, citronella grass and patchouli, while mung bean, maize and rice have been grown for subsistence. Early Indonesian experience with soyabean and rice has been reviewed by Dijkman (1951), and more recent data from an organ-

Fig. 7.3   Intercropping with maize and beans in a Thai smallholding

ized smallholder development scheme in north Sumatra are given in Table 7.7. Yields of wet-season rice and rice interplanted with maize were considered satisfactory under the local circumstances, but in the dry season, largely because of low inputs, yields were low and variable and returns do not appear to justify the high input of family labour. In this particular scheme it was calculated that, for costs and income to break even, dry-season yields of maize would have had to rise to 942 kg/ha and of green gram to 342 kg/ha (Reed & Sumana 1976).

In Thailand, a survey of rubber smallholdings showed intercrop yields ranging from very low to good with poor yields attributed to low plant densities, use of low-yielding varieties, high pest and disease incidence and an almost complete lack of fertilizer use (Reed 1974). Work at the Rubber Research Centre, at Hat Yai, south Thailand, has shown that given appropriate inputs, satisfactory yields of maize, soyabean, mung beans and sorghum could be obtained and also that intercropping need not harm the early growth of the planted rubber (Prawit *et al.* 1975). In one demonstration, various crop rotations were practised in comparison with standard legume covers, and also with natural covers of grass and small shrubs, with and without ploughing – a common practice in Thailand. Results are given in Table 7.8, and demonstrate that at three years after planting the growing of intercrops had not led to any important reduction in growth of the rubber.

In peninsular Malaysia, a survey of 1200 smallholdings showed

Table 7.7 *Intercrop yields, costs and income per hectare of rubber on a north Sumatran smallholder development project*

Crop	Number of holdings	Yield (kg/ha)	Income (Rp/ha)	Return to family labour (Rp/ha)	Net farm income or loss (−) (Rp/ha)	Total costs excluding family labour (Rp/ha)	Cost of family labour (Rp/ha)
1974 wet season:							
Rice	87	869	65,056	49,263	−1,491	15,793	50,754
Rice interplanted with maize:	13						
rice		843	73,341	56,846	3,412	16,495	53,434
maize		205					
1975 dry season:							
Maize	45	462	22,518	16,434	−23,813	6,083	40,247
Green gram	62	261	33,425	23,816	−23,569	9,609	47,385
Maize interplanted with green gram:	14						
maize		150	26,908	19,810	−35,380	7,097	55,191
green gram		220					

Source: Reed and Sumana (1976)

Table 7.8 *Effect of intercropping and cover plants on growth of young rubber, Thailand: mean tree girth in 1974 three years after planting*

	Treatments and mean girth (cm)			
	Four intercrop systems	Natural covers	Natural covers + ploughing	Legume covers
RRIM 600	24.1	21.4	22.3	23.2
GTI	24.7	24.5	23.8	25.4

Detail of intercropping systems

System	1972	1973	1974
1	Mung bean	Mung bean	Mung bean
	Bambara*	Bambara	Groundnut
		Yam-bean†	Rice
2	Groundnut	Rice	Groundnut
			Rice
3	Groundnut	Rice	Sunflower
	Watermelon	Rice	Dwarf castor
			Rice
4	Sweet potato	Rice	Sweet corn
	Sweet potato		Rice

Plus one row of papaya in the centre of each inter-row

Source: Prawit *et al.* (1975) quoted by Watson (1980)
* Bambara: *Voandzeia subterranea* (L.) Thou.
† Yam bean: *Pachyrrhizus tuberosus* (Lam.) Spreng.

that highest family returns were earned by intercropping with watermelon and tobacco. The least profitable crops were bananas and pineapples, yet because the former two crops used far more labour, the differences noted in the returns per man-day were less marked (Table 7.9).

Intercropping with cassava

Early trials in Indonesia (Dijkman 1951) showed that cassava (*Manihot esculenta*) grown as an intercrop could compete very strongly with rubber, and also that it might help to promote the spread of root disease (*Rigidoporus lignosus*). In the Ivory Coast it is thought to promote also *Helminthosporium* and *Colletotrichum* leaf diseases, probably due to the higher humidity found in the mixed planting (Melis 1978). In Malaysia the crop has not been favoured for fear of nutrient competition (Rubber Research Institute of Malaya 1972c), but recent work suggests that it may not be as harmful as had been thought. Provided that a short-term crop such as groundnuts or maize is grown as the intercrop until the rubber has grown to the third or fourth whorl stage, and provided that the cassava is then planted at least 1.5–2 m away from the rubber, any adverse effect on the trees may be countered by use of additional fertilizers (Pushparajah & Tan 1970).

Problems of intercropping with food crops

What is observed under experimental conditions with the different food crops is likely to be quite different from actual results under smallholder conditions. In the case of cassava, any uncontrolled growth would impede access to the planting and pose a weed problem.

At the Aek Nebara smallholder rubber development project in north Sumatra, large-scale intercropping with upland rice during the first 2–3 years after planting was successfully accomplished. Assisted by mechanical cultivation, the area intercropped annually increased from 447 to 8133 ha for a total production of 25,555 t with an average yield of about 1500 kg/ha in the later years. Unfortunately, at the cessation of cropping, because of the consequent loss of interest by the smallholders and withdrawal of their labour, lalang (*Imperata cylindrica*) invaded the area to present an expensive weed control problem (Habib & Rahman 1978; Bevan 1979).

For a smallholding newly planted with clone RRIM 600, Barlow (1978) predicts an increase in internal rate of return from 15.9 per cent to 33.8 per cent over a period of 36 years if intercropping with bananas and pineapples is practised for the first two years of immaturity (Table 7.13). In the absence of any data on the long-term effects of intercropping on rubber tree performance comparable with that for legume covers used during the immature phase,

Table 7.9 *Statistics of some major intercrops, Malaysia**

Crop	Production per crop cycle (t/ha)	Length of crop cycle (months)	Monthly labour input (man-days/ha)[†]	Family returns($M) (per month/ha)	(per man-day)
Bananas	12.7	23.0	4.7	62	12.9
Pineapples	24.0	—[§]	7.2	73	10.0
Watermelons	7.6	2.9	25.2	374	14.8
Chillies	3.0	6.6	23.0	285	12.4
Tobacco	1.6	5.6	34.3	371	10.9
Groundnuts	2.0	3.3	23.0	217	9.4
Sweet potatoes	9.9	4.2	23.7	165	6.9
Cassava	20.8	11.2	15.1	89	5.9
Yams	13.2	10.0	16.5	291	17.6
Maize	3814 cobs	3.2	15.6	122	7.9

Source: Lim (1969) quoted by Barlow (1978)

* Data are from a survey of over 1200 smallholdings; returns calculated on prices at 1966/67 levels.

[†] Man-day defined as 8 hours; in practice a man can usually work 4–5 hours only with such crops.

[§] Pineapple data are per year, not crop cycle.

such predictions must be speculative. From what is known of the interaction between ground covers and the rubber tree, it seems probable that in the smallholder situation, with little fertilizer use, intercropping during the immature phase will inevitably extract a penalty in terms of the long-term yield potential of the rubber.

Intercropping with perennials

Possibilities for intercropping rubber with perennials are even more restricted than with annual food crops. At normal tree spacings there is only sufficient useful light for inter-row crops during the first three years after planting, and thereafter the canopy closes over virtually until time for replanting. Conditions may become lighter towards the end, but never to the extent observed under coconuts, where mixed cropping of old palms with cocoa, pepper and a variety of other perennials is common (Nair & Varghese 1980).

Intercropping with bananas, plantains and yams

Of the shorter-lived perennials, banana is very suitable for small farmers, provided that there is a market for fruit surplus to the farmer's requirement. It can provide a reasonable cash return for a relatively small labour input (Table 7.9), but one Malaysian study of banana intercropped in rubber (Rubber Research Institute of Malaya 1972c) shows that under estate conditions high labour cost will cause expenditure to exceed returns, so making the crop unattractive to the larger grower (Table 7.10). Choice of variety can have a large effect on profitability; in the trial referred to, variety Rasthali had larger fruit bunches, and more saleable fruit bunches per unit cropped area than did Emas. Yield of both varieties declined over the years as competition developed, and as nematodes and disease, described by Graham (1970), took their toll.

Exploratory work in the Ivory Coast (Melis 1978) has shown that yams (*Dioscorea alata* and *D. cayenensis*) and plantain bananas can give satisfactory yields under young rubber (13,120, 13,458 and 4640 kg/ha respectively), while recent work has led to a rotation of yams followed by rice with groundnuts and eventually plantains (Institut de Recherches sur le Caoutchouc en Afrique 1983). Castor (*Ricinus communis*) has been suggested as an intercrop in Malaysia (Chandapillai 1969) but Templeton (1970) warns that only high-yielding, dwarf varieties are likely to prove at all successful. Passion fruit is a high-value crop that can be grown under rubber, but is expensive to grow on account of the staking and wiring required (Watson 1980).

Intercropping with coffee, cocoa, pepper and tea

There have been several attempts to grow coffee and cocoa under young rubber (Dijkman 1951; Blencowe & Templeton 1970; Watson

Fig. 7.4   Six-year-old rubber intercropped with pepper, Manaus, Brazil

1980). Early crops can be obtained from the third year after planting, but unless the rubber is grown as a hedge planting, development of the rubber shade will inevitably suppress the growth of these crops by the fifth and sixth years. Management problems include the need to provide shade for the young bushes at establishment; in early Indonesian work *Leucaena glauca* was used for the purpose, and in Brazil cassava. Both can eventually present weed problems and excessive competition with the rubber, and are not now recommended.

In Belém, Brazil, where traditionally grown monoculture pepper (*Piper nigrum* L.) has been badly affected by *Phytophthora* root rot, experiments have indicated that, possibly because of a lower moisture status, disease incidence may be lessened when grown as an intercrop in rubber planted at 5 × 3 m. Growth and yield have been

Table 7.10 *Costs and returns: bananas intercropped in rubber, Malaysia*

| Year | Costs ($M/ha)* | | | | | |
| | Planting materials | Fertilizers | Labour | | Gross returns: $M/ha (yields: kg/ha) | |
			Weeding†	Other items	Emas variety	Rasthali variety
Immature phase	160	134	256	286	—	
First harvest year	—	442	201	326	726 (5746)	1344 (8637)
Second harvest year		257	183	405	535 (3646)	740 (5788)
Third harvest year§		129	152	432	212 (1984)	311 (2410)

Source: Rubber Research Institute of Malaya (1972c)
* Cost data are common to both varieties: Emas and Rasthali.
† Weeding costs are combined for rubber and bananas.
§ First 10 months of third year only.

satisfactory and the system appears to have potential (Viégas *et al.* 1980).

Favourable results are also reported from a 20-year Chinese study on the interplanting of tea in widely spaced rubber (Feng *et al.* 1982). The tea is said to benefit from the shade of the rubber and come into production 3–4 years earlier than in monoculture. It is suggested that mixed plantations of tea and rubber can be cultivated successfully at up to 1000 m. No economic data are available, however, and the system is likely to find only local acceptance.

Intercropping with perennials: rubber tree density and the economics of tapping

While there is evidence that perennials may be intercropped successfully in avenue-planted rubber for a time, management problems are formidable. Under estate conditions, where profit is to be maximized and labour is a major constraint, a planting density of approximately 400 rubber trees/ha is desirable; tapping and collection costs might be slightly reduced in a semi-hedge planting, but trees will tend to lean into the inter-row areas and progressively suppress the intercrop. For smallholders with family labour available, and a need to maximize family income, a planting density of up to 760 trees/ha will be more appropriate. To obtain such a density in avenue plantings would demand such close planting along the tree rows as to result in excessive inter-tree competition (Barlow 1978; Ng *et al.* 1979).

For these and a variety of other local reasons, intercropping with perennials is unlikely to develop significantly. With a second crop interplanted in rubber it is virtually impossible to provide optimum conditions for both crops, and better returns will be obtained from monoculture. Where a diversity of income is required it would seem more profitable to establish the additional crops on separate, dedicated plots so that all may receive appropriate management.

7.1.3 Livestock in rubber: poultry, sheep and cattle

After the rubber tree canopy has closed over, at three or four years after planting, intercropping with annual crops is no longer possible. Weed control must still be practised to prevent any weed invasion by such noxious species as *Imperata cylindrica*, but some workers have considered the possibility of rearing livestock as a more permanent venture.

In Malaysia, the RRIM have concluded that broiler poultry production under rubber can be economically viable. A smallholder with a family of three or four can raise about 1000 birds per batch.

With proper disease control and supplementary feeding the birds will thrive, providing income for the farmer and benefiting the rubber by controlling weeds and returning droppings to the soil (Mohamed & Chee 1976).

Studies are also being conducted with sheep, grazing on mixed grasses and creepers in immature rubber, and on the permanent vegetation after canopy closure (Mohamed 1978; Mohamed & Hamidy 1983; Arope *et al.* 1985). Exotic Dorset Horn rams were introduced to improve the local breed, and helminthicides used to control parasitic worms. Lambing rates of more than 1.5 have been achieved and in one project, 3 sheep/ha have been grazed on covers originally of mixed *Chromolaena odorata, Imperata cylindrica* and grasses to give profits of $M125/ha from a flock of 300 sheep (Lowe 1969).

In Sri Lanka, following practice in the coconut industry, suggestions have been made to introduce cattle into rubber. Preliminary trials have shown that zero-grazed pastures based on *Panicum maximum, Brachiaria brizantha* and *B. milliiformis*, declined sharply in production after the fourth year. They depressed growth of the rubber trees, but this effect could be lessened by incorporating legumes into the pasture (Waidyanatha *et al.* 1984). Further applied research involving cattle grazing is needed before any practical application can be considered.

Fodder value of ground vegetation

In a study of the potential for sheep production under rubber, a survey of vegetation was carried out in Malaysia over 182 smallholdings and estates (Mohamed 1978). Vegetation was broadly classified under grasses, broad-leaves, ferns and others. The grasses included *Axonopus compressus, Paspalum conjugatum, Ottochloa nodosa* and *Imperata cylindrica*; the broad-leaves: *Mimosa pudica, Mikania micrantha, Melastoma malabathricum, Chromolaena odorata*, and leguminous cover crops; the ferns included *Nephrolepis odoratum, Gleichenia linearis, Lygodium* spp. and others.

Imperata cylindrica and *M. malabathricum*, together with most ferns, are not suitable as animal feed, being nutritionally poor and generally unpalatable, but generally 60–70 per cent of the mixed vegetation was found to be suitable for animal feed and comparable to certain cultivated fodder grasses (Table 7.11).

The performance of sheep reared and raised on this vegetation was found to be satisfactory, and it has been concluded that, with good management, there is a future for sheep production within rubber provided that shade-tolerant plant species can be encouraged and unpalatable material selectively eradicated.

Table 7.11 *Average chemical composition of ground vegetation in rubber plantation and cultivated grass pasture*

Location	Plant	Chemical composition (%)		
		Crude protein	Crude fat	Crude fibre
Rubber smallholdings	Grasses	9.4	1.5	33.3
	Broad-leaves	13.2	1.9	32.9
	Ferns	11.4	1.8	31.9
	Mixed sample	11.4	2.1	28.0
Rubber estate	Grasses	11.4	1.9	33.1
	Broad-leaves	14.4	1.7	31.9
	Ferns	13.9	1.9	27.2
	Mixed sample	12.4	2.0	32.5
Cultivated grass (cut at 6-weekly intervals)	Hamil or Guinea grass (*Panicum maximum*)	12–16	—	31–33
	Napier grass (*Pennisetum purpureum*)	15–18	—	26–30
	Para grass (*Brachiaria mutica*)	15–18	—	24–26
		14–18	—	24–26

Source: Mohamed (1978)

7.1.4 Effects of improved planting methods and intercropping on the economics of replanting

Rubber is in many ways an ideal crop for the humid tropics, but its lengthy period of 6–7 years of unproductive immaturity is a disincentive to investment in the crop. The economics of this situation have been discussed in Chapter 6, but, briefly, they can be greatly improved by use of advanced planting materials, additional quantities and frequencies of applications of fertilizers, mulching, irrigation and corrective pruning for optimum canopy development. Trials (Sivanadyan *et al.* 1973) have shown that, except for irrigation, all these factors contributed to a shortening of the period of immaturity. Largely because of resulting lower weeding costs, overall savings of about $M170/ha and $M470/ha were obtained respectively when planting large polybagged buddings and stumped buddings as compared with the traditional seed-at-stake method, using normal commercial practice with regard to legume cover establishment and other field operations (Table 7.12). When these savings are added to the financial benefit conferred by advancing the time of opening for tapping (Barlow 1978), it is apparent that significant additional expenses in agronomic inputs can be incurred without loss of profitability.

In smallholdings, the cost of replanting may be lessened even further by intercropping during the first two years after planting (Table 7.13). The returns quoted by Barlow (1978) may be thought optimistic, but confirm the value of early income after replanting.

7.2 Management during maturity

As a planting enters its mature or tapping phase, the attention of management is concentrated on exploitation of the trees. The heavy leaf canopy effectively suppresses all but the most shade-tolerant ground vegetation, so weed control requirements are minimal and restricted largely to maintenance of tappers' paths along the tree rows. The same canopy, coupled with the annual return of leaf and branch litter to the ground, affords some protection to the soil against erosion, but on all sloping land soil conservation works must be maintained. On the better managed properties regular fertilizer applications will be made, based on the results of soil and leaf analyses.

With the passing of time the trees grow in height and more light will penetrate the broken canopy, particularly where losses due to root disease or wind damage create open patches, or where secondary leaf fall occurs due to leaf disease. The heavy incidence of *Phytophthora palmivora* in Sri Lanka, and of *Microcyclus ulei* in

Table 7.12 *Effect of planting techniques in reducing immature period in field, and costs, compared with the 'normal' immature period* using seed-at-stake*

Planting techniques	Large polybag plantings	Stumped buddings	Seed-at-stake
	Reduction in period of immaturity (months)		
Advanced planting materials	12–15	18–21	
Refined nursery management	3	3	
Improved field operations:			
fertilizing	6	6	
mulching	3	3	
extra branching	3	3	
Total estimated reduction	27–30	33–36	
Estimated immaturity period (months)	51–48	45–42	
	Costs during immaturity ($M per planted hectare)		
Holing	32	32	36
Weed control (circle and selective)	813	642	1129
Fertilizer	384	294	455
Planting operations	116	155	51
Nursery operations	157	97	—
Total costs	1502	1200	1671

Source: Sivanadyan *et al.* (1973)
* The 'normal' immature period is taken as 78 months (6 pre-budding and 72 post-budding).

Table 7.13 *Effect of intercropping on economics of smallholder replanting*

	Present value of profits – after tax ($M)	Internal rate of return (%)
RRIM 600 on well-managed smallholding:		
(i) over 6 years immaturity and 30 years' tapping	779	15.9
(ii) as (i) but with 2 years' intercropping during immaturity	1512	33.8

Source: Barlow (1978)
Notes: Estimated yield for RRIM 600 on well-managed smallholding, 1193 kg/ha.

Fig. 7.5 A ground cover of shade-tolerant ferns in mature rubber, with clean-weeded tree rows

Bahia, Brazil, are examples of the latter. With increased light, the ground vegetation will develop to create weed problems, particularly as the time for replanting approaches. Weed control can consequently become a major cost in mature as well as immature plantings.

7.3 Weed control

Weed control is a subject in its own right, for in the humid tropical environment weed growth can be vigorous throughout the year and present major problems of access and competition. Satisfactory solution of these problems is dependent on good management backed by a sound knowledge of the biology of weed plants, and the development of an integrated approach for their control (Seth 1977). The subject has been reviewed at length, in respect of oil-palm cultivation by Turner and Gillbanks (1974), and by Hartley (1977), and of rubber by Pillai (1978), Teoh *et al.* (1978), Faiz (1979a, b), Ibrahim (1979), and Liu (1979).

7.3.1 Weeding costs

Actual costs will vary widely between localities depending on management and the environment, but in Malaysia weeding costs over the immature period may range up to more than $M1100/ha, roughly twice the cost of fertilizer (Table 7.12). In mature rubber, a survey shows that annual costs of weeding can still be a significant item, greater than that required for general maintenance, pest and disease control, but rather less than that for fertilizing (Table 7.14). This survey covered only a limited number of estates and so must be viewed with some reservations, but showed greater expenditure on chemical weed control than manual, with more specific attention paid to lalang (*Imperata cylindrica*) control on the larger estates – probably a reflection of management style rather than of the actual incidence of the weed.

7.3.2 Weeds as indicators of ground conditions

The vigour and composition of the vegetation that develops after clearing and planting an area to rubber will reflect the history of the site, the method of clearing, and give a good indication of the relative fertility of the underlying soil. In a new planting on good soils, established after clearing forest by traditional manual felling and burning, there is likely to be a vigorous regeneration of indigenous species provided that rainfall is satisfactory. These can be managed by ordinary slashing to form a satisfactory permanent ground cover. On poor soils, however, and particularly where mechanical methods are used to fell and clear the forest and cultivate the soil, regeneration will be poor and legume ground covers will need to be established to prevent grass weed invasion and to stabilize soil conditions. A similar policy will generally be required in a replanting, regardless of the method of clearing, for ground vegetation will have become impoverished over the previous plantings leaving the site open to grass invasion, particularly by lalang (see Ch. 5).

Table 7.14 Cost of weed control in relation to other expenditures in mature rubber: Malaysia 1979 ($M/ha)

Item	Size group of estates*		
	Less than 200 ha (14)	200–400 ha (11)	More than 400 ha (10)
Harvesting costs, including tapping, stimulation, transport and collection of latex	500	680	744
Fertilizing: Materials	87	89	65
Labour etc.	4	4	6
Weeding: Materials	32	19	24
Labour for spraying	12	10	9
Hand weeding	16	17	10
Lalang control	2	6	10
Mechanical weeding	1	—	2
Maintenance, pest and disease control etc.	17	25	21

Source: Yee and Lim (1982)
* Numbers in parentheses are number of estates surveyed.

Weeds of acid, infertile and poorly drained soils

Where soils are particularly infertile and low in base status, with pH values in the region 3.5–4.0, conditions may be particularly poor. On such soils in Malaysia, regenerating vegetation may be limited to thin stands of Straits rhododendron (*Melastoma malabathricum*), Staghorn moss (*Lycopodium cernuum*) and bracken (*Gleichenia linearis* syn. *Dicranopteris linearis*), probably in mixture with lalang.

Poorly drained sites will be marked by the presence of *Cyperus* spp. and *Scleria* spp. sedges, *Stenochlaena palustris* fern, and *Colocasia* and *Caladium* spp. aroids.

Elimination of these weeds is not easy, for they are adapted to withstand adverse conditions, and in itself will not be sufficient to ensure satisfactory establishment of rubber. On infertile acid soils, specific efforts will be required to raise the base status of the soils, probably by inclusion of rock phosphate and magnesium limestone in fertilizer regimes. Where drainage is poor, it will need to be improved. With the introduction of these measures normal procedures can follow, though establishment of legume covers will not be easy under the conditions described.

7.3.3 Methods of weed control

Traditionally, all weed control in rubber has been by hand labour. Sodium chlorate was introduced for the control of lalang in the late 1930s (Greig 1937), but even up to the 1950s hand forking was used to eradicate lalang, and hand slashing for creepers and shrubby growth. Mechanical methods of chopping or slashing weed vegetation have been useful at times, where terrain has permitted use of equipment in the inter-row areas, and disc cultivation is still commonly used to clear land preparatory to planting on a large scale (Edgar 1958).

In nurseries, and along planting strips, hand hoeing was commonly practised and may still be used in the early months after planting. However, such techniques can cause damage to the stem and roots of the young rubber, and, by scraping soil from around the young plants, will create depressions that can lead to local waterlogging. These disadvantages, coupled with the increasing cost of labour, and the development of herbicide technology, have led to the general adoption of chemical herbicides. Accompanying research has led to an understanding of factors affecting herbicide efficacy, weed succession and the economics of weed control.

Environmental factors affecting weed control

One of the most effective factors helping weed control is shade cast by the developing rubber. In the early years after planting, weed growth can be vigorous, but with increasing shade from the third year onwards growth slows down and can be more easily controlled. Herbicides will give better and longer lasting control under such conditions and lower rates of chemicals are required than under full sunlight. Indeed, where poor management has permitted the development of heavy weed growth, generous fertilizer dressings should be used at the same time as application of weed control measures, to promote vigorous leafing by the rubber that will help to suppress any regeneration.

Weed control in wet, poorly drained areas is difficult, for under these conditions regeneration from the roots is vigorous. Contact herbicides will have only a short-lived effect, and translocated herbicides will be needed to kill the root. *Colocasia* spp. native to such conditions are particularly difficult to kill for their leaves shed any spray application, and hand forking may be needed. Under conditions of moisture stress stomatal closure and toughening of the cuticle, sometimes accompanied by dust collection, can restrict the effects of some herbicides, notably paraquat.

In most rubber-growing areas rainfall can be an additional limitation on herbicide effectiveness. Paraquat, for instance, is adsorbed on to the foliage at spraying and is unaffected by subsequent rainfall, but dalapon and glyphosate, among others, require several hours for full absorption by the leaf and their effectiveness can be impaired by intervening rainfall.

Local choice of herbicide will be dictated by availability and price, the price being not the cost per kilo of active ingredient, but overall cost, including labour, when the herbicide is used in a spray programme over a significant period of time. An intrinsically expensive, but long-lasting, translocated herbicide for instance, may prove cheaper to use in field practice than a cheap contact herbicide with no residual effects.

One further aspect that can influence the use of chemical herbicides is their toxicology. In general, herbicides on sale have been subject to approval under local pesticide regulations and any restrictions on use will be detailed on the label. These must be adhered to, and are likely to refer to use near potable water supplies, proximity to cattle or food crops, and to the volume and method of application and degree of protection required for spray operators. Care must be taken to store them safely (away from children and animals), in their distribution and field use and in the disposal of used containers.

There are three main types of chemical herbicide: contact, translocated and soil-acting. They are sometimes used in mixture, not so much to optimize the weed-killing characteristics of each component, but rather to save the cost of applying them separately.

Contact herbicides

One of the cheapest and most effective contact herbicides is sodium arsenite, for a long time the main control agent for sheet lalang when sprayed at 1–2.5 per cent concentration in 450–675 litres/ha. This material, however, is very toxic to mammals and with the introduction of safer alternatives has been banned in many countries. Sodium chlorate is a safer alternative, with a rapid desiccant action when sprayed at 5–10 kg/ha, but can give rise to a fire hazard under dry conditions. It is often used in mixture with the safer monosodium methane arsonate (MSMA) or with MSMA and diuron or 2,4-D (Wong 1966; Yogaratnam 1971).

Paraquat is another widely used contact herbicide, of particular value as it is rain-fast, is harmless in contact with brown bark, and is inactivated on contact with the soil. Applied at 0.5–0.75 kg/ha of active ingredient it is effective against a broad range of grass and broad-leaved species, but does not give long-term control of perennials. To extend its effectiveness it is often applied in mixture with diuron, MSMA, 2,4-D and other materials (Faiz & Liu 1982).

Translocated herbicides

These are most effective against perennial grasses and creepers, with dalapon (2,2-dichloropropionic acid-sodium salt), and glyphosate the principal herbicides used to control lalang. 2,4-D amine ester is commonly used to control leguminous and non-leguminous creepers, and is often used in mixtures to widen the spectrum of weed control of other herbicides. The butyl esters and isopropyl of 2,4-5-T are sometimes used to control heavy brush growth by wiping on a 1:60 mixture with diesel oil. Picloram (4-amino-3,5,6-trichloro-picolinic acid) is a further material that can be useful against the most resistant vegetation, but must be used at only low dose rates for soil residues can persist to provide a long-term hazard to rubber. The more recently developed fluazifop butyl will give good control of most perennial grasses (although not lalang) and can be used for selective control of grasses in legume covers.

Soil-acting herbicides

Attempts have been made to use soil-acting herbicides such as atrazine and prometryne to maintain clean conditions in the planting strip (Riepma 1965), but invasion across the strip by leguminous creepers, rather than germination of weed seed in the strip, has been a limitation. In consequence, their use is restricted to weed control in nurseries and during legume cover establishment. Formulations with oil or surface-active agents can be phytotoxic, but for direct weed control they are more usually used in mixture with contact herbicides.

Methods of herbicide application

The traditional method of herbicide application has been by manually operated knapsack sprayers, using volumes of application up to 1000 litres/ha in the case of lalang control, but more generally of 300–400 litres/ha. For large-scale operations tractor-mounted spray tanks feeding hand lances are effective (Anderson 1976), but mist blowers have generally proved unsatisfactory because of operator hazard and maintenance costs. Of recent years increased attention has been paid to the use of controlled droplet application (CDA) units with herbicide solution fed by gravity on to battery-operated spinning discs. These units weigh some 2 kg compared with the 11–20 kg of knapsack sprayers; volumes of application are low, in the region of 10–50 litres/ha, and droplet size is uniform so eliminating the very small droplets that normally present a drift hazard (Mathews 1982; Han & Maclean 1983; Jollands *et al.* 1983). Due to the low volume of application, spray penetration into dense vegetation is limited, but such equipment can be useful for spraying operations in difficult terrain, where water supplies for spraying are restricted, and in mature rubber where weed cover is scattered and thin. The system is particularly effective with translocated herbicides such as glyphosate. In experimental rubber strip spraying, Han and Maclean (1983) found that CDA units gave 32 per cent cost savings in comparison with knapsack sprayers, and their use is likely to increase.

Minor equipment which might be useful to smallholders in particular, are rope-wick and sponge-pad wipers for applying herbicide directly to sporadic patches of weeds, particularly lalang.

Control of major weed species

Among the more obnoxious weeds presenting problems in rubber cultivation, *Imperata cylindrica* (lalang, illuk illuk, cogun), *Chromolaena odorata* (*Eupatorium odoratum* – Siam Weed), and *Mikania micrantha* deserve special mention.

Control of lalang

Lalang, with its variants, is a major weed problem throughout the humid tropics (Rubber Research Institute of Malaya 1963; Hartley 1977) and has been investigated in detail. It can be eradicated by hand or mechanical cultivation, or even, under favourable circumstances of soil and climate, by establishing *Pueraria phaseoloides* within the stand. In many areas, however, it is easier and cheaper to use herbicides. In Thailand, application of dalapon at 22.5 kg in 1000 litres/ha of water was at one time recommended, followed by spot treatment. Overdosing immediately round the young rubber was a hazard, and an alternative recommendation was to spray with

Fig. 7.6 Young rubber in Sumatra infested with lalang (*Imperata cylindrica*), *Melastoma malabathricum* and other weeds

an initial application of 9.5 kg dalapon in 375 litres/ha water followed by 3 litres of paraquat in the same volume 2–4 weeks later (Harper 1973).

In a review of the Malaysian situation, Yeoh and Pushparajah (1976) found that dalapon acted faster against lalang than did glyphosate, but that on sandy soil glyphosate at 2.2 kg a.e./ha in 899 litres/ha water gave better control than did dalapon at 16.8 kg/ha in 1121 litres of water. The effectiveness of dalapon could be increased by addition of a wetting agent, present in some commercial formulations, and overall glyphosate was more expensive than dalapon especially on clayey soils, where wet conditions aided regeneration. On sandy soils or in shaded areas, the cost of controlling lalang by glyphosate or dalapon was reduced by 42 per cent and 16 per cent respectively, when compared with the cost on clay soils.

Any sporadic lalang left after cultivation or spraying can be eliminated by forking out or 'wiping', an operation in which the leaf blades are wiped with a cloth soaked in herbicidal aromatic oil containing a wetting agent (Edgar 1958).

Control of Mikania micrantha

The *Mikania* spp. are vigorous creeping and climbing weeds causing serious problems in Asia and the South Pacific. *Mikania scandens* is said to be strictly a North American species, while *M. cordata* occurs widely in Africa and Asia. It is, however, *M. micrantha* H.B.K. which has been identified as the most aggressive of the family (Parker 1972), and to which must be attributed the harmful characteristics described above (alternative non-leguminous covers, p. 258). Once established, *Mikania* is difficult to eradicate, and accordingly every effort is made to suppress the weed at an early stage, either by hand weeding or by spot spraying. In pure stand, *Mikania* can be controlled by application of 2,4-D-amine at 1.1 kg a.e./ha, but when associated with perennial grasses such as *Paspalum conjugatum* a mixture of 2,4-D-amine with paraquat and diuron would prove more satisfactory (Faiz & Liu 1982). In very young rubber, where the hormone herbicide may show phytotoxicity, repeated sprays of paraquat may be preferred (Seth 1969).

Control of Siam Weed

Siam Weed (*Chromolaena odorata* K and R, formerly *Eupatorium odoratum*) is a ubiquitous weed in Thailand, Malaysia and West Africa. It is an erect, bushy composite producing a thicket of stems, and can be very competitive with rubber. In field trials it has shown a depressive effect on rubber (Wycherley & Chandapillai 1969). It can be controlled by repeated hand or mechanical slashing, but treatment with 2,4-D sprays after preliminary slashing is more usual (Kasasian 1971), with triclopyr (3,5,6-trichloro-2-pyridyloxyacetic acid) a recently proved alternative (de Vernou 1981; Rubber Research Institute of Malaysia 1984d). Where 2,4-D may present a hazard to young rubber, mixtures of atrazine or diuron with sodium chlorate or paraquat would be preferred (Turner & Gillbanks 1974).

Weed succession

Nature abhors a vacuum, and as one weed problem is eradicated its place will rapidly be taken by another. Use of grass-killers on a mixed weed population will swing the floral balance to broadleaved species, while elimination of the broad-leaved component will encourage grass development. In the Malaysian work on lalang carried out in 2–4-year old rubber (Yeoh & Pushparajah 1976), *Mikania micrantha* invaded the area cleared of lalang to give a 31 per cent ground cover at 335 days after treatment, with *Melastoma malabathricum* and *Chromolaena odorata* two unwanted new species in the area. To avoid such a problem it might be better to suppress the lalang by establishment of *Pueraria phaseoloides* directly within the stand. However, if this is impracticable, as soon as the lalang is cleared by chemical means, leguminous creeping covers should be established to prevent any invasion by noxious weeds.

Problems can also arise where there is a local shift in weed populations. Teoh *et al.* (1982) report such a situation in Malaysia where the freely seeding *Asystasia intrusa* and *Clidemia hirta*, and the rhizomatous wild ginger *Ellattariopsis curtisii*, have recently developed as potentially serious weed problems. Fortunately the first two can be cheaply controlled by repeated sprays of 2,4-D-amine and 2,4,5-T esters. Control of the ginger has been more difficult, because of the regenerative power of its rhizomes, and may only be achieved by an overall spray of paraquat in mixture with translocated herbicides, followed by spot spraying. Under mature rubber, succeeding vegetation included the relatively beneficial fern *Nephrolepis biserrata*, but the case illustrates the need for close management and flexibility in spray regimes in order to prevent the development of expensive weed problems.

Standard weed control practice

Weed control in nurseries

Traditionally, weed control in seedling, bud-wood and polybag nurseries has been conducted manually in order to avoid damage to the young rubber. However, with the increase in labour costs and development of new herbicides, chemical weeding is now more usual. After preparation of the site, application of an alachlor/linuron mixture, each at 0.5 kg a.i./ha, will give effective control of mixed weeds for up to three months (Yeoh *et al.* 1980). Thereafter, three applications of a paraquat/diuron mixture, at 0.5 kg a.i./ha each, or two of glyphosate at 0.4 g/ha a.e. at two-monthly intervals, will maintain weed control. Care should be taken to avoid herbicide contact with the stem and leaf of the young rubber, and with unskilled labour it may be advisable to use shielded spray nozzles.

Weed control during immaturity and maturity

After planting, the young trees are at first hand weeded to a radius of 0.5 m, supplemented by application of a pre-emergent herbicide such as alachlor, diuron, oxyfluorfen, or simazine. These circles are kept cleared by subsequent herbicide sprays of paraquat, paraquat-diuron mixes, dalapon or glyphosate, to facilitate fertilizer appli-

Table 7.15 *A guideline weeding programme for immature and mature rubber*

Year	Manual weeding circle (rubber)	Selective (covers)	Lalang* wiping	Chemical strip control
Immature rubber				
1	monthly	monthly	monthly	—
2	—	bimonthly	bimonthly	bimonthly
3	—	—	quarterly	quarterly
4	—	—	4-monthly	4-monthly
5	—	—	4-monthly	6-monthly
6	—	—	annually	6-monthly
Mature rubber				
7	Slash inter-rows at 6-monthly intervals		annually	annually
8	As for year 7		annually	annually
9	Slash inter-rows annually		annually	annually
10 and above	As for year 9			

Source: Arope *et al.* (1983).
* Any sheet lalang should first be sprayed to wiping stage.

cation and to minimize competition. Weeds in the inter-row areas are selectively controlled as described on p. 261. With progress of time, the weeding circle is extended until, as the trees come to tapping, the entire planting strip to a width of 2 m should be free of weeds. By this time, the tree shade should effectively be suppressing weed growth, and only discriminatory spraying is needed against the most aggressive weeds. The schedule given in Table 7.15 offers a rough guide to the frequency of weed control rounds required through the life of the planting (Arope *et al.* 1983).

Chapter 8

Nutrition

G. A. Watson

Hevea will grow satisfactorily on the majority of tropical soil types when newly cleared from jungle. At clearing there is some loss of nutrients due to burning, erosion and leaching, but the residual fertility following jungle is generally sufficient to promote early growth, and consequent development of a good root system will help to support the plant through maturity. In areas being replanted, however, the loss of nutrients caused by extraction of old rubberwood, added to those sustained during the previous planting period and in the clearing operations, leaves most sites greatly impoverished. Severe Mg and K deficiencies have commonly occurred at this stage, and major fertilizer inputs are generally required to promote satisfactory growth and yield. In Malaysia, favoured by a long period of stability and staff continuity, it has been possible to study this situation in much greater detail than has been the case for any other tropical tree crop, and a classic pattern of behaviour has been established.

8.1 The nutrient cycle of rubber

From planting to replanting the rubber plantation presents an environmentally acceptable replacement for the native forest, being a 'closed' ecosystem with a constant cycle of uptake and return of nutrients from and to the soil. Initially, ground covers used to prevent erosion take up and recycle nutrients at the immediate soil surface. Thereafter, as the trees come to dominate the situation, their roots exploit the upper soil layers, taking up nutrients for immobilization within the biomass, and returning a proportion to the ground in dead branch litter, or in the leaf litter deposited at the annual defoliation or wintering. Nutrient increments to the system come from rainfall and fertilizer application, and losses arise from minor leaching and possibly surface wash, but more particularly from loss in latex and timber taken off the site. An appreciation of this system is basic to any understanding of the nutritional requirement of the plantation.

8.1.1 Dry matter production

Shorrocks (1965c) has calculated that over the first five years of growth, the annual rate of dry matter production within a rubber plantation will increase from about 1000 to 14,000 kg/ha, and during the period when the canopy provides a complete ground cover the annual rate of dry matter production is approximately 24,000 kg/ha. This could vary ± 25 per cent or more, depending on local conditions and the vigour of the clone concerned (Table 8.1), but is roughly of the same order as that reported for secondary forest in Ghana (24,400 kg/ha/an.), for maximum yield of pine forest in England (22,000 kg/ha), and evergreen forest in Japan (21,600 kg/ha) and Thailand (25,300 kg/ha). Comparison may also be made with results from one of the most detailed surveys of humid tropical forest ever made, that of the Pasoh forest in Malaysia, showing net production of 27,400 kg/ha/an. dry matter (Kira 1978).

Table 8.1 *Dry matter production: clonal variation in shoot dry weight*

Clone group	Estimated shoot dry weight* (kg/tree) at 5 years		at 10 years	
1 PB 86	79.7	(72)†	267.7	(91)
2 GT 1, RRIM 501, RRIM 628, PB 5/63	107.0	(96)	247.7	(84)
3 RRIM 600, RRIM 605, PR 107	111.9	(101)	310.1	(106)
4 PB 5/51	124.9	(112)	278.4	(95)
5 RRIM 607, RRIM 623, LCB 1320	132.1	(119)	362.0	(123)
Mean	111.1	(100)	293.2	(100)

Source: Shorrocks (1965d)
* Estimated from representative girth data using equation: log shoot wt = 2.783 log girth − 2.584. Shoot defined as all above-ground parts.
† Data expressed as percentages with reference to mean values.

8.1.2 Soil nutrient loss after jungle clearing

In Malaysian experiments on a granite-derived Rengam series soil (Typic Paleudult), it was found that with exposure and insolation of the soil surface after jungle clearing, organic matter reserves were rapidly mineralized, with levels of 100 ppm of nitrate-N recorded in the top soil (Watson *et al.* 1964c). Even at 2–3 years after clearing, peak levels of up to 44 ppm of nitrate-N were recorded in topsoil under bare conditions, compared with levels of less than 18 ppm found under covers of leguminous creepers, grasses and mixed shrubs and ferns. A depletion in easily decomposable N reserves under bare soil was indicated in pot-incubation trials.

At four years after planting, levels of nitrate-N under the different covers remained fairly constant at 1–2 ppm down to a

Table 8.2 *Effect of ground cover type on soil nutrient status at 4, 6, and 10 years after jungle clearing and rubber planting*

Depth of sampling (cm)	Treatment during immaturity	Years after planting	pH	% C	% N	Exch Ca me %	Exch K me %	Exch Mg me %
0–15	Legumes	4	4.93	1.74	0.14	0.32	0.11	0.15
		6	4.50	1.55	0.14	0.13	0.07	0.09
		10	4.63	1.39	0.12	0.09	0.06	0.07
	Grass	4	4.98	1.68	0.12	0.56	0.11	0.19
		6	4.62	1.47	0.12	0.20	0.06	0.10
		10	4.65	1.44	0.12	1.10	0.06	0.06
	Naturals	4	4.93	1.68	0.12	0.60	0.14	0.26
		6	4.62	1.35	0.12	0.17	0.09	0.15
		10	4.63	1.20	0.11	0.11	0.05	0.07
	Bare soil	4	4.53	1.26	0.10	0.18	0.05	0.04
		6	4.50	1.32	0.11	0.10	0.05	0.08
		10	4.57	1.16	0.09	0.07	0.04	0.05
	Min. 5% sig. diff.	4	0.28	0.33	0.02	0.25	0.05	0.12
		6	0.17	0.24	0.03	0.18	0.04	0.10
		10	0.05	0.20	0.02	0.04	0.01	0.01
30–46	Legumes	4	4.83	0.85	0.07	0.12	0.05	0.06
	Grass	4	4.88	0.85	0.07	0.15	0.05	0.07
	Naturals	4	4.80	0.77	0.07	0.14	0.06	0.08
	Bare soil	4	4.25	0.65	0.06	0.07	0.04	0.04
	Min. 5% sig. diff.	4	0.20	0.15	0.02	0.10	0.03	0.06

Sources: Watson *et al.* (1964c); Pushparajah (1984)

depth of 100 cm. Under bare soil, however, levels ranged from 10 ppm in the 0–15 cm layer up to 35–44 ppm in the 45–100 cm layer. This continued mineralization, accompanied by a downward movement of NO_3-N, a fall in pH values and in levels of organic matter and exchangeable bases (Table 8.2), confirms the existence of a leaching process set in chain by jungle clearing and soil exposure. At 6 and 10 years from planting, further falls in nutrient status were recorded under all treatments. To what extent this was due to continued leaching or to withdrawal and immobilization in the tree is uncertain, but nutrient levels fell to growth-limiting levels and were particularly low in the erstwhile bare soil treatment (Pushparajah & Chellapah 1969; Soong & Yap 1976; Pushparajah 1984).

8.1.3 The role of trees and cover crops in the nutrient cycle

On commercial plantations the general practice is to establish leguminous covers immediately after planting, so as to protect the soil and minimize the loss of nutrients. These leguminous covers will thrive for up to three or four years, but progressively die out thereafter under the developing shade of the rubber. From this period onward the site is dominated by the tree population with a thin understorey of shade-tolerant plants. The accumulation and decay of N in the leguminous, and other, covers during the first five years after planting, compared with that in the trees, is illustrated in Fig. 8.1. The story can be extended to cover all nutrients, over the total life of the planting, following work by Shorrocks (1965c,d), Lim (1978) and others on the uptake and immobilization of nutrients in the planting as a whole.

8.1.4 Soil, fertilizer and plant components of the nutrient cycle

Representative data on the main components of the nutrient cycle are given in Table 8.3, which quotes whole tree data from two Malaysian studies, one covering a wide range of soils and planting material (Watson 1964; Shorrocks 1965c,d), and a later one concerning the growth of RRIM 600 when planted on two commonly occurring soil types (Lim 1978). Site and clone variations in commercial rubber are such that the data give only a general indication of nutrient levels involved.

Ground cover plants

It can be seen that at two years after planting, leguminous covers mobilize greater quantities of nutrient than those taken up by the rubber trees at that time, and the amount of cations involved represents a significant fraction of the total topsoil content of a Rengam series soil. These nutrients are returned to the soil as the

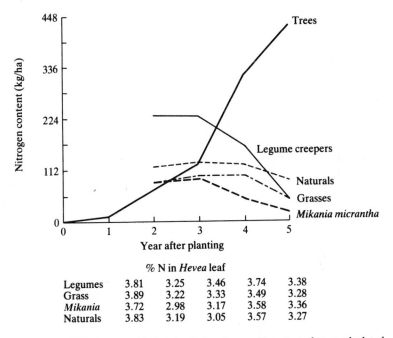

Note: N content of covers calculated as kg/ha of cover; N content of trees calculated on weight of typical RRIM 500 trees

Fig. 8.1 Total nitrogen content of trees and cover plants in first five years after planting, and percentage nitrogen in *Hevea* leaf (*Source:* Watson 1963)

covers die out. In consequence of the favourable effect these legumes have on tree growth the rate of nutrient cycling is enhanced: for example, trees grown with legume ground covers produce greater quantities of leaf litter, richer in N and Ca content, than do trees with grass covers.

Trees

After the first few years from planting, large quantities of nutrients are held in the trees themselves, while significant quantities of N and Ca are recycled in the dead leaf and branch litter. At 8 years after planting, the total Ca held in the trees can be greater than that held in the whole top 45 cm of Rengam soil, and at 16 years the same situation may prevail for K. By 33 years, cations equivalent to the total cation content of the Rengam soils can be immobilized in the trees, with deficits in the case of K and Ca. As the total amount of nutrients in the soil, as analytically assessed, will not all be available to the tree, this deficit in the Rengam soil is an underestimate, and the condition of the Munchong soil nearer to deple-

Table 8.3 *Components in the nutrient cycle of the rubber plantation*

	K(me%) Total	Exch.	Ca(me%) Total	Exch.	Mg(me%) Total	Exch.	% N	Total P (ppm)
Rengam series soil*								
0–15 cm	0.36	0.076	0.37	0.15	0.48	0.06	0.10	194
15–45 cm	0.36	0.055	0.14	0.09	0.49	0.05	0.07	—
Munchong series soil								
0–15 cm	4.51	0.20	0.48	0.15	4.76	0.08	0.11	214
15–45 cm	4.12	0.14	0.31	0.07	4.28	0.04	0.08	131

Total nutrient content(kg/ha)†

	K	Ca	Mg	N	P
Soil 0–45 cm					
Rengam series	840	256	346	5,160	817
Munchong series	10,140	440	3,194	5,790	1,033
Leguminous covers (green material + litter) at:					
2 years after planting	109	114	33	283	25
4 years after planting	24	43	7	107	7
Annual return of rubber leaf litter at 4 years with 408 trees/ha:					
legume covers	1.9	109	6	44	5
grass covers	0.9	31	3	21	4
Annual return of leaf and branch litter:					
at 11 years, 292 trees/ha	20	193	9	114	5
at 31 years, 217 trees/ha	14	58	6	78	3
Annual rainfall of 2500 mm (estimated contribution)	12	38	3	20	0.2

Age	Trees/ha	Clone					
Whole trees:							
2	445	RRIM 501	42	35	14	72	7
4	395	PB 86	62	54	12	72	9
4	395	GT 1	94	96	22	178	18
4	395	LCB 1320	140	115	28	228	19
4	408	RRIM 501	187	169	63	315	30
8	321	RRIM 501	290	415	85	558	49
5	420	RRIM 600 (Munchong soil)	1,170	1,200	285	780	135
16	535	RRIM 600 (Rengam soil)	874	1,301	149	655	134
33	267	Tjir 1	1,233	2,119	417	1,779	277
Fertilizers (125 kg)							
		Potassium chloride	66	—	—	—	—
		Magnesium sulphate	—	—	20	—	—
		Rock phosphate (36% P_2O_5)	—	35	—	—	20
		Ammonium sulphate	—	—	—	26	—

Source: Watson (1964); Shorrocks (1965b); Lim (1978)

* For description of soil types see Table 4.8

† Total soil nutrient content assessed by extraction with $6N$ HCl, and calculated on basis of 1 ha, 45 cm depth of soil, weighing 6000 t

Fig. 8.2   Young rubber growing on Rengam series soil (Typic Paleudult) of the type referred to in Table 8.3

tion. It is easy to understand why severe nutrient deficiencies may develop in the succeeding stand, if this timber is removed from the site at replanting. With increased interest in the utilization of rubber wood this is a significant problem (Tan and Sujan 1981; Nor 1984).

Clonal differences

Of particular interest is the variation in total nutrient content of the trees sampled at four years after budding, with the old and less vigorous clones such as PB 86, GT 1 and LCB 1320 immobilizing significantly less nutrients than RRIM 501 and RRIM 600. Obviously the strain on soil nutrient reserves and consequent fertilizer requirements can vary considerably, depending on the clone planted and the available nutrient content of the different soils.

Fig. 8.3   Removal of old timber at replanting – a major loss of nutrients

Nutrient drain in latex

With early planting material yielding less than 1000 kg/ha dry rubber, little thought used to be given to the question of nutrient drain via the latex. Since modern clones yield around 2000 kg/ha, and perhaps 2500 kg or more with stimulation, nutrient drain has become a significant factor. Since nutrient balance within the tree is biased in favour of the liquid contents of the cell rather than the solid, trees of high nutrient status are subjected to heavier drain of nutrient per unit weight of rubber than are trees of low nutrient status. Nutrient drain will also differ between clones, and will be disproportionately increased by the use of stimulants which reduce plugging and induce longer flow, more latex drip and lower dry rubber content (d.r.c.) (Pushparajah *et al.* 1972; Sivanadyan *et al.* 1972) (Table 8.4).

In the trial quoted in Table 8.4, stimulation of PB 86 raised yields by 85 per cent and K drainage by 169 per cent, with consequent stress on leaf K content. Response to stimulation can vary between clones, and between tapping panels; with very high-yielding material, K drainage equivalent to 76 kg/ha/an. has been recorded (Pushparajah *et al.* 1972). This can only enhance diversion by the tapping operation of metabolites and nutrients from the processes of tree growth, and experience shows that if not compensated for by increased fertilizer application, the normal fall in response to stimulation will be accelerated and yield can eventually decrease to below unstimulated levels (Haridas *et al.* 1976).

Table 8.4 *Effect of stimulation of PB 86 and RRIM 600 on yield and nutrient drain*

Experiment and treatment	Yield (kg/ha/an.)	Nutrient drain (kg/ha/an.)				Leaf nutrient content (%)			
		N	P	K	Mg	N	P	K	Mg
Experiment LF 8, PB 86 tapped S/2.d/2:									
Unstimulated	1390	9.4	2.3	8.3	1.7	3.55	0.325	1.44	0.22
2,4,5-T (1%)	1660	11.9	3.1	11.1	2.1	3.54	0.31	1.41	0.23
Ethrel (10%)	2570	23.9	7.2	22.3	4.1	3.40	0.32	1.26	0.21
Experiment LF 10, RRIM 600 tapped S/2.d/2, Panel A:									
Unstimulated	1819	18.0	3.6	14.6	2.5				
2,4,5-T (1%)	1928	20.0	4.2	16.5	3.2				
Ethrel (10%)	2132	25.4	5.5	23.2	4.0				
Experiment LF 19*, RRIM 600 tapped S/2.d/2, Panel C:									
Unstimulated	2314	22.9	4.6	18.6	3.1				
Ethrel	6955	82.9	18.0	75.6	13.1				

Source: Pushparajah *et al.* (1972)
* Estimated from 10 months' data.

8.2 Effect of fertilizers on tree performance

The above picture of nutrient withdrawal and strain has become clear only within the last 30 years. In earlier, mainly ex-jungle plantings, growth during immaturity had seemed satisfactory with few visual symptoms of nutrient deficiency, and once the trees reached maturity they were commonly thought to be self-sustaining with no need for fertilizer inputs. Even in the early 1900s, however, responses by *Hevea* to fertilizer application were being demonstrated in Vietnam, Cambodia, Sri Lanka, Malaysia and Indonesia (Compagnon 1962), and since then a considerable body of information has accumulated.

8.2.1 Early experiments

In West Java during the 1930s, young rubber on a variety of volcanic ash-derived soils showed responses to NPK fertilizers of up to 29 per cent (Dijkman 1951). In Malaysia, main responses were thought to be to N and P only, until a review (Owen *et al.* 1957) in 1957 of 17 trials laid down during the period 1935–38 and reinstated in 1947/48, led to a broad distinction between rubber growing on coastal marine alluvium where only N might be required, and that on various inland soils where N was only fully effective when applied together with P and K. Initial effects of fertilizer applied in the mature phase were on girth, but measurable effects on yield followed within six years.

8.2.2 Nutrient interactions and their influence on fertilizer effectiveness

A detailed analysis of one of the above 17 trials, sited on a Serdang series sandy loam derived from quartzite (Typic Paleudult), compared the effects of fertilizer applied before and after maturity and concluded that by far the greater response was to the former (Bolton 1960). In this particular experiment, K fertilizer had little effect, possibly because of a low Mg status, while leaf Mg content was depressed to limiting levels by the K application. A similar effect noted in a NPK trial on shale-derived soil of low Mg status (Bolton 1964a) led to a better appreciation of the need to include both nutrients in further trials.

Kieserite (magnesium sulphate) is a quick and effective ameliorant for Mg deficiency in young rubber, but for mature rubber the cheaper magnesium limestone is more appropriate: in one trial with rubber showing deficiency symptoms (Bolton & Shorrocks 1961), three biennial applications of magnesium limestone at 1400 g/tree, gave eventual increases of 8 per cent in girth and 15 per cent in yield, the effects becoming apparent in the first and fifth years respectively after commencement of the treatment. An additional effect was a depression in leaf Mn content, possibly due to a Mg/Mn interrelationship pointed out by Bolle-Jones (1957a), but more probably the result of a local increase in soil pH following the limestone application. Such an interaction would be of potential concern on many alluvial sandy soils, and those derived from si-liceous granites, with low Mn status.

As post-war planting/replanting progressed in Malaysia, more information became available on the profitability of fertilizer use, and on the importance of nutrient interactions. A review of two trials (Pushparajah 1969), with yield levels in the range 1100–1800 kg/ha, showed increases in yield of 100–400 kg/ha and in net annual profit of $M36–203/ha, following appropriate fertilizer application. In one of these trials, SE 62, the N status of the trees was maintained by basal application of ammonium sulphate, while potassium chloride was applied to correct K deficiency; a significant yield increase was recorded in the second year after application, reaching 30–40 per cent in the third and fourth years of tapping, to result in a net profit of $M101/ha (Table 8.5). The N/K inter-action was highlighted in another trial of the series (SE 1/21) where application of ammonium sulphate alone depressed leaf K content, growth and yield to result in a net annual loss of $M349/ha. When ammonium sulphate was applied together with potassium chloride, however, increases in all three items were recorded to give a net annual profit of $M70/ha.

8.2.3 Effects of fertilizer applied during immaturity, and interaction with cover type

Results from the first Ceylonese (Sri Lankan) long-term fertilizer trial on clonal material (Constable 1953), showed that the use of NPK fertilizers during the immature phase and in the early tapping years, resulted in a 27 per cent increase in yield. This was achieved from a low base yield of 560 kg/ha, suggesting that more vigorous and higher-yielding modern clones may show even greater response to fertilizer. Later work with such clones confirms that this may be the case. Jeevaratnam (1969), reporting on a trial with an overall yield of approximately 2000 kg/ha/an., covering a period of NPK fertilizer use over 7 years of immaturity and $7\frac{1}{2}$ years of tapping, quotes a mean yield increase of 49 per cent. In this trial, NPK fertilizer applied during immaturity gave a 30 per cent increase in girth by the time of opening for tapping, and an increase of 22 per cent in early yields. Eventual yield increases amounted to 66 per cent per annum, with fertilizers applied during maturity accounting for 50–60 per cent of the effect.

Table 8.5 *Effect of N/K fertilizer interaction on leaf nutrient content, growth, yield and profitability*

Experiment	Duration* (years)	Treatment	Leaf K content (%)	Girth increment (cm)	Effect on yield (kg/ha/an.)	Income	Costs/returns ($M/ha/an.) Fertilizer costs	Other costs	Net profit
SE 62†	3	k_0	0.76	23.2					
		k_1	0.95	23.6	(of K) 186	190	27	62	101
		k_2	1.06	25.2					
		s.e. +/-	0.02	0.31					
SE 1/21	6	n_0k_0	0.83	17.7					
		n_2k_0	0.76	15.8	(of N) -345	-351	116	-114	-349
		n_0k_2	1.48	20.4	(of NK) 302	306	133	99	74
		n_2k_2	1.25	20.7					
		s.e. +/-	0.006	1.22					

Source: Pushparajah (1969)

* Years of tapping.

† Experiments SE 62 and SE 1/21 have also been referred to as experiments SW and SPME, respectively.

Unit levels of fertilizer application (g/tree/an.):

	SE 1/21 over 7 years	SE 62
Ammonium sulphate (n)	907	—
Potassium chloride (k)	227	114 for 3 years
		227 for 2 years

The importance of early fertilizer effects on tree performance was noted by Bolton (1960) and has been further amplified by two Malaysian studies (Ti *et al.* 1972) concerning the long-term effect of cover plant and fertilizer policies during immaturity. In experiments B and S, previously referred to in Chapter 7, the use of leguminous covers during immaturity gave additional returns over the best alternative cover type, aggregated over a number of fertilizer treatments, of \$M413/ha and \$M511/ha (Table 8.6). Results from an associated trial, experiment SE 60, showed that application of N fertilizer additional to normal would improve yield, but would not be economic where there were leguminous covers. Greater effects on yield were apparent with grass covers, particularly at the higher level of application, but the return to use of grass covers at the highest level of N application was still \$M423/ha less than to legumes receiving no N.

Table 8.6 *Effect of cover plant and fertilizer treatment during immaturity on yield (kg/ha dry rubber) and profitability (discounted returns \$M/ha)*

	Experiment		Legumes	Grasses	Mikania	Naturals
B	Cumulative yield*		9,096	7,371	6,365	7,895
	Cumulative discounted returns		2,014	1,421	1,003	1,601
S	Cumulative yield*		10,619	9,369	9,615	9,028
	Cumulative discounted returns		2,799	2,279	2,288	2,038
SE 60[†]	Additional	n_1	53	386	—	—
	cumulative yields*	n_2	161	1,079	—	—
	relative to n_0	n_3	492	1,749	—	—
	Additional	n_1	540	592	—	—
	cumulative yields	n_2	1,045	1,042	—	—
	required to repay	n_3	1,538	1,534	—	—
	costs of N fertilizer					

Source: Ti *et al.* (1972)
* In experiments B and S yields were for 6 years of tapping, in experiment SE 60 for 3 years.
[†] In experiment SE 60, differential rates of N fertilizer were applied from 13 months after planting onwards, reaching rates of 0, 0.45, 0.90 and 1.35 kg ammonium sulphate/tree/an. at 49 months after planting (Pushparajah & Chellapah 1969).

8.2.4 Fertilizer response of high-yielding, well-managed rubber

In Malaysia, by the late 1950s and early 1960s, large areas of high-yielding young rubber were becoming established, based on Class 1 clonal material and the use of leguminous covers and regular fertilizer application during immaturity. Trials (Mainstone 1963b; Sivanadyan 1983) laid down to study the fertilizer requirement of such material confirmed that, as in earlier work, leaf nutrient status responded quickly, but effects on yield took longer to develop. Responses in annual yield were largely to N, sometimes expressed only in the presence of K and Mg, becoming evident at 3–6 years after commencement of fertilizing, and reaching maxima of 11–58 per cent after 8–10 years (Table 8.7).

Table 8.7 *Effect of fertilizer application* on Class 1, high-yielding clones*

Experiment No., soil type and clone	Year of 1st[†] application	Year of 1st[†] response	Maximum response
SE 101 Durian series PB 5/51	1	5	23% to N in year 10
SE 104 Durian series PB 5/51	1	4	11% to NK in year 9
SE 106 Holyrood series GT 1	4	3	44% to N in year 9
SE 108 Klau series PB 5/51	2	6	12% to NK in year 8
SE 122[§] Munchong series RRIM 600	2	4	58% to NKMg in year 9

Source: Sivanadyan (1983)
* Fertilizer applications were at 3, 3 and 2 levels respectively for N, K and Mg, unit levels being 0.1 kg N/tree/an. as ammonium sulphate, 0.18 kg K_2O/tree/an. as potassium chloride, and 50 g MgO/tree/an. as magnesium sulphate.
[†] Years of first application are tapping years, years of first response are years after commencement of treatment.
[§] The only experiment with non-leguminous covers during immaturity.
Note: All clones of good vigour and giving yields in excess of 1500 kg/ha.

Taking into account the residual effect of treatment during immaturity, and the likely levelling off of nutrient immobilization within the tree with age, it was concluded that fertilizer use could be suspended for four years after the commencement of tapping, and that thereafter only sufficient N should be applied to replace that lost in latex. At the same time, the nutrient status of the trees should be monitored at 3–5-year intervals, and P, K and Mg fertilizer applied as indicated.

The above conclusions are only applicable to well-managed rubber plantings, where leguminous covers have been used during immaturity, and where fertilizer has been regularly applied. Where there has been competition from natural covers and weeds, where intercropping with food crops has been practised, or where fertilizer use has been inadequate, all experimental evidence indicates that a full schedule of NPKMg fertilizer application will be required to sustain tree growth and develop optimum yields.

8.3 Recommended fertilizer practices

8.3.1 Nurseries

Field nurseries and bud-wood nurseries are generally sited on the best soils locally available. Management is directed towards obtaining good growth and in Malaysian practice, at the time of establishment, basic dressings of limestone and rock phosphate are applied and cultivated into the soil. Thereafter, regular dressings of a complete fertilizer incorporating soluble phosphate are applied, according to the schedule quoted in Table 8.8 (Haridas 1981). Similar practice is followed in Sri Lanka and Indonesia.

In polybag nurseries, the polybags are filled with a good class soil, preferably with a fairly heavy clay loam texture with added rock phosphate, and then a complete fertilizer is applied as the young seedling develops. In all cases, care must be taken to avoid scorching the young plants. Soluble fertilizers should not be placed in contact with the stems, and fertilizers containing nitrate should not be used for fear of damaging roots.

With optimum management, advanced planting material such as stump buddings can reach 10–12 cm girth for transplanting by about 15–18 months after budding, or 20–22 months after nursery establishment. Large polybag plants at 6–7 whorls can be transplanted by 7 months after budding or 12 months from establishment. Bud-wood nurseries should be able to yield reasonable bud-sticks by about 5–6 months after cut-back.

8.3.2 Immaturity

In the early years after planting, the rubber tree is a minor component of the plantation, and must compete for soil moisture and nutrients with cover plants and weeds. Fertilizer applied during these years is aimed at promoting vigorous tree growth so as to advance the time at which the trees may be large enough to be opened for tapping (50 cm girth at 150 cm above the union by current standards), and must take into account the role of the cover plants in the nutrient cycle, and in particular the return of N to the soil by leguminous covers in the third to sixth years after planting.

This is a dynamic system in which the trees represent, as it were, a 'moving target' for fertilizer trials, with requirements shifting as the trees age and the cover quality changes. As a result, fertilizer recommendations are based almost as much on general experience and on knowledge of the local soil type as on actual experimental data, with quality and quantity of fertilizers determined by the need to ensure good early growth.

There are two situations where excessive fertilizer use can

Table 8.8 *Nursery fertilizer programmes*

Material	Time of application	Fertilizer type (N:P:K:Mg) and quantity	
Budded stumps and mini-stumps Stumped buddings Bud-wood	Before planting:	magnesium limestone: followed by rock phosphate:	250 kg/ha incorporated in soil 625 kg/ha incorporated in soil † 112 g/point
Budded stumps and mini-stumps Mini-stumps	2,3,4 & 5 months after planting, before budding: 1,3 & 5 months, after cutback:	9:15:7:2 equivalent	60 g/m planting row 60 g/m planting row
Stumped buddings	2,3,4 & 5 months after planting, before budding: 1,3,5 & 7 months after cutback: 9 & 11 months after cutback: 13,15 & 17 months after cutback:	9:15:7:2 equivalent	42 g/point 28 g/point 42 g/point 56 g/point
Bud-wood	2,3,4 & 5 months after planting, before budding: 1,3 & 5 months after cutback: 7 months after cutback: 2 months after pruning:	9:15:7:2 equivalent	42 g/point 28 g/point 42 g/point 42 g/point
Small polybags	At planting: 2 & 3 months after planting, before budding: 4 & 5 months after planting, before budding; 1 month after cutback:	rock phosphate 15:15:6:4 equivalent 9:15:7:2 equivalent	56 g/bag 14 g/bag 22 g/bag 14 g/bag
Large polybags and soil core buddings	Applications up to cutback as for small polybags 1 month after cutback: 2 months after cutback: 3,4,5 & 6 months after cutback: 7,8 & 9 months after cutback:	9:15:7:2 equivalent	14 g/bag 22 g/bag 28 g/bag 42 g/bag

Sources: Haridas (1981)
Notes: Except where stated otherwise, fertilizers are to be broadcast over the soil surface along the rows of young plants.

present problems. In the early months after planting, soluble phosphate fertilizers are often used to promote vigorous growth. On some sites, however, the consequent growth flush may outrun the capacity of the roots to supply trace elements, in particular. This, coupled with some interference with trace element metabolism within the plant and perhaps actual damage to the roots (Mainstone 1963a), can cause tip dieback, followed by bushy side shooting (Middleton *et al.* 1965). The other situation may occur in young plantings with vigorous leguminous covers where the N return from the covers, added to that contained in the fertilizers, may promote heavy leafing and unbalanced canopy development with consequent risk of bending and wind damage. In such cases the application of N fertilizer should be lessened or dispensed with altogether. Fertilizer schedules used in Malaysia that take these factors into consideration and have provided satisfactory growth over a range of planting conditions, are given in Table 8.9 (Pushparajah & Yew 1977).

Fertilizer schedules used in other rubber-growing territories (Pushparajah 1983a) though based on local trials (Angkapradipta 1973; Guha 1975; Hardjono & Angkapradipta 1976; Punnoose *et al.* 1976; Silva 1976; Kalam *et al.* 1980; Omont 1981; Yogaratnam & Weerasuriya 1984), are similar to those used in Malaysia, but pay less regard to the presence or absence of legume covers (Table 8.10). There is obviously room for further research here, particularly in those areas where food cropping is standard practice, when N is likely to be a limiting factor.

8.3.3 Maturity

From the fourth or fifth year after planting, with cover plants thinning out under the heavy shade of the young rubber, the tree becomes the dominant component and latex yield the prime consideration of the planter.

For any particular planting material, latex yield (y) is primarily dependent on the tree girth (g), according to the relationship $y = f(gm)$, where m is between 1.5 and 2.5 (Bolton 1964b; Narayanan & Ho 1970). It follows that to attain satisfactory yields the general health and girthing of the tree must be maintained during maturity. For this purpose, and to compensate for the progressive immobilization of nutrients within the tree, and their loss from the site in latex, the use of 'maintenance' dressings of a complete fertilizer up to perhaps five years before replanting is common practice. Generalized recommendations for fertilizer use during maturity, together with suggestions for use under high-yield situations, have been published by Chan *et al.* (1972) (Table 8.11).

Smallholdings have shown a higher N and P fertilizer require-

Table 8.9 *Fertilizer programmes for immature rubber, West Malaysia*

Time of application, months after planting or budding	Fertilizer type (N:P:K:Mg) Soils with medium K level	Soils with low K level	Quantity, with legume covers (g/tree)	Additional ammonium sulphate required in presence of non-legume covers (g/tree)
In planting hole	Rock phosphate		113	—
1	8:14:7:2 recommended for use on all soils in early months		57	—
2½			85	—
4			85	—
5½			113	—
7			113	—
9			113	—
11,13,15	9:16:3:2	8:14:7:2	170	113
18			170	170
21,24			227	227
27			284	227
30			340	284
34			340	340
38	13:9:4:2	11:10:7:2	340	340
42			454	340
46,50,56,62			454	454

Source: Pushparajah and Yew (1977)

Notes: (a) These programmes are suitable for fertilizer-responsive soils, with high K formulations used on the less fertile, low K soils. (b) The additional ammonium sulphate is recommended to be applied at 1–2 months after application of the complete fertilizer as convenient – it may also be required where *intercropping with food crops* is conducted without specific fertilizer inputs. (c) In the presence of legume covers, young trees may develop heavy canopies and be subject to leaning, in such cases low-N regime should be instituted, using a fertilizer of the general formulation 9:10:13:2 at the rates recommended.

Table 8.10   *A comparison between countries of recommended fertilizer applications to immature rubber (kg/ha/an.)***

Country	Region/soil/cover condition	N	P_2O_5	K_2O	MgO
Brazil	—	205	200	175	45
India	Legume cover	200	200	116	21
	Non-legume	260	220	104	21
Indonesia	East Java	290	309	132	50
	West Java & Sumatra	251	274	217	50
Malaysia[†]	Low K, no legume	640	250	170	50
	Low K, mixed legume	225	250	170	50
	Low K, pure legume	30	250	170	50
	High K, no legume	660	260	90	50
	High K, mixed legume	225	260	90	50
	High K, pure legume	30	260	90	50
Sri Lanka	Lateritic soils	195	260	140	45
	Micaceous soils	205	280	100	30
	Quartzitic and alluvial sandy soils	170	230	195	60
Thailand	Low K soils	250	270	220	50
	High K soils	290	300	130	50
Ghana	—	190	260	190	—
Liberia	—	225	250	200	70

Source: Pushparajah (1983a)
* Applied over a period of 66–72 months after planting, with stand assumed to be 450 trees/ha.
[†] The levels of nutrients quoted for Malaysia are related to the fertilizer quantities given in Table 8.9.

ment than the bulk of estate holdings, because of low input during immaturity, and this is reflected in the relevant recommendations. Similarly in Sri Lanka, where smallholdings have received little or no fertilizer in the past, enhanced rates of application are recommended (Silva 1975).

8.4   Effect of fertilizers and soil nutrient status on latex composition and flow

Both the quantity and the composition of the latex produced by the tree can be affected by soil nutrient status and fertilizer application (Beaufils 1955; Collier & Lowe 1969; Pusparajah *et al.* 1976a). Data from a number of clones in several fertilizer trials in Malaysia showed that application of ammonium sulphate increased the levels of N, K, Mg and the Mg/P ratio in the latex and also slightly increased the content of volatile fatty acids. There was no consistent

Table 8.11 *Nutrient requirement and fertilizer recommendations for estate and smallholding mature rubber, West Malaysia (kg/ha/an.)*

Soil series	Clonal* groups		Nutrient requirement				Approximate fertilizer formula and dose recommended for smallholdings
			N	P_2O_5	K_2O	MgO	
Rengam/Jerangau (granite-derived Ultisols)	Estates	1	16	28	94	10	
		2	20	28	118	10	
		3	8	28	94	10	
	Smallholdings	Wind-fast	38	48	53	—	400 kg/ha 10:12:13
		Wind-prone	18	48	60	—	300 kg/ha 6:16:20
Munchong/Prang (shale-derived Oxisols)	Estates	1	16	21	59	10	
		2	20	21	94	10	
		3	8	21	59	10	
	Smallholdings	Wind-fast	41	38	29	—	350 kg/ha 12:11:8
		Wind-prone	20	43	40	—	250 kg/ha 8:17:16
Selangor/Briah (marine/riverine alluvium Inceptisols/Entisols)	Estates	1	20	—	—	—	
		2	24	—	47	—	
		3	12	—	—	—	
	Smallholdings	Wind-fast	23	—	25	—	50 kg/ha urea + 50 kg/ha potassium chloride
		Wind-prone	12	—	25	—	25 kg/ha urea + 25 kg/ha potassium chloride

Sources: Chan *et al.* (1972); Pushparajah and Yew (1977) to cover the period from commencement of tapping to five years before replanting

* Group 1 All clones except groups 2 and 3. PB 5/51 is in this group, but should receive K appropriate to group 2.
Group 2 RRIM 600, GT 1, with high K requirement.
Group 3 Clones susceptible to branch and trunk snap, e.g. RRIM 605, RRIM 623, RRIM 501. Nitrogen supply is restricted.
Notes: (a) These are put forward as guideline fertilizer programmes, and should preferably be refined in the light of soil and leaf diagnostic survey. (b) Assumed mean yield is 1500 kg/ha. For every additional 1000 kg/ha apply 11 kg N and 20 kg K_2O to compensate for nutrient offtake.

effect on dry rubber content but at higher rates of application the total solids content was sometimes reduced by 1–2 per cent. Rock phosphate raised P and Ca levels and lowered the Mg/P ratio. Potassium chloride increased the K and P contents but lowered those of Ca and Mg and the Mg/P ratio. Magnesium fertilizers increased Mg and decreased K and thus decreased K/Mg ratio and increased Mg/P, the increase in the latter ratio being enhanced if nitrogen was applied with magnesium.

It has been shown that the levels of divalent cations in the serum phase of the latex, particularly of Mg, differ markedly between clones and are inversely related to latex stability (Yip & Gomez 1980). High P content, and low Mg/P and Mg + Ca/P ratios are associated with good latex stability. Poor latex stability tends to result in pre-coagulation on the tapping cut (and probably to more rapid plugging within the vessels), leading to slower average flow rates and reduced yield. In general, clonal differences are more important than variation in nutrient levels, but the latter cannot be disregarded.

As indicated by the data in Table 8.12 from a long-term NPKMg fertilizer trial in Malaysia (Rubber Research Institute of Malaysia 1979b), potassium has the greatest influence on latex flow, increasing rates of flow in comparison with other treatments and also producing the most stable latex concentrates (see p. 490). Work in the Ivory Coast links this effect with the role of K in sucrose translocation and enhancement of production in stimulated trees by the use of K fertilizers (Tupy 1973a).

Table 8.12 *Effect of fertilizer on latex flow and latex concentrate stability**

Treatment	Volume	D.r.c.	Time of flow	Rate of flow	Latex concentrate MST[†]	
	(ml/tree)	(%)	(min)	(ml/min)	Fresh	1 month
N	96.8	29.36	80.6	1.20	65	120
P	93.6	29.84	78.1	1.20	60	95
K	95.4	33.36	72.4	1.32	185	315
Mg	79.2	31.56	92.3	0.86	60	80
NPKMg	92.7	30.68	75.7	1.22	60	100
Control	62.0	33.33	73.7	0.84	65	115

Source: Rubber Research Institute of Malaysia (1979b)
* Values quoted are means of two readings. First sampling done during refoliation after wintering, and second when leaves fully developed.
[†] MST: mechanical stability time.

Magnesium fertilizer application in the Malaysian trial reduced the rate and volume of latex flow and this may be linked with experience in the field showing that trees growing on soils with high

Ca and Mg status may be affected by pre-coagulation on the tapping cut (Rubber Research Institute of Malaya 1964c). If rubber is planted on such soils, Mg fertilizers should be used only to the extent warranted by actual field trials, or by visual diagnosis of deficiency symptoms and leaf analysis. In both Malaysia and Cambodia (Pushparajah 1969; de Geus 1973) the use of high K fertilizers has been recommended to narrow Mg/P ratios, particularly with such clones as Glenshiel 1 which have high Mg content in the latex and are prone to pre-coagulation. The point is illustrated (Liang 1983) by data from China (Table 8.13) which, when related to leaf nutrient data from Malaysia (Table 8.16), seem likely to have come from an area of rubber affected by K deficiency.

Table 8.13   *Effect of K fertilizer on K/Mg ratio in the leaf and latex yield*

Treatment	% leaf content			Pre-coagulation of latex (%)	Yield of dry latex (kg/tree)
	K	Mg	K/Mg		
O	0.89	0.47	1.91	17.2	1.85
N	0.84	0.50	1.68	17.2	2.02
K	1.08	0.45	2.39	5.1	2.22
NP	0.82	0.47	1.74	17.4	2.00
NPK	1.16	0.45	2.57	2.2	2.25

Source: Rubber Institute of the Tropical Crop Academy of South China, quoted by Liang (1983)

8.5 Effect of fertilizer on seed production

Rubber seed is not normally harvested, except when taken from selected mono- or polyclone plantings ('seed gardens') for growing on as high-yielding seedlings or for production of rootstocks. In the Amazon basin, however, native Indians will eat the seed, after first cooking it to destroy cyanic poisons (as in the case of cassava). At times of emergency, seeds have also been used as a foodstuff in most rubber-growing territories and are collected perhaps most generally in Sri Lanka for processing for oil with extracted seed cake used as an animal feedstuff (Nadarajapillai & Wijewantha 1967; Nadarajah 1969). The subject has been reviewed by a number of authors (Wheeler 1978; Stosic & Kaykay 1981; Vimal 1981), and it is conceivable that seed production could become a useful by-product of the plantation provided that collection and processing could be efficiently organized.

Estimates of seed production vary widely, and will depend on the planting material, tree stand, soil and seasonal conditions, but levels of up to 600 kg/ha/an. have been recorded. Malaysian trials (Watson & Narayanan 1965) have shown responses in seed production to the application of NPK fertilizer. In one multi-clone seed garden, on a marine alluvium known to be responsive to N application, annual application of 1.8 kg ammonium sulphate per tree gave an overall 30 per cent increase in seed production over a period of 7 years (Table 8.14, experiment F). In a planting of PB 86, of low N status on shale-derived soil, similar fertilizer treatment produced an 80 per cent increase in seed production, over 3 collecting years. In a further trial, on granite-derived soil, with a relatively high N status but low in K (experiment S G), potassium fertilizer tended to depress seed production and general indications were that seed production is favoured by a wide N/K ratio in the leaf.

8.6 Diagnosis of fertilizer requirement

In the absence of good support facilities, management may have to rely on general experience in preparing fertilizer schedules for local conditions. However, in order to optimize the use of expensive fertilizers, schedules should preferably take into account local soil characteristics and the nature and conditions of the planting material being used. To do this a variety of techniques may be used. Pot culture techniques have been tested, but never developed on any scale (Middleton 1961; Polinière & van Brandt 1969). Replicated fertilizer trials in the field are used, but on their own have limited value. With minimum plot sizes of about 0.20 ha, even a simple 33 NPK trial would require over 5 ha of land. For the preparation of general recommendations, trials need to be sited on a variety of soil types, and since results can only be properly assessed after several years of fertilizer application, a great deal of time and effort together with continuity of management is required before satisfactory conclusions can be reached. In order to make such trial programmes more effective, extensive use is made of soil and leaf analysis to permit extrapolation of results from one site to another, and to help in diagnosis of nutritional problems in the field.

8.6.1 Soil and leaf analysis as a guide to fertilizer requirement

Soil analysis

On the leached soils of the humid tropics there can be obvious links between soil characteristics and tree performance (Wong *et al.* 1977). Riverine and marine clay alluvia (Inceptisols/Entisols) can give good tree growth without fertilizers, provided that drainage is good, but coarse sandy alluvia are generally infertile. Ultisols derived from siliceous granites are coarse-textured, pale in colour, low in pH and nutrients, while those derived from more basic gran-

Table 8.14 *Effect of fertilizer application on seed production*

Experimental details	Treatment	Application rates (kg/tree/an.)		Mean annual* production	Leaf N† (%)	Leaf K (%)
Experiment F	n_0	0	—	83 seed/tree	—	—
Selangor series	n_1	0.9	ammonium	92	—	—
Mixed clonal seed garden	n_2	1.8	sulphate	107	—	—
Experiment SP	n_0	0	—	113 kg/ha	3.04	—
Seremban series	n_1	0.9	ammonium	154	3.25	—
PB 86	n_2	1.8	sulphate	205	3.35	—
	k_0	0	—	159 kg/ha	—	1.85
	k_1	0.22	potassium	147	—	1.87
	k_2	0.45	chloride	166	—	1.96
Experiment SG	n_0	0	—	259 kg/ha	3.33	—
Rengam series	n_1	0.9	ammonium	277	3.47	—
Mixed clonal seed garden	n_2	1.8	sulphate	268	3.51	—
	k_0	0	—	294 kg/ha	—	0.88
	k_1	0.22	potassium	269	—	1.16
	k_2	0.45	chloride	242	—	1.28

Source: Watson and Narayanan (1965)

* Data are means of 7, 3 and 7 years of seed collection in experiments F, SP and SG respectively.

† Leaf nutrient content data are means of 15 and 7 samplings, taken over 3 years, in experiments SG and SP respectively.

ites and granodiorites are likely to be darker in colour due to their higher content of iron, manganese and magnesium minerals, strongly textured, with relatively high K and Mg levels, and give good tree growth. The nutrient content of Oxisols derived from basic volcanic material will vary, depending on age, with low K a limiting factor at times; nevertheless, the deep profile of these soils permits free rooting that can mitigate the effect of a low nutrient status, while the dark, ferruginous colour indicates high P fixation capacity.

For these reasons, and following historical tradition, early attempts to rationalize fertilizer use were centred on soil analysis. Work by Owen (1953), later confirmed by Lau *et al.* (1977), showed a relationship between NH_4F-extractable P in the soil and the yield and leaf P content of rubber, with immature rubber likely to show a response to P fertilizers when the level fell to below 11 ppm. On coastal alluvial soils, however, with high organic P levels, NaOH-soluble P proved a better indicator of availability. In the case of K, both $6N$ HCl extractable K and exchangeable K correlate well with leaf K content, growth and yield of the rubber (Lau *et al.* 1972). Quality and quantity measurement of available K have proved less useful, being too time-consuming for routine use (Mohinder Singh & Talibudeen 1969), and determination of K-supplying power using the electro-ultra centrifuge is still at an experimental stage (Thiagalingam & Grimme 1976). In the case of N, total soil content and the C/N ratio have correlated well with leaf N and tree performance.

Using data from fertilizer experiments on two Malaysian soil types, Sys (1975) has concluded that, with mineral soils at least, the higher the organic matter content the lower the requirement for P, K and Mg fertilizers. In more extensive soil and analytical studies covering 8 fertilizer trials and 350 commercial sites, Pushparajah and Guha (1969) were able to determine the probability of obtaining fertilizer response on four of the main Malaysian soil groups. A simplification of their results (Pushparajah *et al.* 1983) suggests guidelines on the nutrient-supplying capacity of the soils (Table 8.15) that find some agreement with work on similar soils in Sri Lanka (Silva 1969). These guidelines will, however, be influenced by many site-specific factors that need to be considered before making definitive fertilizer recommendations.

Leaf analysis

Soil sampling programmes are laborious and expensive, and following early work by Chapman (1941), detailed studies by Beaufils (1955) and others (Shorrocks 1965b; Guha & Narayanan 1969; Chan 1972), have shown that leaf analysis can be a valuable additional tool in the diagnosis of fertilizer requirement. The method must be used with care, however, for the leaf nutrient

Table 8.15   *Categorization of soil nutrient availability (0–30 cm)*

Nutrient	Low	Medium	Category* High	Very high
Total N%	<0.10	0.11–0.25	0.26–0.40	>0.40
1:1 $H_2SO_4/HClO_4$ ext. P (ppm)	<250	250–350	350–600	>600
NH_4F ext. P (ppm)	<11	11–20	20–30	>30
6N HCl ext. K (me%)	<0.5	0.5–2.0	2.0–4.0	>4.0
6N HCl ext. Mg (me%)	<0.75	0.76–3.0	3.0–8.0	>8.0

Source: Pushparajah *et al.* (1983)
* Low = deficiency
   Medium = likelihood of deficiency
   High = level at which only imbalances with other nutrients need be considered

content can vary depending on position in the canopy, leaf age, season and the planting material involved (Chang & Teoh 1982).

Effect of leaf type on nutrient status

In young rubber, higher levels of N, P and K are found in top-whorl leaves than in those of the lower whorls. In mature rubber, the position of the leaf on the canopy, whether exposed to the light or in shade, can have a significant effect. Leaf N and K contents tend to decrease with age, while Ca will tend to increase; accordingly, leaf analysis will reflect the age of the leaf at the time of sampling. In Malaysia, to achieve some standardization, analytical data is 'corrected' by taking these ageing factors into account, to give an estimate of what the nutrient content would be at an optimum age of the leaf, taken as 100 days after emergence, when maximum photosynthetic activity is assumed to have started (Chua 1970; Pushparajah & Tan 1972; Pushparajah 1975). However, this has its drawbacks, and in India local variability is said to be too great to permit such an approach (Nair 1973). More recently, it is suggested that nutrient content should be determined in tissue water, and sufficiency assessed in relation to supposed satisfactory constant values, obviating the need for any correction factor (Tinker & Leigh 1984).

Leaf nutrient content as an indicator of fertilizer requirement

Other factors that can affect leaf nutrient content include inherent soil characteristics, the type of cover management and fertilizer use, the planting material and the level of yield and system of exploitation. Taking all these into account, the Rubber Research Institute of Malaysia has published a range of nutrient contents that give an indication of fertilizer requirements (Table 8.16), although acknowledging (Pushparajah & Tan 1972) that seasonal variation may extend beyond the limits of sufficiency and deficiency levels (Table 8.17).

Table 8.16   *Range of percentage leaf nutrient content at optimum age in the canopy shade*

Nutrient	Clonal group*	Low[†]	Medium	High	Very high
N	1	<3.21	3.21–3.50	3.51–3.70	>3.70
	2	<3.31	3.31–3.70	3.71–3.90	>3.90
	3	<2.91	2.91–3.20	3.21–3.40	>3.40
K	I	<1.26	1.26–1.50	1.51–1.65	>1.65
	II	<1.36	1.36–1.65	1.66–1.85	>1.85
P		<0.20	0.20–0.25	0.26–0.27	>0.27
Mg		<0.21	0.21–0.25	0.26–0.29	>0.29
Mn (ppm)		<45	45–150	>150	

Sources: Pushparajah and K. T. Tan (1972); Soong (1981)
* For N: Group 1 clones are all clones except those in groups 2 and 3
　　　　Group 2 clones are RRIM 600, GT 1
　　　　Group 3 clones are all wind-susceptible clones, e.g. RRIM 501, RRIM
　　　　513, RRIM 605, RRIM 623, etc.
For K: Group I are all clones except those in group II
　　　　Group II are RRIM 600, PB 86, PB 5/51
[†] Low: well below optimum, tending to visual deficiency
Medium: sub-optimal
High and Very high: levels above which responses are unlikely

Table 8.17   *Seasonal variation in percentage leaf nutrient content*, compared with 'critical' levels*

Soil series	Clone, year of planting		N	P	K	Mg
Malacca	PB 86	1951	3.05–3.33	0.22–0.26	0.96–1.34	—
Serdang	PB 5/51	1962	2.98–3.40	—	1.35–1.91	0.26–0.34
Rengam	GT 1	1957	3.06–3.58	—	1.04–1.44	0.19–0.29
Jerangau	LCB 1320	1958	—	0.17–0.25	0.52–0.88	—
Selangor	Tjir 1	1954	3.43–3.72	0.23–0.27	1.19–1.49	0.28–0.33
Nutrient	level below which response likely:					
leaves in sun			3.20	0.19	1.00	0.23
leaves in shade			3.30	0.21	1.30	0.25
Nutrient	level above which response unlikely:					
leaves in sun			3.60	0.25	1.40	—
leaves in shade			3.70	0.27	1.50	0.28

Source: Pushparajah and K. T. Tan (1972)
* Nutrient content of leaf samples taken from 'control' plots of long-term fertilizer trials, 1964–70.

I Symptoms of nitrogen deficiency in rubber leaves (*Source*: V. M. Shorrocks)

II Symptoms of potassium deficiency (*Source*: V. M. Shorrocks)

III Symptoms of manganese deficiency (*Source*: V. M. Shorrocks)

IV Symptoms of magnesium deficiency (*Source*: V. M. Shorrocks)

V Healthy shoot recovery after transient zinc deficiency symptoms in nursery seedlings

VI White root disease, *Rigidoporus lignosus*, affected root (*Source*: Rubber Research Institute of Malaysia)

VII Red root disease, *Ganoderma philippii*, affected root (*Source*: Rubber Research Institute of Malaysia)

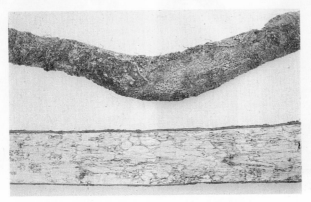

VIII Brown root disease, *Phellinus noxius*, affected root (*Source*: Rubber Research Institute of Malaysia)

IX *Ustulina* collar rot, *Ustulina deusta*, affected root (*Source*: Rubber Research Institute of Malaysia)

X Pink disease, *Corticium salmonicolor*, affected stem (*Source*: Rubber Research Institute of Malaysia)

XI Stem canker, *Phytophthora* sp., affected stem (*Source*: Rubber Research Institute of Malaysia)

XII Black stripe, *Phytophthora* sp., affected panel (*Source*: Rubber Research Institute of Malaysia)

XIII Mouldy rot, *Ceratocystis fimbriata*, affected panel (*Source*: Rubber Research Institute of Malaysia)

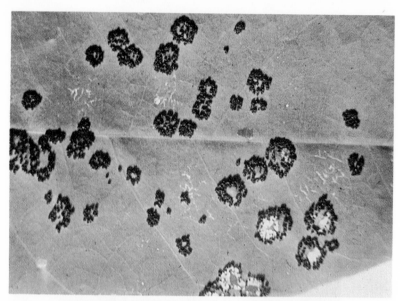

XIV South American leaf blight, *Microcyclus ulei*, affected leaf (*Source*: Rubber Research Institute of Malaysia)

XV Powdery mildew, *Oidium heveae*, affected leaves (*Source*: Rubber Research Institute of Malaysia)

XVI Fallen leaves affected by *Phytophthora* sp. (*Source*: Rubber Research Institute of Malaysia)

XVII Bird's eye spot, *Drechslera heveae*, affected leaves (*Source*: Rubber Research Institute of Malaysia)

XVIII Termites (*Source*: Rubber Research Institute of Malaysia)

XIX Cockchafer larva (*Source*: Rubber Research Institute of Malaysia)

Use of leaf nutrient ratios as an indicator of fertilizer requirement

Alternative methods of using leaf analysis have been explored, notably by Beaufils (1954, 1955, 1957) in Vietnam and Cambodia, who elaborated the use of nutrient ratios in leaf and latex as a guide to fertilizer requirement. Reviewed by de Geus (1973), this work was based on studies of rubber growing on the 'terre rouge' and 'gris' soils of Vietnam and Cambodia. Both soil types, and particularly the 'terre rouge', Oxisols derived from basalts, are deficient in K. 'Normal' levels and ratios of leaf nutrient content were defined, and fertilizer requirements judged on the basis of variations from those norms, diagrammatically represented in Figure 8.4.

Very good results were claimed following use of this method of diagnosis. Attempts to apply the system elsewhere have, however, been less successful and Fallows (1961) determined that the 'normal' values for leaf nutrient content in Cambodia were not applicable in Malaysia. The former reflect the low K and high Mg status of the Vietnamese/Cambodian soils as compared with those of Malaysia (Table 8.16), and lead to differences in the ratio values that indicate nutritional imbalance (Table 8.18).

Table 8.18   *A comparison of leaf nutrient ratios in Vietnam/Cambodia and Malaysia*

Vietnam/Cambodia[*]				
	N	P	K	Mg
'Normal' leaf nutrient levels %	3.40	0.22	0.90	0.40
	'Normal'		'Unbalanced'	
N/P	12.7–16.0		<10.8>18.0	
N/K	3.4–4.3		< 2.9> 4.8	
K/P	3.4–4.3		< 2.9> 4.8	
Malaysia[†]				
N/P	13.0–17.20		<11.0 >19.20	
N/K	1.9– 3.07		< 1.45> 3.61	
K/P	4.69– 7.07		< 3.50> 8.26	

Sources: [*]Beaufils (1957); [†]Fallows (1961)

It might be concluded that the success of the Beaufils' system is dependent on the existence of a gross nutrient deficiency, in this particular case of K, and that where nutrient supply is more in balance, even if at low levels, movements in nutrient ratio values are small and less useful in diagnosing fertilizer requirements.

The complexity of some of the approaches that have been attempted, despite possibilities for computerization (Rosenquist *et al.* 1976; Iyer *et al.* 1977), reflects the difficulty of analysing a biological situation such as that on a rubber plantation, where time

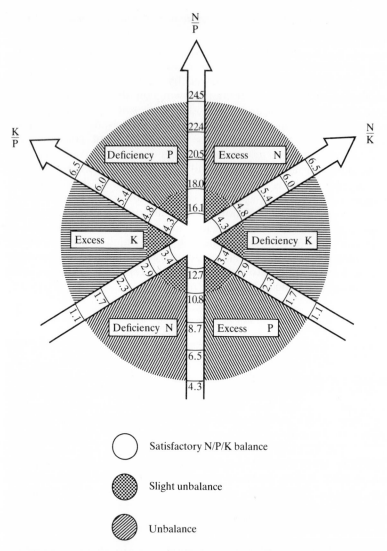

Fig. 8.4 Diagnosis of nutritional status of *Hevea* by reference to N/P/K balance of leaf (*Source:* Beaufils 1957)

of sampling and of fertilizer application, soil type, season, planting material and management can all affect the situation. The most generally satisfactory approach to the problem has been found in the development of joint soil and leaf diagnostic surveys, linked with the results of specific fertilizer trials on known soil types (Guha 1969; Pushparajah & Guha 1969).

8.6.2 Combined soil and leaf diagnosis of fertilizer requirements

Such an approach is used in surveying rubber plantations in Malaysia and elsewhere. Planted areas are first divided into approximately 20-hectare lots, as indicated by soil boundaries and planting material (Chan 1972; Chang *et al.* 1977). A composite soil sample is then taken, comprised of 10 random cores taken over the 0–15 cm and 15–45 cm depths. A composite leaf sample is taken from 30 randomly chosen trees in the same area, taking shade leaves in the case of mature rubber, and light leaves in young plantings. Both leaf and soil are analysed and, taking into account site history and past fertilizer treatment, the data are compared with the results obtained in formal fertilizer trials on the same soil type to arrive at a rational fertilizer recommendation for the area.

Data provided by the above approach have given a much better appreciation of the variation that is possible within a given planting, and led to improved efficiency in the use of fertilizers. A large proportion of Malaysian rubber is now subject to annual or biannual leaf sampling programmes, and the standard fertilizer programmes quoted in Table 8.11 may be modified in the light of the resulting data.

Diagnosis and correction of nutrient deficiencies

Although growing well on a wide variety of tropical soils, *Hevea* will respond sharply to any deficiency or excess of nutrient, in the process showing characteristic symptoms. These are, in general, related to the element in most critical supply, but on occasions a complex of symptoms may be present. This may happen, for instance, where rubber is ill-advisedly planted on deep, acid peat soil newly cleared from jungle. With drainage, these old swamp soils can shrink dramatically leading to root exposure and damage; this handicap, together with the low nutrient status of such soils and the unavailable nature of trace elements held in the organic matter, causes poor tree growth, while leaves show a variety of symptoms linked to deficiencies of Mn, Cu, Fe and the major elements. Under such circumstances correction is likely to be difficult and unprofitable, but where simple nutrient deficiencies on mineral soils are concerned, the proper use of fertilizers can quickly remedy the situation.

Detailed sand-culture work by Bolle-Jones (1954, 1956) and Shorrocks (1964) at the Rubber Research Institute of Malaya, established the characteristic leaf deficiency symptoms shown by both rubber and cover plants. These are illustrated in the book *Mineral Deficiencies in Hevea and Associated Cover Plants* (Shorrocks 1964). Subsequent work in the field has identified levels of fertilizer application required for their correction. The effect of mineral deficiencies on the ultrastructure of bark (Hamzah *et al.*

1976) and of leaf (Hamzah & Gomez 1981a) have been studied in sand-culture, but not yet in the field. Cover plants are convenient for experimental purposes, and work on their nutrition has at times led to a better understanding of aspects relevant to rubber; where appropriate this is commented on in the following paragraphs. Some of the deficiency symptoms in rubber described below are illustrated in Plates I–V.

8.6.3 Nitrogen deficiency

Rubber

On young, unbranched trees a shortage of N will first appear as a pale yellow-greening of the lower whorl leaves, becoming yellow and extending to younger leaves as the deficiency sharpens. On mature trees the yellowing will be most pronounced on leaves exposed to full sunlight. Unless corrected, N deficiency will lead to reduced leaf size and tree girthing, and eventually to stunting of the tree.

The symptoms will rarely be seen in properly managed plantings, particularly where leguminous ground covers are used. However, where young, newly established plants are affected by drought, or where intercropping is practised, there can be a shortage of N. More generally, the deficiency is likely to be seen in young rubber on poor soil infested with a competitive weed such as lalang (*Imperata cylindrica*), or an unchecked growth of bushes and ferns. In such cases, N fertilizer will only be fully effective if the weed is eradicated; subsequent fertilizer use will promote healthy leafing and canopy development.

Table 8.19 *Rates of fertilizer application required to correct acute nutrient deficiencies (g/tree)*

Nutrient in short supply	Fertilizer	Age of planting (years)		
		0–3	*3–5*	*More than 5*
N	Ammonium sulphate	110–220	220–450	900
	Urea	55–110	110–220	450
P	Rock phosphate (36% P_2O_5)	220–450	450–900	900–1350
K	Potassium chloride (60% K_2O)	110–220	224–450	450–900
Mg	Kieserite (26% MgO)	110	220	450–900
	Magnesium limestone (30% MgO)	220	450	900–1350

Source: after Shorrocks (1964)
Notes: (a) In cases of severe deficiency, the above applications may need to be repeated after 6 months. (b) The rates of application of urea are for clay soils, or where incorporation with the soil is possible. If this is not so, and particularly on sandy soils, the suggested rates should be increased by 50%. (c) In the case of soluble fertilizers, dressings of more than 450 g/tree should preferably be applied in two doses, separated by an interval of 2–4 weeks, to minimize possible loss by surface wash.

Two dressings of fertilizer applied within a period of six months should suffice to correct the situation (Table 8.19), and thereafter normal fertilizer dressings should be applied appropriate to the soil type concerned.

Cover plants

Nitrogen deficiency in leguminous cover plants will also be shown by leaf yellowing and stunting of the plant, and may be caused by a variety of factors. Drought and flooding, for instance, can both affect the activity of N-fixing rhizobia. The application of a 'starter' dose of 50 kg/ha of a general NPK fertilizer should stimulate cover growth and thereafter, given adequate rain and phosphatic fertilizers, normal growth should be resumed.

8.6.4 Phosphate deficiency

Rubber

Symptoms of P deficiency have only been clearly defined in sand culture. They consist of a bronzing on the undersurface of the leaf, and the leaf tip frequently dies back. The symptom has not been observed in mature rubber and is only infrequently found in young rubber. It must not be confused with the pale bronze tints sometimes found on the undersurface of very young leaves, nor with the yellowing and bronzing of mature leaves on lower canopy branches as they die and fall, or those present on trees that are dying from the effects of root disease.

Despite the absence of clear symptoms, P deficiency can be the cause of poor growth, but normal basal dressings of rock phosphate at planting and in subsequent fertilizer programmes should maintain a satisfactory P status. Where it is felt that the plants need a stimulus, possibly after a period of neglect, then the levels of fertilizer quoted in Table 8.19 should be applied.

Cover plants

In the case of legume covers, P deficiency will result in marked stunting of the plant and some defoliation. Leaves are very small and dark green, and the stems become very dark and sometimes dark red-purple in colour. Application of a 'starter' dose of NPK fertilizer, followed by regular application of rock phosphate should correct the situation.

8.6.5 Potassium deficiency

Rubber

The characteristic symptom of K deficiency is the development of a marginal and tip chlorosis followed by necrosis. This necrosis, and the absence of a definite herringbone pattern of yellowing, distinguishes K deficiency symptoms from those of Mg deficiency. On young unbranched trees the symptoms will appear first on the older whorls of the plant. In mature plantings, the condition is revealed by the appearance of a butter-yellow colour over the canopy.

K deficiency is most common on leached sandy alluvia and siliceous granite-derived soils. It may also be found on Oxisols derived from basaltic rocks, where K fixation can be a problem, and where any shortage can be accentuated by the drought conditions that can affect these very freely draining, well-aggregated, clay soils (van Diénst 1979).

Potassium deficiency may be corrected by application of one or more dressings of an appropriate fertilizer (Table 8.19) and on the poorer soils regular fertilizer application should consist of high K formulations.

Cover plants

In leguminous covers, K deficiency is marked by a characteristic marginal yellowing, first found on the older leaves. These symptoms can become apparent during the early stages of establishment, and serve as an indicator of low soil K status. Use of NPK fertilizer should eliminate the problem, and in most cases the steady improvement of soil conditions under the developing covers should prevent any reappearance of the symptoms.

8.6.6 Magnesium deficiency

Rubber

Magnesium deficiency is characterized by the development of an interveinal bright yellow chlorosis of the leaf. This chlorosis develops at the leaf margin and is often followed by interveinal and sometimes marginal necrosis. On young unbranched trees the symptoms first appear on the lower, older leaves, and in mature rubber on leaves fully exposed to sunlight. There is usually only little reduction in leaf size, but in extreme cases there can be defoliation and reduction in tree growth. One clone, PB 86, has a characteristically low leaf chlorophyll content, and is particularly prone to showing symptoms of Mg deficiency (Bolton & Shorrocks 1961).

The condition is common in replanted rubber and acute cases can be corrected by application of kieserite and subsequent use of magnesium limestone. As a prophylactic, on soils known to be deficient in the nutrient, dressings of magnesium limestone should be used.

Cover plants

In legume covers, symptoms of Mg deficiency first appear as an overall yellowing of the older leaves, with a tendency to be more pronounced in the interveinal areas leaving the mid-rib and main veins dark green. In severe cases, necrotic patches may develop in the interveinal areas. Correction may be by use of kieserite dressings, or more economically of magnesium limestone.

8.6.7 Calcium deficiency

Rubber

A deficiency of Ca is shown by the development of a tip and marginal, papery, white to light brown, leaf scorch. In young trees the symptom is observed on the younger leaves, and in severe cases the growing points may die back. In older trees, the symptom only appears on leaves low down in the shade of the canopy.

The more acid rubber-growing soils, with pH values of less than 4.5, have only low reserves of Ca, and it is not surprising that deficiency symptoms have been found quite frequently in Malaysian rubber, although so far there is no record of a direct response in growth to applications of Ca fertilizers. Regular use of rock phosphate, and of magnesium limestone where this is used, should prevent the appearance of the problem. Excessive use of limestone fertilizer must be avoided, however, for high levels of Ca may accumulate in the bark, as calcium oxalate crystals. This can sometimes occur on calcareous soils, with deleterious effect on latex formation and flow.

Cover plants

In leguminous cover plants, *Pueraria phaseoloides* shows an irregular patchy yellowing of the leaf, with development of an orange-bronze colour and necrotic patches. In *Calopogonium mucunoides*, yellowing is followed by tip and marginal necrosis, and in *Centrosema pubescens* the leaf tips wilt and fail to expand, resulting in a slightly cupped, deformed leaf with tip necrosis. Regular application of rock phosphate to the covers should prevent the appearance of such symptoms and maintain healthy growth.

8.6.8 Sulphur deficiency

No cases of S deficiency on rubber have been recorded, possibly because the element is likely to be applied regularly to the trees in the form of ammonium sulphate fertilizer. Symptoms developed in sand culture are of a gradual yellowing of the whole leaf followed by tip scorch. Leaf size is not reduced. In legume covers, leaves become pale green and later develop a marginal and tip necrosis.

8.6.9 Manganese deficiency and toxicity

Rubber

At the pH values typical of rubber-growing soils, 4.0–4.5, Mn is mobile and subject to leaching. On acidic, basalt-derived Oxisols with high Mn contents, however, uptake of the element may be excessive, and instances of Mn toxicity have been reported on such soils in Malaysia (Talib *et al.* 1984). Symptoms are similar to those of Fe deficiency, with leaves yellowish-green in colour and showing distinct green mid-rib and veins, and become apparent at levels of more than 500 ppm Mn in the leaf. When necessary, the condition can be corrected by liming to pH 5.0.

On poorly buffered, coarse, sandy alluvial soils, and on acidic granite-derived soils, Mn reserves are low and symptoms of deficiency are commonly found. They consist of an overall paling and yellowing of the leaf with bands of green tissue outlining the main rib and veins. No necrosis occurs and there is little reduction in leaf size. On young unbranched trees the symptoms first appear on the lower leaves but may extend to all leaves in severe cases. On older trees, deficiency symptoms are generally found on low, shaded leaves, but when the deficiency is severe symptoms will also appear on exposed leaves.

Field studies have shown that symptoms of Mn deficiency will appear in leaves containing less than 50 ppm Mn, while severe symptoms are associated with contents of approximately 20 ppm. Two annual applications of 100–200 g/tree of manganese sulphate, broadcast over the soil surface, are recommended to correct the condition (Shorrocks & Watson 1961).

Effect of manganese sulphate application on tree, latex and bark scrap

Results from one experiment on a granite-derived soil demonstrate the sensitivity with which nutrient application to poorly buffered rubber-growing soils can be followed. In this experiment, mixed clones were showing visual symptoms of Mn deficiency, and were given applications of 113 and 170 g/tree of manganese sulphate (Shorrocks & Watson 1961). A visual improvement in leaf colour was obvious at 5 months after application, and at 6 months a significant effect on leaf nutrient content was recorded. Increases in Mn content of bark, bark scrap and latex were also noted, while positive effects on exchangeable Mn in the soil persisted up to 3–4 years after application (Table 8.20).

Table 8.20 *Effect of manganese sulphate application on Mn content of leaf, latex, bark scrap and soil (ppm)*

Item	Treatment*	4/57 (F1)	10/57	5/58 (F2)	6/58	12/58	11/59	11/60	3/61	Total ppm	Exchangeable me %
Leaf Mn	Control	32	29	29	26	17	29	25	—		
	1	38	63	121	99	50	90	60	—		
	2	22	68	132	156	98	147	82	—		
	s.e. ±	11.1	8.2	21	11.7	16.0	13.4	16.4			
Latex Mn	Control	—	1.3	1.8	1.0	1.5	(12/59) 1.6	0.8	—		
	1	—	1.9	1.8	1.7	2.5	2.9	1.7	—		
	2	—	2.3	2.4	2.7	2.8	3.7	1.6	—		
	s.e. ±	—	0.26	0.22	0.47	0.07	0.20	0.09	—		
Bark Mn	Control	43	44	—	—	37	—	—	40		
	1	39	97	—	—	147	—	—	77		
	2	34	117	—	—	224	—	—	92		
	s.e. ±	8.6	9.7	—	—	23.3	—	—	7.8		
Bark scrap rubber Mn	Control	—	—	—	—	16	—	—	13		
	1	—	—	—	—	60	—	—	28		
	2	—	—	—	—	73	—	—	45		
Soil Mn 0–30 cm (Rengam series)	Control	—	—	—	—	—	—	—		18	0.0029
	1	—	—	—	—	—	—	—		33	0.0116
	2	—	—	—	—	—	—	—		42	0.0135
	s.e. ±	—	—	—	—	—	—	—		6	0.00193

Source: Shorrocks and Watson (1961). Data from experiment 1 with mixed clones

* Treatments 1 and 2: 130 and 170 g/tree of manganese sulphate (30% Mn) applied either side of planting strip at two times, F1 and F2.

There is a legal tolerance limit of 10 ppm on the amount of Mn that can be present in commercial rubber, for the element has a deleterious effect in promoting oxidation. At no time, however, in this experiment did the Mn content of latex cause concern, but that of bark scrap rubber rose well above tolerance limits. Such bark scrap is normally soaked and washed during milling; this should reduce Mn levels and so avoid any problem.

Cover plants

In the case of cover plants, Mn deficiency typically results in an overall yellowing of the leaf with the mid-rib and main veins outlined in healthy green tissues. This symptom is very like that of mild Fe deficiency, and analysis is required to distinguish between the two. Correction could be by use of a spray of Mn salts, but in severe cases it is likely that the cover would be abandoned with all efforts (and costs) directed at promoting good tree growth.

8.6.10 Iron deficiency

Rubber

Iron deficiency in rubber only occurs where extremely adverse soil conditions are limiting tree growth. Such situations occur on peat soils, humic and leached sands, and on coastal clays containing calcareous shell layers giving rise to alkaline conditions. The symptoms are generally accompanied by symptoms of Mn deficiency, and correction by fertilizer use, modification of soil conditions or by tree injection with Fe salts, is generally impracticable.

The first leaf symptom is of a general chlorosis resembling that of Mn deficiency. With increase in severity of the deficiency, the entire leaf assumes a pale yellow to white colour and there is reduction in leaf size. Young leaves and those exposed to sunlight are most affected.

Cover plants

In cover plants, early symptoms are like those of Mn deficiency, but in severe cases the younger leaves become very small, pale yellow to white. These symptoms can occasionally be seen when the plant is growing on acid, basalt-derived soils with high Mn status, where Mn uptake may be sufficiently high to interfere with Fe metabolism in the classic Mn/Fe antagonism.

8.6.11 Boron deficiency and toxicity

Rubber

There have been no clear cases of B deficiency in the field. In sand-culture, B-deficient leaves are distorted, reduced in size and somewhat brittle, but there is no loss of colour. On young unbranched plants, stem extension ceases and the terminal leaf whorls are produced without discrete internodes, giving a 'bottle-brush' effect. In severe conditions the apical meristem may die and axillary meristems will develop near the top of the stem. This 'bottle-brush' effect by itself is not a sure indication of B deficiency, for it is quite a common occurrence in Tjir I and GT 1 bud grafts in the field.

Cover plants

In cover plants, B deficiency is shown by stunting in growth and the production of short thick stems carrying small, misshapen leaves that are thick and brittle to the touch. Similar symptoms have been observed as a result of accidental application of 2,4,5-T or picloram.

Great care must be used before applying any B-containing fertilizers in the field, for the element can be highly toxic at anything more than trace element levels. Cases of toxicity following use of fertilizers containing B have been reported from both Sri Lanka and Malaysia. In Sri Lanka, levels of 204 ppm B were found in affected leaves compared with 49 ppm in healthy leaves (Jeevaratnam 1967; Rubber Research Institute of Sri Lanka 1978). Marginal necrosis occurred on otherwise healthy green leaves and some defoliation occurred. In Malaysia (Rubber Research Institute of Malaya 1964c), mistaken application of approximately 28 g/tree of borax to RRIM 513 growing on coarse granite-derived soil caused marginal necrotic scorch and raised leaf boron level to 240 ppm. Apical meristems became dormant, and even at 17 months after treatment growth was still checked and leaf B contents more than 100 ppm. Boron toxicity has also been confirmed in rubber growing on the coastal alluvial soils of Malaysia, and on eroded siliceous granite-derived soils when the rubber seemed to be absorbing B directly from weathering mineral tourmaline. Such cases are not common, but may be ameliorated by application of limestone to reduce solubility of the element.

8.6.12 Molybdenum deficiency

Rubber

Molybdenum is adsorbed strongly on to the soil at low pH levels, and can also be held in unavailable form together with sesquioxides, so that deficiency might be expected on the more acid, ferruginous Oxisols, and particularly those with depleted organic matter reserves. There have, however, been no reports of Mo deficiency in rubber in the field, although in Malaysia there have been a few cases where leaves showing symptoms associated with the deficiency have been found to contain very low levels of the element.

Molybdenum deficiency is characterized by the development of a very pale brown marginal leaf scorch, particularly towards the tip. In sand-culture this symptom has been particularly marked where N was supplied in nitrate form rather than as ammonium (Bolle-Jones 1957b), as might be expected from molybdenum's association with the nitrate reductase enzyme system. In young trees the symptom is found on mid- and lower-leaf whorls.

Cover plants

In sand-culture, deficiency symptoms develop only after growing for many months in the absence of Mo. Growth is only slightly reduced, with leaves of a pale green colour. Young leaves of *Pueraria phaseoloides* and *Centrosema pubescens* may develop a very pale brown papery tip scorch, and marginal-interveinal chlorosis can occur. With *Calopogonium mucunoides* the symptoms appear as a necrotic mottling over the whole lamina, followed by marginal necrosis and eventually defoliation.

While it appears that cover plants require very little Mo in order to maintain growth, the element is important in symbiotic N-fixing processes and as such could become a limiting factor on certain soils.

Effect of liming and molybdate application on nutrient status of covers

In pot culture trials with *P. phaseoloides*, on three inland soil types (Watson 1960), application of sodium molybdate at 1.1 kg/ha and liming to pH 6, increased the Mo and N content of the plants, and also increased dry weight production (Table 8.21). Relatively little response was found on a coastal alluvial clay soil in the same trial. In field trials (Watson *et al.* 1963) two spray applications of 1.1 kg/ha of sodium molybdate increased the Mo content of the covers, but no response in N content or dry matter production could

Table 8.21   *Effect of lime and molybdate application on leaf nutrient content of Pueraria phaseoloides: pot culture trials*

Soil pH	N(%) m_0	N(%) m_1	Ca(%) m_0	Ca(%) m_1	Mo(ppm) m_0	Mo(ppm) m_1	Dry weight leaf and stem (g/pot) m_0	Dry weight leaf and stem (g/pot) m_1
<5	2.77	2.97	0.35	0.29	0.50	0.58	96	117
6	3.66	3.68	1.30	1.28	0.53	0.83	141	137
7	3.74	3.66	1.61	1.62	0.78	3.02	111	121
Mean	3.39	3.44	1.09	1.06	0.60	1.48	116	125

Source: Watson (1960)
Notes: (a) Data meaned over results on 3 soil types (Kedah, Rengam and Malacca series) with initial pH and SiO_2/R_2O_3 values of 4.98, 4.74, 5.00 and 0.95, 0.84 and 0.46 respectively. (b) Treatments: m_1 application of sodium molybdate at 1.1 kg/ha, and soils limed to pH 6 and pH 7.

be determined. It is of interest, however, that the effect on Mo content could be detected at $3\frac{1}{2}$ years after application of the molybdate, and tended to be higher in the presence of rock phosphate application (Table 8.22).

Table 8.22 *Effect of molybdate and phosphate applications on molybdenum content of legume covers (ppm): field trials*

Time of sampling	Treatment	p_0	p_1	p_2	Mean
4 years after planting, $3\frac{1}{2}$ years after spray application of 1.1 kg/ha	m_0	0.23	0.18	0.33	0.25
					s.e. $\pm$ 0.391
of sodium molybdate to the covers (m_1)	m_1	3.88	5.31	7.66	5.61
Means		2.05	2.74	3.99	
		s.e. $\pm$ 0.679			

Source: Watson *et al.* (1963)
Notes: (a) Trial carried out on Munchong series soil (Tropeptic Haplorthox).
(b) p_0, p_1, p_2: rock phosphate applied at total levels of 0, 500 and 1000 kg/ha.

8.6.13 Zinc deficiency

Zinc deficiency has been reported from Brazil, Sri Lanka and Malaysia, all with young plants. In the upper stories, leaf laminae become very reduced in breadth relative to length. They may become twisted and the margins appear wavy or undulating. There is also a chlorosis of the leaf, similar to that observed in fairly severe cases of Mn deficiency, the mid-ribs and main veins remaining dark green in colour. Death of the apical meristem occurs under conditions of very severe deficiency, with side shoots developing.

In cover plants, Zn deficiency generally results in a marked reduction of growth and the formation of very small chlorotic leaves. The leaves rapidly develop marginal and spot necrosis. The situation is transient and no treatment should be required.

Effect of tree burning on soil pH and zinc deficiency
Zinc deficiency is only likely to appear under special circumstances. Zinc solubility in the soil decreases with rise in pH and is very low in the presence of free calcium carbonate. Such conditions arise in localized patches of ground affected by ash left from burning of jungle or old rubber trees (Ling & Mainstone 1983), where pH levels may be above 7.0 (Table 8.23). Leguminous covers or cocoa planted in such areas may thrive, but young rubber plants will quickly display symptoms of Zn deficiency. Also, if excessively high levels of P occur in the young plants following use of soluble phosphatic fertilizer, then Zn metabolism may be affected and symptoms

Table 8.23 Effect of ash from burned rubber trees, at replanting, on soil nutrient status*

Site and soil type	Months		pH	Available P (ppm)	Total K (me %)	Total Ca (me %)	Total Mg (me %)
1 Rengam	31	burnt	6.1 (5.8–6.4)[†]	48 (30–69)	0.47 (0.35–0.59)	3.73 (2.31–5.31)	—
		unburnt	4.7 (4.3–5.0)	10 (6–18)	0.26 (0.13–0.31)	1.02 (0.76–1.46)	—
2 Bungor	21	burnt	6.7 (6.2–7.2)	71 (52–93)	0.44 (0.36–0.56)	5.16 (2.29–7.48)	1.01 (0.62–1.46)
		unburnt	4.7 (4.3–5.1)	13 (10–19)	0.22 (0.15–0.23)	1.10 (0.67–1.46)	0.32 (0.14–0.62)
3 Rengam	3	burnt	6.9 (6.0–7.6)	92 (37–150)	1.82 (1.15–2.96)	6.62 (3.27–8.56)	1.85 (1.08–2.96)
		unburnt	4.3 (4.1–4.7)	13 (6–20)	0.40 (0.29–0.50)	1.08 (1.17–2.31)	0.59 (0.25–0.99)

Source: Ling and Mainstone (1983)
* Soils sampled at 0–15 cm, at times indicated, after felling, stacking and burning of old rubber trees.
† Data in brackets give the range of values encountered.

of deficiency appear (Middleton 1961). In both cases the condition is likely to be only transient, with the plants resuming normal growth as roots extend beyond the affected area or as the high pH and P levels become attenuated by rainfall and time. To hasten recovery, a spray application of 0.5 per cent concentration zinc sulphate solution may be helpful.

8.6.14 Copper deficiency

Beaufils (1957) records a positive yield response to copper sulphate on certain Vietnam/Cambodian soils. In general, however, Cu deficiency is unlikely to present any problem in rubber cultivation, except possibly in freshly cleared peat soils. The first sign of Cu deficiency, under sand-culture conditions, is wilting of the margins on the youngest leaves, with subsequent upward curling of the leaf tip. The marginal wilting develops into a very pale brown scorch which can spread down the lamina. If there is defoliation the apical growing point usually dies, and new shoots give rise to multiple branching. In severe cases the lamina on the new side shoots fail to expand properly, and the shoot may be covered with numerous dead and shrunken petioles.

In the case of cover plants, leaves are thin, become pale green in colour and develop a pale brown to white scorch at the tips. Like B deficiency, Cu deficiency affects the leaf shape but without the leaf becoming grossly distorted.

8.7 The effect of fertilizers on soil nutrient status

The considerable quantities of fertilizer that are applied over the life of a planting can significantly affect soil nutrient status, with effects enhanced by the localized nature of most applications. In immature rubber, for example, NPKMg fertilizers are generally applied at rates of 250–300 kg/ha of rubber over the first year. The zone of application around the young trees is, however, only about 0.006 ha so that the effective rate of application may be as high as 40 t/ha. In mature rubber, with perhaps 250 kg/ha of a complete fertilizer applied annually, but generally broadcast along the planting strips which cover one-third of the total area, the effective rate of application is 750 kg/ha/an. Over a 20-year period, total quantities applied may be equivalent to over 20 t/ha, and will represent a significant proportion of all nutrients in the nutrient cycle. There are obvious implications for the base-deficient and poorly buffered soils on which rubber is grown, and analytical

studies in lysimeter and field trials have been carried out to eluci-
date the situation.

8.7.1 Lysimeter trials

In uncropped lysimeter trials using low rates of fertilizers applied
at 1.7–3.4 kg/ha to a freely draining sandstone-derived Serdang
series soil (Typic Paleudult), Bolton (1961, 1963) showed that the
chlorine anion leaches faster than nitrate and sulphate anions, the
latter showing strong adsorption. Magnesium was more mobile than
either K or Ca. Rock phosphate remained largely unchanged in the
soil over a 26-week period, but some fixation of P in superphosphate
occurred. When superphosphate was applied together with lime
some reversion to less soluble calcium phosphate took place.
Manganese proved highly mobile at the pH level concerned (pH
4.8) and leaching was increased by both ammonium sulphate and
potassium chloride.

In later lysimeter trials with the same soil, but cropping with
rubber seedlings and using much higher rates of application equivalent to the effective rates used during immaturity, Soong (1973)
showed that both ammonium and nitrate ions were leached; that
nitrate anions appeared in the leachates within two weeks of application of urea fertilizer; and that application of potassium chloride
increased leaching of Ca and Mg in the absence of N fertilizer. The
general effect of nitrogenous fertilizer was to increase the leaching
of K by 62 per cent, Ca by 300 per cent and Mg by 300 per cent,
compared with non-fertilized lysimeters, with associated depressions
in pH values. Leaching losses will vary with soil type, being particularly high on sandy soils, with N losses being greater with
ammonium sulphate and nitrate than with urea (Pushparajah *et al.*
1977b). The extent of loss is also more affected by intensity of rainfall than by total volume.

8.7.2 Field effects

Effect of N fertilizers on soil pH and nutrient status

In field trials with rubber, where the soil remains undisturbed for
many years, it is possible to follow the effects of fertilizer application in some detail. A decrease in pH and loss of cations has been
recorded in immature rubber after only six years of ammonium
sulphate application (Bolton 1964a). In another field experiment
with mature rubber, sited on alluvial, coarse, sandy loam (Bolton
1961), eight annual applications of 1.4 kg/tree of ammonium
sulphate broadcast down the centre of the inter-row caused a
decrease in pH from 5.34 to 4.41, and of exchangeable Ca from
0.04 me% down to 0.02 me% (Fig. 8.5).

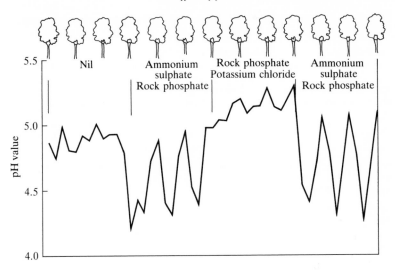

Notes:

(a) ammonium sulphate 1.4 kg/tree, rock phosphate 0.5 kg/tree, potassium chloride 0.3 kg/tree, applied annually for 8 years to inter-row areas.

(b) points at 150 cm intervals across 6 m wide inter-row areas, sampled to 30 cm depth.

Fig. 8.5 Effect of fertilizer application on soil pH (*Source:* Bolton 1961)

In a larger study of seven field experiments carried out on a variety of soil types, and involving fertilizer applications over periods ranging from 6–14 years, detailed soil sampling was carried out along the tree rows where fertilizer had been applied (Pushparajah *et al.* 1976b). Results varied according to soil type, but in general the continued use of ammonium sulphate reduced soil pH and the amount of exchangeable bases in the soil, with measurable effects down to 45 cm. Urea application showed only minor effects. Typically the application of rock phosphate or limestone resulted in an increase in pH, exchangeable Ca, and tended to counteract the effect of ammonium sulphate (Table 8.24). Most soils involved in this particular study showed some retention of K. For kaolinitic soils the build-up was mainly in the exchangeable form but there was indication of K fixation in soils with 2:1 clay and micas.

Residual effects of P fertilizer on soil nutrient status

In the nutrient cycle of the rubber plantation, P is unique in that its level in the soil is likely to increase rather than decrease. The continued use of P in the experiments described above led to considerable build-up of total and available P in both surface soil

Table 8.24 *Effect on soil nutrient status of fertilizer treatment applied over eight years*

Treatment	pH	Total P (ppm)	Available P (ppm)	Exch. Ca (me %)	Total K (me %)	Exch. K (me %)	Exch. Mg (me %)
n_0	4.8	—	—	0.50	—	0.14	0.24
n_1	4.4	—	—	0.09	—	0.11	0.10
n_2	4.2	—	—	0.13	—	0.12	0.08
s.e. ±	0.03			0.03		0.011	0.014
p_0	4.4	121	11	0.13	—	—	—
p_1	4.5	210	38	0.27	—	—	—
p_2	4.5	309	108	0.42	—	—	—
s.e. ±	0.03	18	8	0.03	—	—	—
n_0p_0	4.8	123	6	0.23	—	—	—
n_0p_2	4.9	324	149	0.76	—	—	—
n_2	4.3	130	6	0.07	—	—	—
n_2p_2	4.3	332	94	0.20	—	—	—
s.e. ±	0.05	31	14	0.21	—	—	—
k_0	—	—	—	—	2.52	0.09	0.12
k_1	—	—	—	—	2.75	0.12	0.14
k_2	—	—	—	—	3.71	0.15	0.17
s.e. ±					0.42	0.011	0.014

Source: Pushparajah *et al.* (1976b)
Notes: (a) Total P obtained by 1:1 $H_2SO_4/HClO_4$ extraction, total K by $6N$ HCl, and available P by $0.03N$ NH_4F/HCl. (b) Experimental detail: experiment no. 28/3, on Seremban series soil (Typic Paleudult), unit level of fertilizer used over 8 years of application (kg/effective ha): ammonium sulphate 5525, rock phosphate 4200, potassium chloride 1440.

and at lower depths, with accumulation of P in the silt and clay fractions. Exhaustive cropping in pot culture with *Pueraria phaseoloides* showed that a large proportion of the residual P as determined by NH_4F/HCl extraction was available to plants.

The levels of available soil phosphate found in the above study were generally much higher than those considered as limiting for rubber, and suggest possibilities for economizing in the use of phosphatic fertilizers. This is confirmed by data from a field trial in which a new stand of rubber was established in the planting rows of an old fertilizer experiment at the time of replanting (Pushparajah *et al.* 1977a). Brief details are given in Table 8.25. At 8 years after planting, and 14 years after the last applications of P to the former planting, the girth, yield and leaf nutrient content of the young trees reflected a continued beneficial effect of the earlier phosphate fertilizer.

Additional long-term data are available from a NPK fertilizer trial laid down in 1935 on a heavy inland clay (Mainstone 1963b). Trees that received a double level of rock phosphate in 1935–41

Table 8.25 *Residual effects of phosphate fertilizer on rubber growth, yield and leaf nutrient content*

Treatment	Girth (cm)	Yield dry rubber (kg/ha)	Leaf P (%)	Leaf Ca (%)
Control	47.8	1,592	0.21	0.87
'Residual P'	51.9	1,889	0.23	0.97
'Fresh and Residual P'	54.9	1,869	0.27	1.12

Source: Pushparajah *et al.* (1977a)
Notes: (a) Data taken in 1977 from experiment SE 1/4, sited in a 1969 replanting on Serdang series soil (Typic Paleudult). (b) 'Residual P' plots received 2030 kg/effective ha P_2O_5 over years 1937–1963, as double superphosphate and rock phosphate. (c) 'Fresh and Residual P' plots received an additional 2561 kg/ha P_2O_5 after replanting.

gave in 1959–61 approximately twice the latex yield of trees receiving no P; they also displayed higher P content of leaf and latex, increased height, better girth and superior bark renewal. N and K fertilizer exerted little residual effect.

In replantings, where legume covers have been grown and the recommended fertilizer inputs supplied, the high N and P levels built up during immaturity can be expected to benefit the rubber during the first few years of tapping. Fertilizer applications can then be limited to those indicated as necessary by leaf analysis (Pushparajah *et al.* 1983; Sivanadyan 1983). In those areas where inputs during immaturity were less than optimum, a full fertilizer regime as indicated in Table 8.11 will need to be implemented.

8.8 Fertilizer quality and methods of application

Fertilizers are a significant cost item in the plantation budget (Tables 7.12, 7.14), and for economy care is required in their selection and use. They can consist of 'straight' fertilizers, simple salts supplying one or two of the major nutrients (Table 8.26), that may be applied on their own or in mixture with others. Where there is a difference in particle size, shape and specific gravity, such mixtures may be prone to segregation and caking due to crystal bridging. Preferably the different components should have uniform particle size, when proper mixing will produce an acceptable blend.

Alternatively, granulated mixtures may be used that are physical mixtures of two or more straight fertilizers processed into granular forms. More generally, the granulated formulations in international trade consist of compound fertilizers manufactured by chemical reaction. In Malaysia, hot liquid ammonium nitrate is reacted with

Table 8.26   *Composition of some common fertilizers, as percentages*

	N	P_2O_5	K_2O	CaO	MgO	SO_3	Equivalent* acidity
Ammonium sulphate	21	—	—	—	—	59	110
Ammonium nitrate	33–35	—	—	—	—	—	60
Urea	45–46	—	—	—	—	—	84
Rock phosphate	—	36	—	50	—	—	—
Superphosphate, single	—	14–22	—	24–31	—	25–32	—
Superphosphate, double	—	42–50	—	17–23	—	3.5	—
Monoammonium phosphate	11	49	—	1.5	—	6	65
Potassium chloride (muriate of potash)	—	—	48–62	0–3	0–3	0–7	—
Potassium sulphate	—	—	48–52	0–3	0–2	39–48	—
Potassium magnesium sulphate	—	—	21–30	0–7	6–20	32–56	—
Magnesium sulphate (kieserite)	—	—	—	—	26	30	—
Magnesium limestone (dolomitic limestone)	—	—	—	30–36	20	—	—
Magnesium sulphate (Epsom salts)	—	—	—	—	16	58	—
Multiplication factor to convert oxides to absolute nutrient values		0.437	0.83	0.71	0.60	0.40	—

* Equivalent acidity is the number of parts by weight of calcium carbonate required to neutralize the acidity resulting from the use of 100 parts of the fertilizer material, e.g. it takes 110 kg of calcium carbonate to neutralize the acidity developed in the soil by the use of 100 kg of ammonium sulphate.

dry powdered rock phosphate, potassium chloride and kieserite to produce a range of products with varying NPK composition. Other compounds are manufactured by reacting liquid nitric acid and phosphoric acid with ammonia, dry powdered potassium chloride and kieserite, followed by granulation. Such products are easy to handle, and when applied to the soil give an even distribution of the individual nutrients.

8.8.1 Nitrogen fertilizers

Urea

Urea, NH_2CONH_2, with 45 per cent N, is the cheapest form of N fertilizer, and is often the most easily available, in prill form, as a by-product of petro-chemical operations. It is very water soluble and after surface application is readily washed into the ground where nitrification takes place. The material is not quite as acidifying as ammonium sulphate, but regular application will tend to reduce soil pH in time. In Malaysia, work by the International Fertilizer Development Center (1983) has resulted in the development of a urea-based NPK granular fertilizer, specially designed for aerial application.

Loss of N by volatilization

The prime use of urea is, however, in the rice crop. In plantation crops it has one major disadvantage, in that on application to damp soil it will quickly hydrolyse by enzymatic action to give ammonium carbonate (Pushparajah *et al.* 1982). This salt dissociates and loses ammonia rapidly by volatilization. Sri Lankan work (Silva & Perera 1971; Yogaratnam & Perera 1981) has shown that the urease activity is positively correlated with soil pH, and urea is not recommended for use on soils of high pH (>5.5). Tests in Malaysia suggest that urea can generally be applied safely to clay soil, for adsorption is rapid with little ammonia loss to the atmosphere. Applied to sandy soils with cation-exchange capacities of less than 10 me%, however, or to cover plants in immature rubber or leaf litter in mature rubber, losses can range up to 40 per cent of the applied N, with the bulk of loss taking place in the first four days after application (Watson *et al.* 1962; Chan & Chew 1983). In field trials on a strong-textured Jerangau series soil (Typic Paleudult), urea was less efficient than either ammonium nitrate or ammonium sulphate, annual treatment levels of 1800 g N per tree being required to match the effect of 900 g N per tree applied as ammonium sulphate (Rubber Research Institute of Malaysia 1978b).

In general, urea can be recommended safely for rubber growing on clay soils, but on light soils rates of application need to be increased to compensate for the likely loss of N. In Indonesia, an increase of 15 per cent is recommended, but increases of 30–50 per cent might be more realistic. Unless sufficient labour is available to fork the urea into the ground it should preferably be applied only when rain is expected, in the expectation that the fertilizer will then be washed into the soil with little loss.

Ammonium sulphate

Ammonium sulphate traditionally has been the main source of N used in rubber. It is an effective source of N and also supplies sufficient S to prevent deficiency of this element developing on the more infertile tropical soils. In crystal form it blends easily with rock phosphate, muriate of potash and other powder materials to provide mixtures that can be formulated in relatively small batches appropriate to local requirements.

Ammonium sulphate can have a significant acidifying effect on the soil (Table 8.24). On well-buffered soils with pH above 5.0 this may not be a serious consideration, but on those with pH values down to 4.0 the use of ammonium sulphate will lead to further depletion of bases by leaching, and to deficiencies of Mn and Mo on some soils and of Mn toxicity on others. Use of rock phosphate will tend to counter these effects, but in the more severe situations corrective dressings of magnesium limestone may be necessary.

Ammonium nitrate

Ammonium nitrate is used quite commonly in rubber cultivation, but mainly as a component of granular compound fertilizers. Limited trials work has shown that the material is more effective than urea and roughly comparable with ammonium sulphate. Its high N content makes it an economic fertilizer to use but leaching losses of nitrate-N are likely to be higher than with either ammonium sulphate or urea. This would suggest that, when applied on its own, several light dressings might be more effective than the equivalent dose given in one application.

In temperate agriculture, application of ammonium nitrate to soil with moisture content at or near to field capacity can result in losses of 10–30 per cent of the applied N due to denitrification (Ryden 1984). This can also happen on rubber-growing soils, particularly those with high levels of mineralizable organic matter (Tan & Bremner 1982), and at times of high water table, use of nitrate fertilizers on such soils should be discouraged.

Calcium ammonium nitrate (CAN) is not used in rubber because of its cost and, possibly, also because of a fear that its calcium content might cause nutrient imbalance in the tree.

8.8.2 Phosphatic fertilizers

Rock phosphate

Rock phosphates occur naturally in many countries as calcium fluoro-apatites of the general formula $Ca_3F(PO_4)_2$ (Rubber Research Institute Malaya 1972a; Emsley 1977) and are applied on their own, or in powder mixes with other straight fertilizers. They are water-insoluble, but the P content becomes slowly available to plant uptake on soils with pH values of less than 6.0. The availability of their P content is generally measured in terms of citric acid solubility, and can vary depending upon origin of the rock, its softness and fineness of grinding. Malaysian standards (Pushparajah 1979; Lau 1981) demand a minimum of 30 per cent P_2O_5 by weight with 25 per cent of this being soluble in citric acid, and not less than 95 per cent by weight passing through a sieve of 500 μm (BS Mesh 30) square apertures.

In general, on the typical acid rubber-growing soil, pot and field trials have shown that rock phosphate meeting the above standard will be as effective in supplying P as the more soluble phosphate fertilizers, although the latter may be preferred in the establishment of young plants (Mahmud 1978).

Soluble sources of phosphate

These include single, double and triple superphosphates, the first produced by reacting rock phosphate with sulphuric acid, and the others by treatment with mixtures of phosphoric and sulphuric acids. In addition, there are the P-containing complete fertilizers obtained by reacting rock phosphate with phosphorous and nitric acids in the presence of K and Mg fertilizers. The higher nutrient content of these materials, their standard composition and free-flowing characteristics, offer advantages over straight mixtures, in terms of precision, convenience and handling.

In all cases the P content is readily available for plant uptake on application to the soil, but fixation of P by sesquioxides in the more ferruginous tropical soils can quickly reduce availability. Locally within the soil an equilibrium will be developed between the applied source of P, P fixed in the soil, and that left available for plant uptake. Regular application of either soluble or rock phosphate will help to sustain this availability, as can the maintenance of high organic matter levels particularly by use of leguminous covers.

Soluble sources of P are likely to be more expensive than rock phosphate but they are of value where a stimulant is required, provided that excessive levels of P in the plant are avoided.

8.8.3 Potassium fertilizers

The most used K fertilizers are the naturally occurring potassium chlorides and potassium sulphate, and the double salt, potassium magnesium sulphate (Table 8.26). Potassium chloride (muriate of potash) is the cheapest and most widely used, but rapid leaching of the chlorine anion can enhance leaching of Ca and Mg. On these grounds potassium sulphate would be preferred to the chloride, as the sulphate is more strongly adsorbed and retained by the soil, but its extra cost is a disincentive to use.

8.8.4 Magnesium fertilizer

In case of Mg deficiency, magnesium sulphate, in the form of kieserite mineral, is the preferred fertilizer for it is immediately available for plant uptake. It is expensive, however, and for prophylactic purposes the less expensive dolomitic limestones would be better. Their Mg content is not so quickly available to the plant, but breakdown in the more acid tropical soils is quite rapid and, as well as providing a longer-lasting source of Mg, the Ca content is of benefit. As with the rock phosphates, actual composition will vary according to origin, and effectiveness is dependent on fineness of grinding. The limestone is generally applied on its own, for in mixture with urea and ammonium sulphate the alkaline reaction would cause a loss of ammonia.

8.9 Methods of fertilizer application and formulation

In rubber cultivation, high rainfall and undulating terrain can lead to fertilizer losses by leaching and surface wash, while fixation of P by sequioxides can further reduce the effectiveness of applied fertilizer. Various methods of application and formulation have been tested in order to minimize these factors.

8.9.1 Methods and time of application

Possibly the most effective fertilizer application of all is that of phosphate fertilizer applied in the planting hole at the time of establishment. Rock phosphate is the most generally used material, and is mixed intimately in powder form with the soil, both to aid dissolution and effect maximum contact with the plant roots. In subsequent applications to the young rubber, early policy was to 'pocket' the phosphate in holes dug round the planting point, with the object of minimizing soil contact and presumed fixation. Trials, however, showed that little or no advantage was gained thereby, and when the phosphate is applied on its own, preference is given to broadcasting it over the cover plants in the inter-row area (Push-parajah & Chellapah 1969). Uptake by the covers, and subsequent retention in the organic form, ensures long-continued availability of the nutrient to surface rubber roots.

Complete fertilizers are applied broadcast over the soil surface, initially in the immediate proximity of the young plants, and progressively thereafter along the planting strips and out into the inter-row area. Eventual broadcasting over the whole surface is the norm, subject to labour availability and convenience. Broadcasting by machine spreader and even by air (Rajaratnam 1979) has been tested, but has not yet become general.

In nurseries and young rubber, fertilizers are applied according to the schedules outlined in Tables 8.8, 8.9, and 8.10, with little regard to weather conditions, in order to ensure steady, uninterrupted growth. In mature rubber, however, fertilizer applications are timed to avoid periods of maximum rainfall, and the main application of the year should be at the time of refoliation after wintering, when nutrient uptake is active. In high rainfall areas it is sensible to split annual doses into at least two separate applications.

8.10 Fertilizer formulation

8.10.1 Traditional and slow-release formulations

Traditionally, fertilizers have been applied as 'straight' N, P or K fertilizer in crystalline or powder form, proprietary mixtures of these straights, or as granulated compounds. Their general characteristics have been discussed in section 8.8. There have been attempts from time to time to improve the efficacy of fertilizers by encapsulation in plastic (Soong *et al.* 1976), or in permeable sachets, by coating urea with S, or using nitrification inhibitors in order to slow down mineralization of the ammonium nitrogen content (Push-parajah *et al.* 1982; Tan 1982). These methods are, however, more applicable to horticulture where returns may be high enough to justify the additional cost, than to the plantation industry.

More recently, trials have been carried out with rubber-based and rubber-encapsulated slow-release fertilizers (Soong & Yap 1976, Yeoh & Soong 1977). Promising results were obtained, and at times of low rubber prices this technique might find application. Of more immediate interest is the utilization of factory waste effluents as fertilizer sources.

8.10.2 Factory waste effluents as fertilizer

Effluent types

In processing of rubber large amounts of water are used for washing, cleaning and dilution, with some 22 litres of effluent requiring disposal per kg of dry rubber produced. Discharge of such effluent can have a deleterious effect on local water courses, and in the major rubber-growing areas this has led to effluent control legislation. The industry's response has been to consider possibilities for utilizing such wastes (Collier & Chick 1977; John *et al.* 1977).

In the Malaysian situation, three major types of effluent are identified, originating respectively from factories processing latex concentrate, block rubbers and sheet rubber. The composition of the waste is dependent on the type of process employed, but generally it consists of processed water, small amounts of uncoagulated latex and latex serum containing quantities of proteins, sugars, lipids, carotenoids, inorganic and organic salts. Discharge from concentrate factories may consist of skim and combined discharge effluent (skim plus factory washings), while the bowl sludge remaining in the centrifuges after latex concentration is a further waste product. These former wastes contain significant quantities of nutrients, particularly of N and K (Table 8.27), and various attempts have been made to exploit this hitherto neglected resource.

Table 8.27   *Analysis of rubber factory effluent and wastes*

| | Type of effluent* | | | | |
	Skim	Combined discharge	Block rubber	RSS	Bowl sludge
pH	4.7	5.3	5.6	4.9	—
Suspended solids	1206	1133	540	165	—
Total solids	8365	3672	1410	2650	—
Chemical oxygen demand	8735	5566	2140	3280	—
Biological oxygen demand	5523	3410	1130	2615	—
Ammoniacal N	610	415	60	10	—
Albuminoid N	178	62	30	100	—
Total N	952	512	100	115	2.6
Ca	39	72	—	Trace	0.03
K	724	390	—	63	5.9
Mg	80	47	—	5	7.0
Na	16	8	—	6	—
P	99	58	—	8	13.3
Fe	7	5	—	Trace	—

Sources: John *et al.* (1977); Tayeb (1978)
* Effluent data in ppm, that for bowl sludge as %.

Possible uses for effluents

Bacteria, yeasts and edible fungi have all been cultured successfully on block rubber serum, and *Chlorella* spp. algae can be grown in digested block rubber and latex concentrate factory effluent (John *et al.* 1977). The freshwater fish, *Tricipodus leeri*, *Clarias* sp. and *Tilapia* have been successfully grown in the stabilizing ponds used in treating block rubber effluent (see p. 494) feeding on green algae, but no commercial development has been reported. More successful have been attempts to use the waste materials as sources of fertilizer.

Fodder production

In many rubber-growing areas cattle provide a secondary enterprise for the local farmers, but must rely on marginal areas of grass for fodder. There are indications from peninsular Malaysia that allocation of specific areas for pasture, and their irrigation with factory effluent, could significantly upgrade productivity of this sector.

Both Napier grass (*Pennisetum purpureum*) and star grass (*Cynodon plectostachys*) have responded well to surface irrigation with effluent. Typical results were obtained in one trial (H. T. Tan *et al.* 1976), where application of a mixture of effluents from latex concentrate, cup lump and skim processing plants in the ratios of

Table 8.28 *Effect of application of NPK fertilizer and two types of mixed effluent on the yield of Napier grass (kg/ha/harvest)*

Fertilizer treatment	Ratio * 6:4:1			Ratio *8:4:3		
	Green forage	Dry matter	Crude protein	Green forage	Dry matter	Crude protein
No treatment	9,784	2,113	158	2,509	497	60
NPK, 17:18:17, fertilizer at 6420 kg/ha/an.	25,590	4,683	656	22,328	3,260	535
Effluent:†						
level 1	52,183	6,210	1,105	28,099	2,950	620
level 2	52,183	5,270	1,070	26,342	2,713	577
level 3	46,413	6,127	1,157	27,346	3,008	650

Source: H. T. Tan *et al.* (1976)
* Ratio of effluents derived from latex concentrate, cup lump and skim processing plants (see text).
† Three levels of application:

	Ratio 6:4:1	Ratio 8:4:3
level 1	37. 5 mm r.e.m.	25 mm r.e.m.
level 2	75.0 mm r.e.m.	50 mm r.e.m.
level 3	112.5 mm r.e.m.	75 mm r.e.m.
Harvesting interval	6 weeks	4 weeks

(r.e.m. = rain equivalent per month)

6:4:1 or 8:4:3, gave higher production of total dry matter and crude portion than when 6400 kg/ha of a 17:18:17 NPK fertilizer was applied (Table 8.28). Cattle fed on treated pasture showed no ill effects and liveweight gain was 15 per cent higher than with untreated pasture.

In applied work on estates, surface irrigation of pasture with 6000–7000 litres/ha of skim serum applied at 45-day intervals after each cut, supplying the equivalent of some 5 t/ha of ammonium sulphate, gave dry fodder yields of up to 35 t/ha/an.

Bowl sludge may also be used to advantage as a fertilizer for fodder production. Applied to pastures of *Brachiaria mutica* and *Axonopus compressus* it has increased N content of the grass, and when applied at 2.5 t/ha doubled production of fodder grass with projected green yield of 220 t/ha/an.(Pushparajah *et al.* 1975).

Food crops

Pot trials have shown that anaerobically digested effluent might be a suitable fertilizer for irrigated padi (Tayeb 1978). Bowl sludge has also been shown, in pot culture, to be as effective a source of P as superphosphate in growing maize, soybean and groundnut, but no commercial applications of these findings have been reported.

Oil-palm

Oil-palm is often grown in areas adjacent to rubber and because of its high nutrient requirement might offer a profitable disposal site for effluent. In one Malaysian experiment H. T. Tan *et al.* 1976), furrow irrigation with mixed effluent applied at rates of up to 18.75 mm rainfall equivalent per month, supplying about 24.5 kg/an. N per palm, increased the number and weight of fruit bunches. Over an 8-year period a mean increase of about 17 per cent of fruit bunches per annum was recorded (Lim & P'ng 1984). Overhead irrigation had no harmful effects, and circle irrigation of young palms with skim serum showed a net profit of $M140/ha/an. when compared with normal NPK fertilizer use.

Rubber

The profitable use of factory waste and effluent within the associated rubber plantings deserves particular consideration, for in outlying districts it could help to minimize dependence on imported inputs. Trials have shown, for example, that bowl sludge can be used effectively as a planting hole application in place of rock phosphate, although normal NPKMg surface dressings are still required for satisfactory growth (Table 8.29).

In experimental application of skim serum to irrigate mature rubber, using rates of about 20 litres of serum per tree, one Malaysian estate obtained yield increases of up to 25 per cent. The work

Table 8.29 *Effect of bowl sludge on growth of PB 5/51 selfed seedlings*

Treatment *	Height (cm)[†]	Girth (cm)[†]
Control	76.6	2.82
No basal P + NPKMg	110.4	3.72
Basal CIRP	103.2	3.88
Basal CIRP + NPKMg	141.5	5.23
Basal bowl sludge	105.3	3.66
Basal bowl sludge + NPKMg	153.2	5.22

Source: John *et al.* (1977)
* Basal treatments applied in planting hole, NPKMg applied broadcast on surface.
[†] Height and girth measured at 10 months after planting.

was discontinued, however, because of increased wind damage losses, possibly occasioned by the high N contribution of the serum. In later observation trials (Pillai 1977), mixed effluent from concentrate factory washings, block rubber and skim serum, in the proportions 8:4:3, was used at rates of 12.5 and 37.5 mm rainfall equivalent per month to furrow-irrigate PB 86 and RRIM 513. Over a period of $2\frac{1}{2}$ years, the monthly irrigation regardless of season increased yields by up to 75 per cent, the increases being particularly marked in the presence of 'Ethrel' stimulation although possibly at some cost to girth increment (Table 8.30). Over a period of 8 years, overall increases due to effluent application totalled 11 per cent and 19 per cent for the two clones respectively (Lim & P'ng 1984).

Table 8.30 *Effect of irrigation with effluent on response to yield stimulation*

Treatment	Yield (kg/ha/an.) * PB 86	RRIM 513	Girth increment (cm) * PB 86	RRIM 513
No effluent, no stimulant	2091	—	3.6	—
No effluent + Stimulex[†]	2519	1908	3.4	2.7
12.5 mm r.e.m. + Stimulex	2670	2595	2.8	1.9
37.5 mm r.e.m. + Stimulex	3098	2614	2.4	1.9
No effluent + Ethrel	2872	2366	2.8	2.2
12.5 mm r.e.m. + Ethrel	2947	3301	2.3	1.8
37.5 mm r.e.m. + Ethrel	3728	3339	2.7	1.9

Source: Pillai (1977)
* Data recorded over period June 1974 to November 1976, effluent applied to single plots of 1.6 ha.
[†] Stimulex – a formulation containing 2,4-D.

In one particular area affected by a moisture deficit of 19 cm over 3 months, it has been calculated (John *et al.* 1977) that the drought would have been overcome by the application of 190,000 litres/ha of water overall, or 4500 litres per hectare of rubber planting strip per application, over perhaps 10 rounds of application. To use factory effluent for such a purpose might not be suitable for plantings on soils with impeded drainage, where waterlogging could be an unwelcome consequence. However, on open-textured freely draining soils, and on droughted areas, particularly in the marginal areas referred to in Chapter 4, the use of factory effluent for irrigation would seem particularly appropriate (Tayeb *et al.* 1979, 1982). Reservations have, however, been expressed with regard to the cost of distributing effluent over any area remote from the factory source (Yeow & Yeop 1983), and nearby oil-palm plantings would always deserve preference on account of their greater nutrient requirement (Lim & P'ng 1984). So far no harmful effects have been noted on soil characteristics, but the possible appearance of nitrate-N in drainage water needs further study (Tayeb *et al.* 1982).

Chapter 9

Exploitation of the rubber tree

E. C. Paardekooper

9.1 Conventional tapping systems

The rubber-containing latex, which, as described in Chapter 2, is produced in latex vessels within the phloem tissue of the bark of the tree, was traditionally harvested from wild rubber in South America by slashing the trees with an axe or machete, fresh cuts being made at each tapping. This incision method of tapping commonly resulted in severe injury to the tree and gave poor yields in the long term. As explained in Chapter 1, the development of modern methods of tapping followed the invention of excision tapping in 1889 by Ridley, who showed that *Hevea* can be continuously exploited by removing at regular intervals a thin shaving of tissue from the surface of a sloping groove made into the bark (Ridley 1897). The shavings remove the cut ends of latex vessels that are plugged with coagulated latex from a previous tapping, thus reopening them and permitting a fresh flow of latex from the vessels down the sloping cut and into a container. Ridley also demonstrated that with excision tapping the bark was regenerated above the cut and could be tapped again after several years.

A multiplicity of tapping systems have been employed in an effort to maximize yield or optimize profit. The type of system used is influenced by a number of factors, including the cultivar grown, the age of the trees, the rainfall regime, the availability of skilled tappers and local wage agreements. This section only attempts to mention the basic types of system employed on estates without the aid of yield stimulants. Later in the chapter, after a discussion of the nature and mode of action of stimulants, exploitation systems involving the use of the safe and cost-effective chemical ethephon are briefly described. Tapping systems for smallholders are described in Chapter 12. The international notation used to designate exploitation systems is summarized in Appendix 9.1 (p. 544), which the reader unfamiliar with the subject should read in conjunction with this chapter.

9.1.2 Single low-cut systems

The commonest system used for budded rubber, known as the half spiral alternate daily system and designated as $\frac{1}{2}$ S d/2 by the international notation, will first be described in some detail as an introduction to the main features of tapping. To start tapping young trees on this system in Malaysia, a groove, starting at 1.5–1.6 m above the union of scion and stock and running for half the circumference of the trunk at a slope of 30° is initially cut into the bark. In order to sever as many latex vessels as possible, the high end of the groove should be at the left and should be cut almost as deep as the cambium (see Fig. 2.2) but should avoid damaging the latter as otherwise the bark renewal will be unsatisfactory. Tapping is done on alternate days with a knife or gouge which pares from the groove a shaving of bark just thick enough (1.0–1.5 mm) to re-open the plugged vessels. At each tapping, latex flows upward through the latex vessels to the cut from an area of bark extending some distance below it. It then proceeds along the cut and down a vertical guide-line, cut shallowly in the bark, to a metal spout, inserted in the trunk, which delivers the latex into a cup held in a wire loop attached to the tree (Fig. 9.1). The flow of latex at each tapping is rapid at first but gradually slows down and ceases after several hours.

With successive tappings, the cut gradually moves down towards the base of the tree and the bark above it is regenerated. After about five years, when the lower end of the cut reaches the union of stock and scion, and the first panel of virgin bark (B01) has all been tapped, a second half spiral cut is made at the same height as the first on the other side of the tree to start exploiting the second panel of virgin bark (B02). After a further five years, tapping returns to the top of the first panel (BI1) where the regenerated bark is now thick enough to tap again. When this has been exploited, after another five years, the second panel of renewed bark (BI2) is tapped. Thus, the tapping of the virgin bark and the bark of first renewal is completed in about 20 years, with the rate of bark consumption about 30 cm per year, measured down the trunk. In Sri Lanka, where (partly due to the much more difficult terrain) trees are opened at 1.2 m above the union, bark consumption is commonly as low as 22 cm a year, again allowing 20 years for the exploitation of the first four panels, after which tapping of bark of second renewal, or of higher panels of virgin bark, may be undertaken.

Under this system, a tapper is allocated two tapping tasks, each of, say, 500 trees, and he taps the same trees of each of these two tasks on alternate days. Normally, he starts work at, or soon after, dawn because, as explained later (section 9.2.3), lower yields are

Fig. 9.1 Tapping a half spiral cut (Rubber Research Institute of Malaysia)

obtained from tapping later in the day. At each tree he first removes
the strip of coagulated latex, known as 'tree lace', remaining on the
cut from the previous tapping, and puts it in a bag, along with any
other coagulated 'scrap' which he may find on the spout or on the
bark below the cut. He then pares the shaving of bark off the
tapping cut with his knife and places a clean, empty collecting cup
in position to receive the latex flow. By the time he has finished
tapping his task, or within about an hour thereafter, the flow of
latex from the trees tapped first will normally have ceased and he
then returns to his starting point to begin collecting the latex,
emptying the cup from each tree into a bucket, cleaning it and
replacing it on the wire hanger upside down. As explained later, in
certain circumstances – usually where yield stimulants are used –
latex flow may be prolonged and it may be necessary to replace the
cup in an upright position and to make a second collection later on
the same day. Alternatively, the 'late drip' may be collected the

next day when, as a result of bacterial action, it will have auto-coagulated to form 'cup lumps' which are collected in bags.

Seedling trees are usually tapped every third day on a half spiral cut ($\frac{1}{2}$S d/3). For reasons explained later (section 9.2.1) the first panel is opened at 50–75 cm above ground level but subsequent panels at about 150 cm; this allows three years' tapping on panel B01 and seven years each on panels B02, BI1 and BI2. The d/3 frequency is adopted for seedlings (and for certain clones) because they are susceptible to 'brown bast' and the more intense d/2 tapping leads to unacceptable levels of dryness. The lower frequency reduces tapping costs but is not used primarily for this purpose since on suitable cultivars the value of the extra yield obtained by d/2 tapping outweighs the higher tapping costs and is profitable even at quite low rubber prices.

Full spiral tapping, S, (or SR, a reduced spiral cut leaving 10–15 cm of the circumference untapped) is usually at d/4 frequency and is especially popular in former French territories with PR 107 and GT 1 – two clones whose latex flow pattern is particularly suited to long cut tapping. With lower frequency tapping more bark is consumed on each tapping occasion, but it is still practicable to attain a 10-year interval before renewed bark is tapped.

9.1.3 Single high cut systems

The virgin bark higher up the tree, if exploited at all, is not usually tapped until bark of first renewal on the lower panels has all been tapped out. If downward tapping of a high panel from a ladder is adopted, a $\frac{1}{2}$S cut is not practicable, for the worker has to lean too far to the side to make the cut; a V-cut comprising two $\frac{1}{4}$S cuts from right and left is used instead. Much smaller tapping tasks greatly increase cost of production per ha but the virgin bark of old mature trees is very productive and the operation is profitable even at moderate rubber price levels. Bark consumption can be controlled satisfactorily and the high panels, usually about 1 m in length, may – if tapped d/2 – each be exploited over a three-year period. Renewed high bark is not usually exploited, apart from slaughter tapping (see below).

On old trees with bark renewal on the lower panels too poor for further tapping, upward tapping into the virgin bark above the previously exploited panels used to be practised to a limited extent in order to prolong the useful life of the tree. It was not very satisfactory because tappers soon became tired, the poor standard of tapping caused severe wounding and much latex was lost through spillage, but it could be used for slaughter tapping. Latterly, upward tapping has become more popular in conjunction with yield stimulation with ethephon – see section 9.5.6.

9.1.4 Multiple cut systems

Panel changing

Various systems of alternately tapping two half spiral cuts on low panels on opposite sides of the tree have been employed with intervals between panel change ranging from alternate tappings to periods of months. Panel changing in such systems did not contribute significantly to profitability, but it has assumed considerable importance in systems employing shorter cuts with yield stimulation (see section 9.5.5).

Panel changing between high and low cuts is much more extensively used. In some circumstances the high panel is tapped periodically to allow more time for the renewal of the bark on the lower panel; in others, the high and low panels are tapped alternately with various time intervals. It has also proved useful to split the task, part of it being tapped high, the other low, on each tapping occasion. This eases the work load on the tapper and evens out the amount of crop he has to carry in each day.

Slaughter tapping

For two or three years before replanting, tapping intensity may be considerably increased, with little consideration of bark consumption and no attention to wounding of the cambium, in order to extract as much latex as possible before the trees are felled. High and low cuts may be tapped on each tapping occasion, and as replanting becomes imminent more than one cut, each with its own collecting cup, may be opened at each level and tapping frequency increased. Yield stimulants may also be used.

9.1.5 Rest periods

In areas such as Burma, Thailand and Vietnam the monsoonal rainfall pattern imposes a tapping rest period, tapping being impracticable during periods of intense rain or drought. In China, where the term 'wintering' of the rubber tree has a real rather than a descriptive significance, great importance is attached to cessation of tapping from the onset of the dry, cold weather (well before leaves have fallen) until after the new leaves have completely hardened. This resting of the trees during a period of considerable physiological stress may well be critical. In countries where rainfall is more evenly distributed, benefits of a rest period have been more difficult to demonstrate; properly designed experiments are complex and expensive to conduct, and any benefits – real or otherwise – have generally been outweighed on estates by practical difficulties of 'laying off' the tapping work-force for periods of time.

9.2 Practical aspects of exploitation

9.2.1 Standards for commencement of tapping: height of opening

As explained later (section 9.4.3), when trees are brought into regular tapping the growth rate (and in particular the annual girth increment of the trunk) is reduced. If tapping is started when the trees are too young and slender, subsequent growth and girthing will be poor. Yield is positively correlated with girth and it is, therefore, necessary for trees to have attained a certain minimum girth before they are opened for tapping and for them to continue to increase in girth, albeit more slowly, thereafter.

Seedling trees have tapering trunks and the thickness of the bark and the yield decrease with increasing height. They are opened relatively low in order to take advantage of the greater girth and bark thickness and to obtain an economic yield in the first year of tapping. The trunks of budded trees are practically cylindrical; the yield declines much less with height than it does with seedlings, changing little between 50 and 100 cm above the union, and commonly decreasing as the tapping cut approaches the union, probably because of the limited area of bark remaining below the cut from which latex can be withdrawn. Buddings can, therefore, be opened as high as they can be tapped conveniently by a tapper standing on the ground.

In Indonesia it has been standard to open buddings when they reach a minimum girth of 45 cm at 150 cm above the union and seedlings when they attain a minimum of 50 cm at 50 cm above ground level. Malaysia had a similar standard for seedlings but adopted a more conservative one for buddings, opening them at a girth of 50 cm at 150 cm above the union. Recently, arguments have been put forward in Malaysia for an earlier opening (at 45 cm instead of 50 cm) based on economic considerations. Such earlier opening reduces the immature period by about six months and, even if tapping intensity is initially low, it improves the profitability of estates; cumulative yields, even over a long period, are higher for earlier opened trees (Ng *et al.* 1972) and generally no adverse effects need be feared. Recent experiments in Indonesia (Lukman 1979, 1980a,b) confirmed that opening at 50 cm compared with 45 cm may result in a cumulative loss in yield over a four-year period of the order of 1400 kg/ha. Thus it seems generally justifiable to open buddings at 45 cm, and some companies in Malaysia have changed their practice accordingly, but an exception should be made for wind-susceptible clones in areas prone to wind damage, for which the 50 cm recommendation remains in force. (Further delay in opening, up to 65 cm, has been shown to reduce wind damage, but such prolongation of the immature period is uneconomic.) When buddings are opened at 45 cm, they should be tapped at low intensity (e.g. $\frac{1}{2}$S d/3) for one or two years.

Obviously not all the trees in a field reach the selected criterion at the same time, and company policy usually specifies that at least 50 to 70 per cent of the trees in a field must have reached the required minimum girth before the field is brought into tapping. Subsequently, further trees reaching the tapping standard are brought into tapping, usually at six-monthly intervals. As a result the average girth of the trees when brought into tapping will be greater than the minimum standard by 2–5 cm.

The actual period of immaturity depends mainly on the growth rate of the trees. When opening is at 45 cm girth, a well-grown field of buddings will usually reach the 70 per cent standard $4\frac{1}{2}$–5 years after planting in the field, but in conditions of poor soil or poor husbandry, such as are often found in smallholdings, the immature period may stretch to over 8 years. Ways of shortening the immature period by using advanced types of planting material have been discussed in Chapter 6; the possibility of earlier exploitation through puncture tapping is discussed in section 9.5.8.

9.2.2 Direction and slope of tapping cut

In general, latex vessels in the tree do not run vertically, but spiral from bottom left to top right of an observer as he faces the tree, the average angle to the vertical being 3.7°. It has long been recognized (and confirmed by experiments) that, as a result, tapping cuts sloping from high left to low right, given the same length and angle, cut through more vessels and consequently give a higher yield than cuts sloping to low left. Cuts sloping from left to right have become standard on estates in all rubber-producing countries, but on smallholdings, notably in Thailand, 'inverse' cuts can still be observed.

Gomez and Chen (1967) found that the inclination of the vessels is not uniform; they observed marked clonal differences in the angle of deviation from the vertical. Surprisingly, in three clones (including RRIM 600) a substantial proportion of the trees showed spiralling from low right to high left; such trees would be expected to yield more on a cut sloping to low left. Paardekooper (1968) in Thailand found that almost every sampled tree of clone KRS 13 showed an inclination of the vessels to high left; no such trees were found in other clones. (This characteristic is apparently heritable; of 14 illegitimate seedlings of KRS 13, eight showed an inclination to high left.) An experiment in which the latex from the left and right part of a V-cut was collected separately confirmed that, while with other clones the left arm (corresponding to a normally sloping cut) outyielded the right arm by 8 per cent, in KRS 13 the right arm yielded 8 per cent more.

For a given length of cut measured as a proportion of the trunk circumference (e.g. ½S), the steeper the cut, the more vessels will be cut and the higher will be the expected yield. Nevertheless, for practical reasons tapping cuts should slope steeply enough to allow the latex to flow down them without overflowing. Steep cuts will increase the size of the triangle of bark left untapped at the bottom of the panel; they will also increase the length of the cut disproportionately, and thus increase tapping time. Based on practical experience, most research organizations recommend an angle of 25° from the horizontal for seedling trees and 30° for bud-grafted trees. Because trees slowly expand in girth, after some time slopes tend to become flatter and need adjustment. In full spiral systems, steeper slopes are often recommended to avoid latex spilling over the cut and to increase the vertical distance between the two ends of the cut, thus somewhat reducing the possibly deleterious 'ring-barking' effect of the full spiral. Upward tapping also requires steeper slopes, up to 50° to reduce latex spillage.

9.2.3 Time of tapping

For a given clone and exploitation system, the yield of latex from a tapping mainly depends on the initial flow rate of the latex (see section 9.4.2) which in turn depends on the turgor pressure in the latex vessels. After sunrise the turgor pressure normally falls as a result of withdrawal of water under transpirational stress, reaching a minimum at 1300–1400 hours and then slowly recovering. In consequence, delay in starting tapping after sunrise results in lower yields. For this reason it is usual on estates to begin tapping at 06.00 hours, but smallholders in Thailand often start considerably earlier and complete a large task before daybreak. Their experience that night tapping gives higher yields has been confirmed in an experiment in which the relation between yield and time of tapping, covering a full 24-hour cycle, was investigated (Paardekooper & Sookmark 1969). No differences in yield were found within the period 20.00 hours to 07.00 hours; after that yields decreased gradually to reach a minimum of about 70 per cent of the 'night' yield around 13.00 hours. The dry rubber content followed a reverse pattern and was often four percentage points higher at noon than during the night. The variation in yield was shown to be the result of initial flow rate as a consequence of a variation in turgor pressure, while the plugging index (see p. 372) was not affected. The diurnal variation in yield followed closely the vapour pressure deficit of the air (Fig. 9.2).

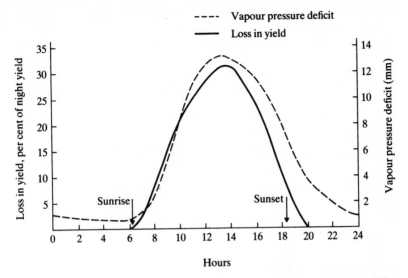

Fig. 9.2 Diurnal variation in vapour pressure deficit of the air (average of days without rain at Haadyai, Thailand) and relation between hour of tapping and loss in yield expressed as a percentage of 'night yield' (After Paardekooper & Sookmark (1969)

Early morning tapping is often subject to interference from rain. Rain in the night or very early morning necessitates postponement of tapping until the panel is reasonably dry, because rainwater causes coagulation on the cut. In these circumstances, estates resort to 'late tapping' or – when rain continues during the morning – to afternoon tapping. If a whole tapping day is lost due to prolonged rainfall, 'recovery tapping' of the task A may be carried out on the afternoon of the following day after the tapper has tapped his other task, B. On an alternate daily system, task A is then tapped again on the following morning after a lapse of less than 24 hours. Since the yield is lower and the tappers have to be paid overtime rates, such afternoon or 'recovery' tapping is not always economically justified. When rain falls during, or immediately after, tapping, its flow down the tree trunks (which amounts to about 25 litres/tree during a 25 mm rainstorm) may wash the latex out of the cups. Such 'washouts' result in the loss of both the crop and the wages, since the tappers have to be paid their full basic wage.

9.2.4 Depth of tapping

As the most active latex vessel rings are located close to the cambium, tapping should be as deep as possible without damaging the cambium, and the major element of skill in tapping is to achieve adequate and uniform depth without causing more than a minimum of wounds. The standard recommendation is to tap to 1.0–1.5 mm from the cambium. Present tapping systems seldom exploit the bark of second renewal, and hence in tapping the bark of first renewal a higher degree of wounding is acceptable than in tapping virgin bark.

It has long been known that deeper tapping can give higher yields, particularly in young trees in which there is a marked concentration of latex vessels close to the cambium. De Jonge (1969a) published results of experiments in Malaysia which showed that deep tapping to within 1 mm of the cambium gave considerably higher yields (up to 50 per cent) than tapping to 1.5–2.0 mm, particularly during the first year on a new panel and with full spiral cuts. It slightly decreased dry rubber content and girth increment but had no effect on bark renewal.

9.2.5 Bark consumption

The thickness of the bark shaving removed at each tapping depends on the skill of the tapper and the degree to which the bark has dried out. Excessive consumption, common on smallholdings, wastes bark capital and so shortens the economic life of the tree. The shaving should be thick enough to remove all plugged vessel ends, but above a certain minimum no increase in yield has been found. Under alternate daily tapping, the recommended consumption is usually 1.5–2.0 mm per tapping for virgin bark, or about 25–30 cm per year, but in Sri Lanka, where the use of the gouge (see p. 359) is predominant, 1.3 mm is considered adequate under d/2 tapping.

In order to keep a check on bark consumption, estates frequently put a dot of paint on the renewed bark just above one end of the cut at monthly intervals. Another effective method is to draw grooves at, say, 12 cm intervals at the time of opening for tapping to mark the expected consumption for each six-month period.

Low frequency tapping results in greater drying out of the bark tissues; good tappers adjust to this automatically by increasing the thickness of the shavings. De Jonge (1969b) suggested that for fourth daily tapping bark consumption per tapping should be 20 per cent greater than for alternate daily tapping. Paardekooper *et al.* (1976) found a large effect of tapping frequency on bark consumption; with tapping every sixth day the consumption per tapping was almost double that for daily tapping. The higher bark consumption is often overlooked in the comparison of tapping systems; full spiral fourth daily and half spiral alternate daily systems (S d/4 and ½S d/2) may result in the same yield per hectare over a given period, but the yield per unit of consumed bark would be 20 per cent less under the former system.

9.2.6 Tools and equipment

The main tools traditionally used in rubber exploitation are knives and sharpeners, latex cups, cup hangers, spouts and collection buckets. In specific situations, additional tools may include ladders, polythene bags, rain-guards, scrapers, brushes and lamps.

Knives

No less than 22 types of tapping knife were used at the beginning of the century (Wright 1912). Modern tapping knives are of two basic types: the 'jebong' knife, adapted from an old English knife used for paring horses' hooves, and pulled downwards along the tapping cut; and the gouge, consisting of a U-shaped blade attached in line to the handle, which is pushed upwards along the cut. The jebong knife is used in Malaysia and Indonesia while the gouge (the so-called Michie-Golledge knife) is popular in Sri Lanka and India (see Fig. 9.3).

There has been a recent development of knives designed to facilitate upward tapping. These include a double-bladed jebong (kanoor knife) for V-cuts, a new gouge with a V-shaped blade (P'ng *et al.* 1976) and a bi-directional knife as an improvement on the ordinary jebong (Abraham & Anthony 1980b). The latter can be pulled or pushed and forms grooves which reduce the spillage liable to occur with upward tapping.

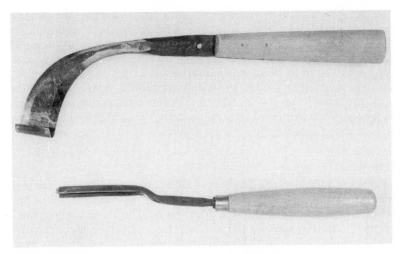

Fig. 9.3 Tapping knives: jebong (above), gouge (below)

Mechanized tapping tool

Collaboration between the Rubber Research Institute of Malaysia and a Japanese company has recently resulted in the development of a mechanized tapping tool fitted with a micro-motor powered by a rechargeable battery and using a control guide to enable a reciprocating blade to cut a uniformly thin shaving of bark. Tests have shown that unskilled workers readily learn to operate the tool effectively, that it gives an even shaving with minimum bark wounding, and that latex yields are comparable with those obtained by conventional tapping. This tool has now reached the stage of commercial production (Anon 1984, 1985). It is hoped that it may alleviate the shortage of skilled tappers now experienced in some areas, but the recurrent cost of providing the tool must surely impose serious constraints on its practicability. In any case, cup cleaning and latex collection take up more time than the actual tapping operation, so only a separation of this unskilled work from that of actually making the cut is likely to have a marked effect on task size and hence on tapping costs. Polybag collection (see section 9.2.7) has not generally been found practicable, but the possibility of automating collection by dripping an anti-coagulant on to the tapping cut to give longer flow times and using a system of closed pipes and a suction pump to collect latex at a central point, is under investigation (Rubber Research Institute of Malaysia, 1984e). However, even if technical problems are solved, it seems unlikely that the necessary capital investment would give an economic return.

Latex cups

The receptacles which receive the latex after tapping are usually made of aluminium, earthenware or modern synthetic material. In Sri Lanka and India, however, coconut shells continue to be used. Although very cheap, they have some disadvantages; being porous they are difficult to clean, which may result in pre-coagulation; also, they are generally too small to accommodate high yields. Aluminium cups are usually of 300 or 500 ml content; earthenware cups used in Vietnam were often of 1000 ml capacity. With high-yielding trees, two or more cups are sometimes required.

Cup hangers

In many countries cups are simply hung on a nail driven into the bark; in others, a hanger is made of wire, or bamboo slings are used, to prevent bark damage from nails. The most sophisticated method, whereby damage to the tree is completely avoided, consists of a hanger on a spring-loaded wire hooked around the tree. Coconut shells are often placed at the foot of the tree; this requires a long channel to convey the latex, and is a frequent cause of spillage.

Ladders

In the period 1950–70, when exploitation of high panels with down-ward cuts was popular for older rubber, ladders were a common tool for tappers. Many different designs were used, light wooden ladders with a small platform being the most popular. With the introduction of upward tapping, ladders are no longer required; cuts on high panels are tapped upwards with knives at the end of long handles.

Lamps

In Thailand, where tapping often starts at midnight, extensive use is made of carbide lamps, usually fixed on the forehead. An electric head lamp powered by a nickel cadmium rechargeable battery was recently developed in India (Rubber Research Institute of India 1983).

Rain-guards

As already explained, rain may seriously interfere with tapping. Rain-guards which protect the cut from the stem flow of rainwater may be useful in certain conditions, and many different types have been tried. In Sri Lanka and India, skirts of polythene sheet found some use, sometimes being stitched to provide frills. Their two main disadvantages were humid conditions under the skirt, leading to panel diseases, and the need to lift the skirt for every tapping. Similar problems were experienced with lampshade types made from paper board. More practical are gutter-type guards which, while not keeping the panel dry during rain, nor always preventing stem flow from entering the cup, at least expedite the drying out of the panel after rain. Such guards, made from compound rubber, were found to be of some benefit but considered uneconomic in Sri Lanka. In India, simple gutter-type guards, made of jute hessian reinforced with aluminium strips, or from split polythene tubes, appear to be promising (Rubber Research Institute of India 1981). At the end of the 1960s, in conjunction with the possible use of polybag collection, a simple polystyrene rain-guard was developed, consisting of a narrow strip of expanded polystyrene glued around the tree with latex (Chang *et al.* 1969; Southorn 1969). Such guards were claimed to be cheap and effective, but susceptible to bird and insect attacks and thus of limited service life. A recent development is the 'Ebor eave', made of flexible bitumen sheet cut to a crescent-like shape and fixed to the trunk with a contact adhesive to form a protective hood above the panel. It is claimed that it can signifi-cantly reduce panel wetting from rain during the previous night or early morning and thus largely prevent loss of tapping days and late tapping (Gan *et al.* 1985).

9.2.7 Latex collection

Latex collection on estates consists of: (a) transferring the latex from cups to buckets and transport of the buckets to a collection point – 'crop lifting'; and (b) the transport of the latex from the collection point to the factory – collection proper. In this chapter only the former activity, which is the responsibility of the tapper or a member of his family, will be considered; transport of latex to the factory is discussed in Chapter 11.

On most estates tapping starts about 06.00 hours and finishes at about 10.00 hours. Traditionally, the tapper then takes a rest and starts 'crop lifting' at 11.00 hours. Latex from each cup is poured into a small bucket. The cup is wiped and replaced on the tree, either upside down to avoid collection of rainwater or, if the tree is still dripping, in an upright position to collect more latex. This 'late drip' forms a lower grade of latex because its longer exposure in the field results in increased contamination with bacteria and dirt. The latex in the small collection bucket is subsequently transferred to one or two larger buckets which are carried to the collection station. At these stations, which may consist of high platforms to facilitate transfer of the latex into tanks by gravity, the amount of latex brought in by each tapper is measured and recorded, and commonly a sample is taken for dry rubber content determination.

As an alternative to the frequent collection of latex from the ordinary cups, the Rubber Research Institute of Malaysia investigated the use of polythene bags in which the latex from several tappings could accumulate and autocoagulate before the bags were collected at intervals of two or three weeks (Collier & Morgan 1969; Southorn 1969). Polythene bags of 0.04 mm gauge and of about 2-litre capacity were found satisfactory. They were attached to the tree by means of a wire stirrup, or hanger, and the tapped latex flowed into the bag down a modified spout, which passed through a small hole in the otherwise sealed bag. Latex from the first tapping, which flowed into a virtually sterile bag, stayed liquid for 24 hours or more before bacterial contamination built up to the point where auto-coagulation occurred, but the second and subsequent tappings coagulated rapidly. The bags were collected in baskets and transported by lorry to the factory, where they were slit and the coagulum dropped into clean water before processing. Originally it was thought that rain-guards would be essential to prevent the bags filling with water, but they proved to be unnecessary because, once coagulation had occurred, the rainwater could be poured out by tilting the bag.

Later, polybag collection was tried out in several countries, and its main advantages can be summarized as follows:

1. There is no need for the tapper to do 'crop lifting' because the bags are collected by unskilled labour. The actual tapping time is also reduced because there is no need for cup cleaning. Hence, task size can be considerably increased, which is a great advantage in places where there is a shortage of skilled tappers.
2. There is a substantial saving in tapping and collection costs.
3. Any 'late drip' which may occur is not contaminated by exposure in the cup and can be included in the first grade rubber.
4. Coagulation in the factory is eliminated.
5. The coagulum is particularly clean; compared with rubber produced from acid-coagulated latex or from cup lump the processed rubber has a very low dirt content, a slightly lower plasticity retention index (see Appendix 11.4, p. 554) but better dynamic properties.

In spite of these marked advantages, polybag collection is presently only used on a large scale by some companies in the Ivory Coast and Cameroon. The main problem is theft of the full bags. A minor problem is the lack of space for the bag between the spout and the ground when the tapping cut is approaching ground level, but obviously this would not prevent polybag collection from most of the panel.

9.2.8 Tapping tasks

Task sizes vary greatly, depending on many factors such as tapping system, planting density, age of the trees, nature of the terrain, method of payment and method of collecting the latex. While a tapper could, in theory, cope with a larger number of trees in a full working day, in practice task sizes are restricted by the decrease in yield which occurs when tapping is extended from early morning to late morning. As a result, an increase in task size, while increasing the yield per tapper, depresses the yield per hectare. De Jonge and Westgarth (1962) found a decrease in yield per hectare of 1 per cent for every increase in task size by 50 trees. Whether this small loss is financially justified depends largely on the way the tapper is remunerated and on the price of rubber. If tappers receive a fixed daily wage only, larger tasks would in general be more economic; if, on the other hand, tappers are paid a fixed rate per kg of rubber, smaller tasks giving the maximum yield per unit area are preferable.

An analysis of the tappers' activity shows that the collection of latex may take up to one-third of the total time spent in the field; substantial increases in task size would thus be possible by eliminating collection. Chuang (1965) reported task sizes up to 1200 trees where collection was done by cheaper, unskilled labour. However, any procedure involving increase in task size will decrease yields per hectare since tapping will continue up to or beyond noon.

Tapping time (excluding collection) has been shown to consist of about 25 per cent actual knife work, 35 per cent walking and 40 per cent cleaning and replacing cups. Walking time can be decreased by more pronounced hedge planting (see p. 191); Barlow *et al.* (1966) showed that savings in time in 20 m hedge plantings were relatively small compared with a conventional planting system, although total walking distance was about half. Cleaning time can be reduced by the procedure adopted on many estates whereby the dirty cup is cleaned during the tapper's walk between trees, and used as a replacement for the cup on the next tree.

The size of the tapping task is greatly affected by the tapping system; longer and multiple cuts increase actual knife time, and thus require a smaller task. (The economic evaluation of different tapping systems consequently depends on the assumed task sizes.) In commercial practice, the following relative task sizes are often found compared with half spiral tapping: 70 per cent for full spiral, 115 per cent for $\frac{1}{3}$ spiral, 120 per cent for $\frac{1}{4}$ spiral and 50 per cent for double cut systems.

The common task size for $\frac{1}{2}$S d/2 tapping in most countries is nowadays around 500 trees. In Indonesia, where pre-war tasks were around 300 trees, tasks were subsequently increased to 400–500. In Sri Lanka, where tasks have traditionally been smaller, 350 trees is recommended. In Malaysia, a set of maximum task sizes under different conditions of terrain, age of trees and exploitation systems was agreed with the Union of Plantation Workers as part of a wage agreement in 1979. Maximum task size varies from a high 600 (young trees, flat or undulating terrain, $\frac{1}{2}$S low cut tapping up to the fifth year of tapping) to a low 166 (old trees on contour terraces tapped simultaneously on high and low $\frac{1}{2}$S cuts). On smallholdings the number of trees tapped is mainly dictated by the size of the holding. Under the share tapping arrangement encountered in the larger holdings in Thailand, task sizes are often very large, with tapping starting at midnight and lasting six hours or more.

To facilitate the keeping of proper records, tasks often coincide with block boundaries. Since not all trees are tappable in the first and second tapping year, and since, on the other hand, old trees require a longer tapping time, periodic change in task size is necessary. Long, narrow tasks facilitate supervision; it is also desirable that all tasks border a road. A useful practice, not always adhered to, is for the tapper to start tapping at opposite ends of his task on alternate tapping days.

9.3 Tapped latex

The latex collected by regular tapping consists of the cytoplasm expelled from the latex vessels and is similar to the latex *in situ* already briefly described in Chapter 2. Apart from water, it contains about 30–40 per cent of rubber and about 3.5 per cent of other substances. The structure and composition of fresh, tapped latex has been elucidated by high-speed centrifugation (Moir 1959). Depending on the method followed, 3 to 11 zones can thus be distinguished (Fig. 9.4). The top fraction consists almost entirely of rubber; the middle zones are made up of the watery phase of the latex, generally called C serum; the relatively heavy bottom fraction, normally yellowish, viscid and semi-liquid, consists mainly of the lutoids, while the yellow, lipid-containing Frey-Wyssling complexes are normally found at the upper border of the bottom fraction.

By careful separation of the three major zones, their chemical

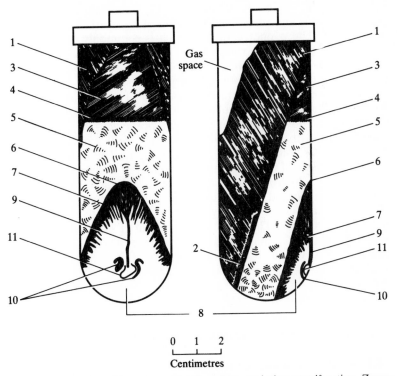

Fig. 9.4 Separation of latex into zones by refrigerated ultra-centrifugation. Zones: 1 – rubber; 4 – yellow layer containing Frey-Wyssling particles; 5 – serum; 6–11 – 'bottom fraction' (After Moir 1959)

composition can be determined. The rubber fraction contains, in addition to the rubber hydrocarbon, the proteins and phospholipids associated with the rubber particle membranes. The serum phase contains most of the soluble substances normally found in plant cells, such as inositols, carbohydrates, free amino acids, proteins, inorganic anions and metal ions, together with the enzymes and intermediates of various biochemical processes, including rubber biosynthesis. The bottom fraction can be studied by repeated freezing and thawing; in this manner the membranes of the lutoids are ruptured and their liquid content, often referred to as B serum, can be analysed. B serum has been found to contain proteins and other nitrogen compounds as well as metal ions. In general, potassium is evenly distributed between both serums; acid-soluble phosphate, magnesium and calcium are mainly concentrated in the B serum, while sugars are predominantly present in the C serum.

The nitrogen, phosphorus and metal ion contents of the latex are variable, being affected by cultivar, soil type, fertilizer treatment and tapping system. Typical average values as per cent by weight of latex are: N – 0.26; P – 0.05; K – 0.17; Ca – 0.003; Mg – 0.05. The magnesium and phosphate contents of latex affect its stability (see p. 490). The copper content, which is commonly in the range 0.1–1.5 ppm Cu, usually shows a positive correlation with yield.

The rubber content of the latex is obviously of economic importance; in addition it has a physiological interest since it reflects the balance between rubber extraction and rubber regeneration within the tree. The dry rubber content (d.r.c.) can be determined by coagulating a known weight of latex with acetic acid, separating the coagulum from the serum, rolling it out to about 2 mm thickness, drying it for 24 hours at 70 °C and weighing it. This determination gives an estimate of the percentage by weight of rubber in the latex, i.e. of rubber hydrocarbon together with the small amounts of protein, phospholipid, etc., associated with the rubber particles. The d.r.c. is influenced by many factors including cultivar, age, girth and tapping system. The total solids content (t.s.c.) can be determined by placing a shallow layer of latex of known weight in a sample dish, drying it to constant weight at 70 °C and weighing the resultant dried film. The t.s.c. includes rubber plus other substances and is usually about 3.5 percentage points higher than the d.r.c., but it varies significantly between cultivars.

9.4 Physiology of latex production and flow

9.4.1 Situation in untapped trees

During the growth of untapped trees a small part of the metabolites is used to generate rubber in the newly formed vessels in the bark and in other organs. In this way, a 10-year-old tree produces around 1 kg of rubber per year, of which about half is in the leaves shed during wintering. Once formed, the rubber, which is mainly in the microgel stage, does not move. Various workers have used the first drops of latex obtained through some form of micro-tapping to estimate the total solids content of the latex *in situ* (Ferrand 1941; Gooding 1952a). This concentration is much higher than in tapped latex; it may vary between 50 and 70 per cent, depending on days, on time of day, and on the side of the tree. These variations in total solids and rubber content *in situ* can be expected to be related to changes both in turgor pressure and osmotic pressure within the vessels, which regulate the flow of water in and out of the vessels. In addition, highly significant differences between clones have been observed (Paardekooper, Rubber Research Centre, Thailand, unpublished).

Buttery and Boatman (1966) devised a method for direct meas-urement of turgor pressure in the bark tissues using simple capillary manometers. They observed early morning pressures in untapped trees between 10–15 atm; pressures decreased during the day, and increased at night, probably as a result of withdrawal of water from the phloem tissues under transpirational stress; such diurnal vari-ation is in the order of 3–4 atm. The same authors found a significant difference in turgor pressure between two clones, but no correlation with the girth of the tree. By measuring the osmotic concentration of the first drops of latex obtained by micro-tapping Pakianathan 1967) estimated the osmotic concentration of latex *in situ* as around 450 milli-osmoles/litre. Osmotic pressures appear to be always larger than the turgor pressures, resulting in positive 'suction pressures'.

Results of these and other experiments suggest that latex vessels behave as a relatively simple osmotic system. Turgor falls during the day as a result of withdrawal of water under transpirational stress. The diurnal pressure changes are positively correlated with the rela-tive humidity of the atmosphere and negatively correlated with changes in temperature, leaf water deficit and stomatal opening. Most significantly, they do not occur in trees without leaves during wintering. The diurnal variation in turgor pressure is responsible for the small diurnal variation in trunk diameter observed by Pyke (1941), Gooding (1952a) and Boatman (1966).

In addition to variations in rubber content, there appear to be differences in latex constitution, both between trees and between different locations within one tree. Consistent differences in chemical composition between untapped trees of different clones have been observed. Latex within the young inner vessels is different from that of outer vessels, as was pointed out by Schweizer (1949); the younger vessels are likely to have a higher lutoid content. With increased height in the trunk, latex has a higher magnesium but a lower nitrogen, phosphorus and perhaps copper content; it is also likely to have a higher proportion of lutoids. Tupy (1973b) found a downward gradient in sucrose concentration in untapped trees.

9.4.2 Response to tapping

When a tree is tapped for the first time the latex obtained is very viscous, and flow ceases rapidly. Successive tappings at regular intervals result in increasing yields of more dilute latex, until an equilibrium between rubber extraction and rubber regeneration is reached. At this time, the rubber content *in situ* is considerably lower than in untapped trees; it is, however, higher than found in the tapped latex; this is illustrated in Table 9.1.

Table 9.1 *Total solids content (%) of latex in situ (within the tree) of untapped and tapped trees and of tapped latex*

Clone	Relative yield	In situ			Tapped latex
		Untapped	Tapped	Difference	
KRS 13	low	59.9	55.8	4.1	43.9
Banglang	low	56.4	50.6	5.8	38.2
Tjir 1	average	55.2	46.3	8.9	39.5
KRS 21	high	56.6	45.5	11.1	40.2
PB 86	high	48.9	38.4	10.5	36.5

Source: Paardekooper, Rubber Research Centre, Thailand (unpublished)

These data show that: (a) the decrease in total solids content *in situ* as a result of tapping is particularly marked for the high-yielding clones; and (b) differences in total solids *in situ* in tapped trees are reflected in differences between clones in the total solids content of the tapped latex.

The degree to which the total solids content *in situ* is lowered through tapping depends very much on the tapping intensity and is reflected in the dry rubber content of the tapped latex. Consequently, the d.r.c. is generally considered as a physiological indicator: a low d.r.c. (say below 30 per cent) indicates that the tree is probably 'over-exploited'. In tapped trees there is a clear relation between girth and d.r.c.; an increase in girth of 3 cm has been found to correspond to an increase in d.r.c. of 1 per cent. Similarly, d.r.c. generally increases with the age of the tree.

There has been much research aimed at elucidating the sequence of events that follows the opening of the vessel ends by tapping and at determining the mechanisms involved. This work has been ably reviewed by Gomez (1983) in a monograph on the physiology of latex production, to which reference may be made for more detail.

On each occasion when a tree in regular tapping is tapped, the rate of latex flow is initially high, but it soon diminishes, at first rapidly and then more slowly, until flow ceases after a period varying from half an hour to three hours. By measuring flow rates at intervals of time after tapping, various workers have determined the exact shape of flow curves (Fig. 9.5). The total yield depends on the initial flow rate and the duration of flow; two trees may have the same yield but widely different flow patterns. Mathematical treatment of flow has been attempted by Frey-Wyssling (1952), Riches and Gooding (1952) and Paardekooper and Samosorn (1969). The latter authors showed that, except during the first half hour, a logarithmic transformation of the flow rates results in approximately linear trends. Their model corresponds to the formula

$$y = b.e^{-at}$$

where y is the flow rate at time t after tapping
   b is the initial flow rate
   e is the base of natural logarithms, and
   a is a constant mainly depending on clone.

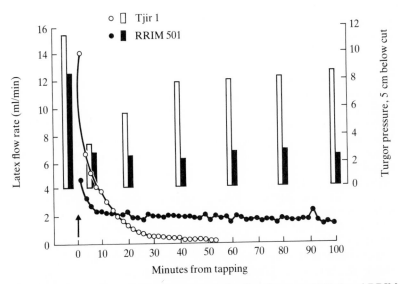

Fig. 9.5 Flow curves and turgor pressures during flow for trees of Tjir 1 and RRIM 501 (arrow indicates tapping) (After Milford *et al.* 1969)

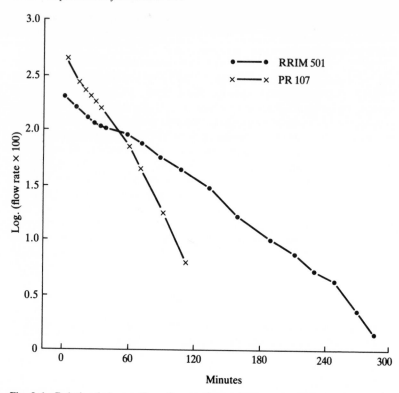

Fig. 9.6 Relation between time after tapping and flow rate after logarithmic transformation of flow rate data (After Paardekooper & Samosorn 1969)

Figure 9.6 illustrates the flow curves, after logarithmic transformation, for two clones. Ninane and David (1971) established a similar relationship between time and flow rate.

Before tapping, latex is contained in the vessels at high hydrostatic pressure, usually between 10 and 15 atm in the early morning. Immediately after tapping, the pressure at the cut ends of the vessels is reduced to ambient and the elastic contraction of the vessel walls under the pressure of the still turgid surrounding cells expels latex at high speed. By use of capillary manometers it has been shown that the turgor pressure of the laticiferous system just below the cut falls considerably and very rapidly, but with increasing distance from the cut the pressure declines less and more slowly (Buttery & Boatman 1967). From an area of bark extending some distance below the cut (see p. 378), latex flows through the vessels along the pressure gradient but the gradient per mm becomes smaller with increasing distance from the cut.

The loss of turgor pressure in the vessels disturbs the osmotic equilibrium, creating a suction pressure which causes an inflow of water from the neighbouring cells into the latex vessels. This movement of water spreads throughout the phloem tissue and, presumably, into the xylem. As a result of this so-called 'dilution reaction', the total solids and dry rubber contents of the latex decline immediately after tapping but usually show some recovery before flow ceases and the rubber content is restored to its usual level by synthesis between tappings. The turgor pressure gradually recovers as the latex flow slows down to approach its normal value about the time when flow stops on a humid day, or a few hours later if the weather is dry (Fig. 9.5).

After tapping, the rate of flow can be expected to slow down with time because latex withdrawn over greater distances from the cut is subject to reduction in the pressure gradient and to the greater resistance of an increasing length of partially collapsed vessels. But these factors are partly counteracted by the movement of water into the vessels from the surrounding cells, which tends to restore turgor pressure and to render the latex less viscous so that it flows more readily. It was suggested that, after the early stages of fast flow, the vessels might become sufficiently constricted to impede or even stop flow, but dendrometer measurements of trunk diameter (Pyke 1941, Gooding 1952a, Boatman 1966) indicated that this could not be a major influence on latex flow since vessel diameter is reduced by no more than 25 per cent, even close to the cut where constriction is greatest.

The concept of latex vessel plugging

A major advance in understanding the mechanism of latex flow resulted from the work of Boatman (1966) and Buttery and Boatman (1966, 1967). By reopening the tapping cut by removing shavings approximately 1 mm thick at 10-minute intervals after tapping it was shown that flow rates recovered markedly after each reopening of the vessel ends, so that stepped flow curves were obtained (Fig. 9.7a). This indicated that the flow rate is markedly decreased within a very short time after tapping by some impediment developing at, or within about 1 mm of, the cut ends of the latex vessels.

The nature of the impediment was revealed by optical and electron microscope studies of longitudinal sections of latex vessels near the tapping cut (Southorn 1968a). In sections made before flow had ceased, internal plugs of rubber particles and damaged lutoids completely blocked some vessels but were entirely absent from others. This suggested that plugging was rapid and complete in a particular vessel rather than that the plugs form slowly at the walls of the vessels and grow inwards. In sections made after flow had ceased there was also an external cap of coagulum, of the same

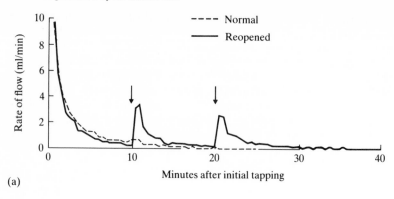

(a)

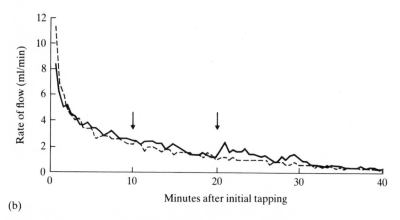

(b)

Fig. 9.7 Rate of flow recorded at half-minute intervals after tapping a Tjir 1 tree. Comparison of normal ½S d/2 tapping (broken line) with reopening the cut, as indicated by arrows, 10 and 20 minutes after initial tapping (solid line): (a) before stimulation, (b) after application of the stimulant 2,4,5-T (After Boatman 1966)

composition as the internal plugs, which sometimes appeared to have invaded the vessels just before flow ceased.

The plugging index

Milford *et al.* (1969) studied clonal variations in the rate of plugging and proposed that plugging behaviour could be characterized by a 'plugging index' derived from the ratio between the initial flow rate and the total volume of latex per tapping:

$$PI = \frac{\text{mean initial flow rate (ml/min)}}{\text{total yield volume (ml)}} \times 100$$

If the flow curve is represented by the experimental model

$$y = b.e^{-at}$$

as proposed by Paardekooper and Samosorn (1969) (see above), the plugging index is equivalent to 100*a*. The steeper the regression, and hence the greater the relative decrease in flow rate with time, the higher the plugging index and the shorter the flow time. It can be shown that the flow time in minutes is approximately equal to 500 divided by the plugging index. Short-flowing clones with a high plugging index, such as Tjir 1, show both the rapid recovery of turgor pressure soon after tapping and the increased flow rates following repeated reopening of the vessels that are characteristic of rapid plugging, while both these features are absent from long-flowing clones with low plugging indices (Figs. 9.5 and 9.8).

The initial flow rate is usually calculated as the average flow rate during the first five minutes. This rate depends on: (a) the number of latex vessels cut, therefore on the length of the cut and the number of vessel rings; and (b) on the pressure in the vessels before tapping. The initial flow rate may vary from less than 1 ml/min to over 5 ml/min, depending on clone and length of cut. The plugging index has been shown to be a clonal characteristic, but also to vary markedly with season, tapping system and stimulation practice. The index can be as low as 1 for very long flows or over 10 for short flows.

The main practical advantage of the plugging index is that it allows the flow behaviour of a relatively large number of trees to be derived from two simple yield measurements. A simplification was proposed by Sethuraj (1968), who measured the flow rate by counting the number of drops falling off the spout within a given time, but investigations in Thailand showed that this method is unreliable since many factors influence the weight of a drop of latex, which may vary from 15 to 17 mg. Sethuraj *et al.* (1978) and Subronto and Harris (1977) introduced two further flow indices, both based on the flow rate during the initial 5 minutes and the flow rate 30 minutes after tapping. Southorn and Gomez (1970), proposed an index based on the flow rates obtained by reopening the tapping cut after certain intervals. The 'intensity of plugging' is calculated as

$$\frac{100 \ (b - a)}{b}$$

in which *a* is the flow rate just before reopening the cut and *b* that just after. Samsidar and Gomez (1982) in their studies of puncture tapping, introduced a 'micro-plugging index' as the ratio of the first minute's yield over the total yield and found that, with a number of clones, there was a high correlation with the usual plugging index.

Many workers have studied the effects of clone, season and tapping system on plugging index and the relation of the latter with

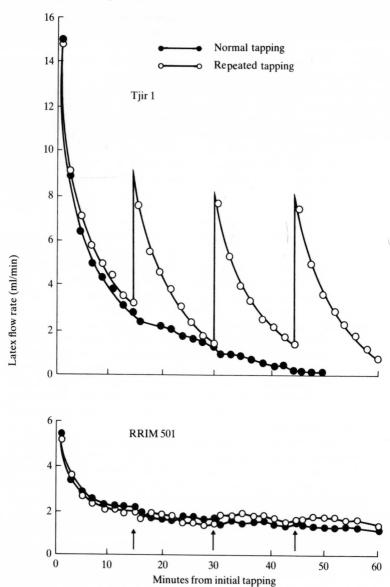

Fig. 9.8 Latex flow patterns of trees of clones Tjir 1 and RRIM 501 when tapped normally and after repeated opening of the cut as indicated by arrows (After Milford *et al.* 1969)

other characteristics of the latex. It was found that the well-known yield depression just after wintering is accompanied by a marked increase in plugging index, while the initial flow rate is not affected. Similarly, the lower yield obtained from cuts at higher levels is related to increased plugging. Plugging increases with progressively shorter cuts. Southorn and Gomez (1970) observed that the 'intensity of plugging' and the initial flow rate increased dramatically for very short cuts. Long cuts, such as full spiral cuts, exhibit a decrease in flow rate per mm of cut and an increase in flow time.

The role of the lutoids in latex vessel plugging

Various workers have built up evidence that a major cause of vessel plugging during latex flow is damage to the lutoids. Microscopic examination of latex from the ultra-centrifuge showed that dilution with water progressively damaged the bottom fraction particles (mainly lutoids) which could then aggregate with rubber particles to form flocs. This was thought to be caused by osmotic shock. Later, evidence was obtained which suggested that changes in the osmotic concentration of the latex during flow after tapping damaged lutoids. During the initial period of fast flow these lutoids were swept out of the vessels before they suffered irreversible damage, although the permeability of their membranes was affected and they subsequently formed aggregates with rubber particles which were found in large numbers in the latex collected at the spout at the bottom of the tapping cut. During subsequent, slower flow the lutoids suffered greater damage within the vessels and aggregated with rubber particles to form flocs which accumulated near the cut ends, thus initiating the plugging process (Pakianathan *et al.* 1966; Pakianathan & Milford 1973).

Following upon the microscopic observation of plugs of rubber particles and damaged lutoids near the cut ends of vessels, Southorn and Edwin (1968) discovered that the fluid contents of the lutoids (B serum) caused rapid and complete flocculation of an aqueous suspension of rubber particles. In whole latex, breakage of lutoids by ultrasonic treatment resulted in the formation of flocs of rubber and damaged lutoids. Thus, it seemed likely that plugging within the vessels is caused by release of B serum from ruptured lutoids.

Fresh latex always contains micro-flocs, the formation of which appears to be confined to the neighbourhood of damaged lutoids where the concentration of B serum is momentarily high, and to be limited by a stabilizing effect of C serum. Southorn and Yip (1968a) envisaged fresh latex as a dual colloidal system composed of: first, negatively charged particles (mainly rubber and lutoids) dispersed in the neutral C serum containing anionic proteins; and, second, a system within the lutoid membranes comprising the acidic B serum with metallic ions (especially Mg^{++} and Ca^{++}) and some cationic proteins. The two antagonistic systems can only exist so long as they are separated by the intact lutoid membranes; release of the B serum results in interaction between its cationic contents and the anionic surfaces of the rubber particles, causing the formation of flocs. Evidence was obtained that, prior to their disintegration, the intact lutoids undergo loss of polarization and increase in permeability.

Following the observation that the enzyme acid phosphatase is released in the B serum when lutoids are damaged, Pujarniscle and Ribaillier (1966) measured the activity of the so-called 'free' acid phosphatase in latex and also the total acid phosphatase, after disruption of all the lutoids with a detergent, and called the ratio of the two the 'bursting index' of a sample:

$$BI = \frac{\text{Free acid phosphatase activity}}{\text{Total acid phosphatase activity}} \times 100$$

They considered that this index, which was inversely related to the osmolarity of latex samples, indicated the percentage of the lutoids in a sample that had ruptured. Pujarniscle *et al.* (1970) showed that the bursting index of the first fraction of latex collected after tapping was higher than in subsequent fractions. This was in agreement with the finding of Pakianathan *et al.* (1966) that damage to bottom fraction particles was greatest in the first flow fraction after tapping. Yeang and Paranjothy (1982), in studying seasonal variations in flow pattern, found high correlations between plugging index, intensity of plugging and bursting index.

Low (1978) noted a positive correlation between total cyclitols in C serum and plugging index and suggested that the higher osmotic pressure associated with higher cyclitol concentration could lead to a faster and greater dilution on tapping and hence to increased lutoid damage and faster plugging.

While these investigations established that lutoids can be disrupted by a sufficient degree of osmotic shock, it is not certain that osmotic changes within the vessels after tapping can alone account for lutoid damage. Pakianathan and Milford (1973) recorded only a drop from 500 to 435 mOsm/litre in vessels 5 cm below the cut, and even in collected latex the minimum they recorded was 340 mOsm, but Pujarniscle *et al.* (1970) considered that lutoids are perfectly stable at 350 mOsm.

Yip and Southorn (1968) produced evidence that lutoids could be disrupted by the mechanical shearing forces to which latex is subjected when flowing through the vessels under high pressure gradients after tapping. Optical and electron micrographs showed distortion of particles along flow lines in the vessels, indicative of shear stress. When fresh latex was forced through glass capillaries of similar internal diameter to that of latex vessels, lutoids began to break down at pressure gradients above 0.2 atm/mm and at gradients of 0.4–1.2 atm/mm the capillaries plugged and flow stopped abruptly. With latex freed from lutoids there was no plugging and flow could be maintained indefinitely even at 13 atm/mm pressure gradient. The evidence of these experiments is not conclusive since the conditions differ from those in latex vessels, but it does suggest that shear of lutoids could be a factor in plugging.

Factors other than lutoids involved in slowing and stopping flow

It is possible that differences between clones in the composition of the protective film on the rubber particles may be partly responsible for differences in the ease with which flocs are formed within the vessels. Rubber particles are strongly protected by a complex film of protein and lipid material. Ho *et al.* (1976) found marked differences between clones in neutral lipids content of the rubber phase; on the other hand, phospholipid contents did not differ very much. These clonal differences are reflected in the values of acetone extract of total solids films (Subramaniam 1976) and there appears to be a negative correlation with plugging index; long-flowing clones such as RRIM 501 have a high neutral lipid content in the rubber phase (as much as five times that of the high plugging clone Tjir 1). Sherief and Sethuraj (1978) found, in addition to a negative correlation between phospholipids of the lutoid membrane and bursting index, a negative correlation between triglyceride (the major neutral lipid) in the rubber phase and plugging index. Premakumari *et al.* (1980) noticed that during wintering, when plugging index and bursting index increase, the neutral lipid content of the rubber particles decreased slightly, as did the phospholipid content of the bottom fraction. These findings suggest that, in addition to the effect of lutoid behaviour, plugging is influenced by the degree to which rubber particles are 'protected' by lipids.

While slowing down and cessation of flow after tapping seem to be mainly due to plugging within the vessels, it is likely that, at least towards the end of flow, coagulation on the cut also contributes. The formation of the external cap of coagulum (the 'tree lace') over the cut ends of the vessels may partly result from the aggregation of rubber particles and damaged lutoids but is at least partly caused by the action of sap released from damaged bark cells. The flocculating activity of this sap was reported by Cook and McMullen (1951) and has been investigated by Gomez (1977) and by Yip and Gomez (1984). High molecular weight substances in the sap are the major active components but the inorganic cations present (Na^+, K^+, Mg^{++}, Ca^{++}) are also essential for the flocculation of the latex particles. Gomez (1977) found that there were clonal differences in the flocculating activity of bark extracts. Paardekooper (unpublished) recorded that the net weight of tree lace of 100 clones varied from 0.2–2.6 g/tree but found no significant correlation with plugging index.

Conclusions on plugging mechanisms

Despite much research, the mechanisms involved in latex flow after tapping are still not fully understood. It is clear that after the initial rapid drop in flow rate the subsequent slowing up and cessation of flow is due to internal plugging of vessels near their cut ends and, perhaps, to the external cap of coagulum that forms over the cut ends. The precise contribution of these two impediments has not been established, but it is likely that internal plugging is mainly responsible for slowing down flow, with the external cap playing a part towards the end of flow and eventually stopping it. The internal plugs consist of aggregates of rubber particles and damaged lutoids formed by the flocculating effect of B serum released from ruptured lutoids. The lutoids are probably damaged both by osmotic effects and by the mechanical shearing forces to which the latex is subject when flowing through the vessels under high pressure gradients. During the initial stage of rapid flow, the latex is swept out of the vessels before there is irreversible damage to the lutoids; further damage occurs after the latex has left the vessels and some of the resulting flocs are retained on the cut. Later, when flow is slower, flocculation occurs within the vessels and an increasing number of them become completely blocked. The external cap may partly result from the destabilizing effect of B serum released from damaged lutoids, but its formation is at least partly due to the flocculating activity of sap from bark cells damaged in tapping.

Drainage area (the area of bark which contributes latex at a single tapping)

It was obvious to early investigators of latex flow that a knowledge of the extent of the drainage area could contribute to the rational development of tapping systems. Frey-Wyssling (1952) and others attempted to calculate the drainage area from anatomical observations and theoretical considerations of the flow process. Based on the volume of latex vessels, the volume of the total tissue in the bark, and an assumed proportion to which vessels are emptied on tapping, an estimate can be made of the area of bark that contributes to the latex at one tapping.

Lustinec *et al.* (1966) defined the part of the bark from which latex runs out during one tapping as the 'outflow area'; and the area of bark from which latex, after tapping, flows towards the cut without actually reaching it as the 'displacement area'. An even larger area, called 're-equilibration area', shows changes in latex concentration, including some slow movement towards the cut. Observations of Paardekooper (unpublished) that tapping lowers carbon content of leaf petioles (in high-yielding clones as much as 20 per cent one year of tapping), suggest that the re-equilibration area extends throughout the tree, including the leaves.

The actual outflow area has been measured by various methods, including micro-tapping, measurements of bark contraction, turgor pressure measurements and the introduction of labelled (radioactive) chemicals into the bark at various distances from the cut. Lustinec *et al.* (1969) measured trunk contraction to delineate outflow areas; shrinkage was greatest 10 minutes after tapping, 5 cm below the top end of the cut. These authors proposed a parabolic relationship between yield and vertical expansion of flow area. For a review of the most interesting work with radioactive chemicals, see Lustinec *et al.* (1966) and Lustinec *et al.* (1968). The outflow area was shown to extend to 40–60 cm below the cut, and to a lesser degree above the cut. Based on various observations, these authors assumed that latex flows directly from the renewed bark above the cut to the tapping cut; Ribaillier (1972) estimated that as much as 30 per cent of the latex collected comes directly from above the cut. These findings are contrary to the generally accepted view that because of the insignificant number of connections between latex vessel rings, there can be no flow from the unsevered latex vessels in the renewed bark, and that the contribution of the bark areas above the panel has to be channelled 'around' the tapping cut, through the undisturbed virgin bark on the other side of the tree. This controversy has yet to be resolved.

Pakianathan *et al.* (1976) studied the drainage area by measuring the difference in turgor pressure in the laticiferous system before and after tapping. Pressure falls on tapping were observed in upward, downward and lateral directions from the cut. The resulting shape and extent of the drainage area are illustrated in Fig. 9.9. Falls in turgor give a measure of the overall drainage area arising from the displacement of latex and water; they do not make it possible directly to separate the 'outflow', 'displacement' and 'equilibrium' areas.

It has often been assumed that two cuts of which the outflow areas do not overlap are independent, so that older mature rubber could most efficiently be exploited by two cuts at a sufficient distance apart. However, in high-yielding clones the displacement area is considerably greater than the outflow area, while the re-equilibration area involves the whole tree; this may explain why the total yield obtained from tapping two cuts simultaneously is generally less than double the yield obtained by tapping either of the two cuts alone.

The yield determinants

Bearing in mind the anatomical features of the bark, described in

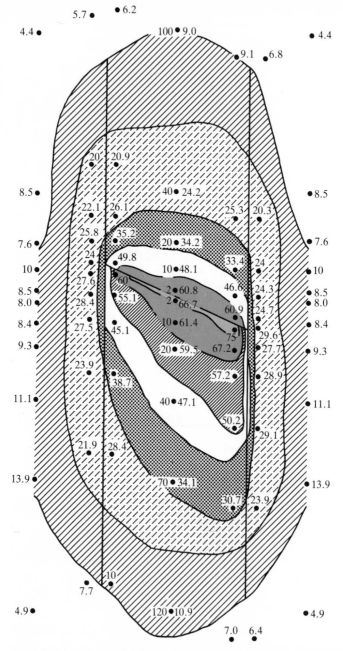

Fig. 9.9 Map of potential displacement area of a tree of clone RRIM 600 as indicated by falls in turgor pressure following tapping (After Pakianathan *et al.* 1976)

Chapter 2, and the above discussion of the flow behaviour of latex upon tapping, the various factors which determine the yield of a tree can now be summarized. The two major yield components are initial flow rate and duration of flow. The initial flow rate per unit length of cut is determined by: (a) the amount of latex-bearing tissue in the bark, which varies between cultivars and with age of the tree; and (b) the turgor pressure in the vessels, which is influenced by time of day and by the tapping system, and perhaps also varies between clones. The duration of flow is determined by the plug formation in the vessels and by the cap formation through coagulation on the cut. It is strongly influenced by cultivar and by the tapping system, long cut systems decreasing the plugging and prolonging flow. The plugging of vessels is influenced by differences in stability of the lutoids, which are disrupted by shear and by changes in osmotic pressure caused by the dilution reaction. It is also influenced by clonal differences in the neutral lipid content of the rubber phase, while differences in the flocculating activity of the bark sap affect formation of the external coagulum cap.

In the longer term, the yield is also determined by the rate of rubber biosynthesis. The balance between latex withdrawal and rubber regeneration is reflected in the dry rubber content of the latex. Between cultivars, an indication of rubber regenerating efficiency is provided by the difference between total solids *in situ* in untapped trees and that in tapped trees. More intensive tapping results in a lower d.r.c. equilibrium.

9.4.3 Rubber production and growth of the tree

Rubber formation in the tree can be expected to compete with growth for the metabolites resulting from photosynthesis. The growth rate of a crop depends on the amount of photosynthetically active radiation absorbed by the leaves, the leaf area index and the net assimilation rate. Few growth studies have been carried out with tropical perennials. The growth and related parameters of a number of such crops, including rubber, have recently been discussed by Corley (1983). Results of growth studies on immature rubber were published by Templeton (1968).

The growth rate expressed as the increase in total dry weight of a plant is normally calculated from periodic dry weight determinations. As, however, it is impractical to measure the dry weight of complete trees, except on a very small scale, various research workers have studied the relationship between tree circumference and dry weight, so that dry weight increments could be calculated from simple girth measurements. For rubber, it was found that the relationship between girth and weight is given by

$$W = a.G^{-b}$$

where W denotes shoot dry weight
   G denotes girth, and
   a and b are constants

Published values for b vary between 2.4 and 2.9. Shorrocks *et al.* (1965) developed the equation

$$W = 0.0026 \ G^{-2.78}$$

which has a subsequently been used by other authors to calculate the dry shoot weight increment of both tapped and untapped trees from consecutive sets of girth increments.[1]

The annual dry shoot weight increment of a stand of untapped trees is of the order of 70 kg/tree, or 30t/ha. Assuming that this rate remains constant throughout the life of the trees, it can be shown that the corresponding annual girth increment of these trees would gradually decline from about 10 cm to less than 2 cm.

When trees are brought into regular tapping, the assimilates, so far used for growth only, are partitioned between rubber production and vegetative growth. Consequently, the growth rate is reduced, to a varying extent, as a result of tapping; part of the potential shoot weight increment becomes, as it were, 'unrealized'. In general, the higher the yield – be it through a high intensity of tapping, through stimulation, or genetic constitution – the greater is the 'loss' in shoot dry weight increment and hence in girth increment, compared with untapped trees; this loss can be as high as 50 per cent. The partitioning between rubber production and vegetative growth has often been expressed as a partition ratio, similar to the harvest index in other crops; various formulae have been used, such as

$$\frac{2.25 \ R}{W + R}$$

in which R = rubber production (kg), and
   W = dry shoot weight production (kg)

The factor 2.25 allows for the fact that the energy content of rubber is at least double that of wood. The partition ratio varies with the intensity of tapping, clone and age, between 9.6 and 44.3 per cent (Templeton 1969a).

Various authors have found that the amount of rubber produced is less than the 'unrealized' shoot dry weight increment, even allowing for the difference in energy content. Hence, the ratio rubber/shoot loss is less than 1 : 2.25 or 44 per cent; this ratio has been found to vary from 10 to 40 per cent: 1 kg rubber would thus be equivalent to 1.0–2.5 kg dry weight increment. One possible explanation for this unfavourable rubber conversion rate is that in tapping many other high energy value products besides rubber are lost and need to be re-synthesized.

Thus, the rubber production of a tree depends upon the total annual biomass increment, the partition ratio, and the rate at which the unrealized dry weight increment is converted into rubber (Simmonds 1982). To this should be added that tapping not only results in a reduced shoot weight increment, but also causes a gradual change in weight distribution between crown and stem, in favour of the former. This has important implications for wind damage susceptibility.

There are marked differences in rubber/shoot loss ratio between cultivars; two clones can have the same yield level but differ in the degree to which their girth increment is reduced through tapping. This will gradually change the apparent relative 'vigour' of a clone; for example, RRIM 501 exhibits a slightly better growth rate during immaturity than RRIM 600, but its girth rate on tapping is sharply reduced, so that after 10 years' tapping trees of RRIM 501 are very much thinner than those of RRIM 600. To avoid confusion, the term 'vigour' is now confined to the girth at the time of commencement of tapping.

There appears to be a most interesting relation between the plugging behaviour (i.e. the latex flow pattern) of a clone and the degree to which girth increment is reduced on tapping. Clones that have a short flow time, a high plugging index and exhibit a high initial flow rate (usually corresponding with a high number of latex vessel rings) have, on average, better girth increments than long-flowing, lower plugging clones. Thus, high index clones have a more favourable rubber/shoot loss ratio than do clones with low plugging indices. For example, Waidyanatha and Pathiratne (1971) found that within a population of clones there was a partial correlation coefficient between plugging index and girth increment, at constant yield, of 0.69.

Numerous exploitation experiments have shown the pronounced effect of tapping system and of yield stimulation on girth increment, and this effect is an important criterion in the choice of the most appropriate exploitation system. As was found with differences between clones, long cut systems, which are characterized by low plugging and long flow, are associated with poor girth increments (Wycherley 1975,1976b); the rubber is produced at greater cost to accumulation of dry weight, the rubber/shoot loss ratio is low and the partition ratio high. To illustrate the effect of tapping intensity on girth increment, data from an experiment with 30 tapping systems (Paardekooper *et al.* 1976) are shown in Fig. 9.10.

Various authors have tried to estimate the physiologically maximum yield that could be expected from a mature stand of rubber, and arrived at estimates varying from 7000 to 12,000 kg dry rubber per ha (Templeton 1969b; Sethuraj 1981; Corley 1983). The maximum annual growth rate that would seem feasible on the basis of leaf area index and conversion efficiency would be of the order

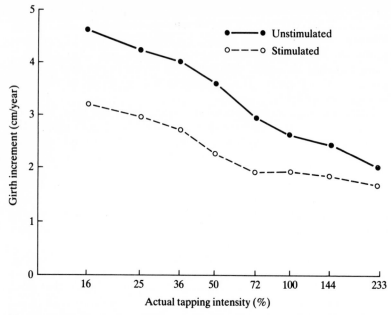

Fig. 9.10 Relation between tapping intensity and girth increment (After Paardekooper *et al.* 1976)

of 40 t/ha of dry matter; if half of the assimilates are used for rubber production (thereby reducing girth increment to 42 per cent of that of untapped trees), and if the rate at which 'unrealized' dry matter increment is converted to rubber equals the theoretical maximum of 44 per cent, the rubber production would be 44 per cent of 20 t, or around 9000 kg/ha. Such trees, however, are likely to suffer from severe wind damage, as a result of the unfavourable crown/stem dry weight ratio.

9.4.4 Dryness ('brown bast') of tapped trees

The absence of any latex exudation, either over part of the cut or over its entire length, is generally referred to as dryness. In most cases complete dryness is associated with a typical brownish discolouration on and below the tapping cut. In severe cases bark cracking and flaking follow, sometimes involving the entire length of the trunk below the cut. All stages of this disorder are commonly referred to as 'brown bast'.

Features, causes and possible treatment of brown bast have been the subject of much research since the beginning of this century, as dryness is often a serious problem on estates and smallholdings; in the latter, incidence of dry trees may be as high as 30 per cent. While from time to time pathogens have been proposed as the cause of brown bast, it is now generally considered to be a physiological disorder associated with excessive exploitation. The underlying mechanisms of the disorder, however, are still not completely clear.

Incipient brown bast trees often show a prolonged flow of watery latex; subsequently, translucent bodies appear on the cut; there is pre-coagulation on the cut, causing spillage of latex over its edge. Once latex flow has stopped, the typical discolouration can be observed, appearing as brownish streaks or patches on the cut. Upon scraping, or in anatomical sections, this discolouration is seen to spread rapidly downwards following the vessels; within 2–5 months the entire panel can be affected. With upward tapping, the symptoms progress upwards.

In young cases, the disorder does not spread to the renewed bark above the cut, nor to the untapped bark on the opposite side of the tree to the panel, but in older cases brown bast spreads slowly throughout the vessel rings, and even to originally unaffected inner rings. In advanced stages of the disorder, many trees show cracking and heavy flaking of the bark; no flow of latex occurs when the bark is punctured and it is found that all latex has coagulated in the vessels. The brown colour is caused by tannin deposition in parenchyma cells around the vessels. Cross walls, tyloses and stone cells also develop, indicating general disintegration of cellular organization.

The work of Chua (1967) appeared to show that an earlier view that dryness resulted from deficiency of assimilates, especially carbohydrates, was incorrect. He could detect no difference in the level of soluble sugars and starch in normal and dry bark tissues and he suggested that an excessive loss of enzymes and other proteins disrupts normal cell metabolism. However, other workers have found that high intensity tapping and/or stimulation reduces the sucrose content of latex (Tupy 1973f; Tupy & Primot 1976) and that bark dryness is associated with low latex sucrose levels. Tupy (1985) considers that inadequate sucrose availability for basic biological functions of latex vessels may result in their premature degeneration and in bark dryness.

From work reported by Paranjothy *et al.* (1976) it seems likely that dryness is initiated by massive lutoid damage and release of B serum, leading to coagulation of latex within the vessels. The initial damage is possibly related to an extreme dilution reaction, as shown by the often watery latex of incipiently dry trees. Recently, Chrestin *et al.* (1984) postulated that intensive tapping or intensive use of stimulants increased NAD(P)H oxidase activity at the lutoid tonoplast; this would lead to damage to the lutoid membrane, causing flocculation through the release of B serum.

Sivakumaran and Pakianathan (1983) induced dryness by puncturing the bark at four or five points along vertical strips on the panel and sealing the punctures with drawing pins. By repeating this operation weekly or fortnightly, sharply reduced flow rates and a high incidence of dryness were induced in several clones. The authors do not attempt to explain this interesting phenomenon; it could be that each puncture induces localized *in situ* coagulation of the latex, which would trigger off further floc formation.

An aspect of brown bast that is puzzling is its non-random distribution in the field; affected trees tend to come in clusters, suggesting that brown bast spreads along the rows. Non-randomness of dryness was statistically proved in Thailand in a field of seedlings (Paardekooper, unpublished) and recently in Sri Lanka for 5 out of 6 clones on commercial estates (de Soyza *et al.* 1983). The non-random occurrence of brown bast suggests the involvement of a particular environmental factor, since it seems unlikely that the disorder would spread through the vessels via rootgrafts to neighbouring trees.

It is well known that brown bast incidence increases with tapping intensity. For a given intensity, long cut systems seem to lead to higher incidence. Increased yield through stimulation is almost invariably associated with higher brown bast incidence. Various workers have induced severe brown bast by intensive tapping; on the other hand, Paardekooper *et al.* (1976) observed only a very moderate incidence in two clones after five years of 300 per cent intensity tapping.

It is also well known that large differences exist between cultivars in susceptibility to brown bast; notoriously sensitive are PB 5/63, PB 28/59 and RRIM 628. Seedling trees are always more susceptible than bud-grafts; this has been confirmed in breeding experiments where legitimate seedlings are compared with buddings derived from them. The reason for this is not clear. Among clones there appears to be a negative correlation between incidence of dryness and plugging index: long-flowing clones tend to develop more brown bast. Paardekooper (unpublished) examined the records of a number of clones and found a positive correlation between incidence of dryness and susceptibility to wind damage, suggesting that a certain form of dryness is induced by strain in the tissues associated with bending under wind stress. Trees which have actually been damaged by wind often develop dryness.

Brown bast incidence can usually be limited to acceptable levels of around 1 per cent by lower tapping intensities such as $\frac{1}{2}$S d/3 (67 per cent); in susceptible clones the use of stimulants on young mature trees should be combined with even lower intensities. The occurrence of partial dryness on a large number of trees could be used as an indicator of incipient brown bast, but in practice partial dryness is difficult to observe. Trees which have developed complete dryness are usually rested for six months, after which tapping is resumed in healthy bark, often on a reduced cut length. However, the fact that in most trees brown bast will gradually spread throws doubt on the value of resting affected trees, and usually only a small percentage of the trees actually recover (Paranjothy & Yeang 1978).

Appreciation of the fact that brown bast is initiated in, and spreads along, the vessels has recently led to a re-examination of an old recommendation to isolate affected bark areas from healthy ones by a system of horizontal and vertical grooves. Detailed recommendations for smallholdings, where brown bast is a particular problem because of the tendency to overtap, have been published by Anthony *et al.* (1981). As a precautionary measure it is recommended that three months before starting tapping all trees should be provided with a single vertical isolation groove to the wood on the back guide line of the tapping panel from the cut to the base of the tree. In the event of dryness developing, this groove should ensure that it does not spread to the second panel. In trees which develop partial or total dryness a further system of isolation grooves is proposed. Tapping should be resumed over the full cut but with periodic rest to minimize the recurrence of dryness.

9.5 Yield stimulation

9.5.1 Historical review

Since the early days of the rubber industry there has been an interest in yield stimulants. Smallholders in various countries frequently applied mixtures of cow dung and clay to the bark to improve yield and bark renewal; later, it was established that light scraping of the bark below the cut, as well as application of palm, or other, oil had similar effects. Chapman (1951) showed that all effective treatments with oils, cow dung, etc., contained plant hormones and that synthetic growth regulators were effective as yield stimulants; in particular, 2,4-dichlorophenoxyacetic acid (2,4-D) applied to scraped bark below the cut induced large increases in yield. As a result of this work, a proprietary yield stimulant 'Stimulex', based on the butyl ester of 2,4-D in palm oil, was developed and soon became popular in Malaya and other countries.

At the same time the Rubber Research Institute of Malaya started extensive experimentation with synthetic growth regulators and obtained particularly good responses with 2,4,5-T (trichlorophenoxyacetic acid) and its use in mixture with red palm oil and petrolatum was recommended (Baptist & De Jonge 1955). The same authors reported for the first time that the application of growth regulators on the tapping panel above the cut increased yield and accelerated bark renewal.

The large number of papers which appeared in various rubber-producing countries in the period 1955–70 reflects the importance attached to yield stimulation, both in research programmes and in commercial practice; a detailed review is provided by Abraham and Tayler (1967b). During this period, commercial yield stimulation remained confined to older mature stands (at least 12 years in tapping) since treatment of younger trees was considered risky. Various factors which influence the response were studied, such as the quality of the bark, the age of the tree, the cultivar, the tapping system, the concentration of the active ingredient in the mixture, and the method and frequency of application.

Meanwhile, the search for more effective and safer stimulants continued. A wide range of biologically active compounds was studied by the Rubber Research Institute of Malaya, but none appeared to be more effective than 2,4-D or 2,4,5-T (Abraham *et al.* 1968). In Vietnam, Compagnon and Tixier (1950) first reported that injection of copper sulphate into the wood produced a similar response to that obtained from growth substances; this was later confirmed in other countries. Mainstone and Tan (1964) reported that the use of copper sulphate in association with 2,4-D or 2,4,5-T in low concentrations had an additive effect. To avoid damage to the tapping panel, copper sulphate pellets were inserted into holes drilled near the foot of the tree, using lightweight petrol-driven drills (Lowe 1964). Nevertheless, stimulation by means of copper sulphate never reached large-scale application, partly because of fear of copper contamination of the latex, which could affect the manufacturing properties of the rubber. (There has been little speculation as to the mechanism through which copper acts as a yield stimulant; in this connection it is of interest to note that high yield, whether the result of cultivar, tapping system or stimulation with growth substances, is invariably associated with a high copper content in the latex.)

Taysum (1961) reported in Malaysia the first use of a gas, ethylene oxide, which increased latex flow dramatically; some years later in Vietnam acetylene gas was found very effective. Abraham, Wycherley and Pakianathan (1968) were among the first to report on the yield-stimulating effect of 2-chloroethyl phosphonic acid (CEPA or ethephon), and to comment on the fact that all known substances with yield-increasing properties appeared to exhibit the capacity to produce ethylene in plant tissues. About the same time, d'Auzac and Ribaillier (1969) in the Ivory Coast reported large yield increases of trees treated with ethylene, acetylene and ethephon. During the early 1970s, a commercial preparation of ethephon 'Ethrel'® gradually replaced the other types of stimulant. Extensive research was initiated into its action, its effects, methods of application and interaction with cultivar, age of the trees and tapping system.

When it was realized that the common factor of all compounds able to increase latex yield was their ethylene-generating property – a hypothesis fully confirmed by Audley *et al.* (1978), who demonstrated by treatment of leaflets and pieces of stem of *Hevea* that the most effective ethylene inducers were also the best yield stimulants – further work on a wide range of ethylene-generating substances was carried out at the Malaysian Rubber Producers' Research Association in England. No compounds of greater effectiveness than ethephon have, however, been reported so far.

Another line of research aimed at a practical method of using ethylene gas by absorbing it on activated charcoal and other potential absorbent materials (Dickenson *et al.* 1976); since ethylene is much cheaper than ethephon formulations, this could bring down the cost of stimulation considerably. Molecular sieves (artificial zeolites) proved the most effective, and a system with a commercial potential named 'Ethad' was developed. However, up to the time of writing, ethephon has remained the only compound widely used on a commercial scale.

In the following sections, the various effects of ethephon application, the factors influencing its effect, and the possible mode of action of ethylene will be discussed briefly. For a more detailed review of the subject, the reader is referred to Bridge (1983).

9.5.2 Effects of yield stimulation

Effects on yield

Yield increases resulting from stimulation are extremely variable. Peak yields of over 1 litre of latex per tree per tapping have been reported, but at the other extreme, over-exploited trees often show a negative response after repeated stimulation. The usual range reported from experiments or commercial practice with ethephon is from 20 to 100 per cent increase in rubber production averaged over one stimulation cycle.

Chapman (1951) observed that stimulation increased yield mainly by prolonging latex flow and that the effect lasted up to three months after the first application. Since then, many workers have shown that the primary response to stimulation is increased flow time due to a decrease in plugging intensity and that the initial flow rate is probably not affected. One practical consequence of the prolonged flow is that either a second latex collection or a collection of the late drip in the form of cup lumps is required to benefit fully from stimulation, which thus usually increases the proportion of lower grades.

Repeated bark application of stimulant at regular intervals results in peak yields during the first 10 or so tappings after application, followed by a decline; modern methods of application therefore aim at obtaining a more even yield pattern. The decline in response corresponds with a gradual increase in the plugging index to its original level. Subronto and Napitupulu (1978) found that the ratio of the plugging index of stimulated to that of unstimulated trees increased linearly with days after stimulation. It has also been generally observed that the response to repeated stimulations becomes progressively lower compared with that obtained in the first cycle.

Dry rubber content

Yield stimulation invariably results in a decrease in d.r.c., varying from 2 to 7 percentage points. The hypothesis that this was a direct effect of the stimulant and that the lower latex viscosity was the main reason for the longer flow time became untenable when it was demonstrated that an increase in flow duration occurs within 24 hours of stimulant application, while the effect on d.r.c. is usually only noticeable one or two tapping days later. It is now generally assumed that the increased latex extraction after stimulation shifts the balance between rubber biosynthesis and extraction to a lower d.r.c. equilibrium. In practice, the decrease in d.r.c. can be avoided by changing to less intensive tapping systems.

Latex composition

Stimulation increases the content of nutrient elements (except magnesium) in the latex; this, combined with the lower d.r.c. and increased yield, causes a marked increase in nutrient drainage, which is proportionately much greater than the increase in rubber yield. For this reason, additional fertilizer applications are recommended whenever trees are stimulated. Of more academic interest is the marked increase in the copper content of the latex, probably related to increased enzyme activity. Immediately after stimulation, the invertase activity and sucrose content of the latex are increased temporarily and the starch content of the bark lowered (Chong 1981), but subsequently both the sucrose and the total cyclitol contents of the latex are decreased (Tupy 1973f; Low 1978). Eschbach *et al.* (1984) found a marked and persistent increase in the level of thiols in the latex; these are known to play a role in the protection of lutoid membranes and in the activation of certain enzymes. Latex obtained from stimulated trees appears to be more stable than that from untreated trees, in the sense that spontaneous coagulation is delayed.

Technological properties of latex and rubber

Latex concentrate prepared with latex from stimulated trees has a slight increase in KOH number (Gorton 1972). The same author found a decrease in magnesium and an increase in copper content of raw rubber films. In processed rubber, copper content is usually normal. Subramaniam (1972) observed a temporary effect on the viscosity of the rubber, but no effect on the plasticity retention index. Sumarno-Kartowardiono *et al.* (1976) reported that stimulation had no appreciable effect on any of the rubber properties.

Dryness

Stimulation often increases the number of wholly or partially dry trees especially in young mature stands. This danger can be minimized by reducing the tapping intensity.

Girth increment

Because of the increase in rubber extraction and the change in flow pattern to longer flow, which has been shown to decrease the rubber/shoot ratio, stimulation usually has a depressing effect on girth increment; by reducing the tapping intensity this effect can be limited. In older trees, to which stimulation is usually confined, the girth increment is less important.

Bark renewal

Application of a yield stimulant above the cut increases the initial rate of bark renewal, but this beneficial effect is confined to the first year of renewal and is of no consequence by the time the renewed bark is due for tapping. The young renewed bark under stimulation has been shown to have no more latex vessel rings than the bark without stimulation (de Jonge 1957).

Leaf composition

Stimulation has lowered the nutrient content of leaves in certain experiments, no doubt as a result of increased latex withdrawal. Similarly, the rubber hydrocarbon content of the petioles has been found to be depressed six weeks after a stimulant application (Ho & Paardekooper 1965).

Long-term effects

Results of a series of experiments in Malaysia with continuous ethephon stimulation during nine years have been reviewed in detail by Sivakumaran *et al.* (1981, 1982, 1983, 1984). With conventional ½S d/2 tapping, there was generally a progressive decline in response, a negative response being common in certain clones in the later years. Low frequency systems showed progressive declines in response as the cut moved down the panel (possibly because of limitation of the drainage area), but a resurgence of response when cuts were changed to a new panel. In some clones, cumulative yields over nine years on d/4 tapping with stimulation were comparable to those on d/2 tapping without stimulation. With three clones (Tjir 1, RRIM 605 and RRIM 623), particularly good results were obtained with $\frac{1}{4}$S d/2 (t, t), a short-cut panel changing system. This system of 50 per cent intensity appeared to have caused minimum stress in the trees and to be most suitable for combination with long-term stimulation starting early in the life of the tree.

In the Ivory Coast, where the standard system was formerly S d/3 6d/7 without stimulation, a series of long-term experiments with several clones, and also 8–9 years' records from commercial plantings, showed that as good, or better, yields could be obtained by tapping $\frac{1}{2}$S d/3 6d/7 10 m/12 (48 per cent) with four applications of 5 per cent ethephon per annum. With the latter system there were less dry trees and lower bark consumption, and the cost of stimulation was more than compensated for by the decrease in tapping cost resulting from a larger task (Eschbach & Banchi 1985).

Pakianathan *et al.* (1982) studied the physiological and anatomical changes associated with repeated stimulation of trees tapped $\frac{1}{2}$S d/2 with 10 per cent ethephon over periods of 3–9 years. Initial flow rates were markedly reduced, while large areas of laticiferous tissue below the tapping panel showed abnormally low turgor pressures, an increase of stone cells in the bark, and various abnormalities in the content of latex vessels. It seems likely that these adverse effects are caused by the excessive extraction of latex (over-exploitation) rather than by ethephon *per se* (Sivakumaran *et al.* 1983). On the other hand, the thickening of the cell walls observed by various workers could be considered as a direct effect of ethylene.

9.5.3 Factors affecting the response to stimulation

The response to stimulation depends on a number of factors, some of which are intrinsic to the tree, while others can be controlled. Among the former, the most important are age and cultivar. Numerous experiments as well as commercial practice have demonstrated large differences in response to stimulation between clones. The traditional practice whereby response is expressed as a percentage of the control yield leads to a negative correlation between response and yield level of the control (see, for example, Sethurai and George 1975); hence, generally high-yielding clones show a lower relative response. However, even if the response is expressed in absolute terms, considerable clonal differences remain; these are mainly related to the flow pattern, as indicated by the plugging index, of each clone: high plugging, short-flowing clones respond much better than low pluggers; this is not surprising since the main effect of the stimulant is to suppress plugging and prolong flow time. Ho *et al.* (1973a) have classified clones in large-scale variety trials in Malaysia, on the basis of absolute yield increase, in three classes:

Good   RRIM 600, 605, 607, 612, 614, 623; Tijr 1; LCB 1320;
          PR 255; AVROS 2037
Medium   RRIM 513; PB 5/51, 5/63; GT 1; PR 107, 261
Poor   RRIM 501, 519, 526, 613, 628, 701, 707; WR 101; PR
          251

Other intrinsic factors which affect the response to stimulation are
the age of the trees, their nutrient status and their tapping history.
In commercial practice hitherto stimulants have mostly been applied
to renewed bark after 10 or more years' tapping as it was found
that, young trees respond poorly, especially to repeated applications. However, it now seems that satisfactory responses can be
obtained from young trees if a low concentration of ethephon
(1.0–2.5 per cent a.i.) is used and tapping frequency reduced to
third daily during the first year (see p. 399). Responses are poor
from trees of low nutrient status, or following intensive tapping.

The major factors contributing to the response to stimulation that
can be manipulated are the type of active ingredient, its concentration, the method of formulation, the mode of application, the
frequency of application and the tapping system adopted.

Method of application

The oldest method ('bark application') involves scraping the bark
directly below the cut over a width usually equal to the amount of
bark to be consumed over a two-month period (Fig. 9.11). Scraping
the bark removes the corky layer and should be done carefully so
that the whitish layer of hard tissue becomes exposed without
'bleeding' of latex. In upward tapping, the bark above the cut is
similarly treated. The main disadvantages of the bark application
method are the cost of scraping and the fact that peak yields
obtained within the first 10 tapping days are followed by lower
yields. Probably the most widely used method is 'panel application'
in which the stimulant is painted at monthly intervals directly on to
a 1.5 cm band of renewed bark, immediately above the cut for
downward tapping or immediately below the cut in the case of
upward tapping. In spite of the more frequent application, this
method is less labour-intensive than bark application; it is very suitable for trees with poorly renewed bark. A third method called
'groove application' was introduced in the early 1970s (P'ng *et al.*
1973) in which a thin layer of the stimulant is applied with a small
brush to the tapping cut, after removal of the tree lace. For monthly
applications, the amount of stimulant used would be about one-third
of that used for panel application, but twice monthly application is
usually recommended in order to obtain an even response pattern.
More recently, a modified groove application method has become
popular, in which the stimulant is painted directly on to the tree
lace, overlapping slightly on to the renewing bark (Fig. 9.12). As

Fig. 9.11   Half spiral tapping cut with stimulant applied to scraped bark below the cut

the tree lace does not have to be removed this 'lace application' method is easier and saves labour. Since the stimulant – (applied on a non-tapping day – would be partly removed by the next day's tapping, there were doubts as to whether there was sufficient time for the stimulant to be absorbed, but Sivakumaran (1982) showed that an interval of about 12 hours between application and first tapping is indeed sufficient. Sethuraj *et al.* (1977), aiming at extending the drainage area, found that stimulation of many separate patches up to the branches gave a much enhanced yield response, but these and other possibly very effective methods, such as long vertical bands, are impractical. A special method of application in relation to puncture tapping is discussed in section 9.5.8.

Concentration

In the early experiments of the Rubber Research Institute of

Fig. 9.12 Application of stimulant to the lace on the cut (Rubber Research Institute of Malaysia)

Malaysia on older mature trees a mixture containing 10 per cent a.i. of ethephon was used, and the commercial stimulant 'Ethrel' was of the same concentration. Subsequently, many researchers found that concentrations ranging from 1 to 15 per cent gave similar responses when applied to the bark (Ribaillier & d'Auzac 1970; de Jonge & Tan 1972; McCulloch & Vanialingam 1972; Ross & Dinsmore 1972). With application to scraped bark, Anekachai *et al.* (1975) similarly found little effect of concentration, 0.5. per cent being almost as effective as 10 per cent, but for application above the cut at least 2 per cent was necessary and the maximum response was obtained in the range 5.0–7.5 per cent. Tan and Menon (1973) showed that, both for above and below the cut, increasing frequency of application to a narrower band was more effective than increasing concentration. They concluded that the optimum amount of ethephon per tree per year was of the order of 1 g.

Results of more recent experiments in Malaysia (Abraham 1981) suggest that where stimulation is effectively supervised, concentrations of 2.5–5.0 per cent will give the best long-term response. With young trees even lower concentrations, down to 1.0 per cent, are usually effective, provided that the frequency of application is increased. Work reported by Sivakumaran (1983) suggests that about 600 mg a.i. per tree year is adequate for a satisfactory response; this amount can be supplied, for instance, by groove application of a 2.5 per cent formulation at the rate of 0.5 g per application at weekly intervals. In line with the current emphasis on lower concentrations, 'Ethrel' is now available with 5.0 per cent and 2.5 per cent of active ingredient.

Formulation

Ethephon is highly soluble in water, but for conventional methods of application a more viscous product is needed. 'Home-made' formulations are usually based on palm oil, but it is difficult to achieve a homogenous and stable mixture. The commercially available 'Ethrel' formulations are based on synthetic gum material; they also include a dye to facilitate supervision. In the groove application method no differences in response between water- and oil-based mixtures have been observed. For bark application, formulations containing bark penetrants have occasionally been used; these make scraping unnecessary.

Tapping system

The interactions between stimulation and tapping system are of the utmost importance. In general, the relative response to stimulation is negatively correlated with tapping intensity; this is illustrated in Fig. 9.13. Since shorter cuts are known to be associated with increased plugging in the vessels, it is not surprising that full spiral systems respond poorly to stimulation. The high response observed with very short cuts has led to the development of micro-tapping, discussed in section 9.5.8. The generally good response to stimulation in high cut tapping is possibly related to the higher magnesium content of latex at higher levels in the tree, which causes early plugging of the vessels.

In younger mature trees, tapped in virgin bark on low panels, it has been shown that with most clones continuous stimulation combined with standard half spiral, alternate daily tapping leads to a gradually diminishing response, to unacceptable levels of dryness and to a poor girth increment. Stimulation therefore has to be combined with a lower tapping intensity, i.e. usually a lower tapping frequency. Used in this way, stimulation is less a means for obtaining substantial yield increases than a method of reducing tapping costs through increased yield per tapping, or of overcoming the problem of shortage of skilled tappers, with the additional advantage of a more efficient use of bark reserves and a potential increase in the economic life of the tree. As trees become older and are tapped in renewed bark, stimulants can be used to obtain moderate increases in yield without incurring panel exhaustion and dryness. The following relative increases are considered judicious in Malaysia (Abraham 1981):

Panel B01 (panel A)   10–15%
Panel B02 (panel B)   15–25%

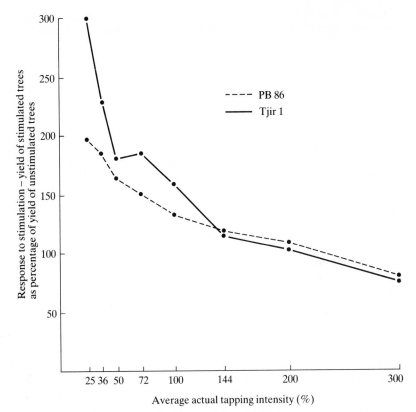

Fig. 9.13 Response (averaged over three years) of two clones to stimulation in relation to tapping intensity (After Paardekooper *et al.* 1976)

Panel BI1 (panel C) 25–40%
Panel BI2 (panel D) 40–50%

The use of stimulants remains most important in old mature trees, where the risk of over-exploitation is least; it has become a standard component of double cut (high level) systems.

9.5.4 The mode of action of yield stimulants

The fact that stimulation delays plugging and consequently prolongs flow was established by Boatman (1966) who showed that, after stimulation, repeated reopening of the cut did not result in a stepped flow curve as was the case without stimulation (Fig. 9.7b). As all known stimulants either release, or induce the formation of, ethylene in the bark, it would appear that this gas acts as an anti-plugging agent. In view of the known role of damaged lutoids in plugging, it is reasonable to suppose that ethylene somehow stabilizes these particles, but the mechanism whereby plugging is delayed has not been clearly established.

As a result of a number of observations, such as the reduced bursting index of latex from stimulated trees and the swelling of lutoids *in vitro* when treated with ethylene, Ribaillier (1970) concluded that ethylene stabilizes lutoids by increasing the permeability of their membranes. He also observed a decrease in the ability of B serum to flocculate rubber particles. The hypothesis that stimulation delays plugging by stabilizing lutoids is supported by the observation of many workers that ethephon increases the stability of tapped latex. An observation which at first sight seems contradictory is that of Pakianathan *et al.* (1966), who noted that latex collected from stimulated trees contained more damaged lutoids and flocculated material than that from untreated trees. However, this mainly occurred in the early stages of flow after tapping and Yip and Southorn (1974) produced evidence that stimulation results in less damage to active lutoids during later flow, probably because of an observed increase in the activity of C serum, which is a stabilizing factor. Coupé *et al.* (1977) suggested that the increase in latex pH which follows stimulation has a direct effect on lutoid membrane permeability.

There have been suggestions that ethephon increases yield by mechanisms other than increasing the stability of lutoids. Various workers have observed changes in the chemical composition of latex after stimulation, but it is difficult to decide whether these are direct effects of the stimulant or secondary effects resulting from biochemical activity in the vessels following increased latex withdrawal. As ethephon application very quickly increases yield – Pakianathan (1977) observed yield increases six hours after application – any changes in latex that occur only after a number of tappings are likely to be secondary effects.

Tupy (1973c) and Tupy and Primot (1976) speculated that the increase in latex pH and consequent enhanced invertase activity after stimulation would cause an intensification of sucrose metabolism and rubber biosynthesis, so that stimulation would increase the rate of latex regeneration. Recently, Low and Yeang (1985) confirmed that in most clones ethephon causes an immediate increase in the invertase activity of C serum and a somewhat slower increase in that of the whole latex. This appeared to be due to an actual increase in the amount of the enzyme present rather than a pH effect. However, with repeated monthly application of stimulant there was a fall in invertase activity, which was not accompanied by a similar decline in the yield response to the stimulant. Except for one clone, correlations between yield and invertase activity and yield and sucrose levels were not significant, suggesting that the role of invertase in rubber production may not be as important as was formerly thought. Other workers have suggested that stabilization of lutoids could enhance rubber biosynthesis because the phosphatase released from bursting lutoids has been shown to interfere with some reactions in the biosynthetic process. Jacob *et al.* (1983) have reviewed the evidence that ethylene directly affects rubber regeneration.

Coupé *et al.* (1977) observed a large increase in polymerization of ribosomes in latex from stimulated trees, but it is not clear whether this is a primary or a secondary effect. Hanower and Brzozowska (1977) found a much reduced level of polyphenoloxidase in stimulated trees; they suggested that polyphenoloxidase released from Frey-Wyssling complexes could play a role in plugging. Gomez (1977) and Yip and Gomez (1984) found that the flocculating activity of bark extracts decreased after stimulation and concluded that ethephon delayed the formation of coagulum on the cut as well as internal vessel plugging.

Various workers have demonstrated that stimulation increases the drainage area, albeit not always in proportion to the yield increase. Sethuraj *et al.* (1976) showed that yield increases are prevented by limiting the drainage area with grooves to the wood, indicating that stimulation enhances yield by increasing the drainage area rather than by increasing production per unit volume of laticiferous tissue. Clearly, the increase in drainage area is the consequence of delayed plugging and increased withdrawal of latex rather than the cause of it.

9.5.5 Exploitation systems with yield stimulation

The availability of ethephon allows not only a sustained increase in yield per hectare but also the possibility of reducing labour inputs by lower frequency tapping. It also presents a much broader choice of tapping system. Given the good supervision to be expected on an estate, there seems no reason why ethephon should not be employed from time of opening provided that, initially at least, concentrations of 2.5 per cent or lower are used, and that stimulation is stopped if the modest increases in yield suggested by Abraham (1981) (see p. 396) are greatly exceeded, or if d.r.c. falls below 30 per cent.

These parameters may be met by reducing tapping frequency to d/3 or d/4 (with substantial savings in labour cost), by shortening the tapping cut (probably to $\frac{1}{4}$S), or by both. A shorter tapping cut extends considerably the life of the bark reserves and offers the possibility of increasing the drainage area by panel changing on alternate tapping occasions. Given this flexibility, ethephon usage can enhance profitability and export earnings whether rubber prices be high or low.

9.5.6 Upward tapping

Upward tapping has become popular in conjunction with the use of ethephon stimulation, which enables good yields to be obtained from $\frac{1}{2}$S or $\frac{1}{4}$S panel changing systems. The introduction of a specially designed long-handled gouge has obviated the need for the ladder formerly required and made larger tasks practicable once the tapper has become accustomed to the operation. Upward tapping provides a drainage area largely separate from the lower panel and this feature may be exploited by systems in which high and low panels are tapped upwards and downwards on alternate tapping occasions.

However, there are several problems with the system. Tappers tend to suffer fatigue. Bark consumption may be excessive. Latex often leaves the cut and flows wastefully down over the panel. To minimize the spillage, the cut may be made at a steeper angle (55°) but this enlarges the triangle of unexploited bark (see Fig. 9.14) and, as the position of the top of the cut is limited by the length of the gouge handle, it means that a $\frac{1}{2}$S cut is only practicable on smaller trees. A $\frac{1}{4}$S cut is therefore more commonly used and four such cuts may be exploited over a period of eight years. Poor quality of tapping usually precludes tapping of renewed bark on upwardly tapped panels.

9.5.7 A modern exploitation system

Abraham and Hashim (1983) have given detailed tapping recommendations for different types of clone, seedlings, and specific smallholder situations. One suggested schedule for normal clones tapped in a typical estate situation in Malaysia may be summarized as in Table 9.2. Such an arrangement (Fig. 9.15) would ensure a tapping life of 25 years. Other schedules are given for situations where upward tapping is impracticable, where wintering is severe, and for clones prone to dryness; these variants would also ensure high sustained yields, and a tapping life of 30–33 years. All schedules are planned to require only moderate labour inputs.

Managers in Malaysia and elsewhere should be able to conduct simple trials, using task-size plots, to select systems best suited to local conditions and work forces. In countries where skilled tappers are plentiful, they might find it useful to tap $\frac{1}{2}$S panels B01 and B03 alternately, followed by B02 and B04. Upward tapping might be introduced, with reduced exploitation of the B panels at an earlier date, and so on.

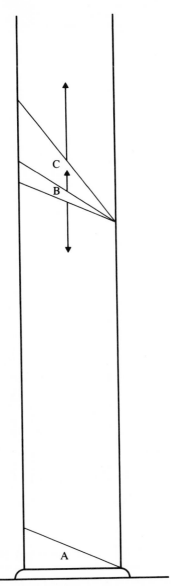

Fig. 9.14 Diagram showing 'islands' of untapped bark resulting from a downward cut at 25° (A), an upward cut at 35° (B), and an upward cut at 55° (B plus C)

Exploitation schedules such as those above will require good standards of field maintenance and careful attention to fertilizer requirements. Their successful implementation should ensure the continued prosperity of the rubber estate industry. In summary, the

Table 9.2 *Exploitation procedures for clonal rubber on estates*

	Panel	*Tapping system*	*No. of years*
I	B01	$\frac{1}{2}$S d/3.ET 2.5%.1/y $\frac{1}{2}$S d/2	2 4
II	B02	$\frac{1}{2}$S d/2.ET 2.5%.1/y $\frac{1}{2}$S d/2.ET 2.5%.3/y	3 2
III	BI1	$\frac{1}{2}$S d/2.ET 3.3%.4/y	5
IV	H01) BI2)	$\frac{1}{4}$S d/2 8m/12.ET 5%.8/y $\frac{1}{2}$S d/2 4m/12.ET 5%.1/y	2
V	H02) BI2)	$\frac{1}{4}$S d/2 8m/12.ET 5%.8/y $\frac{1}{2}$S d/2 4m/12.ET 5%.1/y	2
VI	H03) BI2)	$\frac{1}{4}$S d/2 8m/12.ET 5%.8/y $\frac{1}{2}$S d/2 4m/12.ET 5%.1/y	2
VII	H04) BI2)	$\frac{1}{4}$S +1/2S d/3.ET 10%.10/y	3

Source: Abraham and Hashim (1983)

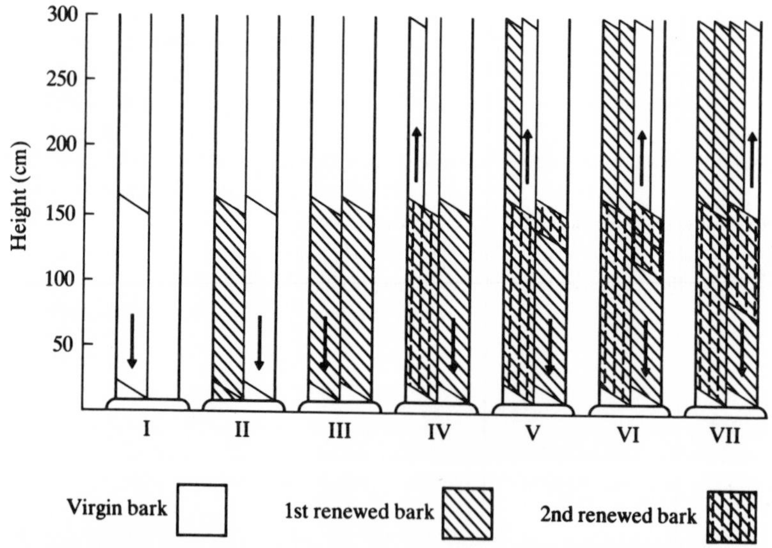

Fig. 9.15 Exploitation procedure for modern clones: for explanation see text (After Abraham & Hashim 1983)

availability of ethephon makes it practicable for estates to gain short-term profits by intensifying exploitation when prices rise, and allowing trees to recuperate reserves when prices fall. This can be achieved with least disruption of work-force schedules by increasing the strength and/or frequency of stimulant applications, whilst monitoring carefully the d.r.c. of latex carried in from individual tasks or fields. It can be used to maintain productivity if skilled labour is scarce or expensive. It can also be used to extend the tapping life of the tree, but the cost of planting and of interest charges on funds used places a considerable premium on judicious enhancement of yield during the early years and discounts to negligible levels any profits from yields that might be obtained if tree life were extended much beyond 40 years.

9.5.8 Puncture tapping

The idea of extracting latex through punctures made in the bark rather than by conventional cuts is not new. However, commercial application only became feasible when it was found that trees tapped by punctures responded particularly well to stimulation. Research workers in the Ivory Coast and Malaysia started serious efforts to develop puncture or micro-tapping systems around 1973 (Tupy 1973g; Primot & Tupy 1976; Leong & Tan 1978). Subsequently, various aspects of puncture tapping were investigated in Indonesia (Basuki *et al.* 1978; Siahaan 1980; Subronto 1982); Thailand (Samosorn *et al.* 1978); Sri Lanka (Waidyanatha & Angammana 1981); and Brazil (Moraes 1978). Reviews and evaluations were published by Tonnelier *et al.* (1979), and by Sivakumaran and Gomez (1980).

In puncture or micro-tapping, an ethephon-based stimulant is applied to a strip or groove of scraped bark; the strip is usually vertical, 2 cm wide, with a length of 50–100 cm. Subsequently, on each tapping day, 4–10 punctures are made with a special instrument, spaced out along the strip, and the latex is collected in the traditional manner. At monthly intervals a new strip adjacent to the old strip is prepared, stimulated and puncture-tapped.

In various experiments the yield under alternate, or third, daily puncture tapping was higher than that of the conventional $\frac{1}{2}$S system without stimulation; on the other hand, yields were usually less compared with conventional tapping with stimulation. Over the limited period of the experiments (up to five years), no decline in yield trends has been observed, but long-term effects of micro-tapping are not yet known. Tonnelier *et al.* (1979) found no difference in yield between bands in virgin bark above an old panel and in renewed bark. However, the same authors observed a drastic reduction in yield when a cycle of high panel puncture strips, having exhausted the tree circumference after two years, was followed by strips on the low panel.

The advocates of puncture tapping argue that, because the phloem transport is not interrupted as is usually assumed for conventional tapping, the stress on the tree is less. This would account for the markedly higher sucrose content of the latex, observed by Tupy (1973f) and Waidyanatha and Angammana (1981). Other workers, however, have reported a decrease in sucrose content of the latex (Samosorn *et al.* 1978; Low *et al.* 1983). Published data on reduction in girth increment suggest that the rubber/shoot loss ratio may be somewhat more favourable under puncture tapping compared with conventional tapping.

Various practical advantages of puncture tapping are generally claimed, such as: less skilled labour required, possibly larger task size; a better girth increment; higher yields (depending on clone); probably no dry trees; suitability for use of mechanized tools; and possible use at an earlier age than conventional tapping. The method has been found to have a number of drawbacks, including: spillage, if latex is not properly guided along the strip; late dripping; need for frequent stimulation; differences in bark thickness that make it impossible for all punctures to be to the correct depth; rather rapid bark consumption; and adverse bark reaction.

Adverse bark reactions to puncture tapping can be classified as external flaking with pitted appearance below the corky layer due to plugs of dead tissue, bark burst resembling patch canker, and uneven swellings on the panel (Sivakumaran & Gomez 1980). The first form of reaction does not pose any practical problems. Bark burst, characterized by a blackened coagulum projecting from a burst, may cause serious wounds. It is mainly restricted to young trees of sensitive clones including PR 107, RRIM 600, PB 5/51, PR 251 and PR 255, and is essentially a non-pathological injury. Although by removal of the coagulum and application of a wound dressing, a reasonable recovery of the bark can be expected, such surgical treatment may not be feasible in commercial practice, and it may be prudent to delay puncture tapping in sensitive clones. The anatomy of renewal bark after puncture tapping has been studied in detail by Hamzah and Gomez (1981b) who found that, although latex vessel initiation is somewhat delayed, bark regeneration in general proceeds smoothly, comparable to the renewal obtained with conventional tapping.

Various factors, such as direction, width and length of the strips, number of punctures per tapping, concentration and dosage of stimulant, and frequency of tapping, have been studied. Yields generally increase with the length of the strip, but strips longer than 80 cm would prevent the exploitation of high and low panels. Although in some experiments yields increased with increasing concentration, it is considered prudent not to exceed 5 per cent concentration. Vertical strips at least 50 cm long and 1 cm wide, with about 6 punctures per tapping, a monthly stimulation with 2.5 per cent ethephon, at about 2 g per application, a needle size of 1 mm and a third daily or lower tapping frequency would appear to suit most clones.

In order to overcome some of the disadvantages of puncture tapping on vertical strips, other methods have been developed. In the spiral puncture tapping method, 4–6 punctures are made on a 45° half spiral cut each tapping day, and once a month 5 mm of bark is removed for reapplication of the stimulant. A promising combination of conventional tapping with micro-tapping is the micro-excision tapping system (Hashim & P'ng 1980). In 'micro-X tapping' a half spiral groove is opened at 40° slope and treated with ethephon. For 9 tapping days (on d/2 frequency) 3–5 punctures are made on the cut; this is followed by three 'normal' tappings, which remove the puncture marks, after which the cut is restimulated. Each cycle thus lasts about one month. Advantages include flexibility, lower bark consumption, less problems with bark bursts, and yields as good or better than with micro-tapping alone. A different combination of conventional tapping with micro-tapping was suggested by Tonnelier *et al.* (1979) in the Ivory Coast. In this method puncture tapping is practised on a 60 cm vertical band above an existing cut; the band and the cut are stimulated alternately every five weeks, and once per week punctures and conventional tapping are carried out together.

Micro-tapping would be particularly useful if it permitted earlier exploitation of young trees. Abraham and Anthony (1982) reported that when trees were brought into puncture tapping at 36 cm girth, yields were not unsatisfactory (about 1000 kg/ha for 18 months), but growth was retarded by 30 per cent. The yield from subsequent conventional tapping over the first year was higher than that of the control trees which had not been subjected to puncture tapping. In Sri Lanka, Waidyanatha and Angammana (1981) puncture tapped trees aged 4 years with a girth as low as 28 cm; yields were, as expected, rather poor, but better than under conventional tapping. Hunt (1983) found in Indonesia that clones AV 2037 and RRIM 600 could be successfully puncture tapped when girth had reached 30 cm by using 5 punctures on a 50 cm strip and tapping fourth daily. After conventional ½S d/2 tapping was started when the trees had reached 46 cm girth, the yield during the first normal tapping year was higher than for trees without previous puncture tapping, but in subsequent years the yields became equal. Although in this manner about 500 kg/ha extra rubber was obtained over a two-year period of puncture tapping, it seems doubtful whether this would be economic when tappers' wages are high. Tonnelier *et al.* (1979) reported experiments in the Ivory Coast in which trees of GT 1 and RRIM 600 were brought under puncture tapping at a girth of about 30 cm; this was about 22 and 29 months respectively before normal opening time at 50 cm girth. Yields from puncture tapping, although not unsatisfactory, were considered uneconomic. Girth increment was reduced by about 30 per cent (GT 1) and 10 per cent (RRIM 600). Recently, Sivakumaran (1984) confirmed in Malaysia that during the first year of $\frac{1}{2}$S d/2 tapping the yield of previously puncture tapped trees was considerably higher than that of trees without a history of puncture tapping. In addition, such trees maintain a capacity to respond to restimulation on conventional tapping.

The development of micro-tapping brought about a search for suitable puncture tools which would allow uniform punctures of a given depth, require little force, be easy to clean, facilitate replacement of needles, and require minimum time. From simple hand tools, semi-automatic instruments were developed (Abraham & Anthony 1980a), such as the 'handy injector' and the 'spiral injector', the latter requiring less force because the puncture is made by a rotating needle. The most sophisticated types, such as the 'rotary injector', are driven by small electric motors powered by batteries.

Research institutes in Africa and Malaysia are still cautious about the possibility of large-scale application of micro-tapping systems. The exploitation of panels regenerated after puncture tapping needs further investigation. In order to make the exploitation of very young trees profitable, improvements in the technique would be necessary. It seems rather unlikely that micro-tapping will be generally adopted, but it may have potential where there is a serious shortage of skilled tappers, or for one or two years' exploitation of immature trees of certain clones before they are opened for conventional tapping.

9.6 Economic aspects of exploitation systems

Whereas the results of experiments on tapping and stimulation give indications of the yield per hectare and the yield per tapper that can be expected for each clone and age, an economic analysis is needed to draw conclusions as to the relative profitability of the different exploitation systems. For smallholders, the situation is relatively straightforward; since smallholders do not cost their own family labour, they naturally aim for the highest yield per unit area. When tapping is done under a fixed share arrangement with outside tappers, a conflict of interest emerges: the owner would still be interested in the highest possible yield from his holding, while an individual share tapper would clearly benefit from the highest yield per tapper per tapping.

In choosing exploitation systems, estates aim at the highest profit per ha, which does not necessarily coincide with the highest yield per ha. Profit per ha is the product of yield per ha and profit per kg, and it is this product which should be maximized. Since, generally speaking, the highest profit per kg is obtained at the highest yield per tapper, the right balance between yield per ha and yield per tapper needs to be found. Unfortunately, measures that increase the yield per ha tend to increase cost per kg. Whether or not a proposed change in exploitation system is financially justified depends on whether the positive effect of an increase in yield per ha outweighs the negative effect of an increase in cost per kg.

The relative profitability of different exploitation systems is further complicated as a result of differences in the ways in which tappers are remunerated. Although arrangements vary, in most countries estate tappers are paid a fixed wage per day augmented by an incentive payment based on the weight of rubber collected in excess of a basic 'poundage'.

Based on work by Paardekooper (1964), Watson (1965) and Ng *et al.* (1969), the Rubber Research Institute of Malaysia has published various formulae for calculating what the 'break-even' yield would have to be at different price levels before it would be worth while switching from conventional $\frac{1}{2}$S d/2 tapping to some other system (Sepien & Lim 1981). A slightly different approach, leading to the same results, will be followed here.

The gross return per ha (disregarding fixed costs per ha such as upkeep, management, etc.) under conventional $\frac{1}{2}$S d/2 tapping is:

$$Y_1 v - N_1 w$$

where Y = yield, kg/ha (including lower grades)

v = value of 1 kg of rubber: this is the net price minus costs fixed per kg (processing cost, incentive rate for tappers etc.)

$$N = \frac{\text{number of tapper days per year, or number of trees per ha} \times \text{number of tappings per year}}{\text{task size}}$$

W = fixed costs per tapper: wage rate plus fringe benefits minus incentive rate × basic poundage (an adjustment necessary because the tapper receives no incentive payment whenever his production is less than the basic poundage)

Similarly, the gross return per ha after change-over to another tapping system would be:

$$Y_2 v - N_2 w$$

In the break-even situation, these two gross returns would be equal:

$$Y_2 v - N_2 w = Y_1 v - N_1 w$$

and, therefore,

$$Y_2 - Y_1 = (N_2 - N_1) \; \frac{w}{v}$$

The difference in yield per ha as a result of a change in tapping system corresponding to an unchanged profit situation is therefore easily derived from the change in the number of tapper days resulting from the new system. (Note that in this model the yield differential is independent of the yield level in the original $\frac{1}{2}$S d/2 situation.) Table 9.3 shows break-even yield differentials compared with $\frac{1}{2}$S d/2 for a number of tapping systems, based on the assumptions shown below the table.

Table 9.3 *Estimated break-even yield differentials for various tapping systems relative to the yield obtainable from $\frac{1}{2}$S d/2 tapping*

System	Task size *	Tappings per year	Break-even yield differential (kg/ha)
$\frac{1}{2}$S d/2 (100%)	530	140	—
$\frac{1}{2}$S d/3 (67%)	530	94	−109
$\frac{1}{2}$S d/4 (50%)	530	70	−167
$\frac{2}{3}$S d/2 (67%)	600	140	− 39
S d/4 (100%)	425	70	−127
S d/6 (67%)	425	47	−194

* 300 trees/ha

Assumptions:

RSS 1 f.o.b. price, $M per kg	2.00
Net price, $M per kg	1.74
Deduction for incentive rate, $M per kg	0.31
Other deductions, $M per kg	0.16
Basic tapper's wage, $M per day	5.10
Fringe benefits, $M per day	3.20
$v = 1.74 - 0.16 - 0.31 = $ ($M per kg)	1.27
$w = 5.10 + 3.20 - (0.31 \times 9.5) = $ ($M)	5.35

The above model ignores the fact that different tapping systems may result in different proportions of lower grades, while the profit per kg is less for lower grades than for latex rubber. To take this into account, the formula can be expanded as follows:

$$Y_2 = \frac{Y_1 (z_1 v_s + (1 - z_1)v_e + (N_2 - N_1) w}{z_2 v_s + (1 - z_2) v_e}$$

where z = proportion of lower grades
v_e = value of latex rubber
v_s = value of lower grades

For Malaysian conditions, tables have been published showing the estimated break-even yields for various tapping systems relative to $\frac{1}{2}$S d/2 yield levels ranging from 1000–2000 kg/ha, at a range of f.o.b. price levels and for assumed proportions of lower grades in the alternative systems of 10, 20, 30 and 40 per cent (Arope *et al.*

1983). For example, at a RSS 1 f.o.b. price of $M3.00/kg, a change from ½S d/2 to ½S d/3 would be profitable if the drop in yield from 1600 kg/ha did not exceed 89 kg/ha; if, however, it is expected that the proportion of lower grades would increase from 10 per cent under d/2 tapping to 20 per cent under d/3 tapping, the drop in yield ought not to exceed 64 kg/ha.

9.6.1 The economics of stimulation

The cost of stimulation varies considerably, mainly depending on the frequency of applications; it should include the cost of the extra fertilizer required. In 1982 in Malaysia it was in the range of $M75–100/ha/an.; depending on rubber prices and percentage of lower grades, this would be the equivalent of 35–120 kg/ha of rubber (Table 9.4). The additional yields required for stimulation treatments to break even can be added to those estimated as explained above to show the break-even yield differential for a proposed change from ½S d/2 to a lower intensity tapping with stimulation. For example, at a RSS 1 f.o.b. price of $M3.00 kg, a cost of stimulation of $M100/ha/an., and an expected increase in the proportion of lower grades from 10 per cent to 20 per cent, a change from ½S d/2 to ½S d/4 with stimulation would be profitable if the yield did not drop by more than $112 - 56 = 56$ kg/ha.

Table 9.4 *Estimated yield increases (kg/ha) required to break even on costs of stimulation*

Cost of stimulation ($M/ha)	RSS 1 f.o.b. price ($M/kg)			
	2.00	2.50	3.00	3.50
75	60	48	42	36
100	90	64	56	48
150	120	96	84	72

Barlow (1978) estimated that stimulation, without a change in tapping system, with four stimulations a year, would require a minimum of 200 kg/ha extra yield to secure a worthwhile return. Yee (1983) analysed data from a large number of fields on Malaysian estates; the average difference in yield between stimulated and unstimulated fields was 32 per cent and mean operating profits were considerably higher in the stimulated fields.

In conclusion, in the absence of stimulation, economic analyses of most tapping experiments show ½S d/2 to be the most profitable system. Lower tapping intensities with stimulation are generally more profitable than ½S d/2 during the first years of stimulation.

9.6.2 Long-term considerations

In the above models, only the effects in the relatively short term have been considered. However, a different tapping system, or the use of stimulants, will probably influence relative yields over a long period and may, in addition, have an effect on the economic life of the tree. Generally speaking, profit made now is worth more than that made in the future, and to a certain extent one would be justified in maximizing yields in the early tapping years and accepting lower yields later on. Low frequency tapping with stimulation conserves bark and would thus lengthen the economic life of the plantation; hence a smaller depreciation charge would be justified. Lian and Cheam (1974) showed that maximizing early yields with ethephon stimulation and shortening the exploitation life would increase the internal rate of return and the benefit/cost ratio of an estate.

9.6.3 The economics of early opening

Barlow and Ng (1966) showed that shortening the period of immaturity by six months through the use of advanced planting material increased the estimated present profit per hectare by 10–25 per cent, depending on the discount rate used. Shortening the immature period by earlier opening will not have the same effect because early yields of trees opened at 45 cm are less than for those opened at 50 cm. Nevertheless, under conditions of high prices, earlier opening appears financially attractive. This was confirmed by Ng *et al.* (1972); their analyses, based on actual records for the clone RRIM 501, and using a discount rate of 10 per cent, showed that advancing the opening time by six months increased the present value of profits by 13 per cent.

9.6.4 Rubber extraction from shoots

From time to time research workers have contemplated a completely different way of harvesting, similar to that employed for other rubber-producing plants, such as guayule and gutta-percha. In such a destructive harvesting method, the shoots and leaves of closely planted buddings would be mechanically harvested after only one or two years' growth, and the rubber extracted from them by suitable solvents. Leong *et al.* (1982) found clonal differences in rubber content, which ranged from 1.4 to 2.1 per cent in petioles and from 1.1 to 1.7 per cent in young stems. At a planting density of 50,000 plants per ha, the annual dry weight production was of the order of 50,000 kg, which would correspond to about 500 kg/ha of rubber. With selected clones, yields of up to 900 kg/ha might be possible, but it is quite unlikely that yields could equal those from conventional tapping. Furthermore, it is not known for how long a stand could be regenerated for successive harvests. The main advantages of destructive harvesting would be an early return on investment capital and a reduction in labour costs. Capital and operating costs of an extraction plant, however, would be likely to be high.

9.7 Utilization of by-products

9.7.1 By-products of latex processing

Tapping removes fairly large amounts of nutrients from the tree which are normally lost during processing. At various times, efforts have been made to use at least part of these nutrients as fertilizer. In the 1960s, one company in Malaysia successfully used the skim resulting from latex centrifugation as a fertilizer for rubber. More recently, the possible use of bowl sludge (a solid by-product of a latex concentrate factory) has been investigated by Wahab *et al.* (1979), who estimated that Malaysia annually produces about 900 tonnes of this waste product. Bowl sludge contains up to 37 per cent P_2O_5 and 20 per cent MgO and could be a cheap source of fertilizer for legume covers and could also replace rock phosphate in the planting hole (see section 8.3.1). The treatment and use of liquid effluents from rubber processing factories is briefly discussed in Chapter 11.

9.7.2 *Hevea* wood

At the time of replanting, a full stand of old rubber trees contains between 150 and 200 t/ha of green wood. When replanting, the old trees may be left in the field to rot or to be burned, but nowadays the timber is often extracted for commercial use. Estate contractors may undertake replanting activities 'free of charge' in exchange for the wood. The total amount of rubberwood available from annual replantings in Malaysia has been estimated at over 8 m. t a year, which is about equal to the present annual timber supply from natural forests (Nor 1984).

The oldest and most widespread use of rubberwood is as fuel in rubber factories, for cooking and for brick making. A change-over from RSS production, which requires a large amount of firewood, to block rubber, as well as a gradual decline in the domestic demand for firewood in towns, has led to a search for other uses. The first alternative use, which has found a fairly large-scale application, is as charcoal for iron smelting and steel production. In the northern states of Malaysia over 1 m. t of wood was used annually for charcoal production in the 1970s. Subsequently, rubberwood has found use on a commercial scale for a variety of products, including cardboard, particle board, fibre board, paper, packing cases, pallets for Standard Malaysian Rubber, building components, flooring and furniture.

The suitability of rubberwood for pulp production and paper manufacture was studied in Malaysia by Peel (1959–60). More recently, the suitability of rubberwood from Liberia for the manufacture of particle board was assessed by Chittenden *et al.* (1973); it was found that rubberwood was not a first choice for particle board manufacture, but its shortcomings could be reduced by mixture with other raw materials. In Sri Lanka, where an annual amount of about 800,000 tonnes of fresh wood is estimated to become available from replanting, the use of rubberwood for paper making was investigated by Jeyasingham (1974). Its suitability was found to be somewhat restricted by the presence of small amounts of latex.

The use of sawn timber of *Hevea* has been the subject of much investigation over the last decade. The properties of the so-called 'heveawood' have been described by the Rubber Research Institute of Malaysia (1974c, 1977d) and by Nor (1984). Heveawood is a homogenous, pale straw-coloured 'light hardwood', with texture and strength properties comparable to other woods, such as *Shorea* and *Palaquium* spp., widely used for furniture making in Asia. Malaysia now has over 50 sawmills specializing in heveawood and a number of factories for the manufacture of 'knock down' furniture, chairs, lounge and dining room sets (Tan & Sujan 1981). Nor (1984) lists a total of 61 articles that have been made commercially from heveawood. Heveawood is also exported to Japan and Europe; one of the products manufactured on a commercial scale from rubberwood in Europe is parquet – the process has been described by Tan *et al.* (1980).

Only about 10–20 per cent of the total wood from replantings could be economically used for sawn timber; nevertheless, the potential availability of sawable logs in Malaysia alone is about 1 m. cubic metres per year. A constraint in developing sawmills for rubberwood is the need for a continuous supply of logs to be collected from scattered individual smallholdings, but the use of portable sawmills may be a solution. Another problem is that fresh rubberwood, because of its high starch content, is susceptible to insect and fungal infestation, and freshly sawn wood must be treated with preservatives, preferably by pressure impregnation (Tan *et al.* 1979). A particular problem is 'blue stain' which has restricted the use of rubberwood for building and furniture in Sri Lanka. Control of the blue-staining fungi by a double dip treatment was described by Gunasekera and Sumanaweera (1974); effective chemicals have been listed by the Rubber Research Institute of Malaysia (1982b) and Nor (1984) has described current recommendations for treatment of fresh logs.

Finally, since the use of wood as fuel for smokehouses and brick kilns is extremely inefficient in its conversion to heat, wood distillation has been proposed as an alternative use of waste wood. This could yield high-quality charcoal, acetic acid, wood spirit and tar (Harris 1979).

9.7.3 Seed

The conventional use for seed in all rubber-growing countries is to grow stocks for bud grafting or to raise clonal seedling trees, the latter being grown from seed from special seed gardens. In some countries, such as India and Sri Lanka, rubber seed is used on a moderate scale for the extraction of oil; whether or not this is an economic proposition depends much on the cost of collection.

Fresh rubber seeds comprise about 45% hard shell and 55% kernel and have a moisture content of about 25%. The dried kernel (5% moisture) contains 40–45% oil and 35–40% can be extracted by simple presses or expellers. The light yellow oil consists mainly of glycerides of oleic (*c.* 17%), linoleic (*c.* 35%) and linolenic (*c.* 24%) acids and somewhat resembles linseed oil in properties. The cake remaining after pressing typically contains 35% crude protein, 45% carbohydrates, 7% oil and 7% crude fibre. The raw rubber seed oil can be refined by traditional processes and used for soap making, or as a rather inferior substitute for linseed oil in the manufacture of paint. It can also be treated with sulphur to produce factice, or epoxidized for use in making adhesives and alkyd resins. The cake, or meal, remaining after oil extraction is used in the production of animal feeds. Feeding trials have shown that cattle, pig and poultry feeds may satisfactorily contain up to 25 per cent of rubber seed meal, provided that its oil content is below 10 per cent. The cake has also been used as a fertilizer.

In Sri Lanka, the use of rubber seed for oil extraction started on a small scale in the 1960s (Nadarajah *et al.* 1973). Proper storage was found to be a problem, as enzymatic breakdown of the oil must be avoided. The use of rubber seed oil for resin manufacture is considered highly promising as it could replace imported linseed oil. It is estimated that Sri Lanka could produce about 4000 t of rubber seed oil and 7000 t of cake; export would probably be feasible. Haridasan (1977) has described the extraction of oil from rubber seed in India. Some 12,000 t of dried kernels are normally crushed in about 100 mills, many of which also process other oil seeds, especially groundnuts. Seeds are mainly collected by children and sold through dealers to the mill owners. The oil is sold to local soap and paint manufacturers, and the cake, formerly used as manure, is now used by animal feed manufacturers. In Thailand, where the total production of rubber seed is probably close to 100,000 t, a limited amount is used for extraction of oil, which is exported (Udomsakdhi *et al.* 1974). In Malaysia, little interest has been shown in the use of seed other than as planting material, although annually between 100,000 and 200,000 t of seed are estimated to be produced and largely wasted (Ghani *et al.* 1976).

Notes

1. It appears to have been generally overlooked that the dry weight production over one year should also include the dry weight of the new canopy, which, according to Templeton, would be around 15 kg per tree. In addition, published data suggest that an indiscriminate use of the above 'general formula' significantly overestimates the difference in dry weight increment between tapped and untapped trees.

Chapter 10

Diseases and pests

A. Johnston

10.1 Diseases

Rubber is affected by several serious diseases, most of which were first noted early in the crop's history. South American leaf blight was first recorded in 1900, brown and white root diseases in 1904–5, and black stripe in 1914. It is unusual that virtually all the diseases so far recorded are caused by fungi, exceptions being red rust, a minor algal leaf spot, and a virus and a bacterial disease recently reported from Brazil.

Useful reviews of rubber diseases have been given in the books by Petch (1921) in Sri Lanka, Steinmann (1925) in Indonesia, and Sharples (1936) in Malaysia, and in the bulletin by Weir (1926) in South America. Although old, these publications contain much detailed information, particularly on symptoms and causal organisms, that is still of value. More recently Rao (1975) reviewed and illustrated the diseases occurring in Malaysia and Gasparotto *et al.* (1984) those in Brazil.

Chee (1976b) produced a comprehensive annotated list of the micro-organisms associated with rubber. It includes details of geographical distribution as well as a list of 351 references. The list mentions over 550 organisms; of these, Chee lists 24 as being of importance.

In this chapter an attempt is made to summarize infomation on the commoner diseases. References are given to publications in which further information, including details of pathogens, can be found.

10.1.1 Root diseases

In Malaysia and adjacent areas root diseases can cause serious economic losses. In decreasing order of severity the three most important root rots in the area are white, red and brown root diseases. These also cause trouble in Africa but an additional serious disease there is *Armillaria* root rot, which does not occur in Asia. These root pathogens are particularly troublesome because they affect the roots of a large number of other plants, which can act as centres of infection from which the diseases can spread to rubber. The pathogens can flourish and spread in a wide variety of conditions of soil and climate and careful control needs to be exercised in all areas except those most unfavourable to their development. Diseases of less importance are stinking root rot caused by *Sphaerostilbe repens*, *Ustulina* collar rot and *Poria* root rot.

In South America root infections are of little importance compared with leaf diseases, particularly the devastating South American leaf blight.

The four major diseases can be recognized by the type of damage they cause to the roots. The differences can be summarized as follows:

White strands on the outside of the root	White root disease
Tough red or purplish skin covering the root	Red root disease
Soil and stones firmly fixed to the root by tawny brown mycelium	Brown root disease
Thick skin of white mycelium under the bark	*Armillaria* root rot

Treatment of these diseases is a classic example of integrated control, depending on crop sanitation, application of chemicals, cultural control, and measures to enable antagonistic organisms to exercise biocontrol.

White root disease

White root disease, first described from Singapore in 1904, has now been reported in virtually all rubber-growing areas in south-east Asia. It does not, however, appear to affect rubber in south India, and does not cause serious damage to the crop in Burma, Kampuchea or Vietnam. It is widespread in Africa and has been reported from Brazil.

The disease is important on young rubber, over 90 per cent of infections occurring during the first five years after planting. In young plantations in Malaysia the losses caused by white root disease are greater than those caused by all other diseases and pests combined.

Over 40 woody plants have been recorded as hosts, and crops such as cocoa, coffee and tea are commonly affected. Many forest tree species have also been found to be attacked and in new plantings these are a source of infection for rubber.

Losses can be heavy. In Sri Lanka the average loss of stand in 15 estates surveyed was just under 10 per cent (Liyanage 1977). If the disease is left untreated large areas of rubber may be virtually eliminated, as is illustrated by Fox (1965) for an area in West Malaysia neglected because of the Second World War.

The most obvious symptom of the disease is the presence of rhizomorphs of the causal fungus on the roots (Plate VI). They typically take the form of stout, smooth cords, up to about 6 mm in diameter, adpressed to the surface of the roots and firmly fixed to them. They are basically pure white but may be discoloured yellowish or reddish by the surrounding soil. They run longitudinally along the roots, branching and anastomosing, and from them hyphae penetrate the roots, although this occurs only some distance behind the advancing tip of the rhizomorph. Growth is comparatively fast, of the order of 30 cm in a month. Wood of newly killed roots is brown and fairly firm and dry. Later the rot is white or cream, usually remaining firm although in wet soils the wood may become jelly-like.

When root infection is well advanced foliage symptoms become visible, leaves developing a slightly off-green colour, then turning yellow and brown and withering. The tree quickly dies. If only the tap-root is affected the process is very much slower, but in any case by the time above-ground disease symptoms are visible the tree is beyond hope of recovery. In West Africa channelling or fluting of the base of the trunk takes place, a symptom not seen in the Far East. Symptoms of this and the other major root diseases are described in detail by Sharples (1936) and Pichel (1956).

The fungus causing white root disease was first regarded as being *Fomes semitostus* Berk. but it was later identified as *F. lignosus* (Klotzsch) Bres. This name was used for many years and is still commonly found in the literature. However, in 1952 the fungus was removed to the genus *Rigidoporus* as *R. lignosus* (Klotzsch) Imazeki and this is the name now most commonly used, although the names *Leptoporus lignosus* (Klotzsch) Heim and *R. microporus* (Swartz) v. Overeem are also in use. For a discussion of the identity of the fungus see Fox (1961).

Rigidoporus lignosus is a typical bracket fungus (Basidiomycotina, Polyporaceae). Fructifications appear on trees and stumps only when they have been dead for a considerable time. The brackets are unstalked, often about 20 cm in diameter. On fallen trunks very much larger compound fructifications may be formed, while on standing trunks the brackets may grow in tiers for a distance of up to about 1 m. The upper surface of a bracket is reddish-brown with a light yellow margin when fresh. Later the colour fades and concentric alternate zones of red-brown and yellow-brown develop. The undersurface is bright reddish-brown, fading on drying, and is covered with pores from which the spores are released. Fructifications appear mostly in wet weather.

Infection of the rubber tree arises through root contact with a disease source, which may be the infected root of another tree, or a dead stump, or a detached piece of infected material. Although capable of causing great damage, the fungus is to be regarded as a weak parasite in that it requires a relatively large food base from which to spread. The fungus is capable of spreading by rhizomorphs from a food base through damp soil but spread by this means appears insignificant. Spread can also be by spores, which are airborne and are capable of infecting the cut surface of tree stumps. Thence the roots of felled trees become infected and can act as a source of disease for living trees.

Once a rubber root has made contact with a source of disease, rhizomorphs spread along the healthy root. Penetration does not take place immediately, so that a considerable length of the advancing rhizomorph is purely epiphytic – a factor having important implications for disease control. Penetration of the uninjured root takes place through natural weak points such as lenticels, but wounding increases the chance of infection, as does loosening of the soil – both factors which are relevant to control. When the rhizomorphs reach the collar they soon penetrate it with mycelia girdling the tree and infecting the tap-root. Infection of the tap-root may proceed to a depth of about 4.5 cm and the tree quickly dies.

The disease is essentially one of moist conditions, causing more damage on coastal and alluvial soils, in ravines, and on fairly flat land than on hilly areas. In areas with a prolonged dry season, as in Kampuchea and Vietnam, the disease is not serious. Similarly in the Ivory Coast losses occur in the rain forest area but not in the savannah.

Infected trees of all ages can be found, but the disease is essentially one of young trees. If it is not controlled during the first five years of growth, before the final thinning, there is great danger of serious losses to the mature stand.

Control of white root disease has naturally been a major preoccupation of pathologists since the early days of rubber planting, and several changes of policy have been introduced over the years, particularly with a view to economy, mainly by reducing labour costs. Control in early plantings, as described for example by Petch (1911) in Sri Lanka, depended on dealing with the disease when an infected tree was recognized by above-ground symptoms. A trench was dug round the infected area, all dead wood removed from within the trench, and lateral roots of the jungle stump that was usually the source of the outbreak were followed up and dug out. Digging the area was advocated and also adding lime – a useless practice, belief in which, however, persisted for many years.

The first major change in control techniques arose from the work of Napper in Malaya (Napper 1932). Recognizing that the rhizomorphs spread along the root surface a long way ahead of penetration, he advocated control based on early location of centres of infection by frequent inspection of the collar region of young trees. Napper also began investigations into two other important factors. In replanting, poisoning of the standing trees can reduce subsequent disease in the young stand by up to 80 per cent. Poisoning of stumps after felling is less effective, but reductions of 40–60 per cent can be achieved. The effects appear to be due to the rapid death of the poisoned trees encouraging the growth of saprophytic fungi at the expense of the parasite. Establishment of creeping cover crops between the planting rows also plays a role in reducing disease losses. There seem to be several factors involved. Physical conditions under the covers (high humidity and soil moisture, prevention of soil insolation) encourage decay and hasten the succession of saprophytes in diseased roots. The cover plants may cause increased development in the soil of micro-organisms antagonistic to the parasite. Infection of the ground covers' numerous fine roots may dissipate the energy of the parasite before it can attack the rubber. The covers should not be allowed to grow around the bases of the rubber trees, as the high humidity engendered would increase the chance of infection of the collar.

In the past, planting of woody covers such as *Crotalaria anagyroides* has been recommended, as indicators of the presence of root disease, but it soon became evident that because of the labour involved in removing infected plants, and the danger of their assisting in disease spread, this was inadvisable. The biological control aspects of tree poisoning and the use of cover crops have been considered in detail by Fox (1965).

The following summary is based on methods of control advocated by the Rubber Research Institute of Malaysia.

1. When old rubber is being replaced, the existing stand may be removed by mechanical uprooting, using heavy crawler tractors, and burning or removal of the old timber. This is possible only for estates and grouped smallholder settlements and has disadvantages in that it is expensive, there is undesirable exposure of bare soil to sun and rain, and there is loss of organic matter. More commonly the old stand is poisoned. Sodium arsenite is now banned in some countries, and so 2,4,5-T in diesolene is used, painted on the bark. As 2,4,5-T is now regarded as posing a health hazard, because of the possible contaminant dioxin, other chemicals are being sought. One alternative material recently shown to be equally effective is triclopyr used at 5 per cent in diesolene or kerosene (Lim & Kadir 1983). Standing trees

may be poisoned or the stumps poisoned after felling. When planting after clearing jungle, uprooting trees and removal of roots, as described in Chapter 5, can aid in reducing the incidence of root diseases, but full root eradication is expensive and would only be done if subsequent mechanical cultivation of the plantation is intended, which is rarely the case.

2. A thick cover of mixed creeping legumes should be established as early as possible between the planting rows. The rows themselves should be kept clean-weeded up to about 1 m on each side of the rubber.

3. About a year after planting, quarterly inspection rounds should be begun to detect diseased trees by leaf symptoms in the canopy indicating drought or wilting. The diseased trees are removed with their roots, and collar inspection of neighbouring trees is done by removing the soil carefully so as to expose a length of the tap-root and the laterals. This process is continued along the row until a healthy tree is reached. Infected trees thus found are treated by exposing all the diseased tissue on lateral roots and collar and cutting it out. However, diseased roots are not followed into the inter-rows but are cut off at the edge of the planting row. Superficial rhizomorph growth is ignored. Cut ends of roots and wounds on the collar are treated with tar, and the collars of all inspected trees are given a dressing with a protectant such as quintozene. Treated trees are marked for reinspection after 12 months and retreatment if necessary. If the source of infection is located in the planting row, it is removed or isolated by digging. Prior to the final thinning out a random tree-to-tree collar inspection is made, perhaps sampling one row in four.

   An alternative method of detecting the pathogen in the early stages of growth, by using infrared aerial photography, has recently been tried but so far without success (Nandris *et al.* 1985).

4. If the disease persists until the plantation is older, infected trees may be isolated with a trench, including within the trench any surrounding trees found by collar inspection to be infected.

There are variations on these methods in other countries. Thus in Sri Lanka it is recommended that before planting an inspection should be carried out to demarcate affected patches. All potential disease sources there should be eradicated, and at planting the soil should be treated with sulphur, about 150 g/m². This treatment, also recommended in Indonesia and the Ivory Coast, reduces soil pH, which may encourage growth of antagonists such as species of *Aspergillus, Penicillium, Trichoderma* and actinomycetes.

In the Ivory Coast planting of *Tithonia* (Compositae) instead of a creeping leguminous cover has been recommended, to reduce soil moisture by transpiration. Another suggested control method is the application of a mulch to detect the presence of the fungus, which occurs almost entirely in the top few centimetres of soil. The mulch induces formation of rhizomorphs on the surface roots and six-monthly inspections are made. Trees found to be affected, and sometimes neighbouring apparently healthy trees, are treated with a fungicide such as quintozene. Recommended more recently is a method involving clearing of a funnel-shaped area around the collar about 10 cm deep; tridemorph emulsion is then poured around the collar (Tran 1982).

Applying often expensive control measures against this and other root diseases is justifiable only if such measures are to result in profit. The measures to be taken can be costed, but it is difficult to forecast the profit to be gained, as the time between application of control measures and the realization of profit from the expected increased yield is likely to be very long. The complex problems involved have been examined in detail by Fox. (1977).

Red root disease

Red root disease has been known for almost as long as white root disease, having first been investigated in west Malaysia in 1915. It has since been reported in most rubber-growing countries in Asia, although not in Sri Lanka. It is also absent from Papua New Guinea. Red root symptoms have been described from several countries in Africa, including Cameroon, Gabon, Ivory Coast, Nigeria and Zaire. The only record for South America is a recent mention from Brazil (Chee & Wastie 1980).

The disease infects trees from about 3 years old but develops slowly and does not usually become noticeable to any extent until the rubber is about 10–12 years old. It is thus most important in older plantations. This slow development is a source of danger, in that treatment of older trees is a major and expensive task. If the disease is allowed to proceed unchecked, large patches may occur in which all the trees have succumbed. However, losses from red root disease are generally less important than those caused by white root disease, although red root can be the most important disease in some areas of Malaysia. This is particularly so where there are large sources of infection in the form of massive stumps of forest trees, which are deep-seated and slow to rot.

The economic importance of the disease varies greatly from area to area. In Indonesia it is most frequently found on heavy clay soils, where there is a high water table, and in hollows, but in Malaysia the disease appears to occur on most types of soil, on both hilly and flat land. As in the case of white root disease, areas with a prolonged dry season are less seriously affected. The disease can cause serious trouble in Malaysia, West Java and Sumatra. In Africa it is widely distributed but is nowhere recorded as a serious threat.

The causal fungus attacks a wide range of woody plants; crops affected include cinchona, cocoa, coffee and tea.

The disease can easily be recognized by the characteristic growth of the causal fungus on the roots. This takes the form of rhizomorphs which begin as blood-red mycelial strands, the advancing edge being creamy white. Later they coalesce to form a membrane on the root surface (Plate VII). The light red of the young rhizomorphs later changes to dark red, and on drying the fungal sheath is almost black; but on moistening a diagnostic reddish or purplish colour appears. In surface roots it is very noticeable that the fungal sheath develops only on the underside.

The rate of extension of the rhizomorphs along infected roots is much lower than in the case of *R. lignosus*, and the length of root showing epiphytic growth in advance of penetration is very much less, of the order of a few centimetres.

The wood of affected roots in the early stages of attack is pale brown and hard. Later it turns to pale buff and degenerates into a soft spongy mass from which water can be squeezed – hence the name watery root rot sometimes used for the disease in the past. Typically the annual layers of wood in the root tend to separate easily, a feature not present in other fungal rots. Another characteristic is the proliferation of adventitious roots from the collar, particularly in mature trees, which takes place as a defence reaction.

Above-ground symptoms of yellowing and dying of the foliage do not distinguish this disease from other root diseases, and so cannot be used for identification.

Specimens of fructifications of the causal fungus from Malaysia were named in 1918 as a new species, *Fomes pseudoferreus* Wakefield. The fungus was later moved to the genus *Ganoderma* as *G. pseudoferreum* (Wakef.) Over. and Steinm., and this is the name that is commonly found in the literature. However, in 1932 it was considered that the fungus was the same as one described in 1881 as *F. philippii*. This name was not taken up by plant pathologists immediately, but the name *G. philippii* (Bres. and P. Henn.) Bres. is now generally accepted and is found in the more recent publications. Identifications from Africa have mostly been based on vegetative characters only, so the species of fungus involved there is uncertain (Steyaert 1975). Fructifications have been found associated with the disease in Zaire (Pichel 1956) but were identified only as *Ganoderma* sp.

The fungus is a member of the Basidiomycotina, Polyporaceae, with typical bracket fruit bodies. These develop at the base of trees killed by the disease, on stumps or on felled logs. They do not appear until 3–4 years after the death of the affected tree, so that if all dead material is cleared away quickly they will not be seen. The sessile brackets are very variable in shape. They are flattish and thin, and grow singly or in groups. Individual brackets may be up to about 30 cm in diameter and up to about 4 cm thick at the base. The upper surface is dull brown with a white margin. It may be grooved concentrically or may be covered with large, uneven knobs. The undersurface is white or buff, with numerous small pores within which the spores are formed. They are brown, and are produced in such large numbers that in layered groups of brackets the lower ones may be covered with a thick layer of the spores, looking as if dusted with cocoa powder.

As in the case of white root disease, infection of rubber trees by *G. philippii* takes place through direct root contact with roots of neighbouring diseased trees or with pieces of infected woody material in the soil. Spread by spores would seem another possibility, but so far no proof of infection of stumps by spores has been forthcoming. The fungus is relatively slow-growing and root infection usually occurs a long time before above-ground symptoms appear, so that trees may be in tapping before the presence of disease is recognized.

Control measures are as for white root disease. The standard poisoning of trees or stumps is not so effective against *G. philippii*. However, if the pseudosclerotial skin covering the root is damaged, saprophytes rapidly take over (Fox 1971), so mechanical clearing reduces sources of infection significantly. For collar treatment tridemorph may be used (Rubber Research Institute of Malaysia 1974d).

Brown root disease

The earliest record of this disease on rubber is that from Ceylon (Sri Lanka) in 1905. It is now known from most other rubber-growing countries in Asia. It is also widespread in Africa. It was tentatively identified in Brazil (Langford & Osores 1965) and is reported as occurring there sporadically.

The disease is generally of less economic importance than either white or red root disease, although in south India it is the most important root disease (Wastie 1975). Even in countries in which it is commonly found, such as Sri Lanka, Indonesia and Malaysia, it causes comparatively little damage, and usually occurs on isolated trees rather than in extensive patches. In Africa it has generally been reported as uncommon or rare, although it has recently caused significant damage in the Ivory Coast (Nandris *et al.* 1985).

Brown root disease occurs on a wide range of hosts besides rubber, and trees in over 50 genera are affected (Pegler & Waterston 1968). These include crops such as cocoa, coffee, kola and tea. The oil-palm is also attacked, but here the disease usually takes the form of a trunk rot some distance above the soil.

Since the spread along attacked roots is slow, aerial symptoms, followed by death, are usually seen only in older trees.

The appearance of affected roots is very characteristic (Plate VIII). The roots are encrusted with a mass of soil, sand and small stones to a thickness of 3–4 mm. This mass is attached to the root by a continuous skin of mycelium which produces mucilage, making the mass adhere so firmly to the root that it cannot easily be washed off. The mycelium is velvety brown, and the overall appearance of the infected area is brown when the attack is recent, although with age a black crust develops. The fluffy brown mycelium can, however, be seen if the crust is broken. In younger trees the encrusting mass is often most developed on the tap-root and it may ascend the stem for a few centimetres. The fungus spreads very slowly, and if the tap-root is first affected it may be completely rotted and disintegrated before the fungus has spread along the lateral roots far enough to cause any foliage symptoms. These are similar to the symptoms caused by the other root diseases, with yellowing and drying out of the leaves.

In the early stages the rot affecting the roots is pale brown. Later the wood becomes soft and friable, and thin sheets of brown hyphae occur throughout the wood appearing, when the root is cut, as a network of fine brown lines. At an advanced stage of decay the brown sheets persist even after the wood itself is completely rotted, at which stage the roots can easily be crushed and broken.

Until 1932 there was confusion concerning the identity of the brown root fungus. In the early days the name applied was *Hymenochaete noxia* Berk. but this was soon recognized as being incorrect, the diagnosis and description having been made from very immature specimens. From 1917 it was considered that the fungus was the same as *Fomes lamaoensis* Murr., but in 1932 specimens from Malaysia and Sri Lanka were recognized as belonging to a separate species which was named *Fomes noxius* Corner. This is the name which has appeared most commonly in the literature; but the fungus was transferred to the genus *Phellinus* in 1965, and the name *P. noxius* (Corner) G.H. Cunn. has been used since then.

Phellinus noxius is a member of the Basidiomycotina, Polyporaceae, and has typical bracket fructifications. These appear on stems, stumps and logs only a long time after death of the tree, so they are not often encountered where good hygiene is practised. The unstalked fruit bodies are up to 13×25 cm across, and 2–4 cm thick, and are very hard. They have a deep reddish-brown to dark brown upper surface, becoming almost black with age. There is at first a white margin, but this later disappears. The lower, pore-bearing surface is dark grey or greyish-brown. Fructifications may grow singly or in layers.

As in white and red root diseases, infection takes place by root contact with diseased stumps, roots or root fragments. Progress of the fungus along the infected root is slow, and epiphytic growth ahead of penetration is only of the order of a few centimetres. Infection can also be initiated by the spores produced in the brackets, in two ways. The cut surface of stumps of felled trees can be infected. The roots are eventually invaded, and spread can then take place to living trees. Or infection can take place, particularly in wind susceptible clones, through broken or pruned stems and branches (Lim 1971) from which the fungus may spread to the roots.

Control is basically the same as for white root disease. As a collar protectant tridemorph is the preferred material (Rubber Research Institute of Malaysia 1974d). To guard against spore infection, cut stumps should be treated with a wood preservative such as creosote. Damaged branches should be treated with a wound protectant.

Armillaria root rot

Although the fungus causing this disease is worldwide in distribution, it occurs in Asia only at high altitudes and so rubber is not attacked. Nor is rubber affected in America. In Africa the fungus can be found virtually down to sea level and *Armillaria* root rot occurs commonly on rubber. It has been particularly noted in Zaire and Nigeria, but has also been observed causing considerable damage in Cameroon and Liberia.

Loss of trees can be serious, particularly in wetter areas with high rainfall and poor drainage. It is virtually absent on free draining, acid, sandy soil. An overall figure for losses of 0.5–10 per cent has been reported in Nigeria (Riggenbach 1966). Attack is usually on young trees in the field but the fungus has been found even on the tap-roots of eight-month-old plants still in the nursery (Fox 1964).

The fungus attacks a very wide range of plants in both temperate and tropical areas. It is an important pathogen of forest trees as well as of plantation crops including cocoa, coconut, coffee, oil-palm and tea.

Two types of root symptom are found. In one case affected bark turns bluish-black and the outer tissues of the root become thickened and somewhat soft. Deep cracks appear and in these the black sclerotial tissue of the fungus appears. In the other type of attack the bark of the healthy root splits ahead of the advancing infection and is surrounded by a mixture of soil and latex which exudes from the splits. The colour of the root and the surrounding material is black. When the outer tissues of infected roots are removed, the white mycelium of the fungus can be seen on the surface of the wood forming a thick skin. No external rhizomorph growth occurs. Characteristic of the disease is the enlargement of the lenticels, which become very obvious.

As with other root diseases, above-ground symptoms appear only when the major part of the root system has been killed. Leaves turn yellow and wither, and the whole tree dies.

The fungus causing this disease is *Armillaria mellea* (Vahl: Fr.) Kummer, a toadstool-like fungus belonging to the Basidiomycotina (Tricholomataceae) (Pegler & Gibson 1972). The name *Armillariella mellea* (Vahl: Fr.) P. Karst. is also to be found in the literature, but the generic name *Armillaria* is preferred.

Fructifications develop in clusters at the base of the trunk of a diseased tree. The pileus is 5–10 cm in diameter, honey-coloured to deep brown, with whitish to pink gills, bearing spores which in mass are light cream. The stipe, 5–6 cm high, is mostly dull yellowish to brown, but above the annulus is white. The fructifications are evanescent and appear only during the rainy season (Fox 1971).

The spores are unable to infect living rubber tissues, but they serve to spread the disease over distances as they are able to invade dead trees or stumps which can then act as centres of infection. In the soil, the fungus is dependent for survival on the presence of tree roots, and spread from one rubber tree to another is by contact of a healthy root with an infected one. Although under some conditions, particularly in temperate areas, thin black rhizomorphs are formed which can extend freely through the soil, this does not occur in the case of rubber.

Control is basically similar to that for white root disease. Where rubber is being planted from forest it is important, before planting the rubber, to ring-bark the forest trees, or at least those known to be potential hosts of *Armillaria* (Fox 1964). Felling should be delayed for 1–2 years after ring-barking until the trees have died and the carbohydrates in their roots, on which survival of the fungus depends, have been exhausted. All stumps should be poisoned after felling. Growth of *Armillaria* is slower than that of *Rigidoporus*, so the disease progresses at a slower pace. Because of this sources of infection may be detected only after the sources of *Rigidoporus* have been found and dealt with.

Stinking root rot

The fungus causing stinking root rot (also known in the past as serpentine disease) was first found on rubber in Ceylon (Sri Lanka) in 1907. Since then it has been reported in most rubber-growing countries in Asia and Africa, although it does not appear to occur in America.

The disease is of minor importance, and the causal fungus is only a feeble parasite. In the past it affected quite large numbers of trees from time to time but avoidance of the conditions predisposing trees to attack has reduced the disease to negligible proportions (John 1964).

The causal fungus is widespread in tropical soils where it occurs as a saprophyte on decaying woody material. Besides rubber it attacks several other perennial crops, including avocado, cassava, citrus, cocoa, coffee, mango and tea.

Affected plants are readily recognized by the characteristic flat, branching rhizomorphs which grow between the bark and wood of affected roots, and are seen when the bark is removed. They are reddish-brown to black with a white advancing edge. When the bark is removed the rhizomorphs are sometimes split in half, part remaining attached to the inside of the bark, part to the wood. This results in the white internal tissues of the rhizomorph being exposed. The wood of affected roots does not become particularly soft but develops a bluish or violet colour (the disease on tea is often known as violet root rot). Most characteristic is the strong foul smell produced by the decaying roots.

The causal fungus is a member of the Ascomycotina (Hypocreaceae), *Sphaerostilbe repens* Berk. and Br.; a recent description is that by Booth and Holliday (1973). It commonly produces two types of fructification on the surface of affected roots or on the base of the stem. The more common, the asexual stage, appears as red stalks, 2–8 mm high, each with a pinkish white pin-like head. The sexual stage consists of very small, red, rounded bodies (about 0.5 mm in diameter and tapering at the top).

Local spread has been reported as being by root contact from tree to tree but this is probably rare, if it occurs at all. The main method of spread is by the asexual spores which infect roots.

As regards control, the key factor is that the fungus is incapable of attacking rubber (or any other crop) except under adverse environmental conditions. The disease can occur only under conditions of poor drainage or where there has been flooding. As the disease is so dependent on such conditions, control must be based on avoiding or alleviating these conditions. Outbreaks are rare now that this is understood.

Ustulina collar rot

The fungus causing this disease is cosmopolitan and probably attacks rubber wherever the crop is grown, but the disease is nowhere of major importance. It usually affects small numbers of older trees, attack being seen mainly on those over 20 years of age. In Cameroon, however, attacks have been more serious, probably due to lack of control of branch infection (see p. 432) in the upper canopy, whence the disease spreads to collar and roots. Progress of the disease is slow, death of a mature tree occurring several years after first infection.

This collar rot affects a wide range of hosts, in both temperate and tropical areas, including arecanut, cocoa, coconut, *Crotalaria*, oil-palm, tea (the most seriously affected crop) and teak.

The characteristic symptom of the disease is a dry rot of the wood, mostly affecting the collar area and lateral roots of older trees. At the collar, attack may result in the formation of a hollow due to rotting and disintegration of the wood. This rot may extend up to about 1 m above ground but even then there may be no sign of deterioration in the foliage, and the tree may still continue to yield latex. Eventually, however, if the tree is not first blown over, the foliage becomes thin, dieback occurs and the tree dies.

Rhizomorphs are formed by the fungus but are not found on the surface of diseased roots. They occur under the bark and are fan-shaped. They are greyish-white or light brown, turning black when exposed by removal of the bark. The underlying wood becomes light brown and disintegrates easily. Characteristic are the thick black lines, often double, which can be seen in the outer layers of affected wood when cut. The lines can be distinguished from those found in brown root disease, being darker and more conspicuous (Plate IX).

The fungus causing this root and collar rot is *Ustulina deusta* (Hoffm.) Lind (Ascomycotina, Xylariaceae). The name *U. zonata* (Lév.) Sacc. is also commonly found in publications. Fructifications occur on infected collars and dead wood. They take the form of thin stromata, 1–3.5 mm thick, adpressed to the bark. They are light grey and velvety at first, later becoming dark grey or black. Spores of two sorts, conidia and ascospores, are produced on these stromata and serve to spread the fungus. A description of the fungus is given by Hawksworth (1972).

The spores are air-borne and are the main means of spread, the collar and lateral roots both being subject to invasion when wounded. Infection can also take place by direct contact of a healthy with a diseased root, but the contact has to be more intimate in this than in the major root diseases, as in *Ustulina* there is no surface mycelium.

As the main means of spread is by direct infection by spores through wounds, damaging of collar and roots during weeding, drain digging, etc., should be avoided. Where wounds occur they should be treated with a wound dressing. Infected trees can be saved by excising affected tissues back to healthy wood. In cases in which the rot has spread to affect a large area of the collar, the gap can be filled with cement to give support and prevent the tree from being blown over. As the disease is spread almost entirely by spores it might be considered worth while to destroy fructifications, but as these are numerous on dead wood as well as on other plants the effort involved is probably not justifiable.

Poria root rot

This disease has been reported most frequently from Sri Lanka

(Petch 1921) but has also been recorded in Brazil, Indonesia, occasionally Malaysia (John 1964) and more recently India (Rajalakshmy & Radhakrishna 1978). In none of these countries is it widespread or serious. Other hosts include cocoa, *Crotalaria anagyroides*, dadap (*Erythrina*), tea and *Tephrosia candida*.

The disease was originally known as red root disease, the disease now known by that name having been called wet rot. *Poria* root rot is now accepted as the common name, although the reason for the original name is obvious from the symptoms. This is particularly so on trees up to about two years old, on the roots of which the fungus forms stout, smooth red strands of mycelium, sometimes joining into a continuous red sheet. Internally they are white and this is seen when roots are dug up and the strands are damaged. Later the fungus becomes black and stones may adhere to the root as in the case of brown root disease, but the incrustation is never so thick. The wood of diseased roots is soft and friable and is permeated with red sheets of mycelium, showing as lines when the root is split.

The fungus causing this disease has generally been known under the name *Poria hypobrunnea* Petch, and this is the name found in the literature. More recently the species has been re-examined and the name *P. hypobrunnea* is regarded as being synonymous with *P. vincta* (Berk.) Cooke var. *cinerea* (Bres.) Setliff. It belongs in the Basidiomycotina (Polyporaceae), but unlike the fungi causing white, red and brown root diseases, the fructification is not a bracket but is supine on the substrate. It is usually annual, about 1–1.5 mm thick, and may occur at the collar or on the undersurface of exposed diseased rubber roots, as well as abundantly on stumps and logs. It is reddish-brown to grey, and is covered with small pores in which the spores are formed.

The fungus spreads by means of these spores to lying timber, roots, moribund stumps, etc., which become infected. As in the commoner root diseases, spread is then usually by root contact. Spread is very slow, and there may be abundant affected material and fructifications in the vicinity without infection of rubber occurring.

Control measures would be as for white root disease.

10.1.2 Stem and branch diseases

Diseases of the stem and branches, excluding diseases of the tapping panel, which are dealt with separately (see section 10.1.3), are of less importance than are diseases affecting other parts of the plant. Only pink disease is capable of inducing significant losses on a large scale. More important than the direct damage caused by the other three diseases covered is their role as sources of infection of other parts of the tree. Thus the fungus causing patch or stem canker is one of the causes of the serious diseases black stripe and abnormal leaf fall; *Ustulina* causes a collar and root rot; and *Phellinus* is the pathogen more commonly associated with brown root disease. White threadblight is included as its characteristic symptoms make it very noticeable.

Pink disease

The first detailed account of pink disease on rubber is that by Brooks and Sharples (1914) but by then the causal fungus was already well known on a wide range of hosts. Plants affected include several tropical crops besides rubber, among them cinchona, citrus, cocoa, coffee, mango and tea. Leguminous plants are also attacked, including *Crotalaria, Desmodium, Indigofera* and *Tephrosia*. A more recent account of the disease is that by Hilton (1958).

Pink disease is found throughout the humid tropics, and occurs on rubber in virtually every country in which the crop is grown, in Asia, Africa and America. It has recently increased in many areas as replanting has been carried out with clonal material less resistant to attack than the original seedlings.

This is a disease mainly of young trees, mostly being seen on trees 3–7 years old. It can, however, occur as early as 18 months and also persists in mature trees. In young trees it can cause a large amount of damage to the upper stem and leading branches, but serious outbreaks usually occur only locally. It is particularly common in inland areas and specially where there are well-marked rainy seasons. It causes virtually no damage on coastal soils, even when the weather is wet.

Often the first symptom noticed is the exudation of latex from a branch, usually near the base where it joins a larger branch or the stem. At the same time silky threads of the fungus grow along the branch, forming the cobweb stage of the disease. The threads dry up as the outer bark is killed, and in their place there develops the light pink incrustation from which the disease derives its name (Plate X). This is generally confined to the lower surface of the affected branch. The incrustation soon dries out and cracks irregularly, the light pink colour fading to off-white. The incrustation does not always appear; in its place there may be only pink pustules protruding in rows through cracks in the bark. By now the diseased tree presents a characteristic appearance, with the affected branch dying, the dried-up leaves remaining hanging on it. The leaves then fall, leaving a bare branch from which latex exudes and coagulates in long black streaks. There may be no further spread, in which case adventitious buds may develop just below the diseased part; or the infection may spread down to the fork and along other branches. In extreme cases, if the main fork of a young tree is affected, the entire canopy may be lost.

The pink disease fungus is *Corticium salmonicolor* Berk. and Br. (Basidiomycotina, Corticiaceae). Recently the name *Phanerochaete salmonicolor* (Berk. and Br.) Jülich has been proposed. The most obvious stage of the fungus is the pink incrustation, and this bears the basidiospores which are dispersed by wind. Another spore form appears on orange-red pustules, about 3 mm in diameter (distinct from the pink pustules mentioned above) which are formed on the bark of diseased branches. The pustules bear numerous spores, probably spread by rain splash. This conidial stage was originally thought to be a separate fungus named *Necator decretus* Mass., and it is still sometimes referred to as the necator stage.

The disease is susceptible to fungicidal treatment but a formulation is needed that is not washed off by rain. For treatment to be successful infection must be recognized at an early stage, since as the disease progresses the affected branch is ringed, and once this happens the branch is lost. Bordeaux mixture is still one of the most effective materials, as it is persistent. It can be applied with ground sprayers fitted with a lance, the nozzle adjusted to give a jet, treatment being repeated weekly during wet weather. Or a paste can be used, brushed on to affected parts. This is more troublesome to apply, but one application may give control.

Other fungicides are continually being tested. Particularly useful is a formulation of tridemorph in latex, applied by brush. One treatment applied at the beginning of the rainy season may be all that is required (Wastie 1975). Other materials showing promise include chlorothalonil and Trifoltan (Bordeaux mixture plus dithiocarbamate) (Rubber Research Institute of Malaysia 1981b). Bordeaux and other fungicides based on copper must not be used on trees in tapping in such a way that the latex is contaminated with this metal.

In areas particularly prone to attack it may be possible to reduce the severity of the disease by choice of suitable clones, although highly resistant clones have not yet been found.

Patch canker or stem canker

Patch or stem canker is one of a group of diseases caused by species of *Phytophthora*. As these fungi are common in all rubber-growing countries the disease is probably widespread, although recent records are few. Countries in south-east Asia in which it has been noted include Burma, south India, Indonesia, Malaysia and Sri Lanka. There are also records from Zaire and Costa Rica.

If left untreated the disease can cause death of the tree, particularly if infection is at the base of the trunk, which may be ringed; but overall the importance of the disease is not great. In West Malaysia, for example, it had not been seen during the 30 years prior to 1968, when it was observed on one clone, RRIM 605, which is sensitive to wounding (Chee 1968).

In the early stages of the disease there is usually little indication of infection. Some discoloration of the bark may be noticed, but often the first thing seen is an exudation of latex (Plate XI) or a dark purplish liquid. Pads of evil-smelling coagulated latex accumulate under the bark and as a result the bark bulges and eventually splits. Boring beetles (Platypodidae and Scolytidae) attack the infected area and although this happens only at a late stage it is often the first sign of damage that is noticed. If affected bark is scraped off, a black layer is seen immediately beneath it. Underneath this the cortex is discoloured, at first yellowish-grey, later becoming dull purple-red (the disease has been known as claret-coloured canker), the colour distinctly different from the clear translucent red of healthy tissues.

Both mature and immature trees are liable to attack. The canker has even been found in bud-wood nurseries in south India (Ramakrishnan 1963). Cankers may occur on any part of the main stem, and even on larger branches, but normally they are found within about 2 m of soil level. They become large and irregular, involving large areas of bark, and continuing to enlarge through the wet season. With the advent of dry weather, spread is halted and cankers may dry out and heal over; but the resultant rough bark makes subsequent tapping difficult.

The fungus causing the disease is *Phytophthora palmivora* (see p. 436), which is also associated with black stripe, leaf fall and pod rot. There is some difference of opinion as to whether the fungus is the primary cause of patch canker, or whether it attacks only after lightning damage. It seems, however, that the fungus can act as a primary parasite, as demonstrated by inoculation tests. But similar symptoms also result from lightning strike followed by invasion by *Phytophthora* and *Pythium vexans* de Bary.

Control with fungicides is effective, particularly if the attack is diagnosed early. Affected tissues are scraped off and the healthy tissues thus exposed are then treated with a fungicide. Bordeaux paste is commonly used. Materials applied against black stripe, such as metalaxyl, Antimucin WBR and captafol (Tan 1979) are also effective. The treated wood should then be protected with a waterproof wound dressing. As the disease occurs only sporadically, prophylactic spraying is not feasible.

In general, clones showing resistance to black stripe have a similar reaction to patch canker.

Ustulina stem rot

Stem rot is another manifestation of disease caused by the species of *Ustulina* that attacks the collar and roots (p. 427), and probably occurs wherever collar and root rot have been recorded. Although not usually of major importance, the disease can be troublesome if neglected.

Any part of the stem and branches may be attacked, but infection can take place only through wounds. These may be caused by careless pruning, wind damage, scorching caused by fire or lightning, cracks in the fork at the junction of stem and main branches, or wounds caused by too deep insertion of spouts (Rao, 1975). Often infection takes place through a wound on the stem resulting from a large branch having been broken off by the wind. A typical result is the copious exudation of latex which coagulates to form a large pad, often foul-smelling and covered by the rotted bark. Affected wood is pale brown and the rot is a dry one. The wood is permeated by black plates of fungus, which appear as black lines when the wood is cut.

The fungus causing this rot is *Ustulina deusta*, described on p. 428. Fructifications occurs commonly on affected parts, often forming an extensive sheet; one reported by Sharples (1936) was 60 cm long and up to 35 cm wide. The fungus may thus be very conspicuous and such fructifications are an obvious sign of advanced infection.

Control should first of all be by prevention, any damaged branches being cut off as soon as possible, the damaged surface smoothed, and a wound dressing applied. If infection has already taken place the diseased parts should be excised and the wound similarly treated. The diseased parts should be burnt to prevent them from becoming sources of infection.

Phellinus stem rot

This stem disease, caused by the brown root fungus, was briefly reported from Malaysia and Ceylon (Sri Lanka) in 1928–29 but has been little mentioned since. There appear to be few records from other countries and the disease in itself appears to be generally of minor importance, although Lim (1971) reported 21–30 per cent infected trees in four areas investigated. In two of these mortality was so high that tapping became uneconomic and the affected areas were prematurely felled. The disease assumes additional importance in that it may act as a source of brown root disease.

Infection can take place only through wounds, such as those caused by wind breakage or lightning strike, or through cut surfaces such as the unprotected stubs of pruned branches. Decay then proceeds slowly from the point of entry into the trunk. Once the fungus has penetrated deeply the distal part of the affected stem or branch dies and breaks off. The rot may also spread down within the trunk, although no external signs occur, and may finally invade the collar and thence the roots, causing the typical brown root syndrome. Within the stem the rotted tissue is at first dry and, when this is cut into, brown lines of mycelium are seen. Later the wood becomes soft and cellular and easily crumbled.

The causal fungus is *Phellinus noxius* (see p. 424) and fructifications may at a late stage be found on diseased stems.

Infection of wounded branches is probably by air-borne spores. The badly affected areas investigated by Lim were all near the forest, from which the spores could have originated.

As entry is only through wounds, prevention of infection, as in the case of *Ustulina* stem rot, depends on removal of damaged branches, careful pruning, and protection of cut surfaces with a wound dressing. Creosote and 20 per cent Izal have been found equally effective.

White threadblight

This disease is mentioned because, although the damage done is slight, its appearance is striking. Occasionally it can be troublesome, under a dense canopy or in particularly humid conditions such as may occur in low-lying areas or near the forest edge.

The typical symptom is the growth of white fungal strands spreading outwards along the branches and twigs and eventually along the petioles on to the leaflets, on the underside of which they spread and ramify. Twigs and branches die back. Leaves die and wither, and remain suspended by white threads of mycelium.

Several species of fungi may be involved, but as only the vegetative phase is usually found the situation remains unclear. However, one species parasitizing rubber in Malaysia has been identified as *Marasmius cyphella* (Dennis & Reid 1957) (Basidiomycotina, Tricholomataceae), whose toadstool-like fruit bodies are formed on the undersurface of dead leaflets.

Spread between adjacent branches or trees can easily occur through movement of dead leaves or fragments of leaflets or mycelium. As the threadblight is favoured by humid conditions, measures to reduce humidity within the canopy should be taken. Diseased branches and leaves should be collected and burnt. If attack is severe, spraying with an oil-based fungicide (such as those used for mouldy rot – see p. 439) is recommended (Rao 1975).

10.1.3 Panel diseases

Many fungi are incapable of attacking plants except through wounds, but once entry has been effected can cause serious damage. The panel is an area which, being continually and repeatedly injured in the tapping process, is an eminently suitable point of entry for fungi which would not otherwise be able to damage the trunk. Despite this vulnerability, the panel suffers from only two serious diseases, black stripe and mouldy rot, but both are widespread and can be damaging, particularly in times of heavy rainfall and high humidity, unless controlled. Few other fungi cause damage and only two minor diseases, panel necrosis and white fan blight, are briefly mentioned here.

Panel infections respond well to treatment with fungicides, and control at present depends on their application. There are, however, significant differences in susceptibility among clones, and the introduction of resistant clones may afford the best solution to control of these diseases in the future.

Brown bast is a disorder of the tapping panel that is non-pathogenic in origin, and is dealt with in Chapter 9.

Black stripe

Black stripe has been known under several other names, including bark rot, black thread, black line canker, cambium rot, stripe canker, and tapping canker. It was first noted in Ceylon (Sri Lanka) in 1909 and has since been found to be widespread in south-east Asia as well as occurring in Africa (Cameroon, Liberia, Zaire) and America (Brazil, Costa Rica, Peru).

The disease can cause serious damage, but economically important outbreaks are more or less confined to areas with long periods of high rainfall and constantly high relative humidity. Such periods occur, for example, in Burma, south India and Sri Lanka during the south-west monsoon, and in Liberia in the wettest period, from August to October. Damage can be severe in these places unless prevention or control is practised. In areas such as most of West Malaysia where hot, sunny mornings with low relative humidity are common even in the wettest months, the disease is much less common, occurring in severe form only in localized areas where the terrain is conducive to the maintenance of damp conditions.

The disease has in recent years been more troublesome than previously because many of the high-yielding clones introduced during the past couple of decades are very susceptible. One hundred per cent infection of particularly susceptible clones has occasionally been found in West Malaysia (Chee 1974).

The first symptoms are not very obvious. They take the form of short, dark grey, vertical, slightly depressed lines, about 1 cm above the tapping cut. The damage is more obvious when the bark is pared away, when dark lines are seen corresponding to the depressions seen on the surface of the panel (Plate XII). These black lines extend internally into the wood, to a depth of almost 5 mm, and develop into stripes of about the same width. If untreated they may later widen further and coalesce laterally into wide lesions which may involve a large proportion of the width of the tapping panel.

The damage done causes uneven regeneration of the panel bark, and when the wounds heal the bark may be irregular. In very susceptible clones protuberances or burrs may be formed. This makes it difficult to tap again, and in extreme cases further use of the panel may be impossible. The stripes can also spread downwards as much as 15 cm into the untapped bark below the panel, which then rots. Latex may accumulate under the bark, causing it to split with exudation of latex. Even if untreated the wounds tend to heal with the onset of the dry season, but when conditions again become favourable infection is renewed.

The disease is caused by species of *Phytophthora* (see under abnormal leaf fall, p. 446). The most common species is *P. palmivora. Phytophthora meadii* is also involved in Burma, (Turner & Myint 1980), India and Sri Lanka (Wastie 1975), while in west Malaysia and Thailand *P. botryosa* is implicated (Chee 1969). In Yunnan, China, the main species involved appears to be *P. citrophthora* (Ho *et al.* 1984). Earlier records suggested that *P. heveae* Thompson was another species capable of causing the disease in west Malaysia, but its pathogenicity to rubber there is open to doubt, although it is listed as the cause of the disease in Sabah (Liu 1977).

Phytophthora palmivora affects a very wide range of woody and herbaceous plants; crops attacked include black pepper, citrus, cocoa, coconut, cotton, mango, pineapple and tobacco. However, the taxonomy of the species is under review, and it seems probable that the occurrences reported in the literature may be of several very similar species. *Phytophthora meadii* is virtually restricted to rubber, although there are a few records of it on pineapple and one or two other crops. *Phytophthora botryosa* is known only from rubber. *Phytophthora heveae* and particularly *P. citrophthora* have wider host ranges.

As the fungi are seldom visible macroscopically on affected plants their appearance in the field is of little use for disease diagnosis. They can, however, sometimes be seen early in the morning in very humid conditions as a whitish bloom on the tapping panel just above the tapping cut.

Spread is by infected tapping knives, by rain splash and by airborne spores.

Development of the disease is very dependent on environmental conditions. Infection can take place only when there is high humidity, and long periods of rainfall and continuous dampness are necessary for an epidemic to be sustained. This suggests steps that can be taken in disease-prone areas to minimize the effects of the fungi. Excessively moist conditions can be avoided by preventing vegetation in the inter-rows from becoming too dense and high. New panels are particularly liable to be infected, so tapping of young trees and opening of new panels on older trees should not be started during wet periods. Complete stoppage of tapping with the onset of the rainy season, as is practised for example in Burma (Turner & Myint 1980) is effective. No very highly resistant clones are known, but particularly susceptible high-yielding material should not be planted in areas where severe monsoon conditions occur.

Besides these preventive measures, direct control with chemicals is widely used. It is effective if the disease is detected early, before it has had time to become firmly established. Treatment should begin with the onset of wet weather and continue until the end of the rainy season. The fungicide is best applied after tapping, preferably soon after latex collection. Tree lace should be removed before treatment. Application should be by spraying, using a fine fan-jet nozzle on a hand sprayer with a trigger release, or with a brush if the number of trees to be treated is small. The aim is to treat an area about 8 cm above and 6 cm below the tapping cut.

Over the years several fungicides have been used with success, those in recent use including captafol and metalaxyl, the latter having been found slightly superior in field trials in Malaysia (Tan 1983). Metalaxyl should be alternated with other fungicides to avoid the possible development of fungicide resistance. Treatment intervals can be extended if an oil-in-water emulsion formulation is used. As a precaution it is advisable to extend treatment to neighbouring healthy trees.

If the disease has been neglected so that cankers and deep wounds have been allowed to develop, these should be excised, all diseased tissue being carefully removed, after which the cut surfaces should be treated with a fungicide and wound dressing.

Mouldy rot

Mouldy rot was first recorded in 1916 in west Malaysia and soon after was observed in Indonesia, Sabah and Zaire. It is now found in most rubber-growing countries in Asia (although not apparently in Sri Lanka) and also in Africa (Cameroon, Nigeria) and America (Brazil, Costa Rica, Guatemala, Mexico). The disease can cause serious damage in certain areas, as the renewing bark on the tapping panel is destroyed.

The fungus causing mouldy rot is common on other crops, causing diseases such as wilt of cocoa, canker of coffee and black rot of sweet potatoes. There is, however, evidence of differences in pathogenicity, and strains of the fungus attacking other hosts may not affect rubber.

The first sign of disease is the occurrence of slightly depressed spots or blotches about 0.5–2.5 cm above the tapping cut. These blotches spread and coalesce to form a very irregular depressed band parallel to the cut. Affected tissues soon darken and become covered in a thick, grey or greyish-white mould (Plate XIII). This is very characteristic of the disease, and is easily seen even from some considerable distance. After 3–4 weeks, if left untreated, the affected cortical tissues are completely rotted and wounds are formed similar to those caused by bad tapping. The underlying wood is exposed, and can be seen to be discoloured greenish-black. The infection remains shallow, penetrating only to about 0.5 cm at most. Unlike black stripe, mouldy rot does not extend below the tapping cut.

The fungus causing mouldy rot is *Ceratocystis fimbriata* Ell. and Halst. (Ascomycotina, Ophiostomataceae). The name *Ceratostomella fimbriata* (Ell. and Halst.) Elliott is found in the literature before 1950, and in very early works the name *Sphaeronema fimbriata* (Ell. and Halst.) Sacc. is used. A description is given by Morgan-Jones (1967).

The fungus is seen on rubber in its asexual state, the thin-walled, colourless conidia being produced in great profusion on the infected newly tapped bark. This is the cause of the typical mouldy appearance of the disease. Thick-walled chlamydospores (another asexual form) are also found on the surface of diseased tissue, or within it. The sexual stage of the fungus, which takes the forms of black, long-necked perithecia, was found on rubber trees in Malaysia only during the first few years after the disease was first identified there in 1916, and has not been seen on rubber in the field since then.

Spread of the disease is by the short-lived conidia, which can be disseminated by wind or in the excrement of insects. Also instrumental in spread are tappers, who may carry spores on their hands, clothing and tapping knives. The thick-walled chlamydospores, being long-lived, are important in the survival of the fungus over periods of unfavourable conditions, and can also assist in spread over long distances, by the same human means.

For its growth *C. fimbriata* requires prolonged periods of high humidity, such as occur during long spells of rainy weather. These occur commonly, for example, in Nigeria, where the disease can cause widespread serious damage. Such favourable conditions can also occur locally. Thus in areas such as much of West Malaysia, where dry, often sunny mornings occur commonly even during the wettest periods of the year, the disease is generally less severe, and attacks often cease after a few days. But even here, under particular conditions such as in shaded or wet valleys, in sheltered areas, or where undergrowth has been allowed to grow excessively, it can flourish.

Because of its dependence on humid conditions, cultural control plays an important part in reducing mouldy rot incidence. Stands should be properly thinned to optimum density, and woody undergrowth kept in check. Wounding of the bark during tapping must be avoided. As the spores can be transmitted by tappers, it is advisable not to transfer tappers from affected to disease-free areas.

Chemical control is essential where the disease is severe. To reduce the risk of spread, tapping knives in infected areas may be disinfected after each tree is tapped by dipping in a fungicide such as 5 per cent Izal. The main application of fungicides, however, is in treatment of the tapping panel, and it is fortunate that as the attack is superficial, chemical control is relatively easy. Prophylactic treatment need not be applied, but treatment should start as soon as possible after the first signs of attack are noticed. It is not necessary to stop tapping to obtain control.

Many fungicides have been used over the years against mouldy rot, and new materials are constantly under trial. Chemicals at present or recently in use include benomyl, carbendazim, cyclo-heximide, captafol and metalaxyl. An added indicator dye helps in checking on where treatment has been properly carried out. The fungicide, which may be water- or oil-based, is best applied by a pressurized hand sprayer with a trigger release, giving a fine mist produced as a fan-shaped spray. Or for small numbers of trees a brush can be used. The technique is, in fact, as for black stripe.

Treatment is best given on the tapping day, after latex collection. It can also be done on the day following tapping. With the newer fungicides, treatment at weekly intervals is often adequate and only four applications may be needed to eradicate the fungus.

All seedlings and clonal materials are susceptible to some extent. Clones below average in susceptibility in Malaysia include RRIM 513 and 623, PB 5/51 and 5/63, and GT 1.

Minor panel diseases

Panel necrosis was first described comparatively recently in West Malaysia (Chee 1971) and appears to be restricted to that territory. Panel guide markings and newly tapped panels are affected. The bark decays and splits open and the rot may spread to the cambium and wood. The cause is *Fusarium solani* (Mart.) Sacc. (Deutero-mycotina, Hyphomycetes) with which is often associated *Botryodi-plodia theobromae* Pat. (Deuteromycotina, Sphaeropsidales). The disease has so far been found only on a few high-yielding modern clones which are evidently particularly susceptible. Spraying with one of the fungicides used against black stripe is effective if the disease is caught at an early stage, and control may be achieved in a couple of weeks.

White fan blight was first described from West Malaysia by Thompson in 1924 but has rarely been seen since. A fan of white mycelium appears on the tapping panel and spreads to cover a large part of it, penetrating the bark down to the wood. The cause is

Marasmius palmivorus Sharples (Basidiomycotina, Tricholomata-ceae), a fungus with mushroom-type fruit bodies which are not, however, found on growing rubber, although they have been induced on infected material in the laboratory. The disease occurs only rarely and under damp conditions, similar to those favourable to the development of mouldy rot, control measures against which would also be effective against the blight.

10.1.4 Leaf diseases

The most injurious leaf disease is South American leaf blight, fortunately still confined to Central and South America. Since 1974 there have been renewed efforts, particularly in Brazil, to control it with chemicals and by breeding for resistance, with some success. Other more widespread leaf diseases of importance are powdery mildew, abnormal leaf fall and *Colletotrichum* leaf disease, part of the secondary leaf fall complex. Particularly in Malaysia but also in other countries, increased incidence of such diseases has been evident in recent years, brought about by the large-scale planting of high-yielding clones into which resistance has not been bred. These clones have often been planted in areas in which conditions are ideal for leaf disease development, without giving consideration to the subsequent control problems. Such disease prevalence has recently been recognized as a major constraint within the Enviromax system (Lim 1982; see section 4.5.1).

South American leaf blight

This is the most devastating disease of rubber, but it is so far of restricted distribution, being unknown outside tropical America. First recorded on *Hevea* in 1900 in Brazil, where it is indigenous in the Amazon and Orinoco basins, it has now spread throughout the South and Central America rubber areas, from Mexico to the São Paulo district of Brazil, where it probably arrived as late as 1960 (Commonwealth Mycological Institute Distribution Maps of Plant Diseases No. 27). A review of this and other aspects of the disease is given by Chee and Holliday (1986) and there is also a summary of information by Chee and Wastie (1980).

SALB has in the past prevented the large-scale development of rubber plantations in the western hemisphere. Attempts were made at establishing plantations. For example, in Brazil the Fordlandia planting of over 3000 ha was begun in 1927, and the 6500 ha estate at Belterra was planted in 1934–42. Both were, however, completely unsuccessful because of destructive epidemics of the leaf blight, and it is only more recently that plantations have been established in different localities and through improved techniques. Should the disease reach Africa, and particularly Asia, the effects on rubber production in these areas would be expected to be extremely serious.

Only species of *Hevea* are attacked. Besides *H. brasiliensis*, species known to be susceptible are *H. benthamiana, H. guianensis* and *H. spruceana*.

The serious nature of the disease, which affects trees of all ages, rests in the fact that it can cause repeated defoliation, leading to branch dieback. At best this results in debilitation of the tree; in extreme cases the whole tree may be killed.

The most conspicuous symptoms are on the leaves. If they are infected very young, grey to black, powdery, angular lesions appear on the lower surface. The leaf is distorted and soon abscisses. On older leaves (up to about 16 days old, when they become resistant to infection) similar spots, up to about 2 cm in diameter, are formed but the leaves are not shed. On the undersurface the spot becomes olive green due to the production of numerous spores of the causal fungus. On the upper surface the spots appear as yellowish areas. As the leaf matures the powdery appearance of the spots is lost. They become brown and the centre may fall out to give a shot-hole effect. Later the upper surface bears many black dots, occurring in clusters or rings; still later the black areas become larger. They are scattered on the upper surface of the leaf or round the edge of the shot holes. They reach full development about 10 weeks after infection (Plate XIV).

Inflorescences, young fruits, petioles and young stems may also be affected. Swellings appear which cause distortion on stems and petioles, which become twisted. Infected flowers and young fruits are destroyed, whereas fruits attacked when older survive but bear swellings.

The fungus causing SALB is *Microcyclus ulei* (P. Henn.) Arx (Ascomycotina, Dothideales). Before 1970 it was known under the name *Dothidella ulei* P. Henn.; *Melanopsammopsis ulei* (P. Henn.) Stahel has also been used.

Three spore forms are produced successively on infected parts, as described by Chee and Holliday (1986). The first is the conidial form, spores occurring in the dark grey or olive green growth on the undersurface of affected leaves. Later pycnidia appear on the upper leaf surface as numerous black dots. Finally the perithecia develop, looking similar to but larger than the pycnidia.

Spread of the disease is mainly by the conidia, which are dry and easily dispersed in air currents, dislodgement from lesions being increased during rain showers by movement of the air and the leaves. Such dispersal is efficient, but the conidia are short-lived and are produced only in the wet season. Carry-over during dry periods and initiation of new outbreaks is probably by ascospores, as they occur throughout the year.

Disease incidence is greatly influenced by climate, most disease occurring with high rainfall, well distributed with no marked dry season, and least with lower rainfall and a long dry season of at least four months (see Ch. 4). Planting in areas of 'disease escape', where environmental conditions are less suitable for disease development, is a possible method of control (Gasparotto *et al.* 1984).

Early attempts at chemical control were not successful, as fungicides and application equipment were not sufficiently effective. Recently, however, research in Brazil on improved chemicals and suitable application techniques has given encouraging results (Chee 1984). Of the fungicides used, good results have been given by mancozeb, benomyl, thiophanate-methyl, triadimefon and chlorothalonil. Various methods of application are in use (Chee & Wastie 1980): ground-based mist blowers (using mancozeb plus benomyl, or triadimefon), thermal fog applicators (with thiophanate-methyl or mancozeb) and aerial spraying (with benomyl, thiophanate-methyl or mancozeb). In the areas where the disease is severest, wintering is usually uneven, so several applications of fungicide are needed. The use of artificial defoliation, as suggested in Malaysia for other diseases, to give rapid and uniform refoliation might be useful against SALB also, and tests of the technique have been carried out in Brazil (Romano & Rao 1983).

When the early plantings were made in Brazil and other countries in the western hemisphere it soon became evident that all the planting material available was highly susceptible to SALB. One method developed to circumvent this problem was to top bud high-yielding clones at a height of about 2.5 m, using a resistant clone with good foliage characteristics. The method achieved some success but the delay in reaching maturity was a disadvantage. In addition the top clones used tended to lose their resistance through mutation of the pathogen. Nevertheless, the technique is again being tested in Brazil on a large scale, using green budding in the nursery to shorten the time to maturity (Chee 1984).

Many resistant clones have now been bred, some with resistance coming from *H. brasiliensis* and some from *H. benthamiana*. Unfortunately the resistance of these clones differs in different localities, indicating the presence of several races of *M. ulei*. It now seems likely that major gene resistance will not provide a long-term solution, and that the most useful type of clone will be one with horizontal resistance (Simmonds 1982), without immunity but exhibiting only light infection, allowing very restricted sporulation of the pathogen, and not subject to leaf fall (see Ch. 3).

As SALB is so far restricted to America, effective plant quarantine measures are essential to prevent spread to Africa and Asia, as all the clones grown there at present are highly susceptible. Soon after the Second World War the need for strict quarantine control was recognized and detailed measures, including intermediate quarantine, were formulated, particularly by the staff of the Rubber Research Institute of Malaya. As a result of these efforts the FAO Plant Protection Committee for the South-East Asia and Pacific Region was set up in 1956. This dealt with all crops, but special consideration was given to rubber, and the quarantine measures to be taken by signatory governments as regards rubber were agreed (Singh 1978). It was also recognized that, should SALB reach any country in Asia or Africa, immediate measures to eradicate it should be set in motion. Rapid defoliation by aerial application of chemicals such as 2,4,5-T or triclopyr, followed by a fungicide, is regarded as the method most likely to succeed (Hutchison 1958; Chee & Holliday 1986). Experiments on alternative chemicals (Lim & Hashim 1977) and techniques, including fogging (Lim & Kadir 1978), continue. It is a matter of some concern that not all rubber-growing countries have the means available to allow the immediate implementation of such an eradication programme.

Powdery mildew

Since secondary leaf fall due to *Oidium* was first seen in Java in 1918, the spread of powdery mildew disease has slowly continued. It has by now appeared in most rubber-growing areas in Asia, as well as in Africa (Zaire, Liberia, Cameroon) and America (Brazil) (Populer 1972; Commonwealth Mycological Institute Distribution Maps of Plant Diseases No. 4). It has been reported for the first time in Papua New Guinea as recently as 1967.

The disease has for many years been particularly serious in parts of Sri Lanka and Indonesia where, as in Malaysia and Zaire, it is the commonest cause of secondary leaf fall. In Malaysia, although common, it has not generally been very damaging until comparatively recently, with the introduction of some improved clones with low resistance to it.

Losses are difficult to assess, but an increase in yield of over 100 per cent by controlling the disease has been recorded in south India (Ramakrishnan & Radhakrishna Pillay 1962).

The few other plants reported as hosts include the weeds *Euphorbia hirta* and *Jatropha curcas*, as well as annatto and rambutan. However, although the fungi on these hosts may be morphologically similar, cross-inoculation experiments suggest that rubber is not normally infected by the fungus from other plants.

The disease is essentially one affecting the young leaves less than two weeks old, and it is thus just after bud burst following wintering that damage occurs (Plate XV). The fungus attacks the leaflets, forming superficial white powdery areas, mostly on the lower surface. A severe attack on very young leaves, when they are still at the shiny brown to yellowish stage, causes the leaflets to shrivel and dry up. Tissues turn black, particularly towards the tip, and leaflets are shed. The bare stalks remain attached to the tree for a time. When this happens the first sign of trouble noticed may be the appearance of a carpet of fallen leaflets. If attack takes place when the leaves are a little older and have become light green, with hardening cuticle, the fungal colonies grow as before but eventually die out without the leaflets being shed. When a large proportion of the leaflets are infected and fall, refoliation takes place in about 2–4 weeks. If this new flush survives unharmed comparatively little permanent damage is done to the tree; but a severe attack may cause further defoliation and the process may, in conditions particularly favourable to the disease, be repeated. This can result in serious debilitation, resulting in branch dieback and even death of the tree.

This shedding of immature leaves, known as secondary leaf fall, may also be caused by other organisms – the fungus *Colletotrichum gloeosporioides* (p. 447), the mite *Hemitarsonemus latus* (p. 454) and the thrip *Scirtothrips dorsalis* (p. 455).

Inflorescences are also susceptible, probably even more so than leaves. Attack results in shedding of flowers and thus considerable loss of seed production.

The powdery mildew fungus is *Oidium heveae* Steinm. It is probably the imperfect state of an Erysiphaceous fungus, but no perfect state has been found, a common phenomenon with powdery mildew fungi in the tropics. *Oidium heveae* is seen on leaves and flowers as the typical white powdery growth in which the colourless, microscopic barrel-shaped spores are formed in large numbers (Sivanesan & Holliday 1976).

Spread of the fungus takes place by these spores, which are easily carried by the wind. They also ensure survival throughout the year. This is facilitated by the presence of minor infections at any time of the year in nurseries, on volunteer seedlings, and on young shoots within the canopy of older trees.

Oidium is greatly influenced by environmental conditions, and is particularly damaging at elevations above 300 m. High rainfall or even occasional heavy showers reduces infection, so the disease is worst in drier areas. It is the weather towards the end of wintering that is crucial, and infection is increased if refoliation takes place at a time of low temperature, with overcast days and cool nights. Continuous mists or very light rain giving prolonged periods of high humidity are ideal for the infection process.

Fungicidal control is widely recommended in areas in which it is known that *Oidium* is likely to be serious. Sulphur dusting, introduced in Indonesia in 1929, is still standard treatment, applied with tractor-mounted or portable equipment. Dusting is normally done at intervals of 5–7 days, beginning at bud break. Three treatments may be effective if timing is correct, but up to six may be needed. More recently fogging with fenarimol or tridemorph in oil has been used to good effect (Rubber Research Institute of Malaysia 1982c; Edathil *et al*. 1984).

Another chemical method that has been found effective and economic in Malaysia is artificial defoliation by, for example, aerial spraying with the contact defoliant sodium cacodylate (Rao & Azaldin 1973). The treatment is carried out a month ahead of normal wintering, to achieve refoliation when the drier weather is less conducive to infection.

Raising the proportion of nitrogen in the fertilizer mixture used immediately after wintering is also recommended, as it increases vigour so that refoliation takes place more quickly.

Some clones are highly susceptible and are to be avoided in high-risk areas.

Abnormal or *Phytophthora* leaf fall

This disease has been recorded particularly from Burma, south India and Sri Lanka, where it has been known since early in the century. In Malaysia it did not cause appreciable damage until 1966, about which time it became apparent that it was also present in Thailand. Similar leaf diseases have been noted in all other rubber-growing countries in Asia, as well as in Africa (Liberia, Zaire) and America (Brazil, Costa Rica, Peru).

The fungi causing abnormal leaf fall are very dependent on environmental conditions, causing economic losses only in areas where these are particularly favourable. Such conditions (long periods of cool, wet, humid weather with few periods of sunshine) occur in south India, where this is considered the most important disease of rubber, with 30–50 per cent loss of crop in severe attacks (Radhakrishna Pillay 1977). It is not only considered the most damaging disease of mature rubber in Burma but is also severe in the nursery (Turner & Myint 1980). In the south of Thailand and the north of West Malaysia (Tan 1979) massive defoliation may take place and serious loss of crop ensue unless control is practised. In Bahia, Brazil, the disease can be as damaging as SALB (Chee & Wastie 1980).

Infection often begins on the young pods, which turn black and remain on the tree, dried up and unopened. From diseased pods infection spreads to the leaves. Presence of the disease is often first noticed by the large number of green leaves which are shed, forming a carpet on the ground. The leaf blades usually appear completely healthy, but typical symptoms of disease are seen on the petioles, in the form of dark brown or black lesions (Plate XVI). In the centre of the lesions appear one or two white spots of coagulated latex. The lesions are most commonly found near the base of the petiole but they can occur anywhere along its length. They also appear on the mid-rib and very occasionally the leaf blade is affected. The lesions are localised but cause premature abscission of the leaves while still green.

At a late stage of an attack young green shoots may also be affected, resulting in dieback. This is characteristic of the variation of the disease in Brazil known as leaf wither (Chee & Wastie 1980).

Abnormal leaf fall is caused by at least five species of *Phytophthora* (Mastigomycotina, Peronosporales). They are *P. botryosa* Chee, *P. capsici* Leonian, *P. citrophthora* (Smith and Smith) Leon., *P. meadii* McRae, and *P. palmivora* (Butl.) Butl. Of these, *P. palmivora* is the most widespread. *Phytophthora meadii* is the main species in Burma, India and Sri Lanka, while *P. botryosa*, recognized as a separate species only comparatively recently (Chee 1969), is the major cause of leaf fall in West Malaysia and Thailand. In Yunnan, China, *P. citrophthora* is implicated (Ho *et al.* 1984), as is *P. capsici* in Brazil (Gasparotto *et al.* 1984). The fungi are microscopic and, as they are not usually visible to the naked eye, are not of use in identifying the disease in the field. They may sometimes be seen as a fine white bloom on affected pods.

Three types of spores are produced. The oospores and chlamydospores are thick-walled and are probably the means of survival between wet seasons, remaining dormant in dry weather in infected pods, shoots and bark, either on the tree or on the ground. The zoospores are motile and are produced in large numbers while the disease is active on pods and leaves. Spread from one part of the tree to another and to neighbouring trees can take place by rain splash.

Prophylactic fungicidal spraying is effective on young trees, and is essential in the worst affected areas, such as occur in Burma and south India. One application not more than six weeks before the onset of the monsoon should be adequate. Various fungicides have been used, best results being obtained with a protectant that is resistant to being washed off by rain. In areas of severe attack aerial spraying of mature trees may be justified, and treatment with helicopters is routinely used in India on the larger estates. (Radhakrishna Pillay 1977). Copper oxychloride in mineral oil is extensively used in aerial spraying and is also applied using low volume ground mist sprayers, as well as with the more effective and economic fogging machines. Captafol or difolatan in oil is also effective, and as these materials contain no copper, latex contamination is not a hazard.

There are marked differences in resistance among clones. Some, such as RRIM 600 and PR 107, are highly susceptible and should not be planted in high-risk areas.

Colletotrichum leaf disease

Since 1905, when this leaf disease was first recognized in Asia, it has been seen in virtually all rubber-growing countries, not only in Asia but also in Africa (first seen in Uganda in 1920), and in America, where it occurred in Brazil by 1926. Details of history and distribution are given in a comprehensive paper by Saccas (1959).

The disease, part of the complex known as secondary leaf fall, is severe only in wet weather. It is particularly troublesome in localities where wet weather occurs during refoliation after wintering. It is also serious where rainfall is more or less evenly spread throughout the year. In such areas infection can take place at any time of the year and repeated attacks can result in debilitation of both old and young plants, dieback being a possible serious consequence. In Java a continuous attack over a long period resulted in losses in yield of up to 45 per cent (Soepadmo 1975).

Many plants, running into several hundreds, are attacked by strains of the fungus causing this disease. They include avocado, banana, citrus, cocoa, coffee, papaya and tea. It is not, however, clear whether cross-infection between other hosts and rubber is common.

As in the case of powdery mildew, which this disease superficially resembles, infection takes place of very young leaves as they expand during the two weeks or so of refoliation after wintering. Leaflets attacked within the first few days after bud burst soon turn black, wither and are shed. The bare petioles remain on the tree for a time, before in turn being dropped. At a later stage the effects of the disease are not so severe, and the affected leaflet remains attached, although bearing spots. Such spots, which may be very numerous, are the only symptoms of attack when more mature leaves are involved. The spots remain small, up to about 2 mm in diameter, are round, and have a yellow halo round the brown margin. The centre dries up and may fall out giving a shot-hole effect. Characteristic is the formation of a small conical protuberance on which the spot is situated. Also characteristic is the appearance under damp conditions of small, soft, pink, rounded cushions composed of the conidia of the fungus. Under conditions particularly favourable to the fungus repeated defoliation may take place. Young green shoots are also susceptible, and dieback occurs.

Plants of all ages are affected, the disease being particularly damaging in the nursery.

The causal fungus is *Glomerella cingulata* (Stonem.) Spauld. and Schrenk (Ascomycotina, Polystigmatales) but the name given to its conidial state, *Colletotrichum gloeosporioides* (Penz.) Sacc. is more commonly used, hence the name *Colletotrichum* leaf disease. Previously the disease was known as *Gloeosporium* leaf disease, as up to the late 1960s the name given to the conidial state of the fungus was *Gloeosporium alborubrum* Petch. Other names that have been applied to the fungus, particularly when associated with the disease commonly called anthracnose, are *C. ficus* Koorders, *C. heveae* Petch and *C. derridis* van Hoof.

The *Glomerella* state is not commonly seen on rubber. However, the conidia are produced abundantly and are very evident on infected parts as the pink, cheesy cushions already mentioned. These consist of masses of the conidia which are sticky and are not easily blown away by the wind. Rather they are dislodged by rain drops splashing on them. They are thus dispersed for short distances in rain splash, but once air-borne thay may travel some distance, and survive for a long time in the damp conditions that induce sporulation.

The disease is essentially one of wet conditions. The disease becomes quiescent in drier periods but spores survive to cause new infections on the onset of the rains.

Fungicides effective in controlling *Colletotrichum* are available. Among those which have been used are copper, captafol and chloro-thalonil. However, although spraying with such chemicals is feasible in nurseries and young plantings, practical application problems arise with mature trees. Dusting with sulphur and mancozeb has been used successfully in Java (Soepadmo 1975). Another possibility investigated in Malaysia is fogging with tridemorph in an oil-based formulation (Lim 1982). This gives good control but the machines are very expensive.

A possible alternative chemical control method when secondary leaf fall is anticipated is artificial defoliation with a contact herbicide, a month before wintering, as mentioned under powdery mildew.

Several widely grown clones, including PB 5/63, LCB 1320, PB 86, PB 28/59, RRIM 527, RRIM 712 and RRIM 728, are susceptible to *Colletotrichum* and these should not be planted in wet areas known to favour the disease. Conversely there are clones such as PR 255, PR 261, PB 217 and RRIM 600 with some resistance to the disease (Soepadmo & Pawirosoemardjo 1977), the resistance being multi-gene (Simmonds 1982). The clone GT 1 has generally been considered resistant, but recent surveys in Malaysia have shown that it can be severely affected under conditions favourable for the disease (Rubber Research Institute of Malaysia 1983c). In Cameroon, both PR 107 and GT 1 are found to be susceptible, and artificial defoliation in advance of the normal period of wintering, by aerial application of 3 litres/ha of 'Ethrel' (ethephon), has proved an effective method of control.

Bird's eye spot

This leaf disease is one of the most widespread affecting rubber and has been widely encountered in all rubber-growing areas in Asia, Africa and America (Commonwealth Mycological Institute Distribution Maps of Plant Diseases No. 270). No plants other than *Hevea* are known to be affected.

The disease affects mainly seedlings; budded plants are somewhat resistant. Once infection has started there is usually rapid spread throughout the nursery, with almost every leaf of every seedling attacked. Its importance lies in the defoliation that can occur, which rarely causes death but seriously weakens plants to be used as rootstocks. Budding may have to be delayed and percentage take may be reduced, particularly in green budding.

The effects of the disease, which has been described in detail by Hilton (1952), are most severe when very young leaves, less than two weeks old, are attacked. Infection then results in the formation of yellow spots. 1–3 mm in diameter, surrounded by a narrow black margin. After about three weeks the typical bird's eye spots will have developed (Plate XVII). The centre of each spot is by then white and the margin is raised and reddish-brown; surrounding the spot is a yellow halo. Several spots may then coalesce, the leaves become distorted, and the whole infected area becomes covered with a dark growth of the causal fungus. At this stage infected leaves are shed. If this defoliation is repeated due to further attacks the upper stem becomes noticeably swollen.

On leaves more than two weeks old damage is much less. Spots remain small, do not coalesce and do not cause defoliation. Leaves more than about three weeks old are immune from attack. When the young stem is at the stage of rapid elongation it also may be affected, with the formation of long, dark lesions.

The causal fungus is *Drechslera heveae* (Petch) M. B. Ellis (Deuteromycotina, Hyphomycetes). It was formerly known as *Helminthosporium heveae* Petch and this is the name used in publications until a few years ago. The fungus is microscopic but is seen on affected parts of leaves as a dark growth, as mentioned above. On it are formed masses of sausage-shaped spores, easily identified under the microscope. The fungus is described and illustrated by Ellis and Holliday (1972).

Spores are disseminated by wind and serve to spread the disease both within and between nurseries. Spread and infection can occur regardless of weather conditions, and thus the disease can persist and spread throughout the year. Very wet conditions, however, seem to reduce attacks somewhat, possibly because such conditions promote better seedling growth. Heavy shade also suppresses infection. As the disease is most damaging where growth is poor, good growth of seedlings should be promoted by adequate fertilizers, although excess nitrogen is to be avoided.

Direct control with fungicides is possible, and weekly spraying with zineb has been used succesfully. Other materials recommended include thiram, copper oxychloride and mancozeb. Fogging, using mancozeb or Kathon 4200 (Octhilinon), has been found to give better and quicker coverage, with less wastage and at reduced cost (Zainuddin & Lim 1979).

Resistant clones have not been identified.

Minor leaf diseases

The same species of *Colletotrichum* that is involved in secondary leaf fall also causes anthracnose, a disease of mature leaves, on which large necrotic areas are formed (Rubber Research Institute of Malaya 1965b). Spore pustules are formed around these dead areas or on concentric rings within them. The fungus causing this disease was formerly considered to be a separate species, known under the names *C. derridis*, *C. ficus* and *C. heveae*, but as mentioned above these are now considered to be synonyms of *C. gloeosporioides*. Affected plants are usually poorly grown with yellowish leaves. This is indicative of the major predisposing factor allowing the disease to develop, which is waterlogging. Improvement of drainage is usually all that is needed to reduce disease incidence. Target leaf spot (Carpenter 1951) occurs in America, where it is common in parts of Brazil. The causal fungus occurs on rubber in some other countries (India, Ivory Coast, Malaysia, Papua New Guinea) but does not cause target spot symptoms. It affects plants in the nursery, where the zonate leaf spots may result in repeated dropping of leaves which weakens the plants. The cause is the ubiquitous and polyphagous fungus *Thanatephorus cucumeris* (Frank) Donk (Basidiomycotina, Corticiaceae). Among several synonyms the one most frequently used in association with rubber is *Pellicularia filamentosa* (Pat.) Rogers. Several fungicides give control, copper being particularly effective. Mist blowing with triadimefon, applied primarily against SALB, is also useful (Conduro Neto *et al.* 1983).

Many other fungi have been recorded on rubber leaves but only a few are pathogenic. Diseases include *Corynespora* leaf spot (Rubber Research Institute of Malaysia 1975c) caused by *C. cassiicola* (Berk. and Curt.) Wei, occurring in Asia, particularly Malaysia and West Africa; scab, caused by *Elsinoe heveae* Bitanc. and Jenkins (Bitancourt & Jenkins 1956) and confined to South and Central America; black crust (Langford 1953) caused by *Phyllachora huberi* P. Henn. known only from South America; and *Periconia* blight (Stevenson & Imle 1945) caused by *P. manihoticola* (Vincens) Viégas in Central and South America. Several fungi are associated with rim blight (Rao 1975), including *Ascochyta heveae* Petch, *Guignardia heveae* Syd. and *Sphaerella heveae* Petch. Rim blight is common but the primary cause is unfavourable growing conditions.

One disease caused by an alga may be mentioned. This is red rust, caused by *Cephaleuros virescens* Kunze, which is of minor importance although widespread in all rubber-growing areas.

Until recently there was no authentic record of a virus disease of rubber, but such a disease has now been reported from Brazil (Gama *et al*. 1983; Junqueira *et al*. 1986). Leaves on affected plants were small and distorted and showed symptoms of mosaic and interveinal chlorosis. Virus-like particles were seen in affected leaf tissues. Transmission appears to be mainly through bud grafting, but transmission from rubber to rubber has also been achieved using *Myzus persicae* as vector.

A bacterial disease has also recently been investigated in Brazil (Junqueira *et al*. 1986). Young plants bear leaf lesions, and severe attacks may result in plant deaths. The associated bacterium has been tentatively identified as a species of *Pseudomonas*.

10.2 Pests

Unlike most crops, rubber is comparatively little troubled by pest attacks, and compared with the effect of diseases the damage they cause is fairly insignificant. Termites and cockchafers are widespread and common, mites and thrips are often associated with secondary leaf fall, and the American leaf caterpillar *Erinnys* causes enough loss to justify extensive applications of insecticidal control. But most other pests appear only sporadically, often in restricted areas, and in outbreaks that quickly die down. No doubt at least part of the reason for the apparent unattractiveness of rubber to insect and other pests is the production of latex by all parts of the plant, which makes it unpalatable. It also seems that many infestations begin in adjacent forest areas, or on covers grown in the plantation, suggesting that rubber is not a prefered host, but is attacked only casually.

Brief accounts of pests are included in the books by Petch (1921), Steinmann (1925) and Sharples (1936), and a well-illustrated account by Rao (1965) summarizes information on the situation in Malaysia. But very few papers have been published over the years on rubber pests, indicating the small amount of research on them and their control that has been considered necessary.

10.2.1 Pests affecting the roots

Termites

Termites attacking rubber trees have been known for many years, and in 1936 Sharples in Malaysia regarded them as the only major pest affecting the crop. They have been particularly troublesome in Malaysia and Indonesia but have also been found in India, Kampuchea, Sri Lanka and Vietnam, as well as in Africa and Brazil (Kashyap *et al.* 1984). The species attacking rubber also infest many other living woody species, but they largely subsist on dead timber.

Serious damage is limited to areas planted from forest in which little clearing or burning of timber has been done. Where clean-clearing is practised, or where the rubber has been preceded by other crops, little damage ensues (Rubber Research Institute of Malaya 1966).

Recently planted budded stumps or seedlings are particularly susceptible to attack, and the infested plant soon dies. In many cases no sign of attack is seen above ground before the plant is killed but when it is uprooted the underground part can be seen to be riddled with galleries made by the termites. Similarly when older trees are attacked there may be no above-ground sign of infestation before the roots are so badly damaged that the tree is blown over by the wind. Above-ground symptoms are sometimes seen in the form of the mud-covered galleries always constructed by termites when they venture out of the soil. These are formed on the base and lower part of the trunk and may be so extensive as to form a complete covering of mud. The trunk beneath the tunnels may be penetrated, when latex exudations become evident.

The species attacking rubber roots in Indonesia and Malaysia is *Coptotermes curvignathus* Holmgr. (Isoptera, Rhinotermitidae) (Plate XVIII). The insects are not normally seen above ground but when exposed, for example by destruction of their runways, they can be distinguished from other termites by the appearance of the soldier caste. The soldiers have large mandibles and when disturbed they grip with these mandibles and exude a drop of white fluid from the front of the head. Other species affect rubber in other areas, for example *C. elisae* (Desn.) in Papua New Guinea, *C. testaceus* (L.) in Brazil *Neotermes greeni* (Desn.) and *N. militaris* (Desn.) in Sri Lanka, and *Odontotermes obesus* (Ramb.) in India.

The termite nests are not associated with rubber but are built underground, in or on logs and stumps. From the nests tunnels extend long distances to sources of food, including rubber trees. Young stumps can be attacked even if healthy. This may also be true in the case of older trees but weakened trees, such as those affected by root disease, are more prone to invasion. Wounding also assists entry.

In areas cleared from jungle, termite colonies may be very numerous and it is impossible to eliminate the nests either by digging out or by chemical treatment. Instead treatment has to be limited to identifying affected trees and treating them. If young plants are involved frequent inspection rounds, say monthly, may be needed. In older trees damage is less serious and such frequent inspection is not necessary. Chemical control is by the application of a liquid insecticide around the base of the tree. Some soil should be excavated from around the collar to facilitate penetration of the liquid down the tap-root. Materials used are chlorinated hydrocarbons such as chlordane and heptachlor.

Cockchafers

Cockchafers affecting the roots of rubber and cover crops have been investigated in Malaysia (Rubber Research Institute of Malaya 1968) and India (Jayaratnam & Nehru 1984) but cause comparatively little damage. Not much attention appears to have been paid to them in other rubber-growing countries although they are known to affect rubber in Papua New Guinea (Smee 1964).

Considerable damage can occur in nurseries. Loss of young plants may also be high if they have been planted out in infested soil, as a single grub can kill a small seedling. More generally, in the field, the first sign of attack is sometimes seen in the leguminous covers, which die off in patches. The grubs then move to the less attractive rubber, and when the tap-root is badly damaged in young plants death follows. Attack on more mature trees is less serious, although repeated destruction of roots leads to debilitation. Although not usually so numerous, it has been estimated that the population of grubs may be as high as 500,000/ha.

Outbreaks occur largely in plantings near the forest edge, as the adult insects feed on jungle plants. They are particularly noticeable in areas of light sandy soil which does not impede the movement of the grubs searching for roots on which to feed. Besides the damage the grubs do, there is an ancillary effect of their presence in the disturbance of the soil caused by wild pigs digging to find them. This can be particularly troublesome in hilly areas of light soil.

Several species of cockchafer beetles (Coleoptera, Melolonthidae) are involved. The main species in Malaysia are *Psilopholis vestita* (Sharp), *Lachnosterna* (*Holotrichia*) *bidentata* (Burm.) (probably the most troublesome and also present in Sri Lanka) and *Leucopholis rorida* (F.). In India the species recently noted are *H. serrata* (F.), *H. rufoflava* Brenske, and *H. fissa* Brenske.

The adult beetles are large, heavily built and of various shades of brown. They are able to fly strongly, which they do in the early evening, travelling to feed on the foliage of forest plants. During the day they burrow in the soil. The larvae are thick, soft grubs, off-white or yellowish. They lie in the soil in a typical curved position (Plate XIX). When dug up they move characteristically on their back and sides (Rao 1965). It is these grubs that damage cover plants and rubber, feeding on the roots. The beetles pupate deep in the soil, the adults emerging during the wet season.

When labour was plentiful, digging out and collection of grubs was practised but this would now be uneconomic, as well as having the disadvantage of causing soil disturbance. Now three other basic control measures can be used, different approaches being appropriate at different stages of the pest and the growth of the rubber (Rao 1969). In young trees control is by the use of liquid soil insecticides, poured into the soil at intervals around the tree. This is aimed at protecting the trees from attack, for which about 1 litre of insecticide should be sufficient for each tree. In somewhat older plantings this treatment would not be practical, and the aim is to kill adults as they emerge from the soil. For this purpose a granular insecticide is used, scattered on the soil surface, and it can also be used on young trees to supplement the liquid insecticide treatment and to prevent reinfestation the following season. Insecticides used in Malaysia have included heptachlor and aldrin.

Another approach depends on the fact that the adult beetles are attracted to near-ultraviolet light and so can be caught in light traps (Rao 1964b). With knowledge of the insects' life history these need be operated for only quite a short period. The light traps are not effective in mature rubber, in which most reliance must be placed on treatment of the whole area with insecticide granules.

10.2.2 Pests affecting aerial parts

Yellow tea mite

Although several species of mites occur on rubber, only the yellow tea mite is an important pest. It has been noted particularly in Malaysia, where it occurs both in the nursery, where it is one of the commonest pests, and on mature trees. On these it is damaging only during refoliation after wintering, when it is involved in the secondary leaf fall complex. It is found on many other plants, including beans, castor, tea and tobacco. An account of the pest is given by the Rubber Research Institute of Malaya (1970).

The mite colonizes tender, actively growing shoots and is found particularly on the undersurface of young leaves at an active stage of growth. Heavily infested leaflets are pale and do not expand, remaining small and distorted, often with wavy margins and downward curling. Leaflets fall, leaving the stalks attached for a time. In severe attacks in the nursery the whole flush may be killed; side shoots then develop. Heavy infestation results in retarded growth. The mites feed mainly along the sides of the veins where typical light green streaks appear as a consequence.

The yellow tea mite is *Hemitarsonemus latus* Banks (Acarina, Tarsonemidae). Individuals are very small, males about 0.15 mm in length, females 0.2 mm. In mass they appear on the undersurface of leaflets as a white dusting of powder. Individually they can be seen with a hand lens; they are pale yellow becoming yellowish-green when older. Eggs are very large compared with adults, being about 0.1 mm long, and are conspicuous because of their characteristic pattern of white tubercles. Development from egg to adult takes only 3–4 days, and the female can then lay about eight eggs a day for about a week. So very rapid increases in population are possible.

Development of an infestation is favoured by dry weather, and when wet conditions return populations quickly decrease. In addition numbers are normally kept in check by predators, including ladybirds and particularly the mite *Amblyseius* (*Typhlodromus*) *newsami* (Evans), so chemical control is rarely required. If for some reason the balance between pest and predator becomes upset, and a very heavy infestation in the nursery occurs as a result, spraying with a pesticide such as endosulfan can be practised, particularly to the undersurface of the leaves.

Thrips

Thrips are associated with secondary leaf fall in Malaysia but are by far the least important cause of the trouble (Rubber Research Institute of Malaya 1962a). They are also occasionally found in the nursery.

They feed by puncturing the tissues on the lower surface of young leaves, particularly during refoliation after wintering. Leaves 2–3 days from bud burst are favoured. As a result of their feeding the leaflets become distorted and curl downwards, becoming concave. If many insects are present the leaflets are cast, followed by the petioles. The same symptoms are seen in the nursery.

The species of thrips causing the damage is *Scirtothrips dorsalis* Hood (Thysanoptera, Thripidae). Eggs are laid in or on leaf tissue. Two nymphal instars follow, the insects at this stage being wingless, white at first becoming yellowish-orange and 0.5–1 mm in length. The winged adults are orange-brown, 1 mm in length, and are very active, flying off when disturbed. Because of this the nymphs are more commonly seen on the leaves than are the adults.

Insecticides are effective against these thrips but it is not generally necessary to take control measures, as outbreaks tend to be short-lived, developing only in dry weather and dying down with the onset of wet conditions.

Leaf-eating caterpillars

The most important lepidopterous pest of rubber is *Erinnys ello*. It occurs in Brazil and Guyana and is the insect causing most damage to mature trees in America (Winder 1976). It also occurs on many other Euphorbiaceae, as well as crop plants in other families, including papaya, tobacco and cotton.

Damage is done by the feeding of the larvae. Young caterpillars are able to feed only on young leaves, but older larvae can attack more mature leaves as well as the bark of young branches. Because of the restriction on feeding of young larvae, outbreaks occur during refoliation after wintering and usually last only until refoliation is complete. When large numbers of the insect are present the effect of the tree is to reduce latex production. Complete defoliation may result, and new leaves may not be produced until the following year.

Eggs of *Erynnis ello* L. (Lepidoptera, Sphingidae) are laid on the lower surface of the leaves. Caterpillars, after five instars, grow to a length of 8–9 cm. Their colour is variable, but on rubber they are usually green or greenish-grey. Feeding continues for about two weeks, after which the insects pupate in the soil or surface litter. Adult moths have a wing length of 34–48 mm. Forewings in the female are grey, in the male darker grey or brown with a black band stretching from base to apex. The moths may be found during the day resting among the leaves.

Although more than 30 parasites and predators have been recorded, they have not been exploited, and the main control method is the use of insecticides. Many have been tested, and in Brazil several, including carbaryl and trichlorphon as dusts, and lindane and naled as emulsions give good control (de Abreu 1982). Light aircraft and helicopters have been used, but cost and hilly terrain are limiting factors. Bacterial insecticide containing *Bacillus thuringiensis* has also been tried experimentally but not for commercial use.

Other lepidopterous pests are of minor importance. Sometimes, as in the case, for example, of *Tiracola plagiata* (Walk.) (Noctuidae), which occurs in Indonesia, Malaysia, Papua New Guinea and Sri Lanka, severe but localized damage can occur from time to time on rubber. But this insect, like others such as *Mocis undata* (F.) (Noctuidae) feeds primarily on cover plants (Rao 1965).

Other insect pests

Although many other insect pests are found on rubber (Smee 1964; Rao 1965) few cause more than occasional damage. Some are mentioned below.

The weevil *Hypomeces squamosus* (F.) (Coleoptera, Curculionidae) feeds on the tender leaves of immature rubber, outbreaks often arising from outside the plantation. Legume covers are also affected.

Of the scale insects (Hemiptera, Coccoidea) the most important in Malaysia are *Saissetia nigra* Nietn., *Pulvinaria maxima* Green and *Lepidosaphes cocculi* (Green). In north Sumatra *Laccifer greeni*

(Chamberlin) is troublesome but, like other scales, it can be controlled with white oil emulsion, or with insecticides to deal with its attendant ant *Anoplolepis longipes* (Jerd.). Mealybugs, including *Ferrisia virgata* (Ckll.) and *Planococcus citri* (Risso), occasionally infest young plants, but predators and parasites are usually effective in keeping infestations to a minimum.

Members of the Orthoptera are common in plantations but generally only in small numbers. However, the grasshopper *Valanga nigricornis* (Burm.) (Acrididae) may occur in swarms, consuming cover plants as well as rubber. It can be controlled by aerial spraying with methamidophos (Ng 1980). The cricket *Brachytrypes portentosus* (Licht.) (Gryllidae) damages seedlings, severing the stem a short distance above ground and eating the upper part.

In Papua the tip-wilt bug *Amblypelta lutescens papuensis* Brown (Coreidae) is the most serious pest, causing the young leaves at the tips of affected shoots to wilt and die. It can be controlled with insecticides, dieldrin having been used with success.

The tunnels of several boring beetles of the families Scolytidae (such as species of *Xyleborus*) and Platypodidae (*Platypus* etc.) are often conspicuous because of the shot-hole effect of their tunnel entrances, often made more noticeable by the strings of wood dust ejected from the tunnels. They bore in bark and wood but are not dangerous pests as attack occurs only on trees that have been affected by disease or have been damaged in some other way, for example by fire, flooding or lightning strike.

Non-arthropod pests

Slugs (Mollusca) are capable of causing considerable damage to young rubber, particularly where there is ground cover in which they can shelter. They are harmful in that they climb the stem and eat the terminal bud. When side shoots develop their buds in turn are destroyed, and the result can be the development of clusters of small shoots. The major pest species in Malaysia are *Mariaella dussumieri* Gray and *Parmarion martensi* Simroth (Rubber Research Institute of Malaya 1967b). Cover plants may also be eaten by slugs and snails. A certain amount of biological control takes place through predation by carnivorous snails, beetles, birds and pigs. Chemical control with metaldehyde, either made up as a bait or applied as a slurry or paint to the base of the stem of young plants, is effective.

Information on mammals which can be pests of rubber has been summarized by the Rubber Research Institutes of Malaya (1964d) and Ceylon (1959). Of the many species that can be troublesome, the most important are rats, which can cause severe damage in nurseries and recently planted areas. Seed kernels and young shoots are eaten. Buds of older plants are also taken, and the bark at the base of the plant, up to about 1 m from the ground, is gnawed. The bark is stripped and the stem may be girdled. Anti-coagulant baits can be used for control. Alternatively tanglefoot can be applied as a deterrent to the base of the stem.

Damage similar to that caused by rats may be due to squirrels, but in addition they attack the bark of older plants. A characteristic result of attack by the common red-bellied squirrel in Malaysia is the gnawing of spiral incisions in the trunk from which the squirrel is said to drink the latex. Control is as for rats, or they can be trapped or shot.

The larger mammalian pests are usually invaders from adjacent areas of primary or secondary jungle, and so damage is worst around the edge of plantings. They include deer, against which fencing, repellent substances and game scares can be used; wild pigs, which may be shot or poisoned; monkeys, against which shooting or trapping is practised; porcupines, which may be trapped or kept off by protective dressings as for rat control; and flying foxes, which might be deterred by netting. Elephants can also be troublesome, through trampling and breaking of trees. Methods of control, including trenches, barricades and electric fences, have been described by Blair and Noor (1981).

Chapter 11

Processing and marketing

J. E. Morris

11.1 Introduction

This chapter aims to outline the essentials of the technology of converting the latex from the tree into forms of raw rubber suitable for export from the producing countries It does not attempt to go into all the practical details of the processes involved, to describe the layout of factories, or to give technical details of the considerable range of equipment now used in preparing different types of rubber. For more detailed information on these matters the reader is referred to general accounts by Edgar (1958), Barney (1968), Verhaar (1973), Pillay (1980), Rubber Research Institute of Malaysia (1980b) and to papers on specific topics quoted later in this chapter.

Tapped latex consists of a suspension of rubber particles (and some other particles – see Ch. 2) in a liquid serum. In addition to the rubber hydrocarbon, the rubber fraction contains other substances, mainly proteins and phospholipids, forming a protective surface coating around the rubber particles. The serum phase is mainly water, but contains small amounts of a variety of soluble compounds including inositols, carbohydrates, amino acids, proteins, inorganic anions and metal ions. Table 11.1 shows the composition of a typical latex. For export the latex is converted either to a liquid latex concentrate of about 60 per cent dry rubber content by removing part of the serum (usually by centrifugation)

Table 11.1 *Composition of tapped latex*

Component	%
Rubber	35.62
Acetone extract (lipids, resins, waxes)	1.65
Proteins (N × 6.25)	2.03
Carbohydrates	0.34
Ash	0.70
Water	59.62

or to various forms of dry rubber of 99 per cent d.r.c., to which the rubber hydrocarbon contributes 96 per cent.

11.2 Preparation of dry rubber

The essential steps in all methods for the preparation of dry rubber from latex are as follows. First, the latex is diluted and coagulated with acetic or formic acid which separates much of the serum from a coagulum comprising a network of rubber particles with a certain amount of entrapped serum. Second, the coagulum is washed with water and passed through various machines in order to remove serum and acid and to convert it into various types of sheets or particles. Third, the material is dried.

In addition to liquid latex the tappers also bring in rubber which has coagulated in the field. This consists of cup lumps formed from late dripping occurring after collection of the bulk of the flow, and rubber which has coagulated on the tapping cut, bark and spout, referred to as tree lace and bark scrap. There may also be some 'earth scrap' – latex which has overflowed or has been spilled and has become contaminated with soil. These materials are also converted into saleable dry rubber by washing, machining and drying. The composition of the dry rubber, particularly the amount of non-rubber it contains (which affects its technical properties and behaviour on vulcanization), depends on whether it is made from latex or field coagulum and on the method of preparation. This can be seen from Table 11.2 by comparing the compositions of ribbed smoked sheet (RSS) and pale crepe, which are made from latex, with those of brown crepes prepared from cup lump and other field coagulum.

For a prolonged period after the development of the rubber plantations in the East the consumer, who obtained his supplies through brokers and dealers in terminal and primary markets, had little contact with the producer. There was virtually no technical discussion between the two, particularly as the consumer tended to be secretive about his processes. As a result there grew up a system of grading natural rubber which was not based on technical properties but on visual appearance. Once this was firmly established it was difficult to change either the grading system or the forms in which dry rubber was prepared and both remained unchanged until 1965 when, as explained later, the production of natural rubber in forms better able to compete with synthetics, but which could not be visually graded, necessitated the introduction of technical specifications.

Prior to 1965 almost all dry rubber was sold as 35 grades of sheet and crepe which conformed to visual grading specifications set by

Table 11.2 *Comparison of properties of different types of rubber*

	Cup lump and tree lace	Field latex	Concentrated latex	RSS	Pale crepe	Thin brown crepe	Thick brown crepe
D.r.c. at time of collection or preparation (%)	45–50	30–36	60.16	99.6	99.6	99.4	99
Composition on dry basis % wt							
Ash	1.0	1.5	0.48	0.5	0.3	0.6	0.9
Protein*	2.5	3.5	1.81	2.5	1.7	2.0	2.0
Fats†	3.0	4.0	3.97	3.0	2.0	2.5	2.5
Calculated rubber hydrocarbon§ content % wt	93.5	91	93.74	94	96	95	94.5

* Nitrogen × 6.25
† Acetone extract – fats
§ 100 – (protein + fats + moisture + ash) ≈ rubber hydrocarbon

the International Rubber Quality and Packing Committee and published as international standards in 'the Green Book'. A considerable, but steadily diminishing, proportion of current rubber production is still prepared and graded in this way. The specifications have been updated from time to time and an abridged version of those in use since 1979 is given in Appendix 11.1. From this it will be seen that, for example, the visual grading of RSS into the six grades from RRS 1X to RSS 5 depends on the extent to which the following defects occur:

(a) Dirt – specks of bark, sand, etc., coagulated in the sheet.
(b) Moisture – due to insufficient drying, or wetting after preparation.
(c) Rust – a resinous surface deposit caused by yeasts and bacteria.
(d) Bubbles – caused by fermentation of latex during coagulation.
(e) Blisters – resulting from gas formation usually caused by overheating.
(f) Mould – fungal growth favoured by high moisture content.
(g) Colour – preferably uniform, without stains, spots or blotches.

Of these, (c), (d) and (e) are of little or no technological significance and (f) is increasingly tolerated in lower grades (Rubber Research Institute of Malaya 1962b).

11.2.1 Preparation of sheet rubber

The preparation of sheets by smallholders is described in Chapter 12. For preparing ribbed smoked sheet (RSS) on estates, an anti-coagulant is added to the field latex in order to prevent natural coagulation before it reaches the factory. The material used varies with local conditions; sodium sulphite, formalin or ammonia, whichever is the most suitable, is added in aqueous solution to give, respectively, a concentration on the latex of approximately 0.05, 0.03 or 0.02 per cent by weight. The addition is sometimes made to the tapping cup, but more often to the tapper's bucket, and may be increased at the collection stations to up to double the amount given above if this is thought to be necessary because of a marked tendency to pre-coagulation. Each tapper's latex is weighed at the collection station and the d.r.c. determined by hydrometer or by the reasonably accurate Chinese 'chee' method. The latex is then taken to the bulking tanks at the factory, diluted to 15 per cent d.r.c., stirred and transferred to coagulating tanks. Bulking of the whole day's latex is desirable to give uniformity of coagulation and of the final product but if no bulking tanks are installed the latex is poured directly into the coagulating tanks and the d.r.c. adjusted by adding water.

In the coagulating tanks, which are usually 300 cm long by 90 cm wide and 20 cm deep, a 2 per cent aqueous solution of formic acid is added at the rate of about 175 cc per kg rubber in the latex. Froth and bark are removed from the surface of the latex, partition plates are inserted and the latex allowed to coagulate overnight. The vertical partition plates are cut away on alternate sides of the tank so that one long sheet of coagulum is produced to go through the sheeting machines. The spacing of the partitions is important as it partly controls the thickness of the coagulum after machining which, as explained below, affects the time that the sheets take to dry. Normally, a 300 cm tank has 72 or 90 partitions.

Fig. 11.1 Coagulum passing through sheeting battery

When coagulation is complete and the partitions are being removed, the tank is flooded with water and the coagulum floated off to the sheeting battery. The latter consists of at least four, preferably six, pairs of rollers, the last pair being grooved to give the

Fig. 11.2 Sheets on trolley prior to drying in smokehouse

sheet the ribbing from which its name derives. Each pair of rollers is fitted with a water spray to wash serum and acid off the sheet. The roller pairs are geared and adjusted so that the coagulum, which increases in length and breadth but decreases in thickness as it passes through them, flows evenly and emerges with a thickness of about 3 mm. On emerging from the final roll, the coagulum is cut into sheets of an appropriate length to fill the vertical hanging space between the horizontal bars on the trolleys on which they are placed for drying.

On coagulation the serum content of the latex is reduced from about 65 to about 55 per cent by weight of the coagulum. Machining further reduces the serum content and the machined sheet is then able to exude water spontaneously by syneresis. After being hung on the trolleys the sheets are left to drip for about four hours in the open air but under cover. After this the moisture content is down to about 30 per cent but when the trolleys are removed to the smokehouse and the temperature raised to about 55 °C the rate of syneresis increases, dripping is renewed and there is a rapid loss of moisture down to 10–15 per cent. At this point the normal migration of water through the rubber to its surfaces takes over and the rate of drying becomes proportional to the temperature and to the square of the thickness of the sheet (Gale 1959).

In the smokehouse the sheets are dried in hot air containing smoke which acts as a preservative and helps to prevent the subsequent formation of moulds. The essential features of a smokehouse are a furnace (usually burning wood), flues to conduct the hot air from the furnace provided with outlets from which the emission of smoke can be controlled, and ventilators. Old types of multistorey smokehouses were replaced in the 1940s by houses of the Subur type which consisted of four separate, parallel chambers each capable of holding two trolleys of sheets (Moore & Piddlestone 1937). The wet sheet was pushed into the first chamber, kept at about 55 °C, and then moved each day, on rails and via turntables on a verandah, to the next chamber, the temperature increasing with each chamber until it reached about 70 °C on the fourth and final drying day.

Later, tunnel-type smokehouses were developed (Rubber Research Institute of Malaya 1940; Graham 1964) and most modern smokehouses are of this type. With these, drying and smoking are continuous. Trolleys carrying wet rubber enter the cooler end of the tunnel and move straight through the building on a single track, being pushed along daily by those entering behind them until after four days they emerge from the hotter end with the rubber dry. The Woods-GEC drying tunnel, which developed from the above, can be used for drying with or without smoking. It accommodates 12 trolleys and is commonly used for making air-dried sheet but a smoke generator to introduce smoke into the heated air is available for making RSS. The air is heated by a fully automatic, oil-fired heat exchanger and circulated with the aid of an aerofoil fan, the temperature being thermostatically controlled.

From the smokehouse the trolleys are pushed to the packing shed where the sheets are inspected, bubbles, etc., are clipped out of them and they are graded and pressed into bales. The latter are usually of 120 kg, are wrapped in rubber of the same type and painted with an approved bale-coating solution.

Technique Michelin sheet

This RSS, which was developed to provide the raw material for Michelin tyres, is made by coagulating undiluted latex in cylindrical drums of about 1 m diameter and then rotating the coagulum against a band-saw type of knife to obtain a sheet of normal thickness. Ordinary coagulating tanks may also be used for controlled high d.r.c. (22.5 per cent) coagulation. The sheet is then processed in the normal way through the sheeting battery and hung on the trolleys, but is not put into the smokehouse until the following day. This delay, or maturation step, is important as it results in a relatively fast-curing rubber with a high Mooney viscosity which has advantages in the manufacturing process. The delay often causes rust and bubbles to appear in the final sheet. Although these are of no technological significance, they are considered defects by the visual grading system, under which TM sheet would be graded below RSS 3, whereas in fact it commands a premium on RSS 1. This illustrates the illogicality of the visual grading system.

Oil-extension

For the manufacture of some products it has long been the practice to add mineral oils to raw rubber as a plasticizer to improve processability, or as an extender to cheapen the product. In recent years there has been a considerable use of oil-extended natural rubber (OENR) in winter tyres since it was discovered that it gives better resistance to wear, groove cracking, and to skidding on icy roads. At the plantation an oil-extended master batch containing 20–40 per cent oil can be prepared by mixing an aqueous emulsion of approximately 70 per cent aromatic, naphthenic or paraffinic processing oil with the diluted latex in the bulking tank, followed by acid coagulation and processing to sheet in the normal manner. The procedure can also be used for the oil-extension of the block rubbers mentioned in section 11.4 (Chin & O'Connell 1969), but nowadays it is more usual to add the oil after coagulation by mixing, or extruding, it with the crumb rubber.

Superior processing (SP) sheet

This type of sheet has been made for some years by preparing a mixture in which 80 per cent of the d.r.c. is provided by normal field latex and 20 per cent by vulcanized latex. To prepare the latter, ball-milled vulcanizing agents, including zinc oxide, sulphur and an accelerator, are stirred into field latex which has been stabilized with an anti-coagulant (see p. 462) in a metal bulking tank. The temperature of the latex is raised by admitting live steam, usually for two hours at 100 °C, which results in a highly cross-linked, vulcanized latex. The mixture of the two latices is processed to sheet in the normal manner. The rubber is particularly useful in the manufacture of products which must have exact dimensions after extrusion (Rubber Research Institute of Malaya 1957; Sekhar & Nielsen 1961).

Air-dried sheet

Air-dried sheet (ADS) is characterized by its light colour and cleanliness and is used for white rubber shoes, white side-walls of tyres, elastic bands, etc. It is sometimes called 'pale amber unsmoked sheet' (PAUS) in order to avoid confusion with smallholders' partly dried unsmoked sheet (USS) (Rubber Research Institute of Malaya 1953). To produce this rubber, sodium sulphite as 0.05 per cent aqueous solution is used as an anti-coagulant in the field, although occasionally ammonium borate may be used instead. When the latex arrives at the factory, sodium metabisulphite is added in the bulking tank prior to coagulation, in order to remove dissolved oxygen and thus prevent the formation of dark pigments by the reaction of oxygen with traces of phenols and polyphenol-oxidase enzymes which are sometimes present in field latex. This pigment formation occurs at the coagulum–air interface and is particularly pronounced for a period of a few weeks during wintering and refoliation and also in certain clones, such as AVROS 49 and PR 107. In order to preserve the light colour, latex for ADS is usually coagulated at 12.5 per cent d.r.c. The subsequent steps in its preparation are the same as for RSS except that it is not smoked.

11.2.2 Preparation of crepe rubbers

The distinguishing feature of the preparation of crepe rubbers is that the coagulum is passed through a crepeing battery consisting of a number of paired, driven rolls which machine it to a crinkly, lace-like sheet. Pale crepes and sole crepe are made from field latex, while brown crepes are produced from field coagula, such as cup lump and tree lace, and from RSS clippings.

Pale crepe

Latex for preparing pale crepe is preserved with sodium sulphite in the field and diluted to 20–25 per cent d.r.c. in the bulking tanks at the factory, after which sodium metabisulphite is added. Pale crepe rubber is used for the manufacture of white, pale-coloured or translucent goods and it is advantageous if the latex used for its production contains little or no yellow pigment. However, the latter is present in the majority of clonal latices and two methods are available for neutralizing or removing it.

One method is chemical bleaching. The bleaching agent originally developed was xylyl mercaptan, which was formulated with a wetting agent in an inert hydrocarbon solvent and emulsified with water to form a product known as 'Emulsion A Concentrate' (Rubber Research Institute of Malaya 1952b, 1954b). This is added to the selected latex in the bulking tank, care being taken not to exceed the prescribed amount of the emulsion, as excess will produce crepe of lowered viscosity and unattractive appearance. Newer, equally effective, bleaching agents are available. After addition of the bleaching chemical, the latex is coagulated with formic acid but the partitions in the coagulating tanks are either omitted or placed in every second slot.

The other method of removing the yellow material is known as fractional coagulation and consists in coagulating the latex in two stages. After dilution of the latex and the addition of sodium metabisulphite, sufficient 1 per cent acetic acid is stirred in to cause 12–15 per cent of the rubber to coagulate in about two hours. This fraction contains most of the pigments, almost all the high-viscosity, high-molecular-weight gel and much of the protein and other nitro-genous compounds. It is milled and dried separately and either sold as an off-colour crepe or manufactured into block rubber, when it is best blended at about 10 per cent into 20 or 50 grade. The latex remaining in the fractionation tank is passed through a 40 mesh screen to remove coagulum particles and coagulated with formic acid in the normal way. Fractional coagulation has the disadvantage that it is time-consuming and that it converts only part of the crop to first quality crepe.

Fractional coagulation may be combined with bleaching of one or both fractions. In the latter case half the normal quantity of bleaching agent is added to the bulked field latex before the first coagulation with acetic acid and half to the residual latex after-wards. In the former case bleaching agent is added only to the residual latex after removal of the first fraction.

In the crepeing battery the coagulum is passed through three types of paired rollers. The first machine, the macerator, has rollers geared to the same speed, or with a low differential, which are grooved with a coarse diamond pattern and which break down the

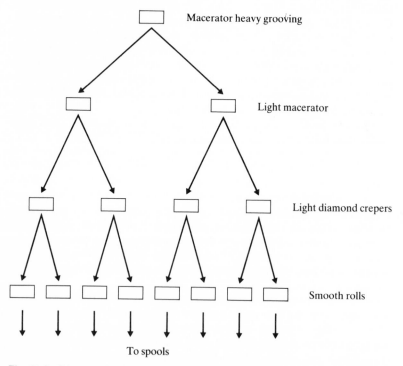

Fig. 11.3 Diagram showing arrangement of crepeing battery

coagulum to form a thick blanket. The second, or intermediate creper, which has more finely grooved rollers with a differential of about 1.0 : 1.5, converts the coarse blanket to a lacey sheet. The finishing machine has smooth rollers with a higher differential of about 1.0 : 2.0 and produces a thin, compact sheet. The minimum requirement is one machine of each type but usually there is a battery of machines, the product of one machine at each stage being split between two machines at the next stage, as illustrated in Fig. 11.3, although other arrangements are used.

Fig. 11.4 Removal of thin pale crepe from last, smooth crepeing roll, up and into a Wood's-type drier in one continuous length (Sime Darby Plantations)

After leaving the finishing machine the lengths of crepe are wound on to spools from which they are rolled off for hanging in the drying shed. Drying sheds are usually tall, corrugated iron buildings capable of accommodating three 5 m tiers of crepe hung on racks spaced 10 cm apart, and having flap windows which are kept open to provide ventilation by day but closed at night. Natural air drying takes 6–8 days in reasonably dry weather and longer under humid conditions. Drying time can be reduced to 3–4 days with heating from wood-fired furnaces with flues, from hot water pipes, or by the use of an oil-fired heat exchanger and distribution ducts. The tunnel driers mentioned earlier can be used for drying crepe after it has been left on the trolleys to drip for $1\frac{1}{2}$–2 hours at ambient temperature.

The above procedures are used for the preparation of both thick and thin pale crepes, the latter being either sold as such or laminated into sole crepe. There are four grades of each thickness, as indicated in Appendix 11.1, but grades 1X and 1 are usually the only grades deliberately made, the others being off-grades.

Thick pale crepe

Thick pale crepe has been made only in Sri Lanka, usually with latex from selected clones which give a pale-coloured rubber, although bleaching is normally done. A rather thick, but lacey, crepe is obtained by light machining which, after drying, is taken to the packaging shed where lengths of about 8 m are inspected, impurities clipped out and the crepe built up layer by layer to a thickness of about 1.0–1.5 cm. Consolidation is achieved by passage through heavy, grooved laminating rolls, after which the crepe is cut into lengths of about 70 cm and pressed into bales which are wrapped in hessian and painted with a bale-coating solution (Garnier 1922).

Thin pale crepe

Thin pale creep has been made extensively in Malaysia, Indonesia and India. The procedure is the same as for thick crepe up to the machining stage. The material is then passed a number of times through a series of crepers achieving a good blending and production of a thin, non-lacey crepe. For shipping as thin pale crepe the thickness is 1.0–1.2 mm but if it is to be made into sole crepe it will be 0.8–1.0 mm. Baling is as for thick pale crepes.

Sole crepe

Sole crepe was developed at the request of the manufacturers and has been used for many years as a completely raw rubber, mostly in children's shoes but from time to time, as fashion has dictated, in all types of footwear. The market for sole crepe is very exacting in regard to the colour and thickness of this product. The first stage is the preparation of thin pale crepe by the procedure already described, with the proviso that fractional coagulation is always done and a bleaching agent is added both before and after the first coagulation.

Sole crepe is built up by hand laminating pieces of thin crepe and successful production depends upon being able to draw on skilled workers, who are normally women, and require about four months to train. After inspection and clipping out of defects, each piece of thin crepe is slightly stretched and laid on a padded table about 70 cm wide and 5 m long where it is patted with the palm of the hand and rolled with a hand roller. When a sufficient number of lengths have been built up the pads are checked for weight and

Fig. 11.5 Cutting off lengths of crepe for building into sole crepe (Sime Darby Plantations)

thickness by a supervisor. They are then left to shrink for a short period, warmed slightly and passed through even-speed laminating rolls. After 24 hours cooling and relaxation the laminated pads are cut to the required size which has traditionally been 12×36 in. or 28×36 in. (30×91 cm or 71×91 cm). Sole crepe is made smooth on both sides and ranging from $\frac{1}{16}$ to $\frac{3}{4}$ in. (0.16–1.90 cm) thick, or corrugated on one side and $\frac{1}{8}$ to $\frac{3}{4}$ in. (0.32–1.90 cm) thick. Corrugated sole crepe has the top layer of the laminated pad made of cuttings which have been soaked, milled and dried, and laid on top of the pile of crepe before lamination. Sole crepe is packed in plywood cases holding about 100 kg.

Estate brown crepes

These crepes are made from cup lumps and other field coagulum collected at the time of tapping. These raw materials are soaked in water and, if necessary, put through a scrap washer before being milled by a creeping battery into a thin crepe about 1.5 mm thickness. The crepe can be dried in about six days. It is then graded into three grades, mainly distinguished by whether the colour is light brown, brown or dark brown, and packed in bare-backed bales. Estate brown crepes have now almost disappeared as the raw material from which they were made is now made into block rubbers, mostly 10 grade (Morris & Nielsen 1961).

Thick and thin brown crepes

These grades were made from wet slab, unsmoked sheets, lump and lace, particularly in Singapore which became a major processing centre for rubber from all over the East and imported large amounts of slab from Indonesia, Burma and other producing countries. After cutting the slabs, the raw materials were washed, macerated and then fed into power-wash mills for about 40 passes. The resulting crepe was usually hung in sheds for natural air drying which took 20–30 days depending on the thickness of the sheet. It was inspected and graded, usually as 2, 3 and 4 thin brown crepe (remills) and B, C and D thick brown crepe (blankets). These grades have virtually disappeared as the raw materials are now made into block rubbers, mostly of 20 grade (Aramugam & Morris 1964).

11.3   The development of dry rubber technical standards and of block rubbers

Rubber manufacturers aim at producing uniform, satisfactory products as cheaply as possible. For most products the raw rubber is mixed with sulphur and other ingredients and heated to produce a vulcanizate suitable for the article being manufactured. The rate at which the rubber 'cures' during this process and the properties of the vulcanized rubber, especially its tensile strength (measured as the load in kg/cm^2 required to break a standard specimen when stretched) and its modulus (load required to stretch the specimen to a given elongation) are affected by the composition of the raw rubber. For many years manufacturers experienced difficulty in achieving uniformity in manufactured products, partly because compounding and vulcanizing was little understood but also because the composition of the raw rubber they received varied considerably due to differences in origin, clone, tapping intensity and processing procedure. Differences in processing cause variation in the extent to which the serum is removed and hence in the amounts of protein and other nitrogenous compounds left in the final raw rubber which, in turn, cause variation in the rate of cure and in the properties of the vulcanized rubber.

An attempt to help consumers to overcome difficulties arising from variation in natural rubber was made by the introduction of a technical classification scheme in 1950 (Rubber Research Institute of Malaya 1952a). This was based on an internationally agreed test in which a rubber sample was cured under prescribed conditions with a specific compound known as the American Chemical Society Compound No. 1 (see Appendix 11.2). Three ranges of cure using this compound were chosen, all measured by the strain value. For each range of cure three Mooney viscosity ranges were specified (see Appendix 11.2) to aid the manufacturer with problems arising from variation in the viscosity of the raw rubber. However, this scheme was largely ineffective, partly because the changes that occur in the viscosity of raw rubber were not properly understood and partly because manufacturers began to use a variety of chemicals to accelerate vulcanization, which increased the rate of cure and reduced the importance of its variability. But from the early 1950s the research organizations in Indonesia and Malaya began to play a more active role in obtaining an understanding of consumer requirements and in endeavours to establish more satisfactory technical and visual standards for natural rubber.

It became increasingly clear that the visual grading system and the conventional methods of preparing and packing natural rubber placed it at a serious disadvantage compared with synthetic rubbers. The visual grading system was based on subjective judgements and there was a considerable overlap with some grades between which subtle distinctions were made which had no technological significance. The latter led to some marketing malpractices. The packaging of natural rubber in large bales, which were cumbersome to handle, transport and store and which required cutting or other treatment before being put into the manufacturers' mixers, also contrasted unfavourably with the presentation of technically specified synthetic rubbers in block form in smaller, easily handled, polythene wrapped bales. By the early 1960s, several producers were beginning to make raw natural rubber by processes which, instead of converting the coagulum to slow-drying sheet or crepe, crumbled or granulated it into particles which could be dried rapidly and then compressed into blocks. Thus, while a new grading system was desirable for conventional type of rubber, it was essential for the new block natural rubbers which could not be graded on the visual system but must be marketed to some kind of technical specification. To avoid an undesirable multiplication of such specifications by different producers it was clearly desirable to introduce national or international standards.

11.3.1 The Standard Malaysian Rubber Scheme

The stage was thus set for studying and introducing new methods of processing, grading and packing in order to improve the product received by the consumer, to ensure that the producer received a fair return for his efforts and to compete better with synthetics. The developments are here illustrated by describing the Standard Malaysian Rubber Scheme, because almost all the modern developments in technical specification and the production of block rubbers followed the Malaysian initiative. The introduction of the scheme in 1965 was preceded by a world-wide enquiry to ascertain consumers' views on which properties should be specified and how the product should be packed and presented. At the same time a study was made of the properties and variability of Malaysian rubber of all grades. The specifications formulated aimed at providing consumers with satisfactory criteria of quality while setting standards with which the majority of Malaysian producers could comply. The scheme embraced both new and conventional types of rubber; any rubber that satisfied the specifications and the packing requirements could be sold as Standard Malaysian Rubber (SMR). It was a permissive scheme but an essential feature of it was that any producer registered under it with the Malayan Rubber Export Registration Board guaranteed consumers that every bale shipped with the SMR symbol conformed to the appropriate specifications. Discipline was exercised by the Export Registration Board with advice from the RRIM, which exercised technical control (Rubber Research Institute of Malaya 1965c).

Essentially, the scheme provided for three grades based on the impurity or dirt content of the rubber instead of the wide range of visual grades. The dirt content was important to the consumer on several counts, especially in relation to the technological performance of his finished goods. The maximum dirt level allowable in the premium grade (SMR 5) was set at 0.05 per cent. Research had shown that with raw rubber containing more than 0.2 per cent dirt there was a tendency to tear in the vulcanized product and hence the dirt limit for rubbers traditionally used in tyre manufacture (SMR 20) should not be above this figure. A dirt limit of 0.5 per cent was provided for a third grade, SMR 50, suitable for lower grade uses. A limit was placed on the ash content of each grade to prevent undue contamination and no more than 0.7 per cent nitrogen was permitted in any grade in order to exclude excessive protein which would usually be derived from skim rubber addition. As one of the main complaints about visually graded rubber was of undue moisture content, a limit of 1.0 per cent volatile matter was set up for all grades. Some of the limits have been altered in periodic reviews of the scheme and the current specifications are given in Appendix 11.3.

The original specification also limited the content of copper and manganese to 8 and 10 ppm respectively with the intention of limiting the proneness of rubber to oxidation. However, research subsequently showed that proneness to oxidation was not necessarily related to the content of these elements but could, for example, arise from prolonged soaking, washing, exposure to radiation or lack of natural antioxidants. A suitable test was found to be the use of the Wallace plastimeter to measure the plasticity of a sample before and after heating for 30 minutes at 140 °C (see Appendix 11.4). The plasticity retention index (heat-aged value as per cent of that of unaged control) was found to correlate well with tyre wear and other properties of compounded rubber (Bateman & Sekhar 1966).

Colour was not considered an essential part of the grading scheme but, as it is important in rubber to be used in making light-coloured articles, a sub-grade of the best quality was introduced – SMR 5L. To test for colour a sample is pressed into a mould, heated at standard vulcanizing temperatures and times, and then compared with a Lovibond colour standard.

In order to compete with synthetic rubber and to satisfy the consumers' desire for small, easily handled bales free from bale coatings but uncontaminated by dirt in transit, the scheme prescribed that any rubber exported as SMR must be packed in polythene wrapped bales not exceeding 50 kg in weight.

In the early stages of the scheme Singapore producers used the SMR designations and much work was done to develop the production of block rubber from low-grade coagulum materials previously processed into thick and thin brown crepes by the Singapore remillers. In 1970 Singapore left the Malaysian scheme and Singapore Standard Rubber (SSR) grades came into being. They are similar to the Malaysian grades, those produced being mainly SSR 20, SSR 50 and SSR 5. Schemes almost identical with the Malaysian one were introduced in Indonesia in 1971 and in Sri Lanka in 1972.

11.4 Methods of producing block rubber

11.4.1 Comminution and drying

The introduction of the technical specification scheme in Malaysia coincided with the development of several processes for converting latex or coagulum into blocks of rubber. These were initially pioneered by French and Belgian companies. As described later, the new processes involved some modification of the coagulating procedure and some pre-treatment and cleaning of coagulum materials, but the essential novel features were the conversion of coagulum to particulate form and the method of drying it.

Several methods are now in use. One consists in comminuting the coagulum with Cumberland and Blackfriars granulators before drying with through-bed hot gas. Care is needed to prevent the development of wet patches in the final bales due to agglomeration resulting from overloading the drying bed but this process has been used with satisfactory results for almost 20 years.

The 'Heveacrumb' process was developeed by the Rubber Research Institute of Malaysia as a result of the observation by Sekhar that the addition of a small quantity of castor oil (0.7 per cent by weight) to latex coagulum followed by passage through a crepeing mill caused the coagulum to crumble (Sekhar *et al.* 1965). The oil acted by forming a thin layer on the surfaces of particles of coagulum torn apart by shear on passage through the rolls set at friction speed and prevented them from joining together again on emergence from the machine, as would normally happen in crepeing. The process could be used to process cup lump, tree lace and bucket lump (Graham & Morris 1966) and to make oil-extended and constant viscosity rubbers (Chin 1966). This method is still in use but the quantity of castor oil used has been much reduced, only 0.12 per cent is usually required with two crepeing passes and one pass through a creper hammer-mill.

A third method, which can be used for estate or smallholders' scrap coagula, is first to hammer-mill or granulate the scrap and then to pass it through an extruder fitted with knives at the die-plate which cut the emerging spaghetti-like rubber into small pellets which are suitable for rapid drying. Extruder drying, which is extensively used in the synthetic rubber industry, has not been found satisfactory for natural rubber because of the high die exit temperature and the high energy and maintenance requirements.

Drying of granulated or crumb rubber could not be done with the facilities used for sheet and appropriate driers had to be designed for the larger factories built to carry out the new processes. Drying is done by passing hot air at 100–110 °C through a bed of crumbs resting on a perforated plate in either trolley-type or box-type driers. Basic designs produced by Graham in 1965 have been developed and improved to give a range of effective driers of both types.

Fig. 11.6 Crumb rubber boxes being pushed into Guthrie-type drier (Sphere Corporation Sdn. Bhd.)

11.4.2 Preparation of latex grades

SMR L

A typical process for the preparation of SMR L may be summarized as follows. If necessary, the field latex is preserved with an anti-coagulant (usually ammonia but sodium sulphite with formalin may be used), but so long as the product is of uniform light colour the onset of fermentation and pre-coagulation does not matter as much in block rubber production as it does in making RSS. To ensure uniformity of the product, the bulking tanks should be big enough to take a whole day's crop and the clonal composition of the latex should be similar from day to day. Sodium metabisulphite is added to the bulking tank at a concentration of 0.05 per cent on the latex to inhibit darkening by enzymic discolouration. Sufficient 2 per cent formic acid to coagulate the latex in the bulking tank is placed in an adjacent tank which is of such dimensions that the latex and acid can be mixed in the right proportions by a matching flow process. When the discharge ports of the two tanks are opened together, the latex and acid flow through appropriately sized pipes which join near the bottom of the coagulating tank or trough. If a coagulating tank is used, only every other partition is inserted. After coagulation, trough coagulum, which is about 30 cm × 30 cm × 30 m or more long, is passed into the crusher rolls. The latter consist of a pair of rolls of about 40 cm diameter with grooves cut into the centre portions but left smooth for a few centimetres at the ends so as to control the width of the coagulum, which they reduce to a thickness of about 5 cm. The crushed trough coagulum, or that direct from coagulating tanks if these are used, then floats to the first of two in-line crepers which reduce its thickness to 2–3 mm before it is fed into a creper hammer-mill which converts it to crumb. The crumb is then transferred, by mechanical conveyors or by hand in baskets, to the trolleys for drying.

The trolleys move through the drier at regular intervals dependent on the number of trolleys required to fill the drier. A nine-stage unit is common and can dry more than one tonne an hour. After leaving the drier the rubber is quickly transferred manually with a hook to a table while it is still warm, as this facilitates baling and packing. It is inspected for any abnormality, contamination or wetness by the baling operative, who cuts out any suspect rubber and weighs out lots of 33.3 kg, each of which is placed in the baling box of an hydraulic press and pressed into a compact bale. The bales are conveyed to an inspection table where they are checked and a proportion, usually 1 in 10, have samples removed to be sent for testing. Each bale is then enclosed and sealed in a plastic wrapper bag, on which the grade, shipping and packing numbers, factory identification, etc., are already printed.

SMR 5

The specification for SMR 5 was originally based on the properties of RSS 1, which was regarded as a quality yardstick against which all other grades could be measured. However, the expectation that SMR 5 would be accepted as a direct equivalent of RSS 1 has not been fulfilled. It is made from latex much as described for SMR L and any SMR L which does not meet the colour specifications becomes SMR 5. Inadvertently, the 5 grade has come to be associated with rubber failing to meet approved specifications and it usually sells at a lower price than RSS 1.

Constant viscosity rubber, SMR CV

Scattered along the polyisoprene chain of natural rubber are a number of aldehyde and methylene groups which, during processing and subsequent storage, can react together to form cross-linking bonds, thereby increasing the molecular weight and inherent viscosity of the rubber (Sekhar 1961). This process is commonly called 'storage hardening'. The cross-linking can be achieved by reacting a bi-functional reagent such as hydrazine with the reactive groups, resulting in a rubber of high viscosity. On the other hand, treatment with a mono-functional reagent such as hydroxylamine blocks the reactive groups and inhibits cross-linking so that the viscosity of the rubber is stabilized and storage hardening does not occur.

The preparation of a constant viscosity rubber such as SMR CV simply involves the addition of hydroxylamine normal sulphate to the field latex in the bulking tank at the rate of 0.5 per cent by weight followed by processing it to dry rubber in the same way as that described above for SMRL. The chemical addition tends to give CV rubber a somewhat darker colour. If a light colour is required, it can be achieved by using particularly light-coloured latex. Rubber prepared from different clonal latices varies considerably in viscosity but most rubber produced from mixed latices has an intrinsic viscosity of 60 ± 5 Mooney units and this became the standard range for CV rubber – CV60. Consumer interest in high or lower viscosity rubbers for certain purposes led to the production of additional commercial grades of CV50 and CV70.

Low viscosity rubber, SMR LV

Low viscosity rubber contains 4 per cent of non-staining extender oil added as an emulsion to the field latex before coagulation. Four per cent by weight of oil was chosen because this gives a 5–10 unit drop in viscosity without loss in physical properties on vulcanization and also because in some countries rubber with 5 per cent or more of added oil attracts punitive duties. Viscosity is stabilized as with

CV, from which LV differs only in that the viscosity is 5–8 units lower than CV made from the same latex. Interest in LV is small.

Packing

The wrapped bales of block rubber may be packed singly in polythene lined, multi-ply, paper bags or in thick-walled polythene bags but usually they are packed in pallets containing 1 t or 30 bales. An argument as to whether 5 layers of 6 bales giving a standard crate of 1000 × 1400 mm or 6 layers of 5 bales giving a standard crate of 1000 × 1200 mm should be used, was decided in favour of the former, which costs the industry several million US dollars a year in extra packing materials. Soon after this decision some packers were making 1.2 t pallets to enable 18 t net of rubber to be packed in an 8 × 8 × 20 ft (2.4 × 2.4 × 6.1 m) container. A recent development has been the shrink-wrapping of bales on a pallet base without side strengtheners or top boards. This type of packing is likely to increase in popularity as it is very suitable for LASH (lighter aboard ship) barge shipments and it reduces the amount of wrapping material needed.

Fig. 11.7 Pallets being consolidated and marked for shipment (Malaysian Rubber Development Corporation)

Now that many plantations are locally owned and large quantities of rubber are being shipped by producers with close ties with shipping companies, the time is ripe for a reassessment of the packing, shipment and delivery from the producer to the consumer. This reassessment should consider the quantities of shipments to be made to various destinations and the cheapest methods by which they might be packed so as to allow for maximum savings and minimum inventories along the supply line. Technical staff should be directed to work out details of bale size and pallet size and consignment without preconceived ideas interfering. New methods, materials and shapes could benefit all in the industry. The recent introduction of box rates for containers, instead of freight by weight, also encourages a serious review of packaging methods for natural rubber.

11.4.3 Coagulum grades – SMR 10, 20 and 50

The main grade produced from coagulum is the 20 grade which meets the requirement of most tyre manufacturers. The 10 grade has been slowly increasing in popularity as consumer product standards have become more exacting but so little of the 50 grade is now produced that it is unlikely to be more than a curiosity in the future. The specifications for these three grades are given in Appendix 11.3, but there is also a requirement that for any property the mean value of the bales supplied in a consignment, plus three times the standard deviation, must be below the specification maximum. As a result of this requirement it has become standard practice in the plantation processing industry to blend the raw materials for uniformity and this has been of great benefit to the consumer.

The equipment to produce the 10 and 20 grades is virtually the same and the processing steps involved are blending, cleaning, comminution, drying, pressing and packing, the methods for the last three being similar to those already described for latex grades. The raw materials for 10 and 20 grades normally consist of cup lump and tree lace but may also include lower grade RSS, USS and clippings. These materials are very variable in their properties and in the extent to which they are contaminated with bark shavings, leaves, dirt, sand, etc. Thorough blending is essential. The amount of cleaning needed varies; estate materials are normally considerably cleaner than those from smallholders, which are usually first sold to local dealers, then to larger dealers and finally to a processing factory.

The need for control and economy has led to the formation of large scrap processing units. Rarely now does a scrap factory process only the scrap from the estate where it is located; a group with a number of estates will set up a factory for processing scrap from an area with a radius of 100–150 miles. Formerly, it was customary to store the scrap under water before blending and cleaning it and this is still done in some factories. Soaking the rubber helps in the process of dirt removal and also softens it, which helps in blending and subsequent processing. However, with modern machinery there is no real need for soaking as an aid for cleaning and processing and nowadays the scrap is usually left to drain and dry under cover until it is collected and taken to the central factory. This saves on transport costs.

A typical scrap processing line is shown in Fig. 11.8. On arrival at the processing factory the scrap is weighed and discharged into storage bins or compartments. Several collections may be placed one on top of the other with a light spray of water to prevent undue massing. When several bins are full, a front-end loader or back hoe lifts scrap from one bin at a time and piles it together. The pile of mixed scrap is then carried by drag conveyor, fork-lift truck or other means to a wet pre-breaker. Copious water is run through while this machine cuts up the large lumps into pieces not larger than 5 cm diameter and at the same time removes a great deal of dirt, leaves,

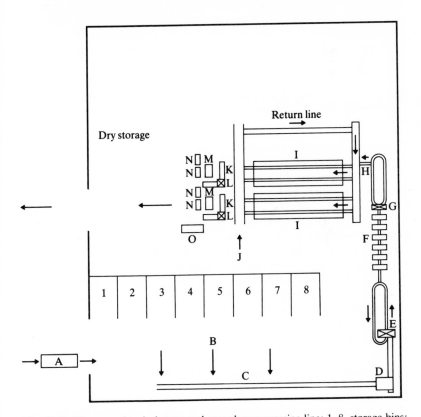

Fig. 11.8 Diagram of typical scrap and coagulum processing line: 1–8, storage bins; A, weighbridge; B, blending area; C, conveyor; D, twin-screw wet breaker; E, creper hammer-mill; F, at least six crepers; G, creper hammer-mill; H, bucket elevator; I, twin-trolley driers; J, transfer trolley; K, table; L, weighing scale; M, baling press; N, pallet; O, reprocessing unit

Fig. 11.9 The dry end of a rubber factory showing a Sphere Corporation drier together with weighing scale and packing area (Sphere Corporation Sdn Bhd)

sand, etc. From the pre-breaker the coagulum falls into water to allow this dirt to be washed away.

It is then passed through equipment to effect further size reduction and cleaning. In Fig. 11.8 this is shown as a line comprising a creper hammer-mill followed by six crepers and a second creper hammer-mill, which should be adequate to deal with even grossly contaminated scrap. But various other combinations of creper hammer-mills, crepers, granulators, shredders and extruders are used. For clean, uniform estate coagulum it may only be necessary to use a granulator followed by a extruder. The rubber from each unit falls into a trough containing water to remove dirt and it may then be recycled through the unit. After leaving the final cleaning and comminuting equipment, the rubber is dried as described for latex grades and then tested, graded and packed.

11.4.4 General purpose rubber – SMR GP

In order to meet consumer demand for a large volume, viscosity stabilized, general purpose rubber, a new grade, SMR GP, was introduced in 1979. This is made from 60 per cent deliberately coagulated latex from an SMR factory or a group processing centre (see p. 524) and 40 per cent field coagulum material. The specification for SMR GP is similar to that for SMR 10 but the former grade is viscosity stabilized and controlled at production to fall within the range of 58–72 Mooney units. The GP rubber should be produced in large quantity from good quality materials to give an homogeneous product with consistent vulcanization properties. Since its introduction there has been a steady, but modest, increase in the offtake of this grade.

11.5 Latex concentrate

In the 1930s significant quantities of preserved field latex of 38–40 per cent d.r.c. were regularly exported from producing countries in 18 litre tins and 200 litre drums. This was obtained by using trees of clones which naturally produced a latex of high d.r.c. and increasing this natural tendency by low intensity tapping. Efforts were then made to find ways of reducing the water content of the latex so as to avoid the high cost of shipping water in the form of low d.r.c. latex and so that normal tapping could be done and a cheaper, more satisfactory product could be marketed. This led to the production of latex concentrates of about 60 per cent d.r.c. The products made from this material world-wide now fall into eight categories, namely: dipped goods such as gloves and balloons (36%), moulded foam (17%), adhesives (15%), thread (10%), carpet backing and underlays (9%), rubberized coir (5%), leather-board (3%) and miscellaneous (5%). Three methods are now used for concentrating latex – evaporation, creaming and centrifuging. An electro-decantation method was developed but is no longer used.

11.5.1 Evaporation

The evaporation process developed by the Revertex Company and used for many years to make their latex concentrates employs a special plant in which latex is circulated through a tubular heat exchanger and then passes to a chamber where evaporation occurs under reduced pressure. The products, which contain the whole of the serum solids originally present in the field latex, find specialized markets of their own and do not compete directly with concentrates made by creaming and centrifuging. The evaporation process has not been taken up by other companies.

11.5.2 Creaming

On standing for a time, field latex, like milk, separates into a cream of mainly rubber particles, and the serum. The rate of creaming and the d.r.c. of the cream can be increased by adding certain chemicals to the latex, of which the most satisfactory, and the only one now used, is ammonium alginate. Field latex preserved with ammonia is bulked at the factory and, if the magnesium content is excessive, is treated with diammonium hydrogen phosphate to precipitate the magnesium, before adding the alginate. There is an optimum level of alginate which maximizes the d.r.c. of this cream; adding more increases the rate of creaming but lowers the d.r.c. Soap, as ammonium oleate or laurate, may also be added to aid in maximizing the d.r.c.

After mixing, the latex is passed through a clarifier or a centrifuge to remove sludge, which can be a source of rapid bacterial deterioration as well as of magnesium ions which destabilize the latex. The ammonia level is adjusted and the latex heated to 40 °C by live steam before being transferred to a tank. After one or two days the watery serum is run off and the cream (of about 55 per cent d.r.c.) is moved to a larger storage tank. About 30 days after filling this tank, serum is again run off and latex is then drummed or shipped in bulk. Creaming can produce a concentrate of 66–68 per cent d.r.c. It tends to have a somewhat lower non-rubber content than centrifuged latex, which can be an advantage for certain applications. Creaming can be repeated several times after dilution and this gives a very pure hydrocarbon concentrate (Duckworth 1964).

11.5.3 Centrifuging

The centrifuge process, which now accounts for about 93 per cent of latex concentrate production (Table 11.3) is based on the modification of milk separators for use with rubber latex. Two manufacturers, the Swedish Alfa-Laval Company and the German Westfalia Company, have put considerable effort into improving the capability of their centrifuges for the production of latex concentrate. This is indicated in Table 11.4, which shows the improvements in the rate of throughput and overall efficiency which have been effected in successive models of the machines. The efforts of these companies and the back-up services they provide have been the backbone of the centrifuge industries.

Processing

Field latex is collected as soon as possible after tapping and at the collection station the requisite quantity of preservative (almost invariably ammonia) is added in order to inhibit bacterial growth in the latex which would cause an increase in its content of volatile fatty acid (VFA), eventually leading to spontaneous coagulation in 6–12 hours (Cook & Sekhar 1955). As soon as practicable the preserved latex is taken to the factory, normally in tankers, and discharged into reception tanks, which should be capable of holding

Table 11.3 *Production of latex concentrate (tonnes)*
(a) Worldwide in 1982

Country	Centrifuged	Creamed	Other	Total
Malaysia	211,300	7,275	6,113	224,688
Indonesia	31,400	5,500	—	36,900
Sri Lanka	1,200	—	—	1,200
Liberia	22,200	—	—	22,200
Total	266,100	12,775	6,113	284,988

(b) Malaysia 1978–84

Year	Centrifuged	Creamed	Other	Total
1978	206,534	3,476	7,445	217,455
1980	216,236	4,750	4,931	225,917
1982	211,300	7,276	6,113	224,688
1983	—	—	—	224,300
1984	—	—	—	227,300
Per cent average	94.9	2.3	2.8	

Table 11.4 *Development of centrifuges*

Model no.	Approx. rate of throughput in litres per hour	Approx. no. of units 1.1.85 Malaysia	Elsewhere	Approx. overall efficiency (%)
Alfa-Laval				
LRH 1900	270–360	9	10	84
LRH 210	360–410	25	15	88
LRH 310	360–410	60	15	88
LRH 410	450–500	514	100	92
Westfalia				
MKA 8001	300	30	—	84
MKA 9001	360	40	—	88
LTA 110–00–006	500	<20	12	92
Westlake				
LX 460	450–500	30	10	92

two days' crop and which serve as a reservoir for the centrifuges.

A high level of magnesium in the latex will decrease mechanical stability (see p. 490). The magnesium content of field latex varies seasonally during the year, reaching its maximum as leaves senesce and are shed during wintering but declining almost to zero during refoliation. After this refoliation it is variable for about two months and then remains constant for about six months before steadily increasing again prior to wintering. If abnormal levels of magnesium are suspected, the field latex from each area should be tested and any excess magnesium precipitated out. This is done, either at the collection point or in the reception tank, by adding to the latex an appropriate amount of diammonium hydrogen phosphate to form magnesium ammonium phosphate which settles in the tank with any natural sludge.

Fig. 11.10 A modern gravity-flow centrifuging factory showing field latex reception tanks in right background, two lines of centrifuges and washing tables (Malaysian Rubber Development Corporation)

When sufficient latex is available, the centrifuges are started and run continuously until the following day, separating the latex into about 50 per cent cream and 50 per cent skim latex by weight, the composition of the two fractions being as shown in Table 11.5. Any sludge found at the bottom of the field tanks or reception tanks is washed out and centrifuged immediately before the centrifuges are shut down for cleaning, thus maximizing the recovery of rubber. The tanks are then throughly cleaned, for which purpose scouring and washing with water has been shown to be as effective as using bactericides. Epoxy, or other, coatings of the tanks facilitate cleaning and also prevent uptake of iron by the latex, which tends to produce blue or black discolouration.

The concentrate is passed through tanks where additional ammonia or other chemical preservative is added. These tanks are of two types, A and B, as illustrated in Fig. 11.11. In type A the

Table 11.5   *Typical composition of concentrate and skim fractions produced at 90% centrifuge efficiency*

Property	Concentrate		Skim	
	Latex	Total solids	Latex	Coagulum
% wt of field latex	50		50	
Total solids, % wt	61.5		8.5	
Dry rubber content, % wt	60.0		4.0	
T.s.c. − d.r.c., % wt	1.5		4.5	
Nitrogen, % wt		0.32		3.19
Protein (N × 6.25), % wt		2.0		19.94
Ash, % wt		0.50		0.22
Acetone extract, % wt		2.73		7.84
Total non-rubbers, % wt		5.23		28.00
Approx. rubber hydrocarbon content, % wt	95.7			72.0
Magnesium, ppm		40		—
Phosphorus, ppm		460		—
Copper, ppm		1.5		3.5
Manganese, ppm		0.5		0.6
Mooney viscosity ML 1 + 4 at 100°C		—		100
Plasticity retention index		95		45
Strain (ACS1), % elongation		—		14.5

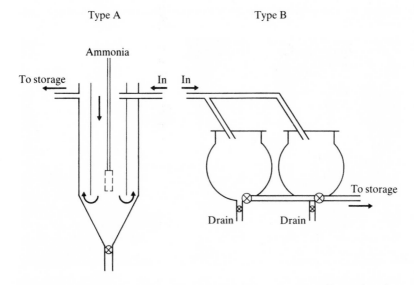

Fig. 11.11   Diagram of ammoniation tanks for latex concentrate (for explanation, see text)

concentrate flows into an annulus past the ammoniation or chemical addition unit and then upwards and out to storage. Type B consists of two calibrated pressure tanks, of 200–500 litres depending on the size of the factory, which serve as mixing tanks for the addition of chemicals.

The concentrate is then conveyed to storage tanks which should be of at least 100,000 litres capacity so that a large uniform bulk is obtained and the amount of testing required is not excessive. Each tank should be fitted with a gentle, efficient, tunnel-type stirrer which will not agitate the latex sufficiently to cause coagulation. When filling the tank, the stirrer is switched on as soon as the level of concentrate rises above the blades. On each day until the tank is filled the latex is sampled for determination of its d.r.c. volatile fatty acid (VFA) number and mechanical stability. This enables any necessary adjustments to be made to the centrifuges before the next day's centrifuging begins, and the correct d.r.c. and preservation levels to be maintained. When the tank is full it is stirred for 12–24 hours, again sampled and tested and, if the properties are correct, the latex is drummed or sent for storage at the port. If the properties are not correct, they are adjusted and the concentrate again stirred for 12–24 hours.

Preservation of latex concentrate

The key to the preservation of field latex and of latex concentrate during its production and storage lies in monitoring the concentration of the preservative in the water phases. For example, the

Fig. 11.12 Latex concentrate storage tanks, showing stirrers and (on floor to left-hand side) discharge pipes to tankers or drumming area (Malaysian Rubber Development Corporation)

minimum amount of ammonia required for complete, long-term preservation is 1.6 per cent by weight on the water phase. Hence, for latex of d.r.c. 20, 30, 40, 50, 60 and 70 per cent (assuming normal total solids content) the necessary ammonia in the water phase is equivalent to 1.22, 1.05, 0.90, 0.77, 0.62 and 0.46 per cent weight of ammonia on the weight of the latex.

In practice, the amount of ammonia which needs to be added to the field latex depends on how soon after tapping it is added and on the time which will elapse before the latex is centrifuged. Ammoniation as soon as possible after tapping is desirable in order to check bacterial growth at an early stage. The longer the delay before centrifuging the larger will be the amount of ammonia required to suppress bacterial growth and VFA development. It is advisable to add the requisite amount in one dose as early as possible as step-wise additions can lead to proliferation of ammonia-resistant bacteria and increase in the VFA number. If field conditions are clean and the ammonia is added in one dose as early as possible, then adequate preservation is obtained with 0.3 per cent and the field latex may be centrifuged up to 6 p.m. on the day of collection. With 0.5 per cent addition the latex may be centrifuged up to midnight but if it is necessary to keep it overnight before centrifuging, then 0.6 per cent is required. If field conditions are bad, d.r.c. low or weather conditions abnormal, these amounts may need to be increased by up to 20 per cent.

Prior to centrifuging, a sample of the latex bulk is taken and its preservation checked by determining its volatile fatty acid content as measured by the VFA number (a low number indicating satis-factory preservation) and further preservative added if necessary. Similar checks are made at later stages during production and storage to ensure that the final product has a VFA number of 0.20 per cent or less, as required in the specification for latex concentrate given in Appendix 11.5.

Systems currently used for the preservation of latex concentrate are summarized in Table 11.6. High ammonia (HA) latex should not normally contain secondary preservatives but, with the

Table 11.6 *Preservation systems for centrifuged concentrates*

Type	Preservatives
HA	0.7% ammonia
LA-SPP	0.2% ammonia + 0.2% sodium pentachlorophenate
LA-BA	0.2% ammonia + 0.2% boric acid + 0.5% lauric acid
LA-ZDC	0.2% ammonia + 0.1% zinc diethyldithiocarbamate + 0.5% lauric acid
LA-TZ	0.2% ammonia + 0.013% tetramethylthiuram disulphide (TMTD) + 0.013 zinc oxide + 0.05% lauric acid

consumers' agreement, small quantities of other preservatives are sometimes used. Various preservatives have been used together with a low level (usually 0.2 per cent) of ammonia. Sodium penta-chlorophenate and boric acid are less popular than formerly because of possible ill-effects on health. Zinc diethyldithiocarbamate has been found to be rather less effective than other preservatives. The LA-TZ system is currently the most favoured one but there have recently been suggestions that nitrosamines are formed from tetra-methylthiuram disulphide (TMTD) during vulcanization. However, it seems most unlikely that 10 parts per billion of nitrosamines in non-ingested rubber products could cause problems and as both TMTD and zinc oxide have been used as vulcanizing ingredients for a long time, apparently without ill-effects, it is expected that LA-TZ will continue to be used as a satisfactory preservative system. Nevertheless, new preservatives are undergoing development and test.

Mechanical stability

The stability of the suspension of rubber particles in latex is dependent upon the negatively charged film of proteins, phospho-lipids, etc., that surrounds these particles and the lutoids and the effect of various chemicals on these can alter the stability of the latex (Philpott & Westgarth 1953). An increase in the VFA content will decrease the stability of the latex. The mechanical stability is quantified by a test which measures the time in seconds taken for a recognizable degree of flocculation to occur when the latex is stirred, at specified temperatures and concentration, in a specially designed apparatus. As shown in Appendix 11.5 a minimum mechanical stability time (MST) of 650 seconds is specified for latex concentrate. A normal high ammonia commercial latex concentrate which has a VFA number of 0.03 and an MST of 1200 seconds will on increase of VFA number to 0.10 or 0.20 suffer reduction in MST to 950 and 600 seconds respectively. As described above, secondary preservatives may be used to ensure that there is no decrease in stability due to increase in VFA.

The presence of magnesium ions decreases stability. A magne-sium level of 30 ± 20 ppm in latex concentrate is generally considered desirable. When the latex reaches maturity the effect of this level on MST is a reduction of 6.5 seconds for every ppm of magnesium. Thus, if a latex has an MST of 1000 seconds with 50 ppm of magnesium present, at 30 or 10 ppm the MST will be 1130 or 1260 seconds respectively. As already mentioned (section 11.5.2) if an abnormally high level of magnesium is present in the field latex, it is precipitated out by addition of diammonium hydrogen phosphate, but the magnesium content of the final concentrate should be checked and adjustments made, if necessary, to give the acceptable level of 30 ± 20 ppm. If further upward adjustment of MST is needed, it can be achieved by the addition of lauric acid soap, which is best added to the concentrate to avoid loss in the skim. The usual rate of addition is 0.03 per cent on the weight of the latex but an increase to 0.05 per cent is necessary if zinc oxide or boric acid are used as secondary preservatives.

Although mechanical stability is spoken of as a relatively uniform property, it undergoes marked changes during the first three months after preparation of latex concentrate. An average latex will leave the centrifuge with an MST of about 80 seconds; this will increase to about 500 seconds after one week, rising to 1000 seconds after one month and to 1200 seconds after three months. After six months it will begin to fall and do so slowly for many years. Stability is maintained particularly well if the latex is gently stirred at intervals and its properties adjusted periodically as necessary. Generally, increasing amounts of lauric acid soap increase the mechanical stability pro rata but as it is not yet possible for producers to be informed of each consumer's requirements, the former produce only a uniform product to specification. If the MST of a well-preserved latex (i.e. with VFA number 0.03) is about 1200 seconds one month after preparation, then the latex can be expected to be satisfactory for all normal uses.

Twice-centrifuged, or multi-centrifuged, latex

There is a demand for a grade of latex purer than normal centrifuged latex for use in special applications where the presence of non-rubbers in the final product is undesirable. In a factory equipped with blow tanks this grade can be made by blowing the freshly concentrated latex back to the field latex storage tanks, diluting it to 30–35 per cent d.r.c. and quickly recentrifuging. This process can be repeated several times, each time losing about 6 per cent d.r.c. by weight as skim. The non-rubbers decrease with each centrifuging and the MST initially increases but falls with repeated centrifuging. Soaps can be used to displace proteins from latex particles, giving a low protein latex. A partially recentrifuged latex, which is acceptable for many purposes, is commonly made by recentrifuging 50 per cent of the latex and adding it back to the remainder, and recentrifuging.

Vulcanized latex

Vulcanized latex is prepared by heating latex with vulcanizing ingredients followed by centrifugation. It is made by several companies, of whom Revertex in Malaysia is by far the largest. As each of these vulcanized latices is normally custom-made, a detailed description is not appropriate. A supply of this latex allows the small manufacturers of latex products to omit the vulcanizing stage in his process, but large manufacturers prefer to carry out their own vulcanization.

11.5.4 Shipment of latex concentrate

Prior to shipment in bulk, latex may be stored at port installations, which have facilities for 500–3500 t, depending on the size of the production units in their area. From the port tanks, latex is usually sent by air-pressure along 15–20 cm pipes to the ship's tanks, which have a capacity of from 80 to 500 t. The larger tanks are mostly in parcel tankers, whereas the smaller ones are usually side or bottom tanks in break bulk ships. The latex in ships' bulk tanks is mostly carried to ports in Europe, the USA and Japan where it is discharged into port installation tanks with a total capacity of 100–500 t, from which it is delivered by road or rail to consumers' factories at short notice. At the ports the d.r.c. of the latex is normally tested but apart from occasional stirring no further treatment is carried out.

Ships with bulk tanks for latex are rapidly disappearing and it is becoming common to ship latex in large containers. The usual method is to use a large plastic bag which holds as much as can be accommodated in the container, which varies from 18 to 22 t net. Several companies manufacture the bags and one company has developed a pallet pack for shipping in containers, using a heavy-weight plastic bag to retain latex within the pallet. Container tanks are becoming more plentiful and bags in containers are cleaned and returned in a deflated condition. This form of shipment is rapidly increasing.

About half of the world exports of latex is shipped in drums. The main users of drummed latex are the eastern European countries and the USSR to which shipments are only made in the summer months to avoid freezing and consequent coagulation of the latex. Normally, new 200 litre drums are used, coated with a bituminous lining to prevent iron contamination.

11.5.5 Skim rubber

Skim rubber is a by-product of latex concentration by centrifuging. About 10 per cent by weight of the dry rubber content of the field latex separates in the skim latex which is about half the volume of the field latex and has a d.r.c. of 4–5 per cent. The high ammonia content of skim latex makes it difficult to coagulate and it is first de-ammoniated by trickling over slats down a tower against an upward flow of air. This reduces the ammonia content of skim latex from high ammonia concentrate, 0.5, to less than 0.1 per cent. One of two processes can then be employed, namely acid coagulation or biological auto-coagulation.

Considering all aspects, the simpler method is de-ammoniation followed by coagulation with dilute sulphuric acid. Coagulation is done in troughs and the coagulum passes through a crusher and creper hammer-mills or extruders as described for preparing block rubber from normal coagulum. The final product contains, by weight, about 70 per cent rubber hydrocarbon, 16 per cent protein (N × 6.25), 13.5 per cent acetone extract and 0.5 per cent ash. The rubber has a very fast rate of cure and is also very 'scorchy', i.e. tends to become prematurely partly vulcanized before the product is moulded, or otherwise fabricated. However, if it is of a light colour and is to be employed for making articles such as toys, its use can save considerable vulcanizing time.

For biological coagulation, some of the serum from a previous skim coagulation is kept and added to the fresh skim latex, repetition of this produure many times building up a very active population of acid-producing and protein-digesting bacteria. The rubber is produced with a loss of about 6 per cent by weight due to digestion of protein and fats. Care must be taken to maintain a satisfactory colour and to prevent the formation of bad odours. Auto-coagulation becomes very difficult if preservatives such as zinc oxide, sodium pentachlorophenate, boric acid, TMDT, or even high levels of ammonia have been used in the field latex. There is a ready market for well-processed skim rubber, typical properties of which are included in Table 11.5.

Apart from the normal processing of skim rubber into crepe or block forms, some incorporation of regulated quantities of skim has taken place in the preparation of other rubbers in Liberia and Malaysia. Such processes, in which the nitrogen content is usually controlled at 0.75 ± 0.05 per cent, can only be satisfactorily accomplished with the strictest factory control and where any off-specification production is reprocessed to bring it within the specification limits. The consumer must also be aware of the composition of such rubbers and agree to accept them. Production of these types of skim and coagulum grades reached more than 1000 t/month in both countries and still continues at a somewhat variable rate as demand requires.

11.6 Factory effluent treatment and disposal

In the early 1950s the rubber research institutes began to give some attention to the problems caused by the effluents discharged from rubber factories. There had been some ill effects from discharges into water courses for many years but these became particularly evident as the production of latex concentrate increased, because of the highly pollutant nature of the effluent from the centrifuging process. This effluent contained considerable quantities of proteins, sugars and other organic materials and in many cases also contained sulphates resulting from the use of sulphuric acid to coagulate the skim rubber. Having a high biological oxygen demand (BOD), these effluents rendered small rivers anaerobic, giving rise to conditions which released hydrogen sulphide and other sulphur compounds with very unpleasant odours so that the streams became polluted, black and evil-smelling, often for many miles away from the factory. These circumstances led to work on methods of reducing the pollutants in rubber factory effluents. In some countries they also led to the introduction by governments of legislation prescribing strict limits for the content of solids, nitrogen, BOD, etc., in effluents to be discharged into waterways.

Factories producing sheet, crepe and block rubbers give a readily biodegradable effluent and it has been found that the classical treatment of 20 days' anaerobic digestion followed by 40 days' aerobic digestion gives an effluent that meets specifications for discharge into a waterway. The effluent from latex concentrate factories is more troublesome to treat, partly because of the sulphates present. A joint programme, set up by latex concentrate producers in Malaysia, to use anaerobic/aerobic systems for effluents following coagulation with sulphuric acid was not particularly satisfactory. The use of oxidation ditches for treating effluents from such factories has also proved unsatisfactory. Some companies have solved the problem by avoiding the use of sulphuric acid and auto-coagulating their skim latex, as the effluent can then be treated in the same way as that from factories producing dry rubber. Where auto-coagulation is done it is usual to delay adding strong bactericides, such as zinc oxide, TNTD, etc., until after centrifuging. Such a procedure requires good hygiene in field and factory.

Many producers now prefer to apply the effluent to their land rather than to treat it to conform to the specification required for its discharge into waterways. The effluent, which contains appreciable amounts of nitrogen and potassium and smaller amounts of phosphate, magnesium and calcium, can be analysed and the rate of application needed to replace wholly or partly the fertilizers normally applied to a plantation can be calculated. Experiments so far have shown no ill effects on soil or ground water even when effluent from latex concentrate factories was applied at rates equivalent to high fertilizer dressings for several years. The effluents from rubber factories have been shown to increase rubber yields by 10–20 per cent and oil-palm yields by up to 30 per cent. Provided that they are situated close to a rubber factory, oil-palms are likely to be a more suitable crop for effluent application than rubber. Relatively high rates of effluent application can be given to replace the high fertilizer inputs required by the palms, whereas mature rubber has a low nutrient demand so that effluent dosage is low and an extensive area is required to accommodate the daily factory discharge, thus increasing the cost of application.

Various methods have been used to apply the effluent, including furrow irrigation by gravity flow, pipe irrigation with sprinklers or trickle nozzles and application by spray guns from mobile tankers. Commonly, the effluent is allowed to settle for three days to remove residual rubber and other particles that might block pipes. Where the topography permits, effluent can be cheaply applied by pumping it to the top of gentle slopes and allowing it to gravitate down herringbone-type channels, one man being sufficient to control the distribution to the channels of a day's discharge from the factory.

11.7 Marketing

Natural rubber has long been marketed through international commodity markets largely operated by dealers and brokers. As described earlier, it was, and to a considerable extent still is, sold in grades distinguished by visual features which bear little or no relation to technological properties. Trading in these grades gave opportunities for malpractices which, despite the introduction of safeguards, cannot, even now, be stopped completely. By contrast, the competing synthetic rubber, which supplies 65 per cent of the world's total rubber needs, has been marketed by the producers in technically specified grades with the support of modern sales techniques developed by the oil and chemical industries. With the development of the technically specified block natural rubbers some changes have begun to occur in marketing. These will be mentioned later, but, first the essentials of the traditional system, by which most natural rubber is still marketed, will be summarized.

Both the domestic marketing organizations in producing countries, which collect rubber from numerous producers for export, and the marketing and distributive systems in consuming countries, are linked to an international commodity marketing network. The latter comprises primary markets in Singapore and Kuala Lumpur, close to the main producing areas, and terminal markets in a number of consuming countries.

11.7.1 The primary markets

Singapore, which still handles about 40 per cent of the exports of its neighbouring territories, was the only primary market until 1962, when the Kuala Lumpur exchange was established. Supervision of the trading activities of these two markets is provided, respectively, by the Rubber Association of Singapore and the Malaysian Rubber Exchange and Licensing Board. Both these bodies are legally empowered to make rules and by-laws to regulate trading and both provide facilities and services for their members, particularly for the settlement and clearing of contracts, for futures trading and for the determination and dissemination of prices, which is done daily by an appointed panel of brokers and dealers (Ng & Sekhar 1977).

The brokers who bring buyers and sellers together in these markets are mainly large firms, with contacts or branches in the other international markets, who can handle physical and futures transactions in all types of rubber. The big dealer-exporters carry out most of the trading activities and generally hold stocks from which they can meet consumers' requirements at short notice. Most of them are also engaged in processing, remilling, packing and shipping. They play an important role in risk-bearing and in providing credit to middle dealers in the producing areas who buy rubber from the smaller estates and from the large number of primary dealers who purchase from smallholders.

11.7.2 The terminal markets

The main terminal markets are in London, Hamburg, Amsterdam, Paris, New York, Tokyo and Kobe. Apart from London, the other European markets deal only in the importation and sale of rubber required by local consumers. The London market developed as a result of the early connections of the businessmen of the city first with wild rubber and, secondly, with plantation development in Asia. It has remained the main market in Europe, providing futures trading and clearing-house facilities and trading physical rubber in Russia and the Middle East as well as in Europe, but it has experienced a decline in business in recent years, partly due to a marked reduction in rubber consumption by manufacturers in Britain. The by-laws under which it operates necessitate all producers, selling agents and importers dealing through brokers with the dealers, who then sell to consumers.

In New York there are no longer facilities for futures trading and the physical trade has been much influenced by the 'big five' tyre manufacturers, who account for the bulk of rubber consumption in the USA and who maintain representatives in the primary markets in the East. The Tokyo and Kobe exchanges handle the physical rubber requirements of local manufacturers, which are mainly met with lower grade sheet rubber from Thailand, but these markets also have a large volume of futures transactions.

11.7.3 Operation of the system

Because of the geographic location of the main international markets and the consequent time differences, price determination and rubber trading can proceed continuously around the clock in different parts of the world. At the end of the trading day in Singapore and Kuala Lumpur information on available rubber and prices is passed to London and similarly from London on closing to New York and from New York to Japan and back to the primary markets.

International marketing is facilitated by the existence of regular and reliable shipping services to transport rubber from producing to consuming countries. Because of this, neither dealers nor consumers normally need to hold large stocks, as the ships act as floating warehouses supplying rubber at regular intervals.

Rubber is normally sold f.o.b., i.e. free on board at the point of shipment in the producing country with all charges for collection, packing, loading, transport to port, port handling, storage, insurance, duty payments, etc., for the seller's account. If the rubber is sold c.i.f., then the seller also pays insurance and sea freight to the port of discharge. As the producing industry has developed and progressed, the tendency has been for the producer or exporter to control the product further along the distribution path. He can then safeguard his own interests by ensuring that the required rubber is available at the right time, thus satisfying the consumer while at the same saving on freight costs by consolidating large parcels or even chartering vessels. Provided that an efficient distribution system at the destination can be developed, either within existing agencies or with a separate system, there are nowadays many possibilities for joint consolidation of consignments from producing countries which could be particularly appropriate for the large producers of south-east Asia.

11.7.4 Changes in marketing

Although the marketing system described above involves several types of operators and a somewhat lengthy chain of transactions and operations, it has generally been considered an effective system. This is primarily because it provides a comprehensive range of services, including establishment of prices, settlement of contracts and claims, provision of facilities for stock-holding, warehousing, credit, hedging and risk-bearing. Originally, almost all rubber from large and small producers was sold through the market, but over the years changes have occurred which have resulted in less being traded in this way and more being sold directly by producers to consumers.

Direct trading

At an early stage, some of the large consumers appointed buying agents in producing countries, who bought some rubber through the market but soon made arrangements to buy the total output of large estates and to supervise the production to ensure that it met the requirements of their tyre factories. The large plantation companies have also long been accustomed to sell at least part of their rubber directly to consumers overseas. There has been an increase in this practice as a result of the development of more large production units. In Indonesia and Sri Lanka estates were nationalized, amalgamated and placed under the management of state corporations. More recently, in Malaysia there has been a consolidation of estates into larger units as a result of the take-overs and acquisitions through the stock market by government agencies and private enterprises. A considerable proportion of smallholders' production in Malaysia is now from land development and settlement schemes (totalling 268,000 ha under rubber in 1983) which have central management and marketing. It can be expected that any future substantial increase in the smallholder area will be in this form rather than as individual smallholdings (Ng 1984). Furthermore, a decision by the Malaysian government to take control of the processing of part of the smallholders' rubber led to the formation of the Malaysian Rubber Development Corporation, which by 1975 was producing 100,000 t of block rubber per annum and had also become one of the largest exporters of latex concentrate, most of its production being sold direct to consumers.

More direct selling is also to be expected, because technically specified rubber is forming an increasing amount and proportion of total rubber production and this trend is likely to continue at the expense of the conventional, visual grades. Technically specified rubber, being in large supply and of consistent properties, is increasingly being sold directly by producers to consumers. Thus, in 1982 the estate sector in peninsular Malaysia directly exported 91.4 per cent of its total production of SMR but only 13.1 per cent of its total production of RSS (Ng 1984).

The development and control of block rubber production has meant that the aim of delivering natural rubber to the consumer, undamaged and to specification, has been achieved. The means by which natural rubber producers can continue to supply satisfactory products at fair and stable prices have still to be resolved. Ideally, the marketing should be taken over by the producing industry in the same way as in the synthetic rubber industry and such a scheme, allied to the International Natural Rubber Agreement buffer stock scheme (see Ch. 1) might give considerable benefit to producer and consumer alike.

Chapter 12

Organization and improvement of smallholder production

J. W. Blencowe

12.1 Introduction

Most of the world's rubber is planted on smallholdings only a few hectares in size. Statistics are less reliable than for estates – the figures in Table 12.1 are an approximation drawn from various recent sources – but they indicate that, of almost 7m. ha planted in the six major producing countries, some four-fifths is on smallholdings. More than half of world production – perhaps about 2.4m. t – comes from smallholdings, indicating a low average productivity.

Table 12.1 *Areas of smallholder rubber plantings in six countries, 1986 ('000 ha)*

	Smallholdings*	Estates	Totals
Indonesia	2020	485	2505
Malaysia[†]	1530	470	2000
Thailand	1520	80	1600
Sri Lanka	125	65[§]	190
India	60	180	240
China	80**	320[§]	400
Total	5335	1600	6935

* Areas of less than 100 acres (40 ha) under single ownership.
[†] Includes some 400,000 ha of land settlement schemes and smallholder block plantings.
[§] National plantations and state farms.
** Rubber grown by communes.

Early rubber plantings by smallholders usually emulated pioneer activities on nearby large estates, although in some areas, such as southern Thailand, the idea spread from family and social contact with smallholders across the border in Kelantan, Malaya. Plantings were usually made on jungle land, often without official permission or land title.

Planting rubber as a follow-on to shifting agriculture in forest clearings provided a future cash crop and, as modern attitudes to land ownership developed, was a step towards establishing a legal rather than a traditional title to the land. Such plantings were particularly widespread in Indonesia and have often been termed 'sleeping rubber' for, as shifting cultivators moved on, the rubber received little further attention and became choked with self-sown rubber seedlings and secondary jungle growths. It was only tapped occasionally when cash was needed or when rubber prices were attractively high.

Most smallholders, within the constraints imposed by dependence on family labour, have copied the early planting methods of estates, but in an unorganized way, using low-yielding, unselected seedling trees planted at a very high density. Rows were often aligned directly up and down steep slopes, no cover crops were planted and fertilizers were not used during immaturity or subsequently. Smallholdings in Malaysia are mostly in the 1–5 ha size group, in Indonesia and Thailand the majority would be between 1–3 ha and in India and Sri Lanka would commonly be less than 0.5 ha. Such small production units are inefficient and often too small to provide a family livelihood; the situation deteriorates further as farms of modest size established by pioneer planters are subdivided between heirs. Where farms have been subdivided, or have been planted casually on former forest land after a few years of subsistence farming, the farmer often lives elsewhere; these unoccupied farms are likely to be tapped irregularly and maintenance limited to occasional clearing of undergrowth along the tapping rows. Tapping of older trees was destructive, with multiple cuts made through to the wood on any bark within reach of the ground; young volunteer seedling trees growing in or near the original planting row would be tapped at a girth of 35–40 cm, cuts sometimes being made from right to left at the whim of the tapper. Yields from pioneer smallholder plantings have usually been one-third or a quarter of those from post-war rubber on well-managed estates, and have often been less than one-tenth of what might be obtained in optimum conditions.

The importance of smallholders in rubber production increased during the 1950s when many old rubber estates were sold, or acquired by national authorities, and fragmented. Sale of estates occurred at times when financial or political considerations made replanting old rubber appear commercially unattractive. Rubber is one of the most labour-intensive crops grown on a large scale in the tropics; as wage rates have risen it has been possible to maintain its profitability by changing tapping systems, but as profits per hectare have fallen below those of other crops, estates have diversified interests by replanting substantial areas with oil-palm and cocoa. As this attrition of rubber estate land continues, a continued supply of natural rubber will depend increasingly on smallholders whose efficiency can, at best, equal that of a modern estate company.

12.2 Replanting

The biggest single step towards improvement of rubber smallholdings has been the replacement of old low-yielding trees with selected seedlings or clonally propagated cultivars. Smallholders and estates face completely different problems when assessing the need to replant. An estate company can estimate the increasing costs and diminishing returns from old rubber, and plan to replant a regular proportion of its total area. Thus, with a 6-year immature period and 27 years of tapping, 3 per cent of the land would be replanted annually and 18 per cent be immature at any one time. The smallholder cannot do this, his trees are likely to be all the same age and equally in need of replanting; even if he could afford to forgo some income from his farm there is a minimum size of field that can be replanted successfully. Plots smaller than about 0.4 ha suffer an excessive edge effect; young rubber is overshadowed by surrounding mature trees and does not grow well. On farms of 1.5 ha or larger, it is practicable to replant in two or three stages, although it is more efficient to do it all at once. The farmer has to forgo rubber income for 6–7 years, and is likely to delay replanting as long as his trees yield a small return. Even if he finds it unprofitable to continue tapping, he will be unable to afford the inputs required to replant, and may be reluctant to accept credit if replanting appears risky.

12.2.1 Assistance for replanting

Rubber on both estates and smallholdings had either been neglected during the Second World War or, in areas not under Japanese control, very heavily exploited. When the war ended many plantings were brought back into production, yields from rested fields were often quite good, and little replanting was done. Even where trees had become exhausted, there were uncertainties about the practicability and profitability of replanting – on some estates the incidence of white root disease caused severe losses among young rubber trees, and price levels during the immediate post-war years were rather low. The outbreak of the Korean War led to an immediate demand for assured supplies of natural rubber, prices rose to levels that – in real terms – have not since been surpassed, and trees everywhere, on estates or smallholdings, were tapped intensively. It soon became apparent that without extensive replanting or new planting, rubber production would decline irreversibly. Whilst the practical problems of replanting were being solved, financial difficulties remained, and there were further uncertainties as to whether natural rubber could compete in price and quality with the synthetic rubbers whose large-scale production was increasing. National and commercial interests of major rubber-producing countries required replanting programmes, and reluctance or financial inability to undertake the work implied that incentives would be essential.

Establishment of replanting funds

In Malaysia and Ceylon (Sri Lanka) schemes to assist replanting started in 1953 with finance from a cess levied on rubber exports. The Ceylon Rubber Replanting Subsidy and Malayan Rubber Industries (Replanting) Board schemes were planned to encourage applications for replanting assistance and, for smallholdings, to provide some physical inputs. In Thailand, an Office of the Rubber Replanting Aid Fund (ORRAF), established by an enactment of 1960, provided a service modelled closely on the Malaysian scheme.

Replanting in Malaysia

In Malaysia, replanting cess was (until 1973) reimbursed to estates on the basis of reported rubber production. Rubber from smallholdings (defined legally as land under single ownership and below 100 acres in size), traded through a multitude of village dealers and could not be identified with individual producers; thus a different system was devised. A separate Fund 'B' was created to receive cess collections not specifically identified with estates and to be used to finance smallholder replanting.

Officers of the Fund were required to encourage smallholders to apply for replanting assistance, check land titles, and confirm that rubber was over-aged. Successful applicants would then be given a replanting grant (so-called, although it was actually a reimbursement of replanting cess), comprising material inputs, such as fertilizers and planting materials, and cash allowances in partial payment for the farmer's own labour. Gardens were planted and nurseries established to supply clonal seed or seedling stumps and, later, for production of bud-grafted plants. Planting materials and fertilizers were delivered to the farm, and progress payments made, after Fund inspectors had certified that work had progressed satisfactorily. If replantings were found to be unsatisfactory, the Smallholders' Advisory Service of the Rubber Research Institute of Malaysia (RRIM) was notified and 'rubber instructors' would visit to help improve cultural practices. When replanted trees reached tappable size (50 cm girth) the farmer was invited to attend a one-month residential course organized by RRIM for 'completed replanters'. These courses, usually attended by sons of replanters, provided standard training programmes on field upkeep, marking trees for tapping, instruction on tapping systems, and on making good clean sheet rubber. The scheme proved extremely successful and by 1980 more than 600,000 ha of Malaysian smallholdings were replanted with high-yielding rubber (Malaysian Rubber Research and Development Board 1983). The inspectors and rubber instructors, trained by RRIM, constituted an effective corps of extension workers for the smallholder sector of the industry.

Special assistance for replanting very small farms

Owners of larger holdings, who were able to replant in stages, and those relatively well-to-do farmers who had bought land from fragmented estates, were the most active participants in the scheme. It became apparent that farms so small as to need replanting in a single operation, belonging as they did to the poorest sector of the community, presented a special problem; the owners just could not forgo rubber income whilst the replanted trees were immature and the cash element of the grant was too small for family subsistence.

A number of moves have been made, in Malaysia and elsewhere, to give special assistance to owners of very small areas of rubber. The Fringe Alienation Scheme in Malaysia offered replanting finance to owners of farms smaller than 2 ha; farmers were allocated an area of jungle land equal to that of their farm and the replanting grant used to plant additional new rubber. The award was conditional on agreement that, when the new trees reached tappable size, the farmer would then replant the original old farm. He would be entitled to a further replanting grant for that land; in effect, he was being offered a replanting grant at double the regular rate. The scheme suffered from several limitations:

(a) It attempted to alienate (allocate legally) a block of jungle land as close as practicable to the villages of participant farmers, but this was not always available or conveniently located.
(b) The replanting grant was inadequate to clear jungle and plant rubber and required labour inputs from participants which, partly due to location, were not readily available.
(c) The schemes were generally too small to afford full-time management inputs and, although assisted greatly by local rubber instructors and replanting staff, were not uniformly well maintained.
(d) The system, in attempting to be equitable, gave the smallest assistance to the owners of the smallest blocks of land.

In 1973 Malaysia established the Rubber Industry Smallholders' Development Authority (RISDA) as a statutory body combining the administrative functions of the Rubber Industries (Replanting) Board Funds 'A' and 'B' with the Smallholders' Advisory Service of RRIM, and with the aim of broadening the rural development activities of the Replanting Board. Refund of replanting cess to estates ceased to be automatic and became conditional on defined programmes of replanting. RISDA made special efforts to encourage owners of very small farms to replant; higher rates of grant and other incentives have been offered, but about 190,000 ha has not been replanted, and substantial areas need a second cycle of replanting. Some 400,000 ha has either been abandoned or is of low productivity; most is in lots smaller than 1 ha, often with aged or absentee owners dependent on share tappers reluctant to work on low-yielding fields (Malaysian Rubber Research and Development Board 1983).

In Sri Lanka most smallholdings are less than 1 ha and the success of the Replanting Subsidy Scheme there has been achieved in face of this disadvantage. In Thailand ORRAF is required to allocate 70 per cent of replanting funds to 'small plantations' (less than 8 ha), 20 per cent to 'medium plantations' (8–40 ha) and 10 per cent to 'large plantations' (although unused funds in one category may be passed to another). Average area replanted is about 1.5 ha, but the most common size is 0.8 ha (Office of the Rubber Replanting Aid Fund and Food and Agriculture Organization 1980). Farmers in Thailand are believed to prefer to replant their land in three or more portions; those with a total of less than 1 ha will thus face the same difficulties as in Malaysia. An arrangement very similar to Malaysia's Fringe Alienation Scheme exists; however, whilst it permits owners of very small farms to use replanting grant entitlement to assist new planting of a similar area of rubber it does *not* provide additional funds for subsequent replanting of the original farm. In consequence, the arrangement has not been popular but, in spite of the lack of special assistance, the problem of replanting small farms has not so far slowed a replanting programme currently running at 50,000 ha per year.

Infrastructural support for replanting, including extension

Replanting inspectors administering replanting schemes fulfilled many of the functions of extension agents, actual replanting being planned to varying degrees to be done by the farmer himself. In Malaysia, to meet the needs of the rapidly expanding programme of land settlement schemes, quite sophisticated contracting services had developed. These contractors were able to undertake land clearing using heavy mechanical equipment, cut terraces, do bud grafting or planting – and at fees that large estates found to be competitive with the costs of doing the work by direct labour. It was argued that, because replanting was a once-in-a-lifetime operation, there was little point in teaching farmers to do the skilled work themselves and many found that they could afford, with the assistance of replanting grant funds, to use the specialized services of contractors instead. In Sri Lanka, estates and smallholdings were smaller in size and there was less opportunity for contractor services to develop, but smallholders benefited from nursery facilities and other services from nearby estates.

In Thailand, where 95 per cent of rubber was planted on small-holdings, no planting contractors were available, and it proved necessary to teach farmers how to do the work; some of this was done individually by ORRAF staff, some by group training at schools operated by the Rubber Research Institute (RRIT). Planting material production was at first centralized on a 1000 ha nursery operated by the Government Rubber Estates Organization, whence budded stumps were distributed by road to all replanters in the country. This worked very well until the rapid increase in replanting raised annual requirements beyond 8 million stumps. The logistical problems of pulling, trimming, packing and shipping more than this number from a single nursery, and within the time-frame of a short planting season, proved insoluble. A number of additional smaller nurseries were established at government 'rubber stations' and private growers were also encouraged to produce budded stumps for sale through the agency of ORRAF, and these, together with the bud grafting done by farmers themselves, are currently producing rather more than 40 million bud-grafted plants annually – supplying both the new planting and replanting programmes.

Teaching farmers to do their own bud grafting was one of the more interesting development programmes executed by RRIT and ORRAF, initially with the assistance of the Food and Agriculture Organization. The aim was for a quarter of all replantings to be done by bud grafting seedlings grown from seed planted at stake in the field. The proposal was somewhat controversial; distribution of budded stumps had been expected to reduce the immature period by a year, and the budding skill taught to the farmer might be used only once in a lifetime. In practice, fields planted with budded stumps had required an allowance of 33 per cent of surplus plants as 'supplies' to replace plants that did not survive transport to or transplanting in the field; in some areas planting material took more than four days to reach the farm and casualties were often much higher than 33 per cent, and additional supplying was often necessary in the second season after replanting. Growth of trees budded in the field was very uniform and at time of opening for tapping a much higher proportion of them were of tappable size. This far outweighed any delay in reaching maturity. Started at first at the request of a group of farmers, a programme was devised that by 1977 was training about 2500 budders each year. At the end of five days' tuition at simply constructed residential schools, more than 80 per cent of trainees were able to achieve better than 70 per cent success with a test task of 100 buddings. Skill and speed of working increased rapidly with practice; the small fingers of children and women proved particularly suited to the task and their work soon added usefully to family earnings. The scheme has been successful beyond all expectation and has had a number of valuable side effects:

(a) Sale of budded stumps produced by trainees whilst under tuition went a long way towards defraying the total cost of the programme.

(b) Gathered together at the schools, the trainees were excellent 'target groups' for other rubber development activities.

(c) Farmers had formerly been most reluctant to uproot budded stumps where the clonal scion failed to grow; they were content to let a seedling tree develop. Doing the bud grafting themselves, the significance of the operation was clear and they would replace or re-bud the plants where scions failed to grow.

(d) Substantial new areas of rubber were being planted in Thailand, most of this being done with unselected seedlings; once budgrafting skills were available, replanters produced surplus budgrafted plants – either for their own use on new land or for sale to neighbours.

12.2.2 Block replanting

It has been suggested that, if groups of smallholders could be persuaded to replant their farms simultaneously, large blocks of land could be cleared at one time and efficient supervision provided. The contractor would have mechanical equipment, and provide full services. The need to teach the farmer skills needed for the supposedly once-in-a-lifetime operation would be obviated. Replantings might be taken over and maintained to estate standards, and handed back to smallholder operation only as maturity approached.

In one interesting programme in Malaysia, farmers have been persuaded to hand over land to RISDA control; contiguous blocks of 30–100 ha are replanted and managed for RISDA by contractors and participants receive a share of the proceeds when the 'mini-estates' come into production. There are arrangements for owners of very small plots to receive a monthly subsistence allowance until the rubber is tapped, which is repayable (with interest) in later years; by 1983 26,300 ha of mini-estates had been established, and RISDA planned further expansion of the programme (Malaysian Rubber Research and Development Board 1983). Whilst the mini-estates provide a method of consolidating and replanting groups of small farms that might otherwise remain unproductive, their size would appear too small to support the cost of as efficient a management as on larger estates or settlement schemes; they do not appear to have attracted direct smallholder participation, but provide an income to farmers unable or lacking the incentive to work the land themselves. RISDA has also established rubber (and oil-palm) on jungle land, and offered settlement to owners of uneconomically small rubber farms, in return for which they would relinquish title to the original holding, which could then be consolidated with others and replanted.

In Thailand, anticipated difficulties of persuading groups of farmers to replant simultaneously, coupled with cost of supervision during immaturity, appeared to outweigh possible advantages. However, in the absence of replanting fund facilities such as those of Malaysia, Thailand and Sri Lanka, Indonesia has implemented the method – on a credit basis – on a substantial scale.

The Indonesian Directorate-General of Estates (DGE) has established Project Management Units (PMUs) to undertake block replanting (and new planting) for smallholders. Farmers are provided with all material inputs during immaturity, given loans covering 25–30 per cent of their labour costs, plus a cost-of-living allowance of Rp15,000/ha for each family. Intercropping is permitted in the early years and provides food or additional income. Typically each PMU would cover about 3000 ha (exceptionally up to 10,000 ha) and be staffed by a manager, administrative assistant, three technical assistants and six field extension workers. Each extension worker would supervise about 100 farmers in the year of planting and perhaps double that number subsequently. During the first three years loan risks are carried by DGE itself but – where rubber growth is satisfactory – loan responsibility then passes to a bank; the farmer commences loan repayment in year 7. Some problems have been encountered because land clearing has been partly by hand, or when *Imperata cylindrica* has invaded land after intercropping. Nevertheless, the success rate has been encouraging: of 25,000 ha at the third-year stage, 20,000 ha was deemed acceptable to the bank, 3000 ha had weed infestation problems and only 2000 ha needed more drastic rehabilitation. In 1983 DGE planted 30,000 ha under PMU schemes and proposes to raise this to an average of 70,000 ha annually by 1990 (Food and Agriculture Organization 1983); (the proportions of replanting and new planting are not specified).

12.3 Planting materials and procedures

Methods developed for estates have usually been recommended unchanged for smallholders. Fertilizer recommendations issued by RRIM have often been adopted in other countries with changes only of units of measurement. Terracing is recommended on sloping land, and cover crop planting wherever practicable. On terraced land, and where intercropping may be practised, inter-row spacing of the trees of up to 9 m is recommended; where rapid ground cover and improved weed suppression are required, it may be reduced to as little as 5 m. Trees spaced closer than 2.50 m within the row may lean excessively across the inter-row and suffer wind damage, but there is considerable latitude in choice of planting density. In simplest terms, the optimum number of trees per hectare is a function of yield and cost of tapping. An estate or smallholder paying for the tapping to be done can optimize profit by increasing yield per tree and thus productivity per tapper. A farmer tapping his own trees may wish to increase yield per hectare by planting the trees more densely. Paardekooper and Newall (1977) suggest that, whilst under normal estate conditions densities higher than 400 tapped trees per ha would not result in appreciable extra profit, smallholders might aim for a mature density of around 500 trees per ha. At closer spacings, poor bark renewal might shorten the tapped life of the field, and they consider that tapped stands of 600–650 per ha could only be acceptable if trees were to be exploited only on virgin bark – perhaps by combination of low frequency tapping, yield stimulation, and reduced task sizes. This would appear to preclude the possibility of planting the smallest farms at extremely high densities so that there are sufficient trees to provide tapping tasks large enough for an alternate daily instead of a daily tapping system to be employed.

12.3.1 Choice of planting material

Clonal seedling versus bud-grafted plants

Selection of planting material for smallholders presents problems different from those faced by estates. Clonal seedlings offer obvious attractions: seed is easy to transport, even to remote areas, and can be stored if necessary for limited periods of time. Seedling trees are easy to establish, usually reach maturity a year earlier than budded trees, and – because of inherent variability – are often more adaptable to variations in soil, climate and methods of husbandry. These advantages were often considered to outweigh their lower yield potential in comparison with budded trees and most countries with public sector planting programmes established special gardens for seed production; some were simply monoclone blocks (often of Tjir 1), other more sophisticated ones attempted to emulate polyclone gardens of the type planted by commercial seed producers. Alternatively, uniform monoclone plots on estates were inspected, registered, and dealers licensed to collect and sell seed to individual replanters or to officially established organizations.

A number of important disadvantages of clonal seedlings have been recognized:

(a) Lower yield potential, often more than 30 per cent below currently available clones.
(b) Smallholders usually tap $\frac{1}{2}$ S d/l (periodic), perhaps $\frac{1}{3}$ S d/l; whilst they might be persuaded to tap alternate daily, it is unlikely that they would ever be persuaded to adopt the $\frac{1}{2}$ S d/3 system recommended as best for seedling trees.
(c) Seedling trees are inherently variable but smallholders have been reluctant to thin out runts. Subsequently they tend to tap selectively; with a reduced number of trees perceived as suitable for tapping, the tendency to tap more frequently is accentuated.
(d) Although seed suppliers are expected to deliver adequate surplus seed to allow for filling vacancies in field plantings, farmers have tended to use the extra seed to plant extra land and – failing to perceive future yield deficiencies – have filled vacancies with unselected seed collected locally.
(e) Unscrupulous dealers have supplied unselected seed under the guise of a reputable certificate or label; a fraudulent practice that would remain undetected until trees came into tapping.

Thus, whilst many plantation companies continue to include selected seedlings as part of estate planting programmes, pointing out that earlier and better rates of return to investment more than offset the lower yields, seedling plantings have virtually disappeared from smallholder planting and replanting programmes.

Choice of clones

Given the choice, a smallholder will opt to plant the highest yielding clone available. After years of waiting for a field to mature, he is eager for a cash return and is strongly biased towards clones that are 'early starters'. The yield obtained by neighbours who have replanted earlier is his most convincing source of information, and it is there to see; the wet sheets hung out to drip each day can be counted by every passer-by. The would-be replanter is naturally most interested in the yields of recently opened fields and clones giving the best precocious yield are the ones most talked of in the village.

Extension workers are more concerned with the element of risk that choice of clone may impose on a smallholder. Large estates can spread the risk by planting small areas of newer, more speculative clones, but most farms are too small to permit planting more than one or two that are known to be reliable and suited to the district. The extension worker would seek the ideal cultivar: fast-growing, resistant to locally prevalent diseases, wind-fast, with thick bark that renews well and does not wound easily, and good performance under daily tapping with the periodic rests that occur when the smallholder is engaged in other activities. If a majority of these attributes are to be found in a clone that is reasonably high yielding, then it will be recommended to small farmers. If only reliably tested material is to be used, farmers are likely to be restricted to a very short list of rather old-fashioned clones. For example, for a number of years planters in Thailand were asked to choose only from RRIM 600, GT 1 and PB 5/51 and, because of the prevalence of *Phytophthora* leaf and panel diseases on the west coast, only GT 1 (which was moderately tolerant of infection) could be recommended for that region. Farmers knew well that RRIM 600 was likely to be at least 30 per cent higher yielding than GT 1 and were very reluctant to accept the latter. In an area where many were doing their own bud grafting it was possible to exert some concealed pressure by increasing supplies of GT 1 bud-wood and reducing RRIM 600, but this led to considerable dissatisfaction. Farmers pointed out quite forcibly that – even when damaged by leaf disease and/or black stripe – RRIM 600 was still substantially higher yielding; when it became apparent that some were so disillusioned by the 'slow starting' performance of GT 1 when they started tapping that they sold their farms, freedom of choice had to be allowed to revert to the farmer.

Elsewhere, on Thailand's south-eastern seaboard, farmers are more willing to accept the lower initial yields from GT 1 in return for its reliability in later years. Meanwhile, increasingly higher priority has been attached to the search for better, more disease-resistant clones than RRIM 600; and owners of larger farms have been encouraged to plant newer clones in task-sized plots that act as demonstrations to owners of smaller farms nearby.

12.4 Intercropping

Smallholders replanting their farms are usually required, as a condition for progress payments from replanting grants, to establish and maintain a leguminous cover, as is accepted practice on estates. However, many will wish to grow food or cash crops between the rows of young rubber, and – on flat or gently undulating land where dangers of erosion are minimal – intercropping is usually encouraged.

Choice of intercrop is limited by a number of factors:

(a) The crop must be easy to grow and have no detrimental effects on the rubber.

(b) Labour requirements, which are often seasonal, must be within the capabilities of the family unit or, if additional labour is to be hired, it must be freely available at a cost appropriate to the value of the crop.

(c) The crop, if to be produced in surplus to family consumption, must be marketable.

Overshading of young rubber can be minimized by planting the rows from east to west, but even so, only low-growing crops are acceptable in the first year. In Thailand upland rice is very popular as the first intercrop. Groundnuts are commonly grown subsequently; soybean or mung bean are occasionally grown. Pineapple may be planted at the same time as the rubber; although it will not produce its first crop until the following year, it can be productive into year 3. All the above may be planted to within 1 m of the rubber row. It may be unwise to plant taller crops such as banana, maize or sunflower until year 2, and they should not be closer than 1.5 m from the rubber. Harvesting sweet potato might cause excessive soil disturbance or root damage and, as with cassava, unharvested roots could serve as a food base for *Fomes* or other root-disease fungi attacking the rubber.

Upland rice requires a lot of hand labour at harvest, but this falls at a time of year when other work (tapping, field maintenance) is less demanding; by contrast, harvesting groundnuts – planted at times carefully chosen to meet such market demands as Chinese New Year or for the cakes made for the Eighth Moon Festival – can prove beyond the ability of the farmer and his immediate family. Bananas are probably the most popular intercrop in Malaysia just because labour requirements are relatively modest (Lim 1969).

Marketability is perhaps the most difficult criterion to meet, for a number of interacting reasons. Crops that are bulky are expensive to carry to market, while those that are perishable may be greatly in excess of demand during peak seasons. A high-value, low-volume crop like tobacco may be extremely profitable for smallholders familiar with its cultivation, but the market is very specialized, and buyers are usually uninterested in visiting new areas with inexperienced growers. An economic study of costs, seasonality and amount of labour inputs and market prices could well conclude that water melon was an outstandingly good choice; this might be true for a relatively small number of growers, but not if too many attempt to emulate the success of a neighbour – prices will fall to levels below costs of moving the fruit to market. A similar study would suggest that upland rice is quite unsuitable, because cost of inputs and returns to family labour are very low; but the popularity of this intercrop in a rice-eating community has a social explanation – a full rice barn gives a stronger feeling of security than money in a bank, and the flavour of the cooked rice is highly prized.

Comparatively little upland rice finds its way into the market. Satisfactory returns may be gained from maize sold as fresh cobs, but not if sold at the lower price commanded by dried grain. Bananas planted in double rows between rubber may be quite profitable, but a single row – desirable from the point of view of the rubber – might be less than satisfactory.

All these problems of selecting and growing crops, added to the fact that a great deal of rubber land is steeply sloping, explain why most young rubber is not intercropped; possible competitive effects of intercrop on rubber add a further dimension to the problems. These are most likely to arise when the farmer attempts to optimize intercrop at the expense of the rubber. Recommendations must always specify that adequate fertilizer be applied to intercrop as well as to rubber; but this rarely happens when the farmer himself does the intercropping, the fertilizers supplied for the rubber usually being shared between both crops. Farmers who lease their inter-row land to others in return for upkeep of the young rubber are more likely to insist that their trees are correctly fertilized and that the intercrop is given additional nutrients.

Replanters who have practised intercropping are usually required to establish a legume cover afterwards. This is not easy to do, and the grower has little incentive, for any remaining replanting grant instalments to which he may be entitled are likely to be small. Invasion of undesirable weeds, especially of *Imperata cylindrica*, is all too common after intercropping. It might be argued that the various ill effects of intercropping outweigh any advantages, but it must be remembered that – in assessing economic returns – *any* income during the early years of investment in replanting has a disproportionately large effect on internal rate of return (IRR). This can be a factor as important to decision-makers in replanting organizations as are the food produced or cash earned to the farmer himself.

It is fortunate that rubber trees are themselves strongly competitive for nutrients. Cassava, with its reputation as a grossly exhaustive crop, is not to be recommended as an intercrop, and is often specifically prohibited by replanting authorities. However, it must be remarked that rubber on farms intercropped with cassava is often in better condition than on farms with no intercropping, if only because the farm is visited and tended more often. It is also interesting to record that tapioca farmers in eastern Thailand have converted thousands of hectares of cassava to rubber, simply by omitting a cassava row at 8 m intervals and intercropping the cassava with rubber! With neither crop receiving optimal fertilizer, the rubber has nevertheless grown quite well, and follow-on crops of cassava planted in year 3 were over-shaded by the trees and did not produce a worthwhile yield.

12.4.1 Other sources of income on the farm

Although many smallholders plant a few fruit trees in or around their rubber, the competitive nature of the latter virtually precludes its use in any system of permanent mixed cropping; rubber will – if it is growing well – over-shade the companion crop.

Livestock in smallholdings

Smallholders have limited opportunities for raising livestock. Grazing cattle may trample young rubber or, as it matures, tapping panels may be damaged irreparably by horns rubbing against the bark. Cutting the cover crop and carrying it to cattle as fodder partially negates the purpose of planting the legumes. Chickens and ducks may be raised, and do well under the shade of mature trees, but in large numbers the latter may puddle the surface soil excessively.

Sheep can be run under mature rubber and help reduce weeds. Arope *et al.* (1985) report from Malaysia that a cross between local ewes and imported Dorset Horn rams thrived under rubber estate conditions, with but little supplementary feeding and veterinary attention. Weed control costs were reduced by 15–25 per cent and, with an imputed value to sales of manure, a return to investment of 15 per cent was achieved. Thai people (and many Chinese) find mutton unpalatable, but rearing sheep offers a valuable opportunity for ethnic Malay smallholders to improve both their diet and the upkeep of their farms – especially if the hybrids bred by Arope *et al.* are 'fixed' as a locally adapted breed.

12.5 Land settlement schemes

Since the 1950s substantial areas of rubber have been planted in officially sponsored land settlement schemes. These have usually been assisted by a central planning agency and financed by national or international loan funds, on-loaned to would-be settlers at concessional interest rates. Loan recovery has been ensured, usually after a grace period of 7 years and spread over some 15 years, by deductions made at point of sale of the crop and usually facilitated by central purchasing and processing of the settlers' crops. Schemes have been provided with management or other forms of supervision, and with varying levels of infrastructure. A cost-effective management service usually requires that schemes be about 1000–2000 ha in size, but methods of selecting settler families, size of plot allocated, range of inputs provided, and social infrastructure have varied widely.

12.5.1 Site selection, land clearing and crop establishment

After sites have been selected on the basis of soil and land suit-ability surveys, the best land has usually been allocated to more prof-itable crops such as oil-palm or cocoa; rubber schemes have often been planted on only moderately good soils and with rather steep slopes. Clearing by felling and burning jungle growth has usually been preceded by extraction of all valuable timber, but some schemes have deliberately been located in areas where loggers have exploited and abandoned jungle land to secondary regrowth, and other areas have been so remote that opportunities for commercial logging have been forgone.

In some instances settlers have been encouraged to clear the land and plant the rubber themselves but this has rarely been satisfac-tory. With minimal mechanical assistance, the work has been extremely laborious and, with no foreseeable income from rubber for a number of years, the need for subsistence food cropping has taken precedence over rubber planting. In extreme cases would-be rubber planters have never progressed beyond the stage of being cassava farmers. It has been much more usual to farm out the initial land clearing to commercial contractors, who would also prepare the land, cut any necessary terraces, plant rubber and cover plants, and maintain the crop during the early years. Settlers might be brought in to occupy the scheme well before the rubber reached tappable size, and paid a minimal subsistence wage whilst working as labourers on the scheme they were eventually to occupy. In other instances, whilst all the land has been cleared of heavy jungle, only a portion has been planted with rubber, the settler moving in when it is ready to tap; he has then been expected to plant up the remainder himself, using materials provided by the executing agency. However, whilst a ready-to-tap portion of the farm ensures an immediate income, it has often distracted the settler from planting up the remainder of the land, and any not used for food production has reverted to secondary jungle. Moreover, costs of initial land clearance, access roads and other infrastructural support are incurred against the total area in the early years of project finance, and loan repayments have to start when crop is harvested from perhaps only a quarter of the land. The most successful schemes have been those where rubber has been established, and the settler given some paid employment before the rubber has reached tappable size. Although this procedure has raised settler indebtedness, this has not been in excess of the ability to repay.

12.5.2 Size of holding

Land allocations to settlers have usually been determined by national policy, influenced by area available and numbers of people requiring resettlement. Allocation of different sizes of plot has created considerable inequalities between settlers on different schemes within the same country. Inequalities have also arisen within schemes: some settlers have been allocated conveniently located plots whilst others received inaccessible and perhaps less fertile land. It has not often been practicable to set a target income, determine crop area required to provide it, and confirm that family labour would be adequate to operate the unit. Perhaps the simplest economic model assumes that a man and wife could together tap (alternate daily) a total of up to 2400 trees; these could be planted on 5–6 ha. Family income from that number of trees would be ample for subsistence and loan repayment. Where land was scarce, planting density could be increased, although Paardekooper and Newall (1977) argue against mature densities higher than 500/ha. In Malaysia, Barlow and Chan (1969) considered that a family unit of 2.4–3.2 ha was suitable, the average allocation on Federal Land Development Authority (FELDA) schemes being 3.0 ha, and on State schemes 2.4 ha. Lim (1972) suggested that 4 ha of land planted with about 500 trees/ha would be an optimum size for a family-operated holding. In Indonesia a size of 2 ha of rubber per family has been adopted. With allocations lower than 2 ha per family, income from rubber may be such that, to earn a living and continue loan repayments, settlers resort to daily tapping. This in turn depletes bark reserves and may bring forward the future need to replant. The internal rate of return (IRR) to national investment in a scheme may be entirely satisfactory with a planned tapping life of 25 years, but not if tapping life proves to be only 15 years.

12.5.3 Social facilities

Settler loans have usually included a share of the costs of land clearing, planting, husbandry during the immature period, management support, field access roads, and provision of central processing facilities. Assistance with housing may be limited to provision of building materials but more usually has comprised a wooden house of standard pattern; in either case, cost would be recoverable. Costs of more sophisticated facilities such as schools, post offices, police stations and main roads would be absorbed by central government. (However, *all* development costs must be included in financial and economic analyses of project viability.)

Houses have been built on individual farm lots, in 'hamlets' of 10–20 families, or in nuclear villages with 300 or more homes. A balance has to be struck between practical and socio-political advantages of the different arrangements; this will be influenced by an assessment of increased efficiency if families live on the farm, social advantages of life in small groups, and the administrative convenience of larger communities. In larger units it is cheaper to install piped water and electricity, and easier to provide schools, shops and other infrastructure. However, the daily journey from home to field is lengthened, and farm wives contribute less to field activities for they are reluctant – if there are small children – to venture far from the house lot. On the other hand, the family unit will be much more stable if young people have some of the modern amenities that might otherwise tempt them to forsake an agricultural life and move to urban surroundings.

12.5.4 Selection of settlers and rights of ownership

Land settlement schemes aim to give a better life to the rural poor and provide economic benefits for the nation. Candidate settlers may be families from over-populated regions, inshore fishermen impoverished by depletion of traditional fishing grounds, or specific groups such as unemployed youths. The easiest success comes when candidates are young, married and already have rubber farming experience (perhaps as share tappers, or as family members on uneconomically small farms). Other types of settler might derive greater social benefit, but the scheme will incur higher charges for training and supervision, and run greater risk of technical failure.

Settlers may be allocated a clearly demarcated plot, with promise of ownership after repayment of development costs (as in earlier FELDA schemes), or nominated for a specified share of a larger unit, or simply given secure employment with entitlement to a daily wage. In Malaysia the Kedah State Land Development Board has purchased and operated estates – or developed them from jungle land – using the services of a commercial estate management agency. Employees are required to be citizens of the state; because many workers had no prior rubber experience, training costs were somewhat higher than normal, otherwise standards of operation and management have been the same as those of other estates in the region. Further south in Johor, RRIM has experimented with a scheme to make estate-type management even more attractive by offering participant workers 20-year security of tenure in a field of rubber on an estate, an 'incentive wage' for work done, a bonus share of estate profits, and the option to nominate an 'heir' to the rights held. It was claimed that, by what was effectively usufruct ownership, participants earned more than tappers on commercial estates or settlers on FELDA schemes. Moreover, production costs were only marginally higher than on comparable small commercial estates. In spite of these benefits, a number of participants left after a few years to join FELDA schemes (Nor & Nayagam 1982).

Choice of system: some practical aspects

Practical considerations, as well as social and political objectives of government, must influence choice of system. Rubber farms are much more suited to individual family operation than, for example, oil-palm, but once a settler has been allocated a specific plot it is difficult, even before he has acquired full title to the land, to enforce a desired standard of upkeep. Moreover, in many communities it may be a customary or legal requirement for property to be sub-divided when the owner dies. Surviving children may each receive a plot that is uneconomically small, and that may be further sub-divided in succeeding generations. The Government of Malaysia has recently been advised (Malaysia Rubber Research and Develop-ment Board 1983) to model its rubber land settlement schemes more closely on commercial estates, but there seems to be little doubt that it is the aspiration for a clear title to land ownership (once loan repayments have been completed), that makes the original FELDA concept so attractive to a settler. It may be that the nucleus estate concept will eventually prove the most popular with governments, lending organizations and participants. A number of oil-palm schemes have already been developed, in Indonesia and elsewhere, and some rubber schemes are being planned. A nucleus estate is planted and developed on commercial lines with adequate processing facilities for a surrounding area of individual smallhold-ings, usually of equal or larger size. The estate, usually operated by government or a statutory body, plants the smallholder area and supervises its upkeep during immaturity. Subsequently, estate staff give technical guidance to smallholders and regulate harvesting to optimize quality and productivity of the central factory. The aspir-ations of settlers to land ownership are satisfied, some measure of supervision provided, and settler families provide some of the labour force for the estate itself.

Land for food crop production or alternative sources of income

Settlers will always expect to have some land for food crops. A small garden lot of about 0.1 ha next to the house is a favoured choice. Swampy land may be developed communally for paddy planting or fish ponds. However, food production on a large scale faces several constraints:

(a) If the land is suitable for rubber, it will usually be more prof-itable to plant it with this crop and to buy food with part of the income from the rubber than to use part of the land to grow food crops for subsistence or sale.
(b) Whilst a few farmers might be able to sell surplus food crops, a few hundred face marketing problems, especially as schemes are often located in remote areas.
(c) The hilly nature of land chosen for rubber often presents unacceptable risks of soil erosion if intercropped.

In spite of this, large areas of land on FELDA rubber schemes were allocated for production of food or alternative cash crops; as much as 1 ha per family was set aside, but in most instances was eventually planted with rubber. It is rather surprising that, on rubber and oil-palm schemes in Indonesia, 1 ha of land is still being allocated for food production to each settler family – even though only 2 ha of rubber per family is planned.

Intercropping might be encouraged but, as the most successful schemes have not brought settlers on site until year 3, large-scale intercropping by settlers has been uncommon.

12.6 Tapping

12.6.1 Improved exploitation of old trees

Large areas of old seedling rubber have been abandoned after being tapped until renewed bark became too thin and scarred for further tapping to be practicable, or yields had become too low to be of interest. Some such rubber may be slaughter tapped before replanting but this is rarely done in a systematic way, numerous cuts of varying shape being made through to the wood.

In Indonesia, smallholders have often been prepared to climb and tap unexploited virgin bark higher up the trees; branches as high as 10 m have been tapped, leading to the term 'telephone pole tapping'. Elsewhere, smallholders have been less motivated to tap higher panels, especially because high-level tapping of inherently low-yielding trees gave returns perceived locally as inadequate recompense for the extra work. The advent of ethephon stimulation has changed this and even old seedling trees can give good yields from virgin bark above old, tapped-out panels.

High-level tapping with stimulation

A high panel may be tapped by gouge or tapping knife, from a simple ladder or from the ground with a long-handled gouge; neither system uses the conventional half spiral cut. A tapper working on a ladder taps downwards, making a V-cut to avoid having to lean too far to one side; with a long-handled gouge, a part-spiral cut is made upwards instead of downwards. Upward tapping presents many new problems (Anthony & Abraham 1981): latex flowing from an upward cut tends to flow down wastefully over the tapped panel and a more steeply sloping cut (55°) is needed if most of it is to reach the cup. The procedure not only involves the use of an unfamiliar implement but also unfamiliar muscles, and is at first extremely tiring. Even when the tapper learns to minimize raising his arms, keeping his elbows near the waist, it is still much more tiring than conventional downward tapping. Control of depth of cut and of bark consumption (which distinguish the procedure from slaughter tapping) is difficult. However, smallholders do not need to be concerned with quality of bark renewal, for sustained responses to repeated stimulation would require fertilizer – something that few smallholders would be willing to apply to old trees. Upward tapping is most likely to be planned for about a year before trees are replanted.

Introduction of high-level tapping and stimulation to smallholders requires considerable extension effort. Farms must be surveyed, rejecting those that, because of infertility or neglect, appear unlikely to give a worthwhile response. Stimulation and trial tappings of a few trees on each of a number of candidate farms will allow further selection of those whose well-grown old trees respond well. Training programmes for groups of perhaps 20 farmers can be held on a centrally located farm; it will take several days to teach the unaccustomed technique, and follow-up visits (and some further demonstration tappings) will be needed on each of the trainee's farms.

Abraham and Anthony (1980c) give detailed recommendations but possible systems are many and varied (see Ch. 9) and, if practicable, demonstration farms should also be used for *ad hoc* trials to determine the best methodology for the district. Ethephon might be applied at different strengths (perhaps 2.5, 5 or 10 per cent), to the panel or to the bark above the cut. Applications might be more effective on a band of scraped bark just above the cut or on to vertical strips scraped into the bark using the long-handled gouge. Frequency of application might be at one- or two-month intervals. A half spiral cut made upwards will only be practicable on smaller trees, for the top of the cut may otherwise be too high from the ground. Farmers with large old trees might be advised to open two ¼S cuts and tap them on alternate tapping occasions or by panel changing at each new stimulation. Tapping at d/2 frequency would probably be recommended, increasing to d/1 as replanting time approached. In favourable circumstances an additional cut reopened (and stimulated) on the low panel might be justified during the last month or so. Clearly, the variety of procedures is large but trial results may be assessed quite rapidly and extension agents and research workers have rewarding opportunities to be of immediate and very practical assistance to farmers.

Stimulation and high-level tapping of well-grown unselected seedling trees will produce yields as good as those when the trees were at their best, perhaps 600–800/kg/ha. Farmers may thus expect to gain at least 400–600 kg/ha more rubber than with conventional tapping. In spite of this, they have often been discouraged, not only by the hard physical work, but because much of the additional crop – especially from the first few tappings after stimulation – consists of late drippings. Moreover, not all late dripping is harvested as relatively low-value cup lump: because of the location and character of the tapping cut, some is further down-graded to earth or bark scrap or may never be collected. A further, socio-economic problem arises if the farm is share tapped. The share tapper resents extra effort on yield stimulation that is not wholly for his own benefit; the farmer finances the work and sees much of the profit going to the tapper who traditionally retains all the scrap, in addition to his agreed share of latex.

Any successful programme will depend on enthusiastic and sustained efforts by extension workers. The best chance of success will lie with farmers who do their own tapping and whose old rubber has not been too badly neglected. Farmers will need to help to devise the best systems for their trees. If long-handled gouges prove too tiring to use, ladder tapping might be advocated instead. A V-cut made downwards would be less productive but might be easier to do, incur less spillage of latex, and seem generally a less unorthodox procedure.

12.6.2 Tapping strategies for young rubber

Ideally, a smallholder opening a field for tapping would be given the same recommendations as an estate owner, but these might in some circumstances be unwise and in others unacceptable. Whilst estates generally would now choose to emulate Indonesian procedures and open trees for tapping at a girth of 45 cm (see Ch. 9), smallholder advisers prefer still to advocate a 50 cm girth criterion. This is wise for, once a farmer starts tapping he tends to tap the whole field rather than just those trees that have attained the 'qualifying' size. When advocating moderate exploitation systems the important constraint to recognize is the number of trees available for tapping. A man and wife together can tap up to 1200 trees each day; if they own between 1200 and 2400 trees they may accept an alternate daily system. Owners of larger farms will employ tappers rather than use a lower-frequency $d/3$ system (this is a strong argument against planting smallholdings with seedlings or other cultivars that respond best to low-frequency tapping). If farmers own fewer than 1200–1300 trees they will tap daily. Of course, if they have land planted to other crops, or other opportunities for off-farm employment, then 'daily' tapping will effectively be daily-periodic. In Thailand, for example, it is customary to stop tapping during wintering; this is not only because of extremely dry weather, but also because farmers are busy with the paddy harvest. Similarly, during the monsoon, farmers may try to tap on the few dry days, but they are often occupied with paddy planting. In effect, farms are tapped about 20–22 days per month during 10 months of the year.

Although climate and seasons vary between countries, this frequency of tapping is probably fairly typical and the intensity of ½S d/1 (20d/30, 10m/12) tapping is (in the old-fashioned notation) only 111 per cent – much less intense than a 'd/1' system implies. The Thai farmer commonly reduces tapping intensity by tapping ⅓S instead of ½S; this has given good yields during the early years, and appears to conserve bark reserves, but the yield of panel B03 has generally been most disappointing. If the intensity of ½S d/1 periodic tapping results in declining yield trends and depleted bark reserves, these effects can be mitigated by short-cut systems with yield stimulation. Abraham and Anthony (1980c) recommend dividing the ½S panel and tapping ¼S d/1 periodic with panel changing on each tapping occasion. In Malaysia 3.3 per cent ethephon would be applied to the cut each month from June to January each year, with no application during or immediately after wintering (applications would also cease during periods of monsoonal rains). Smallholders could minimize day-to-day variations in yield, and in proportions of lower grades, by stimulating half the farm each fortnight and, if driven by economic necessity, tap both panels 2 × ¼S d/1 without stimulation during wintering.

There has been understandable caution in recommending yield stimulation procedures for smallholders. Abraham *et al.* (1981) have reported declining yield trends, reduced d.r.c. and increased incidence of dry trees on smallholdings stimulated regularly with ethephon. They urged restraint in exploiting the trees and made extremely cautious recommendations, especially for smallholders who find it necessary to tap daily. However, as more experience of stimulation was gained, Abraham and Hashim (1983) felt able to recommend a ½S d/1 2d/3 tapping system with twice yearly application of 2.5 per cent ethephon during the first six years tapping on panels B01 and B02. During the next six years the concentration of ethephon used could be increased to 3.3 per cent, applied three times yearly on panel B03 and six times a year on panel BO4. During the next five years panels B11 and B12 would be tapped ¼S d/1 (t,t) 2d/3 8m/12 with four applications of 5 per cent ethephon and 2 × ¼S d/1 2d/3 4m/12 (during wintering) without stimulation.

Where bark consumption has been excessive, and bark renewal poor, it may be necessary for the farmer to adopt high-level tapping and stimulation. Depletion of bark reserves is most likely to arise on smaller farms and a daily system will be preferred; Abraham and Hashim (1983) suggest tapping upwards ¼S ↑ d/1 2d/3 8m/12 with four applications of 5 per cent ethephon; the poorly renewed bark of panels B13 and B14 could be tapped downwards 2 × ¼S d/1 2d/3 4m/12 with a single application of ethephon. The most important proviso with these recommendations was that, although tapping was nominally d/1, there should not be more than 20 tappings per month.

12.7 Processing and marketing

In Malaysia, Thailand and Sri Lanka most smallholders make their latex into sheet; some is dried in a smokehouse and sold as ribbed smoked sheet (RSS), but most is sold off the farm as partially dry unsmoked sheet (USS) and processed to RSS elsewhere. The procedure is quite simple; latex is diluted with an equal volume of water (reducing the d.r.c. to 15–20 per cent) and passed through a sieve to remove dirt. It is poured into shallow coagulating pans, each holding about 4–6 litres of diluted latex, and mixed with acid; froth is skimmed from the top of the pans and the mixture left to coagulate. Formic acid is recommended; 6 ml of acid, diluted in 0.6 litres of water and added to each 10 litres of diluted latex, would coagulate it in a few hours. An estate planning for coagulation to take place overnight would use less acid, a smallholder – who would not wish to wait more than half an hour for coagulation – would add rather more (and contrary to advice would probably use the cheaper and more readily available sulphuric acid, purchased at battery acid strength and diluted an arbitrary amount). When coagulation is complete, the slabs are removed from the pans and squashed to a thickness of 2–3 cm by pressing with a smooth, round stick. They are then passed through a hand mangle with smooth rollers, the nip being tightened progressively so that much of the serum fluid is expelled and the slabs thinned down to about 3 mm. Finally, it is passed through a second mangle with opposing diagonal grooves on the rolls, imparting the characteristic ribbed pattern and – by increasing the surface area by some 60 per cent – speeding the subsequent drying process. At each stage the coagulum is washed with copious amounts of water, and the wet sheets subsequently hung in the shade to drip and be partially air-dried before being sold or taken to the smokehouse for the drying process to be completed.

12.7.1 Influence of processing on quality of the product

Whether it is produced on small farms or large modern estates, latex tapped from the tree is potentially a premium grade product. Nevertheless, the quality of rubber marketed by smallholders usually falls below that of estates. To understand why this is, and to seek improvements, it is necessary to examine each stage of collection

Fig. 12.1 Smallholders passing sheet coagulum through a hand mangle (Rubber Research Institute of Malaysia)

and processing and see how faults arise on smallholdings, and the simple steps needed to correct them.

Crop collection

Down-grading of the product may start if the latex flows down a tapping spout improvised from a fallen leaf and into an old and dirty half-coconut shell. Spillage from the collection bucket may be reduced by floating a handful of leaves on top; share tappers – who usually receive all lower grades collected (cup lump, tree lace and earth scrap) as well as half the latex – will maximize their earnings by floating the scrap in the bucket so that it soaks up as much latex as possible. Before coagulation and processing, leaves, twigs and scrap can easily be removed by straining the latex through a sieve, but the farmer often has to make do with a handful of rice straw. At worst, latex may be poured into a hole in the ground and co-agulated with tapioca extract. However carefully raw rubber is processed, be it into sheet or block form, dirt present at time of coagulation is extremely difficult to remove – and dirt content is the most important single property that concerns the consumer. Improving quality is easy, but requires an investment that seems large to a poor farmer; to get the latex to his farmhouse in clean condition he needs:

1. Tapping cups of earthenware (or varnished coconut shell).

2. Metal spouts and wire hangers to hold the cups in place.
3. A number of good, clean buckets.
4. A pair of fine-meshed Monel metal sieves.

If he is starting to tap trees that have matured after replanting, he will be at the end of a period of minimal income; unless he receives financial help (perhaps as part of the final instalment of a replanting grant), he may be quite unable to afford all these items.

Processing on the farm

If the farmer is planning to make his rubber into good sheet he will need a source of clean water, a clean work-bench (preferably of concrete covered with glazed tiles), coagulating pans (preferably of aluminium), and a pair of mangles. His work-place should have a concrete floor and this is best located under a corrugated metal roof; he will need somewhere shady to hang the sheet to drip. In practice, he will probably make do with old outworn equipment and utensils, and do the work on an earthen floor sheltered by a thatched roof. He will probably hang the sheets in the sun (with possible detriment to their physical properties) and, if his home is near a paved road, he may lay them there to be further dried by the passage of vehicle tyres.

Group processing

Considerable efforts have been made in Malaysia and Sri Lanka to assist smallholders to improve rubber quality by organizing groups to share the simple facilities needed. A typical group processing centre (GPC) would have a roofed area of 12 × 6 m with a concrete floor, a nearby well, concrete work-benches covered with glazed tiles, a supply of coagulating pans, and two sets of mangles (each set comprising two with smooth and one with grooved rollers). The work-shelter may be protected against theft by wire-mesh walls. In Malaysia the first GPCs were constructed and operated privately, participant farmers paying a small fee (in cash or kind) in return for use of the facilities. Most also provided smokehouse services at a charge (in the 1970s) of $M35–40/t. Public support for GPC development came initially through technical assistance by RRIM; about 20 farmers would share the facilities and take turns at caring for the equipment. During 1962–64 interest-free loans for GPC construction were provided by the Rural Industries Development Authority (RIDA, in translation Majlis Amanah Ra'ayat or MARA). Subsequently a grant of up to $M2500 was made available by MARA – usually adequate in the 1960s to build and equip a GPC. By 1974 1050 GPCs had been built at a development cost of over $M2 million; they were used that year by 22,000 farmers who processed 18,400 t of rubber produced from an estimated 40,000 ha of smallholdings (Table 12.2; Lim 1976).

Fig. 12.2 A group processing centre for smallholders (Rubber Research Institute of Malaysia)

Sheet produced at GPCs was usually of quite good quality and could easily be graded, after smoking, as RSS 2 or 3. (It is not easy to produce RSS 1 from smallholders' USS, for the relatively high concentration of acid used tends to cause bubbles to form in the coagulum; these do not affect physical properties but detract from visual appearance of the finished sheet.) Members of GPCs get a better price for their USS, but it is unlikely that the improvement would be more than the differential between RSS 2 and 3 and thus inadequate to justify the fully amortized cost of GPC construction. However, in addition to the financial benefits to the primary producer, a nationwide upgrading in quality of smallholder rubber to the same level as the 4 per cent of Malaysian smallholder production processed through GPCs, would bring further economic benefits. Rubber quality would be less variable and the consumers' 'image' of natural rubber as a product of peasant agriculture removed.

In 1974 RISDA assumed responsibility for smallholder development activities and sponsored the expansion of a number of GPCs to more ambitious 'village development centres' with broader development objectives. Some of these were equipped with small motorized sheeting batteries, but although these reduce time and effort required for making USS, many of the beneficiaries are share tappers rather than the smallholders themselves. The social benefits have proved popular with villagers, but the expense of equipment

Table 12.2   *Smallholders' production by mode of processing and marketing, 1974*

Mode of processing and marketing	Production (t)		Proportion of total (%)	
Group processing:				
Public GPCs*	18,400		2.30	
Private GPCs[†]	12,000		1.50	
Total		30,400		3.80
Central processing:				
MARDEC factories[§]	68,700		8.58	
FELDA factories**	6,000		0.75	
Private factories[‡]	55,000		6,87	
Total		129,700		16.20
Individual processing[§§]		640,900		80.00
Total production		801,000		100.00

Source: Lim (1976)
* Over 90% was USS.
[†] About 85% was RSS.
[§] Includes part of FELDA production.
** SMR grades processed in FELDA factories.
[‡] Includes RSS, latex concentrate and SMR grades. It also covers the greater part of FELDA production.
[§§] Includes scrap and sheet rubber, predominantly USS, processed by smallholders in the traditional manner.

and infrastructural support is unlikely to be cost-effective in improving quality of rubber.

Production of ribbed smoked sheet

It is the need to smoke the sheet that makes it difficult for small-holders to gain the full benefit of improvement in quality of the product. Many individual farmers and GPC members wish to sell their rubber as quickly as possible; smoking takes a week or more, so they sell USS to a village dealer (who, in Malaysia, may also operate a smokehouse). Even if farmers are able and willing to wait for their money, it is difficult to produce the better grades of RSS in small, individually owned smokehouses. Larger group smoke-houses have been built but, whilst capable of producing good quality RSS, have rarely operated successfully.

A survey in west Malaysia by Rama Rao *et al.* (1982) of 135 smokehouses of 400 kg capacity revealed a number of practical problems. The smokehouses had been built during 1965–73, at a cost of nearly $M15,000 each, with grants from RIDA, subsequently MARA; most had ceased working in less than two years. They concluded that such smokehouses could not operate economically, even without capital amortization, either on-farm or as part of a GPC. A 6 t capacity was suggested, but such a smokehouse would cost $M101,000 and require 350 ha of rubber area (perhaps 140 smallholders or 7–10 GPCs) to feed it. Groups as large as this are less likely to operate harmoniously than smaller processing groups; credit would be needed, not just for smokehouse construction, but also for farmers needing cash whilst their sheet dried. The work of operating (and guarding) the smokehouse would be too difficult to organize with shared labour – paid labour would be needed – and the organization, accounting systems and legal status of a properly constituted cooperative would become necessary. One other serious hazard exists; smokehouses are very liable to be destroyed by fire. Whilst insurance is possible, it is an expense that only some small-holders will be willing to incur; unanimous agreement of a group is difficult to obtain. If, however, a smokehouse does burn down, an acrimonious dispute ensues and it is probable that no cooperative has ever resumed operation once a smokehouse fire has occurred.

12.7.2 Marketing

Processing and marketing are interdependent activities; most small-holder sheet is sold unsmoked. In Malaysia smoking is usually done at the village level; in Sri Lanka, most rubber is smoked before being sold to official government rubber purchasing centres located in the rural areas. In Thailand the marketing system is quite different, the market chain extending from itinerant traders who may actually 'buy at the farm gate', through village and small town dealers to the end of the chain where almost all sheet is smoked – by the trader who operates very large smokehouses, grades, packs and exports the sheet as RSS.

Group marketing

Attempts to establish GPCs in Thailand were unsuccessful for reasons that appeared to be social rather than practical. Thai farmers appeared to be less willing than those in Malaysia or Sri Lanka to process their latex with cooperatively owned equipment; they preferred to do the work at home and sold their USS and lower-grade rubbers at intervals of 10 days or more. At village level there was no market for RSS, and problems of improving quality and getting better prices were at least as difficult as elsewhere. Dealers could not easily distinguish between 'good' and 'poor' USS, and – as they did no smoking – were unwilling to offer any premium for quality. Worthwhile progress has been made by encouraging the formation of group marketing organizations (GMOs). These are simple associations of 20–40 farmers who agree to meet on prearranged days, usually at a conveniently located member's house, pool their produce, and offer it for sale by private tender to the various dealers in their nearest village. No credit is provided, although in some instances equipment is provided on loan; platform scales and a simple calculator are all that is needed. The Rubber Research Institute arranges regular radio broadcasts of rubber prices and GMO members have been taught to calculate the discounts they may expect to face for transport cost, water content of USS, etc. A small immediate benefit is gained because errors from weighing many small lots of rubber (always in favour of the dealer) are eliminated. Farmers grade their own USS before sale and, offering it in quantities of 2–4 t, find that they are able to bargain for a small premium. Those whose produce is of poor quality see themselves placed at a disadvantage and quickly learn how to improve processing methods. Groups are soon able to offer batches of rubber for sale that are of a uniform high quality, and dealers are interested in making competitive bids.

Fig. 12.3 Smallholders carrying in sheet for bulk sale (Rubber Research Institute of Malaysia)

One of the main attractions of the GMO movement has been its informal simplicity – almost no capital investment required, no credit requirements and a minimum of book-keeping. Disadvantages have included the need for technical assistance during the first 6–12 months of operation, some unwillingnesss of farmers to join a group where all members gain a knowledge of personal finances and – an unexpected factor – once a local market for premium quality rubber had been created, those who did not wish to join often found that, if they produced USS of similar quality, they could secure privately the same prices as those offered to GMO members.

Central processing and marketing

Central processing of smallholders' rubber utilizes facilities similar to those of estates or large remilling companies (Ch. 11), but purchase and collection of latex from smallholders requires considerable infrastructural support. Collecting stations need to be established and equipped with means of weighing latex and determining d.r.c.; agents need to be paid, and vehicles and drivers provided. A method of paying producers must be devised that allows for spoilage of batches after purchase, and offers some protection against problems of adulteration. This is probably only justified if the central factory can produce and market a premium grade product such as latex concentrate or viscosity stabilized rubber – which is more difficult to achieve with variable smallholder latex than with more stable and consistent raw material from estates.

Most smallholder rubber in Indonesia is coagulated and sold in slab form; the procedure has certain merits: if the coagulating pit is lined with cheap plastic sheet, the latex is passed through a sieve before coagulation with acid, and the coagulated slab is kept clean during transit, there is no reason why a factory should not be able to process it into SNR 20 block rubber. Collection and processing is simpler and cheaper, and savings may well offset price advantages of premium grade rubbers. A major disadvantage lies in the difficulty of determining d.r.c and of using it as a basis for payment that is equitable to both producer and processor.

In a recent review, Lim (1985) advocates a unified and more extensive collection of USS from Malaysian smallholders, the sheet to be smoked at central smokehouses or – subject to market demand – comminuted and incorporated in a general-purpose block rubber. A quasi-government organization such as MARDEC would be expected to equal or exceed prices offered by village dealers and possibly absorb as a social cost the higher transport expenses of rubber produced by smallholders in remote areas.

12.8 Investment in rubber planting and replanting

Any discussion of smallholder rubber development would be incomplete without mention of the financial implications of investment in such a long-term crop. The subject is far too complex to deal with in a few paragraphs, but agronomists must be able to understand the questions that will be raised by economists when planting proposals are appraised.

Internal rate of return to investment (IRR) is derived from calculations of net present value (NPV) and net present cost (NPC) of future investments. At interest rates of 15 per cent per annum, $1000 invested today will be worth $2011 in 5 years' time; conversely the discounted value today (NPV) of $1000 profit expected in year 5 is $497, the NPV of a similar profit in year 32 is only $11.42. The IRR is the average per cent discount which, applied to a series of costs and returns, yields a NPV (or NPC) of zero. To be financially attractive the IRR needs to be several percentage points higher than prevailing interest charges. It is easy to see why rubber, with an immature period of 5–7 years, and much of its yield harvested two to three decades later, can only give a satisfactory return to investment if it is both high yielding and cheap to produce. Costs of actual field establishment may be only half the total; staff recruitment and training, equipment, civil works and other infrastructural support for the project will also be included and are incurred before any benefits start to accrue. The value of a short immature period and of high yield during the early years of tapping are apparent. Although the discounted value of returns during the latter years of a project are low, the IRR of a rubber project may be reduced by as much as one percentage point if tapping life is curtailed by five or six years.

In assessing benefits of any project it is necessary to consider what might happen if no investment were made. In the 'without project' situation would the rubber continue to be tapped with low and diminishing return (if so, for how long?) or would it simply be abandoned? Does the old rubber have any residual value as timber or firewood, could the land be used for other crops if the rubber is abandoned? Even if much of the planting is to be done by family labour, in appraising the project, some cost must be attributed to these inputs; this might be estimated at full commercial rate or a lower 'opportunity cost' if there is much rural unemployment. It is necessary also to make a clear distinction between the financial and economic costs and benefits. Financial costs and benefits are assessed from the point of view of the individual making the investment; in the economic analysis we look at costs and benefits from the point of view of the society as a whole. For example, the farm-gate price a farmer gets would have a financial value lower than the economic value if export taxes were levied or if the crop were sold to a central purchasing agency at a controlled price. Similarly, the financial cost of fertilizer would be lower than the economic price if the farmer receives a subsidy; the economic price of a tractor might be lower if the farmer had to pay very high import duties and taxes.

Some of these problems are illustrated by yield projections made during preparation and appraisal of proposals for a second phase of a replanting programme in Thailand (Office of the Rubber Replanting Aid Fund and the Food and Agriculture Organization 1980). With clones such as RRIM 600, tapped ½S d/2 10m/12, fertilized according to recommendations, and stimulated during the last five years, yields were expected to average 1500 kg/ha over 25 years of tapping, or – with a more realistic allowance for a decline in number of tapped trees per ha – to average 1350 kg/ha (Table 12.3). In the event that smallholders might not apply fertilizer to mature rubber, yields might be expected to fall to an average of 1200 kg/ha, and – if tapping were intensified to ½S d/1 20d/30 10m/12 – tapping life could be expected to be reduced to 19 years and average yields to 970 kg/ha.

Financial implications of these yield projections are illustrated in a simplified way if the PV of each is calculated and expressed in terms of total quantity of rubber produced over the tapping life of the trees. One hectare, tapped and fertilized according to recommendations throughout maturity might yield 37.5 t of rubber during a tapping life of 25 years; at an interest rate of 15 per cent this would have a discounted value of 8.7 t at time of opening for tapping. The 18.4 t expected in the most pessimistic scenario would effectively be worth only 6.3 t. However, the extra rubber produced would be worth less – in PV terms – than the discounted cost of the fertilizer required to achieve the higher crop and longer tapping life. Whilst lower inputs and more intensive tapping might benefit the farmer more (for as long as his trees are in tapping), the government would appear to lose quite substantial benefits in terms of export earnings and taxes. It would be fanciful to suggest that the smallholder understands the implications of discounted cash flow calculations, but the inherent shrewdness of the peasant farmer should not be underestimated. Whilst there is a strong likelihood that expenditure on fertilizer would increase yields in both the immediate and long-term future, some uncertainties exist (trees may not respond as expected, or may be killed prematurely by disease or wind damage); if a farmer has any spare cash available he may make higher and more immediate profit by lending it at prevailing interest rates to less fortunate neighbours, and he may perceive this as less risky!

Actual yields of replantings made on about 400 farms have been assessed (Rubber Research Institute of Thailand 1983) and used to prepare yield projections that may be more soundly based than the theoretical yield profiles used for project preparation (Table 12.3). It appears that yields obtained during the early years (mainly by

Table 12.3   *Projected yields of a field of replanted rubber*

Year	(1)	(2)	(3)	(4)	(5)
8	500	500	500	500	1,040
9	800	750	800	750	1,400
10	1,190	1,000	1,030	1,000	1,530
11	1,400	1,200	1,250	1,200	1,660
12	1,600	1,400	1,420	1,350	1,670
13	1,800	1,500	1,520	1,450	1,670
14	1,800	1,600	1,560	1,500	1,670
15	1,800	1,700	1,580	1,500	1,670
16	1,800	1,800	1,560	1,375	1,670
17	1,800	1,800	1,530	1,225	1,580
18	1,600	1,750	1,350	1,100	1,470
19	1,600	1,700	1,310	1,000	1,380
20	1,600	1,650	1,270	900	1,280
21	1,600	1,590	1,230	800	1,190
22	1,600	1,530	1,200	700	1,110
23	1,500	1,470	1,150	625	1,030
24	1,500	1,410	1,110	550	960
25	1,500	1,350	1,070	475	1,450
26	1,500	1,300	1,030	400	1,120
27	1,500	1,250	990	0	820
28	1,500	1,200	1,230	0	0
29	1.500	1,150	1,200	0	0
30	1,500	1,100	1,170	0	0
31	1,500	1,050	1,050	0	0
32	1,500	1,000	950	0	0
(a)	1,500	1,350	1,202	968	1,369
(b)	37,490	33,750	30,060	18,400	27,370
(c)	8,669	7,972	7,461	6,345	9,041

Notes:
Planting is done in year 1.
(1) Yield of currently recommended clones fertilized according to
    recommendations; planted in an average situation and tapped $\frac{1}{2}$S d/2 10 m/12,
    150 tappings per year, stimulated for the last 5 years.
(2) A more conservative estimate of (1), based on projected yields in terms of
    gm/tree/tapping, trees tapped as in (1), but with tappable trees declining from
    500 to 288 per ha during 25 years of tapping.
(3) As for (1) but with no fertilizer applied during maturity.
(4) A more conservative estimate of (2), trees tapped $\frac{1}{2}$ S d/1 20 d/30 10 m/12,
    200 tappings per year. Tappable trees declining from 500 to 250 per ha during
    19 years of tapping.
(5) A revised forecast based on a survey of results of the first 10 years of tapping
    399 replanted fields (Rubber Research Institute of Thailand 1983).
(a) Average yield per year of tapping.
(b) Cumulative yield (kg rubber per ha) for duration of tapping life.
(c) NPV (kg rubber per ha) at time of opening of cumulative yield (a) discounted
    15 per cent per annum from time of opening for tapping.

tapping ⅓S d/1) have been much higher than expected. Averaged over a shorter 20-year tapping life, returns to the farmer – whether discounted or not – should be better than forecast. The implications of a shorter replanting cycle for governments and their replanting organizations are, however, rather far-reaching – benefits of quicker returns from cess levies and export taxes having to be balanced against the much larger proportion of rubber area that might be immature at any one time.

12.8.1 Success and failure of settlement schemes

Not all settlement schemes have been successful; failures have occurred, perhaps when settlers have been allocated uneconomically small plots, have been brought on to the land too soon, or have lacked the resources necessary to plant their rubber or bring it to maturity. They may have been located in remote areas, lacked prior experience of the crop or received inadequate technical support. In Malaysia the Federal Land Consolidation and Rehabilitation Authority (FELCRA) was established to remedy such situations. In effect FELCRA has taken over some 42,000 ha of rubber planted on fringe development and other unsatisfactory schemes, rehabilitated the plantings and operated them on the lines of commercial estates. Hired labourers have been employed, most of whom are not original participants – the latter receiving a share of profits earned by the enterprise.

The success of this programme has led to the suggestion (Malaysian Rubber Research and Development Board 1983) that future schemes might be operated *ab initio* on the semi-commercial lines adopted by FELCRA. On the other hand, Lim and Noor (1974) have compared the economic implications of a *laissez-faire* policy for unsatisfactory schemes (anticipating several years' delay in the trees reaching tappable size, coupled with long-term reductions in yield potential) with those of an effective but expensive rehabilitation programme. They concluded that, if finance for land development was a limiting factor, it was better to allocate it to new, properly planned and managed schemes than to put further investment into rehabilitation programmes. Settlers on unsatisfactory schemes might be seriously disadvantaged in comparison with those on better run, new ones, but would be far better off than if they had not been assisted at all.

The general principle of allocating public money to land development schemes has sometimes been criticized; Peacock (1981), discussing the large proportion of national development funds allocated to FELDA in Malaysia, has argued that such programmes have been too expensive. By 1972 the marginal cost of settling each family had risen to $M33,935; he considered that these high costs precluded the expansion of such programmes to make any real impact on rural poverty. The pace of settlement was lower than the rate of rural population increase. Benefits might be more broadly based if a more liberal policy of land alienation permitted settlement by individual farmers on their own free initiative. Criticisms such as these may be applied to programmes in more than one country and illustrate the difficulties of optimizing social benefits and national economic development. However, arguments against successful though costly land development schemes tend to ignore or minimize the other economic benefits of such development on other sectors of the rural community and on national economies as a whole, such as improved communications, health, education and welfare facilities, the development of local service industries, and increased government revenue from export duties.

12.9 Development programmes for smallholders

Most of this chapter has been concerned with replanting old rubber; together with the attention given to rubber on land settlement and development schemes of various kinds, a disproportionate amount of attention is concentrated on problems of young rubber. There are few places where owners of small farms with mature high-yielding rubber are able to call on adequate extension service support. In most instances this is because there are just not enough skilled men; those that are available must concentrate on the immature rubber, if only to safeguard the enormous public sector investments in replanting or new planting. Additionally, those responsible for supervising replanters or participants in settlement schemes are usually members of organizations quite separate from the body designated to provide an extension service. In some countries supervision of rubber affairs has been transferred from a ministry of agriculture to a body more directly concerned with industrial or export crop production; whilst ensuring that specialist rubber staff are available, this has sometimes separated the rubber producer from modern developments in agricultural extension methods. On the other hand, where agricultural extension is provided by a specialized service, the extension agent is expected to assist development of all crops grown in his region, and is hard pressed to acquire enough technical knowledge to supplement his expertise in extension methodology.

12.9.1 Extension services

It has been argued that rubber presents unique problems and requires a specialized extension service for the following reasons:

(a) The propagation, planting and maintenance of rubber during its immature period requires specialized advice.

(b) Pest and disease problems are generally unique to the crop.

(c) Tapping requires skills that are unique to the crop.

(d) The crop is not harvested seasonally, tapping continues almost throughout the year, and extension service support may be needed at any time.

(e) Rubber is an industrial crop requiring daily processing on the farm, and consequently expertise in elementary rubber technology is needed.

(f) Marketing raw rubber presents special problems, especially as most of the crop is exported in its raw form and must conform to internationally recognized standards.

The pace of technological change has been such that there are strong arguments in favour of placing an extension service under the direct control of a multi-disciplinary single-crop research institute. This arrangement prevailed at the Rubber Research Institutes of Malaysia and Sri Lanka, but in both instances arguments for a specialized extension organization, coupled with problems of administering both a research institute and an increasingly large extension division have led to administrative separation of extension from research (although care has been taken to maintain lines of communication between advisory and research staff). Created in 1973 to combine the administrative functions of the Malaysian replanting organization with the extension activities of the research institute, the Rubber Industry Smallholders' Development Authority (1983) reported a staff of 1520 extension agents, allowing allocation of one to every 350 smallholders. However, it appears that problems are encountered in concentrating the activities of such a large staff on transfer of innovations to farmers; Tugiman and Said (1983) report that much effort is expended on providing credits and subsidies, and a recent government policy review has proposed that serious consideration be given to making the Rubber Research Institute of Malaysia, once again, the main agency for providing an extension service to smallholders (Malaysian Rubber Research and Development Board 1983).

12.9.2 Project-oriented activities

Some of these problems may be overcome by channelling innovative techniques to farmers through project-oriented activities; research findings may be confirmed by pilot-scale trials executed on farms by research institute staff, then expanded to a larger scale by direct involvement of the extension service. This system gives useful down-to-earth experience to the researcher and keeps the rural extension worker in touch with developments elsewhere. Choosing topics for development projects is deceptively difficult and success may depend as much on the intuitive skills of the extension or research worker as on any formal socio-economic evaluation. The latter method might reveal the differing chances of success of GPC and GMO programmes in Malaysia and Thailand, but would not have foretold the entirely unexpected enthusiasm with which the Thai farmer learnt green budding – a procedure whose anticipated benefits are not gained for a number of years.

By contrast, persuading Thai smallholders to adopt high-level tapping and use of ethephon yield stimulant has been much more difficult – even though the evidence of increased yield can be seen in the tapping cup the day after treatment is applied. A development project of the Rubber Research Institute of Thailand (1984) showed conclusively that yield of old rubber could be increased by 50 per cent; yield stimulant applications cost about $US0.95/ha month whilst additional crop was worth 30–40 times that amount. Demonstrations on nearly 4000 farms have persuaded large numbers of farmers to accept the technique, but the project has taken several years of carefully planned effort by scores of staff, and many farmers still find the new tapping method too unfamiliar and arduous – even though the potential increase in income is about equal to the local average daily wage.

12.9.3 Training

Training programmes should be tailored precisely to the development project being promoted; they need much forward planning, prior study of the target audience, and field testing of methods to be used. At least 70 per cent of the training should be practical work in the field; when lectures are necessary, reliance on the printed word should be avoided. Audio-visual materials are essential but need to be selected with care. Movie film is popular but is generally expensive and too inflexible – it cannot be revised or updated. A cheap and effective substitute is provided by lectures based on sets of colour slides supported by tape-recorded commentary. Lessons can be changed by substituting individual new slides, and the commentary changed at will. An additional advantage is that identical slide sets may be provided with commentaries in different languages or dialects, and quality of presentation of the lecture does not depend on the presentational skills of the instructor; the latter needs only to be conversant with the technicalities of the subject so as to be able to answer questions and lead discussion when the slides are run through a second time to reinforce the lesson's message. Local radio can be very useful for delivering seasonal advice to farmers, and for regular reporting of market information, but colour television is perhaps the most popular and effective audio-visual medium. Operating costs of videotape equipment are low, once the initially expensive cameras and editing machines have been bought, and regional television stations are usually eager to accept short, locally produced instructional items. Whilst many farmers lack electricity and cannot afford television sets, the local coffee shop is usually quick to introduce a new attraction for customers; those who come to watch a football or boxing match will happily stay to watch and discuss short instructional films afterwards.

Training programmes at residential schools have proved very effective, especially when a single specialized topic such as budgrafting or tapping has been presented. Farmers are unlikely to be willing to attend courses that take them away from home for periods longer than one week, and programmes that can achieve useful results within so short a time need very skilled preparation. Buildings and furnishings need not be elaborate, although an electricity supply is essential if audio-visual equipment is to be used. Enthusiastic attendance will usually require public funds to pay for bus fares and to buy food for the participants. Social and educational benefits arising from bringing together farmers from widely separated and isolated communities are difficult to quantify, but are very real; most participants retain lasting memories of their experience and group photographs taken at the end of a course will still be found, years later, displayed in many farmhouses. A certificate of attendance is also popular, but any examinations given should be of an oral and practical nature and should be designed to test the performance of the trainers and not used to make invidious comparisons between the trainees.

For very remote areas, or communities where the farmers find it difficult to leave home, mobile training units have proved a valuable alternative to residential schools. Each vehicle, crewed by two or three extension agents, can be equipped with all the necessary audiovisual hardware and software stowed in specially constructed lockers. With a 4 kW generator, a field classroom or open-air cinema can be almost as effective as a school building. The vehicle also provides all the resources needed at agricultural shows or village fairs. Before selecting the basic vehicle it is important to assess just how often four-wheel-drive (f.w.d.) facilities will be used. Three conventional vans can be purchased for the cost of two f.w.d. vehicles; bearing in mind that the rainy season is unsuitable for outdoor classroom work, the cheaper vehicle may be adequate and on good roads its better springing may be kinder to the electronic apparatus it is carrying. In Thailand, extensive field testing of equipment and of vehicle modifications to house it, proved well worth while; vehicles have carried projectors, amplifiers (and even videotape equipment and television monitors) over many thousands of kilometres with few problems.

12.10 Conclusion

Forecasting future trends in smallholder rubber development may be almost as difficult as forecasting rubber prices. However, it does appear that – just as the smallholder share of rubber production is steadily becoming more important than that of estates – the role of the individual smallholder may be on the decline. Some settlement schemes continue to envisage individual ownership of farm lots, perhaps located around nucleus estates, but the trend towards more centralized management accompanied by shared land ownership and profit seems clear. The introduction of the 'block system' on FELDA schemes in Malaysia – where groups of 24 settlers work on, and eventually assume joint title to, 100 ha of land – looks like being a socially acceptable compromise between outright individualism and the type of public management used to operate FELCRA schemes or RISDA mini-estates. The 'block system' has interesting parallels with the operation of production brigades on state farms in China, where the new 'responsibility system' is giving considerable freedom of action to individual brigades whose members are spurred on by profit incentives short of actual land ownership.

The problems of improving small, individually owned rubber farms continue to be intractable. Block replanting, currently proceeding on a substantial scale under the guidance of project management units in Indonesia, may prove an effective (although expensive) way of returning large numbers of small producers to prosperous rubber production. In some countries, however, a really small farm planted even to high-yielding rubber will not give a large enough income for family aspirations, and conversion to other crops or a drift away from farming to more urban activities may continue. An increasing number of farms may be planted outside the traditional rubber-producing areas; Thailand, for example, plans to replace 130,000 ha of cassava with rubber in the east of the country (Krisanasap 1984). Where climate is less than optimum, rubber may require better soils to grow satisfactorily, but lower yields may still be adequate for family needs if individual farms of 4–6 ha are practicable, and especially if rural poverty has made the hard daily work of rubber tapping and processing appear acceptable and the farmer perceives the potential income as attractive.

Appendix 1.1

Literature of the history of rubber to 1945

This appendix supplements those parts of the text which are too condensed to contain all the relevant works cited in the References.

General

History of the rubber industry (Schidrowitz & Dawson 1952) contains a bibliographic chapter which covers rubber literature, including very early history. The United States Department of Commerce and the American Rubber Association published major surveys of the industry in the 1920s and 1930s (see below).

Amerindian civilizations and colonial South America

Many articles by Schurer cover these subjects. Schultes (1956, 1977a) covers some botanical aspects of early exploitation.

Eighteenth-century investigations

Many scientists other than those mentioned in the text were involved. The work of de La Neuville (1723), Barrière (1743), Macquer (1770), Aublet (1775), de Fourcroy (1791), Grossart (1791) should be mentioned, de Fourcroy, Grossart and Macquer for their laboratory research. Porritt (1926), Schidrowitz and Dawson (1952), and Schurer in several papers, review the subject.

Manufactures

Schidrowitz and Dawson (1952), Allen (1972), Stern (1982) and other writers mentioned in the text, cover this subject. Encyclopaedias are useful for additional detail. United States Manufacturers were described, for the US Department of Commerce, by Barker and Holt (1939, 1940).

Wild rubber

This subject is documented in the text; Schurtz *et al.* (1925), Whitford and Anthony (1926) and Wren (1947) are important for the Amazon and Africa. For the scandals of the 1900s (concerning inhuman exploitation of gatherers) refer to Morel (1919) who worked with Roger Casement of the Foreign office on the Congo atrocities up to 1905, to Inglis (1973), and to Hardenburg (1912) and Parliamentary Papers (1912–13) on comparable atrocities committed in the Rio Putamayo region by the Peruvian Amazon Company (London), reported in the press by the American explorer Hardenburg in 1909 and also investigated by Casement.

Statistics

Nineteenth-century statistics were reviewed by Wallace (1952) and early twentieth-century statistics have been assembled by McFadyean (1944), McHale (1964), Drabble (1973) and Barlow (1978).

The first uniform world statistics were collected by the Rubber Growers' Association in its *Bulletin*. This work was transferred to the International Rubber Regulation Committee in 1934 and to The International Rubber Study Group from 1944. The source book for 1900–37, covering absorption, stocks, production and prices, is a US Department of Commerce publication by Barker and Holt (1938).

Local data on areas planted, labour, etc., were originally published in government publications. A formidable review of these data was made by Figart (1925) for the US Department of Commerce and later by Whitford (1928–32) for the Rubber Manufacturers' Association of America.

Appendix 2.1

Clone identification and nomenclature

Identification of clones

Clones differ in characters of economic importance – such as yield level, growth vigour before and during tapping, bark thickness, colour and dry rubber content of latex, wind and disease resistance, but these are of little value for identification purposes. Each clone has a characteristic seed shape and pattern of markings on the seed coat which enable it to be identified with certainty if its seeds can be compared with those in a reference seed collection. Obviously, this is only possible if the seed of a clone in question is included in the standard seed collection and if the trees are old enough to bear fruit, whereas the need for clone identification commonly arises with young, immature buddings. With the latter, identification depends on differences in a number of botanical features and can only be done by trained clone-inspectors who have knowledge of these features and considerable experience in recognizing differences of detail between clones.

The following are examples of the kinds of variation occurring in different parts of the plant which are useful in clone identification.

1. The main stem may be straight, leaning, bending or twisted.
2. Bark: shade and glossiness of the green bark; occurrence, prominence and colour of lenticels; colour and roughness of the 'brown' bark; pattern left by the cracking and scaling of the corky bark.
3. Axillary buds: sunken or protruding; shape of leaf scars.
4. Density and shape (e.g. hemispherical, conical) of topmost whorl of leaves.
5. Leaves:
   (a) Petioles: length, shape (straight, bow-shaped, sigmoid), inclination (upward, downward, horizontal).
   (b) Pétiolules: shape, length, width, orientation to petiole (upwards, downwards, parallel).

(c) Leaflets: standing free of each other or overlapping, depending on length of petiolules and angle between them.
(d) Laminae: shape (elliptical, obovate, broadly lanceolate, etc.); shape of apex (acuminate, aristate, cuspidate) and of base (cuneate, attenuate, obtuse); colour; glossiness; texture; presence or absence of pubescence on veins on underside; margins (wavy or straight).

In identifying mature, tappable trees, other features that may be useful include the cross-section of the trunk (circular or oval) and its colour (brown, reddish, grey); the shape of the crown (conical, spherical, oval) and the density of its canopy; the spacing and angle of the branches; the thickness of the bark and the colour of the latex.

Nomenclature

Clones

There is no official international registration authority for *Hevea* cultivars, but cooperation between breeders and others in various countries has led to the general adoption of a system whereby a clone is designated by letters indicating its place of origin and a serial number assigned to it by workers at that place. Names of more than one word are abbreviated to the initial letters of each word in capitals without full-stop points between them, e.g. RRIM for Rubber Research Institute of Malaysia. Single word names are abbreviated by giving the initial capital letter only, by two letters if both are consonants, e.g. Ch for Chemara, or occasionally by three or more letters, e.g. Pil for Pilmoor. Some exceptions to these conventions are in use, either because they are long established, or to avoid confusion between similar names such as Tjiomas and Tjirandji. A space is left between the abbreviation of the name and the serial number. If the latter is prefixed or suffixed with a serial letter, this is run on with the number, e.g. Pil A44. Where a subdivision of a series is indicated by a number, this is shown by an oblique stroke, e.g. PB 5/63.

The following list gives the names and abbreviations of the places of origin of the better known clone series.

AVROS	Algemene Vereniging Rubberplanters Oostkust Sumatra
BD	Bodjong Datar
Ch	Chemara
Ct	Cultuurtuin
Ford	Ford
FA	Ford Acre

FB	Ford Belém
FX	Ford Cross
Gl	Glenshiel
GT	Godang Tapen
GyT	Goodyear T series
GyX	Goodyear Cross
Har	Harbel
HAPM	Hollandsh Amerikaansche Plantage Maatschappij
IRCI	Institut des Recherches sur le Caoutchouc en Indochine
IAN	Instituto Agronomico do Nórte
LCB	Lands Caoutchouc Bedrijven
Lun	Lunderston
MDF	Madre de Dios Firestone
MDX	Madre de Dios Cross
MAP	Malayan American Plantations
Nab	Nabutemme
PPN	Perusaha'an Perkebunan Negara
Pil	Pilmoor
PB	Prang Besar
PR	Proefstation voor Rubber
RRIC	Rubber Research Institute of Ceylon
RRII	Rubber Research Institute of India
RRIM	Rubber Research Institute of Malaysia
TR	Terres Rouges
Tjiomas	Tjiomas
Tjir	Tjirandji
WR	Wanggo Redjo

Clonal seedling families

Most plantations of clonal seedlings are established with seed resulting from selfing or cross-pollination in isolated seed gardens which have been planted with clones of buddings selected for their ability to give high-yielding seedling families when crossed in all combinations (see Ch. 6). Seed collected from a mixture of clones in such a garden is designated by letters indicating the name of the garden, e.g. PBIG/GG 1 for Prang Besar Isolated Seed Garden, Gough Garden 1. Seed collected from budded trees of one clone in a polyclone planting is denoted by the abbreviation for that clone, e.g. PB 5/51 seed. Seed obtained from an isolated monoclone planting, i.e. resulting from selfing of, or crossing between, budded trees of one clone, is designated by the abbreviation appropriate to the clone, with the suffix M for monoclonal, e.g. PB 5/51 M seed. Such seed may be useful for raising rootstocks for budding (see Ch. 6).

Appendix 9.1

Exploitation notations

The first international notational system describing tapping systems was introduced in 1940 and was used widely, although various research centres introduced minor variations. With the development of many new tapping systems and the inclusion of stimulation in exploitation of the tree, the existing tapping notation became increasingly inadequate. Between 1974 and 1982, a revised system was developed in a series of international meetings. The final version was published by Lukman (1983) and will be briefly described. The new international notation distinguishes tapping notation, stimulation notation, and panel notation. .

Tapping notation

The tapping notation has the following elements:

Type of cut: S = spiral cut, V = V-cut, Mc = mini-cut.

Length of cut: the relative proportion of the tree circumference embraced by the cut is indicated by a fraction preceding the type of cut symbol (for full circumference cuts no fraction is written; for mini-cuts the length in cm is shown after the Mc symbol). Examples: V; $\frac{1}{2}$S; Mc 2.

Number of cuts: If more than one cut is tapped on the same day and these cuts are identical, the number of cuts is shown before the length of cut notation, for example, $2 \times \frac{1}{2}$S.

Direction of tapping: upward tapping is indicated by an upward arrow ($\uparrow$) written immediately after the cut notation. When one of two cuts is tapping upwards and the other downwards, the notation becomes $2 \times \frac{1}{2} \uparrow \downarrow$.

Frequency of tapping: the 'actual frequency' is a fraction of which the denominator indicates the tapping interval: d/1 for daily tapping, d/4 fourth daily (one day in four), etc. Note that twice-a-day tapping should be shown as d/0.5.

Periodicity: often one or more tapping cycles may be followed by a rest period. This is shown by a further fraction, of which the numerator indicates the number of days, weeks or months of the tapping cycle and the denominator the length of the tapping cycle plus the succeeding rest period. The most common example is that of a 'Sunday rest': d/2 6d/7 represents alternate daily tapping for six days (hence three tappings) followed by one day rest. Similarly, d/1 2d/3 indicates tapping on two successive days followed by one day rest (the old notation would have been 2 d/3). Other common examples of periodic tapping systems are d/3 6m/9: third daily tapping during six months followed by three months rest, etc.

Change-over tapping: when there are two (or more) cuts on the tree they are often tapped on alternating tapping days; this is indicated by (t,t) after the frequency. Therefore ½S d/2 (t,t) refers to alternate daily tapping of two half spiral cuts, one cut being tapped on day 1, day 5, etc., and the other on day 3, day 7, etc. It is also possible to tap one cut for a longer period before changing to the other cut; this is indicated by w (weeks) or m (months). For example, (w, 2w) represents one cut tapped for a week, followed by the other for two weeks.

Combination tapping: when two cuts have not the same length or type, a more complicated notation is necessary. This uses '+' to join the two different cuts if tapped on the same day, and ',' for cuts tapped on alternate tapping days.

Tapping intensity: It has been traditional to indicate the tapping intensity after each tapping notation by indicating the relative intensity compared to the standard ½S d/2 = 100% system. The relative intensity is calculated by multiplying all fractions in the formula by 400. Therefore, ⅓S d/1 = 133%; ¼S d/2 = 50%; 2 ⅓S ↓ ↑ d/1 2w/4 6m/9 = 89%; ½S d/3 (t,t) = 67%, etc. In the new revised exploitation notation, the relative intensity is no longer included; this is perhaps unfortunate since it has been shown (Paardekooper *et al.* 1976) that the tapping intensity has an overriding effect on yield. It should be noted that the actual intensity of a given tapping system is often lower because of days lost through rain, sickness, and holiday. In this case the relative intensity can be easily converted into 'actual intensity' by multiplying with the ratio of the actual number of tapping days over the theoretical number in one year.

Puncture (micro-) tapping: The revised international notation does not include puncture tapping or 'micro-X' tapping; a suggested notation for these systems was published by Hashim and P'ng (1980).

Stimulation notation

The revised international notation for exploitation systems also specifies formulae for indication of the type of stimulant (e.g. ethephon = ET), concentration (%), method of application (Pa = panel application, La = lace application), quantity of formulation and band width (e.g. 2(3) = 2g on 3 cm band), and frequency of application (e.g. 8/y(m) = eight applications per year at monthly intervals). An example of a complete stimulation notation would thus be: ET 5%. Pa2(2). 3y(4).

Panel notation

At the time half spiral cuts were standard, the usual notation was to use A and B to denote virgin bark, and C and D for renewed bark. With the introduction of shorter cuts and high cut panel, various other notations came into use. The notation of tapping panels has now been standardized as follows: B indicates low panels, H high panels; 0 virgin bark, I bark of first renewal, II bark of second renewal. The number of subsequent panels of the same type is indicated by 1, 2, 3 or 4 (depending on the length of the cut). In this manner, the first panel opened at, say, 130 cm would be B01, followed by B02 on the other side of the tree, or – in the case of $\frac{1}{3}$S – B02, followed by B03. If subsequent tapping were on a renewed bark, the sequence would continue with BI1, BI2, etc. On the other hand, if a high panel were to be opened, this would be shown as H01, etc. (Although this system appears logical, it could still give problems if the first panel were opened at a lower height, as in seedlings; the third panel would then be half in virgin bark and half in renewed bark which, according to this new notation, should be shown as H01, followed by BI1.)

Appendix 11.1

The international standards of quality and packing for natural rubber grades (The Green Book)

Established by the International Rubber Quality and Packing Committee under the Secretariat of the Rubber Manufacturers Association Incorporated (USA)

The following general prohibitions are applicable to all grades of natural rubber:

1. Wet, bleached, undercured and virgin rubber and rubber that is not completely visually dry at the time of the buyer's inspection is not acceptable. (Except slightly undercured rubber as specified for No. 5 RSS.)
2. Skim rubber made of skim latex shall not be used in whole or in part in the production of any 'Green Book' grade nor for marking patches for such grades.

Ribbed smoked sheets

Nothing but coagulated rubber sheets, properly dried and smoked, can be used in making these grades; block, cuttings or other scrap or frothy sheets, weak, heated or burnt sheets, air-dried or smooth sheets not permissible.

RSS 1X. The grade must be produced under conditions where all processes are carefully and uniformly controlled.

Each bale must be packed free of mould but very slight traces of dry mould on wrappers or bale surfaces adjacent to wrapper found at time of delivery will not be objected to provided there is no penetration of mould inside the bale.

Oxidized spots or streaks, weak, heated, undercured, over-smoked, opaque and burnt sheets are not permissible.

The rubber must be dry, clean, strong, sound and evenly smoked and free from blemishes, specks, resinous matter (rust), blisters, sand, dirty packing and any other foreign matter. Small pin-head bubbles, if scattered, will not be objected to.

RSS 1. As for RSS 1X, deleting 'evenly smoked' and 'specks' and adding after 'foreign matter' – 'except slight specks as shown in the type sample'.

RSS 2. As for RSS 1 but – 'slight resinous matter (rust) and slight amounts of dry mould not exceeding 5 per cent of sheets not objected to' – and – 'small bubbles and slight specks to the extent as shown in the type sample not objected to'.

RSS 3. As for RSS 2 but delete 'clean'. 'Mould and resin allowable to not more than 10 per cent.'

RSS 4. As for RSS 3 but – 'mould and resin allowable to not more than 20 per cent. Medium-sized bark particles, bubbles, translucent stains, slightly sticky and slightly oversmoked rubber are permissible to the extent shown in the sample.'

RSS 5. As for RSS 4 but – 'mould and resin allowable to not more than 30 per cent. Large-size bark particles and small blisters, stains, oversmoked, slightly sticky rubber and blemishes of the amount shown in the type sample are permissible. Slightly undercured rubber is permissible. The rubber must be dry, firm, free of blisters, except to the extent shown in the sample. Dirty packing, sand and all other foreign matter other than specified above is not permissible.'

White and pale crepes

These grades must be produced from the fresh coagula of natural liquid latex under conditions where all processes are carefully and uniformly controlled. The rubber is milled to produce crepe in thickness corresponding approximately to the piece in the respective samples of 'thin white and pale crepes' and 'thick pale crepes'.

1X thin white crepe. Deliveries must consist of dry firm rubber of very white uniform colour.

Discolouration, sour or foul odours, regardless of cause, dust, specks, sand or other foreign matter, oil or other stains, or evidence of oxidation or heat, not permissible.

1X thick pale crepe, 1X thin pale crepe. As for 1X thin white crepe.

1 thin white crepe. As for 1X thin white crepe with very slight variations of shade permissible.

1 thin pale crepe, 1 thick pale crepe. As for 1X thin pale crepe but the colour should be light with very slight variation permissible.

2 thick pale crepe, 2 thin pale crepe. As for 1 thin pale crepe but slightly darker and slight variation in shade permissible.

Slightly mottled rubber to the degree shown in the type sample not objected to provided the condition does not exist in more than 10 per cent of bales on delivery.

Brown crepes

Estate brown crepes. These grades are made from fresh lump or other high-grade rubber scrap generated on rubber estates. Tree bark scrap, if used, must be pre-cleaned to separate the rubber from the bark. Power-wash mills are to be used in milling these grades to form a crepe of thickness corresponding approximately to the samples of 'estate thin brown crepe' and 'estate thick brown crepe'. Use of earth scrap, smoked scrap and wet slab is not permissible in the preparation of estate brown crepe.

Compo crepe. These grades are made from lump, tree scrap, smoked sheet, cuttings and wet slab. Earth scrap is not permitted.

Thin brown crepe (remills). These grades are manufactured on power-wash mills from wet slab, USS, lump and other high-grade slab generated on estates and smallholdings. Tree bark scrap, if used, must be pre-cleaned. Earth scrap is not permitted. The thickness of the crepe should be that of the type samples.

Estate brown crepes, 1X thin brown crepe, and thick brown crepe. Deliveries must consist of dry, clean rubber light brown in colour. Discolouration, regardless of cause, specks, sand or any other foreign matter, oil or other cause of oxidation or heat, strong sour or foul odour, is not permissible.

Specifications for other grades are in a similar vein.

General

Reference to the Green Book will give the complete descriptions for the 35 grades of natural rubber. Compo crepes, thin brown crepes, thick blanket crepes are similar but vary in colour, contamination and thickness. Type samples are available for most of the important grades. The Green Book also describes other grades of non-international types of NR and gives packing specifications and a glossary of terms used in the rubber trade.

Powder specifications for bale coating

The powders considered as acceptable shall be white and water-insoluble inorganic substances. They must meet the following specifications: 100 per cent penetration through a standard US sieve No. 100; 93 per cent penetration through a standard US sieve No. 325; they must disperse uniformly and without agglomeration on milling. Mineral powders containing calcium sulphate are not acceptable.

Bale-marking requirements

At least the following marks, in addition to those required by law, must appear on each bale:

(a) Grade marks – on two sides of the bale should be 8 in. characters.

(b) Firm marks – letters identifying the shipper's firm on two sides of the bale should be 5 in. characters.

(c) Lot identification marks – numbers appearing immediately below the firm marks on not less than two sides of the bale should be 5 in. characters. These numbers must be the same on all bales covered by the same bill of lading. Different numbers must be used for each separate bill of lading covering lots shipped by a single firm and loaded on the same vessel. Marking paints used on white and pale crepes, thin brown crepes and compo crepes shall not penetrate below the wrapper sheets.

American Chemical Society Compound No. 1 (ACS1) formula and test

The compound consists of the following, in pph (parts per hundred of rubber):

zinc oxide 6   stearic acid 0.5   sulphur 3.5   mercaptobenzthiazole 0.5

In test laboratories master batches of the compounding ingredients and of accelerators are often made. The rubber test sample is banded on a 12 in. × 6 in. mill for six passes, the compounding ingredient master batch added with similar mixing followed by the accelerator master batch. Three bands are rolled off and remilled, sheeting off at 26/1000 in., the whole procedure taking only three minutes.

The compound is allowed to relax for one day and is then cured in dumb-bell moulds for 40 minutes at 140 °C. The dumb-bells are again left for 24 hours and the strain is measured under a load of 5 kg/cm^2 after 1 minute.

Mooney viscosity

The rubber test sample is mill-blended and then a 25 g sample is removed in two parts; one part is placed in the viscometer above the rotor disc and the other part below. The sample is then heated for one minute at 100 °C before spinning the rotor disc for four minutes. The reading obtained at this time is the viscosity in Mooney units, i.e.

$$MV = ML \ 1 + 4 \text{ at } 100 \text{ °C}$$

where MV = viscosity in Mooney units
          ML = Mooney large rotor
           1 = pre-heat time in minutes at 100 °C
           4 = rotation time in minutes at 100 °C

Appendix 11.3

Standard Malaysian Rubber Specifications

Standard Malaysian Rubber Specifications Mandatory from 1 January 1979

Parameter	SMR CV	SMR LV[†]	SMR L	SMR WF
	Latex			
	Viscosity stabilized		—	
Dirt retained on 44 μ aperture (max., % wt)	0.03	0.03	0.03	0.03
Ash content (max., % wt)	0.50	0.50	0.50	0.50
Nitrogen content (max., % wt)	0.60	0.60	0.60	0.60
Volatile matter (max., % wt)	0.80	0.80	0.80	0.80
Wallace rapid plasticity – minimum initial value (P_0)	—	—	30	30
Plasticity retention index, PRI (min., %)	60	60	60	60
Colour limit (Lovibond scale, max.)	—	—	6.0	—
Mooney viscosity ML 1 + 4, 100 °C	—[§]	—[**]	—	—
Cure	R[§§]	R[§§]	R[§§]	R[§§]
Colour coding marker[***]	black	black	light green	light green
Plastic wrap colour	transparent	transparent	transparent	transparent
Plastic strip colour	orange	magenta	transparent	opaque white

Source: Chen *et al.* 1978

* Testing for compliance shall follow ISO test methods.
† Contains 4 phr light, non-staining mineral oil. Additional producer control parameter: acetone extract 6–8% by weight.
§ Three sub-grades. viz. SMR CV50, CV60 and CV70 with producer viscosity limits at 45–55, 55–65 and 65–75 units respectively.

SMR 5	SMR GP	SMR 10	SMR 20	SMR 50
Sheet material	Blend		Field grade material	
	Viscosity stabilized			
0.05	0.10	0.10	0.20	0.50
0.60	0.75	0.75	1.00	1.50
0.60	0.60	0.60	0.60	0.60
0.80	0.80	0.80	0.80	0.80
30	—	30	30	30
60	50	50	40	30
—	—	—	—	—
—	—[‡]	—	—	—
—	R[§§]	—	—	—
light green	blue	brown	red	yellow
transparent opaque white	transparent opaque white	transparent opaque white	transparent opaque white	transparent opaque white

[**] One grade designated SMR LV50 with producer viscosity limits at 45–55 units.

[‡] Producer viscosity limits are imposed at 58–72 units.

[§§] Cure information is provided in the form of a rheograph (R).

[***] The colour of printing on the bale identification strip.

Appendix 11.4

Plasticity retention index (PRI)

Procedure

A bale sample of approximately 300 g is blended on a 12 in. × 6 in. laboratory mill, using six passes with a nip of $\frac{1}{10}$ in. Twenty-five grams of this sample are then sheeted out to 1.6–1.8 mm thickness. The sheet is doubled and six test pieces 1 cm in diameter are punched out. Three pellets are placed in an oven for 30 minutes at 140 °C, then allowed to cool to room temperature. The control pellets and the heated pellets are then tested together on a Wallace plastimeter. The mean aged value expressed as a percentage of the mean unaged (control) value gives the plasticity retention index.

PRI of SMR 5 and visually graded rubber

More than 100 samples taken, 1 sample per tonne on over 100 tonnes

Grade	Maximum	Minimum	Mean
SMR 5	110	70	97
RSS	110	70	91
2X thin brown crepe	65	18	38
Remilled thin brown crepe	70	20	45
4 thick brown crepe (amber)	50	5	19
2 smoked blanket crepe	68	20	41

Effect on PRI of exposure to sunlight

Effect on PRI of exposure to six hours direct sunlight, wet rubber and dry rubber

Grade	Unexposed	Exposed wet	Exposed dry
RSS	94	92	63
Estate brown crepe	88	84	49
Remill	30	26	10

Appendix 11.5

Latex concentrate specifications: ISO 2004, 1979 (E)

Requirements

The latex shall, if required, conform to the requirement for total solids content, and shall conform to all the other requirements given in the table.

If the latex contains preservative(s) other than ammonia or formaldehyde, the chemical nature and approximate quantity of such other preservative(s) shall be stated. The latex shall not contain fixed alkali added at any stage in its production.

The latex shall be sampled by one of the methods specified in ISO 123.

Specifications

The specifications are shown in the table overleaf.

Characteristic	Limits					Method of test
	type HA[5]	type LA[6]	type XA[7]	type HA creamed[8]	type LA creamed[9]	
Total solids content,[1] % min.	61.5	61.5	61.5	66.0	66.0	ISO 124
Dry rubber content, % min.	60.0	60.0	60.0	64.0	64.0	ISO 126
Non-rubber solids,[2] % max.	2.0	2.0	2.0	2.0	2.0	—
Alkalinity (as NH_3), % on latex.	0.60 min.	0.29 max.	0.30 min.	0.55 min.	0.35 max.	ISO 125
Mechanical stability,[3] seconds, min.	650	650	650	650	650	ISO 35
Coagulum content, % max.	0.05	0.05	0.05	0.05	0.05	ISO 706
Manganese content, mg/kg of total solids max.	8	8	8	8	8	ISO 1655
Copper content, mg/kg of total solids max.	8	8	8	8	8	ISO/R 1654
Sludge content, % max.	0.10	0.10	0.10	0.10	0.10	ISO 2005
Volatile fatty acid number (VFA)	As agreed by the interested parties but not to exceed 0.20					ISO 506
KOH number[4]	As agreed by the interested parties but not to exceed 1.0					ISO 127
Colour on visual inspection	No pronounced blue or grey					—
Odour after neutralization with boric acid	No pronounced odour of putrefaction					—

1. Total solids content is an optional requirement.
2. Difference between total solids content and dry rubber content.
3. A minimum mechanical stability may be required which is greater than the minimum value specified.
4. If the latex contains boric acid, the KOH number may exceed the specified value by an amount equivalent to the boric acid content as determined by the method specified in ISO 1802.
5. Centrifuged latex preserved with ammonia only or with formaldehyde followed by ammonia, with an alkalinity of at least 0.60% on the latex.
6. Centrifuged latex preserved with ammonia together with other preservative(s) with an alkalinity of not more than 0.29% on the latex.
7. Centrifuged latex preserved with ammonia together with other preservative(s), with an alkalinity of at least 0.30% on the latex.
8. Creamed latex preserved with ammonia only or with formaldehyde followed by ammonia, with an alkalinity of at least 0.55% on the latex.
9. Creamed latex preserved with ammonia together with other preservative(s), with an alkalinity of not more than 0.35% on the latex.

References

Abbas, B. S., Ginting, S. 1981 Influence of rootstock and scion on girth increment in rubber trees, *Bulletin Balai Penelitian Perkebunan Medan* **12:** 145–52

Abraham, P. D. 1981 Recent innovations in exploitation of Hevea, *Planter, Kuala Lumpur* **57:** 631–48

Abraham, P. D., Anthony, J. L. 1980a Implements for micro tapping, *Planters' Bulletin of the Rubber Research Institute of Malaysia* No. **164:** 124–31

Abraham, P. D., Anthony, J. L. 1980b Improved conventional tapping tools, *Planters' Bulletin of the Rubber Research Institute of Malaysia* No. **165:** 158–65

Abraham, P. D., Anthony, J. L. 1980c Exploitation procedures for mature rubber of smallholdings, *Planters' Bulletin of the Rubber Research Institute of Malaysia* No. **164:** 132–41

Abraham, P. D., Anthony, J. L. 1982 Prospect of micro-tapping immature rubber, *Proceedings of the Rubber Research Institute of Malaysia Planters' Conference, Kuala Lumpur 1981*, pp. 93–116

Abraham, P. D., Anthony, J. L., Arshad, N. L. 1981 Stimulation practices for smallholdings, *Planters' Bulletin of the Rubber Research Institute of Malaysia* No. **167:** 51–66

Abraham, P. D., Anthony, J. L., Gomez, J. B. 1979 Towards automated micro-tapping in Hevea, *Proceedings of the Rubber Research Institute of Malaysia Planters' Conference Kuala Lumpur 1979*, p. 182

Abraham, P. D., Blencowe, J. W., Chua, S. E., Gomez, J. B., Moir, G. F. Y., Pakianathan, S. W., Sekhar, B. C., Southorn, W. A., Wycherley, P. R. 1971 Novel stimulants and procedures in the exploitation of Hevea. I. Introductory review, *Journal of the Rubber Research Institute of Malaya* **23:** 85–9. II. Pilot trials using (2-chloroethyl)-phosphonic acid (ethephon) and acetylene with various tapping systems, *Journal of the Rubber Research Institute of Malaya* **23:** 90–113. III. Comparison of alternative methods of applying stimulants, *Journal of the Rubber Research Institute of Malaya* **23:** 114–37

Abraham, P. D., Boatman, S. G., Blackman, G. E., Powell, R. G. 1968 Effects of plant growth regulators and other compounds on flow of latex in *Hevea brasiliensis*, *Annals of Applied Biology* **62:** 159–73

Abraham, P. D., Hashim, I. 1983 Exploitation procedures for modern Hevea cultivars, *Proceedings of the Rubber Research Institute of Malaysia Planters' Conference, Kuala Lumpur 1983*, pp. 126–58

Abraham, P. D., Tayler, R. S. 1967a Tapping of *Hevea brasiliensis*, *Tropical Agriculture, Trinidad* **44:** 1–11

Abraham, P. D., Tayler, R. S. 1967b Stimulation of latex flow in *Hevea brasiliensis*, *Experimental Agriculture* **3:** 1–12

Abraham, P. A., Wycherley, P. R., Pakianathan, S. W. 1968 Stimulation of latex flow in *Hevea brasiliensis* by 4-amino-3, 5, 6-trichloropicolinic acid and 2-chloroethane phosphonic acid, *Journal of the Rubber Research Institute of Malaya* **20:** 291–305

Akers, C. E. 1914 *The rubber industry in Brazil and the Orient*. Methuen, London

Allen, G. C., Donnithorne, G. 1957 *Western enterprise in Indonesia and Malaya. A study in economic development*. Allen and Unwin, London

Allen, P. W. 1972 *Natural rubber and the synthetics*. Crosby Lockwood, London

Allen, P. W. 1979 Natural rubber a changing scene, *Shell Polymers* **3**(2)

Allen, P. W., Thomas, P. O., Sekhar, B. C. 1974 *A study on competition between natural and synthetic rubber*. Malaysian Rubber Research and Development Board, Kuala Lumpur, Malaysia

Amma, C. K. S., Nair, V. K. B. 1977 Relationship of seed weight and seedling vigour in *Hevea, Rubber Board Bulletin (India)* **13**: 28–9

Andaya, B. W., Andaya, L. Y. 1982 *A history of Malaya*. Macmillan, London

Anderson, J. 1976 The I M U (integrated mechanical up-keep) concept of herbicide spraying, *Planter, Kuala Lumpur* **52**: 161–71

Andrews, E. H., Dickenson, P. B. 1961 Preliminary electron microscope observations on the ultra-structure of the latex vessel and its contents in young tissues of *Hevea brasiliensis, Proceedings of the Natural Rubber Research Conference*, Kuala Lumpur 1960, pp. 756–65

Anekchai, C., Sookmark, S., Langlois, S. J. C. 1975 Search for the best concentration of Ethrel, *Rubber Research Centre, Hat Yai, Thailand* Document No. 58

Ang, B. B., Shepherd, R. 1979 Promising new Prang Besar rubber clones, *Proceedings of the Rubber Research Institute of Malaysia Planters' Conference, Kuala Lumpur 1979*, pp. 219–40

Angkapradipta, P. 1973 Leaf analysis as a guide for a fertilizer programme of *Hevea, Symposium of the International Rubber Research and Development Board, Puncak, Indonesia 2–4 July 1973*

Anon, 1931 A vital rubber question. Bark reserves on small holdings in Malaya, *India Rubber Journal, London*, February 21

Anon, 1983 Rubber trees thrive in S. China, *China Pictorial News* **5**: 2–5

Anon, 1984 Tools for the future, *Rubber Developments* **37**: 20–22

Anon, 1985 Mechanised tapping now a reality, *Rubber Developments* **38**: 84

Anthony, J. L., Abraham, P. D. 1981 Approaches to minimise constraints with upward tapping on smallholdings, *Planters' Bulletin of the Rubber Research Institute of Malaysia* No. **167**: 67–75

Anthony, J. L., Paranjothy, K., Abraham, P. D. 1981 Methods to control dryness on smallholdings, *Planters' Bulletin of the Rubber Research Institute of Malaysia* No. **167**: 37–50

Aramugam, G., Morris, J. E. 1964 The effect of tree lace and cup lump storage conditions on the properties of brown crepe, *Planters' Bulletin of the Rubber Research Institute of Malaya* No. **74**: 144–54

Archer, B. L., Audley, B. G., Bealing, F. J. 1982 Biosynthesis of rubber in *Hevea brasiliensis, Plastics and Rubber International* **7**: 109–11

Archer, B. L., Barnard, D., Cockbain, E. G., Dickenson, P. B., McMullen, A. I. 1963 Structure, composition and biochemistry of *Hevea* latex. In Bateman, L. (ed.) *The chemistry and physics of rubber-like substances*. Maclaren, London, pp. 41–72

Arope, A. bin, Ismail, T. bin, Chong, D. T. 1985 Sheep rearing under rubber, *Planter, Kuala Lumpur* **61**: 70–7

Arope, A. bin, Nor, A. bin M., Tan, P. H. 1983 *Rubber owners' manual*, 2nd edn. Rubber Research Institute of Malaysia, Kuala Lumpur

Aublet, J. B. C. F. 1775 *Histoire des plantes de la Guiane Française*. Paris

Audley, B. G. 1966 The isolation and chemical composition of helical protein microfibrils from *Hevea brasiliensis* latex, *Biochemical Journal* **98**: 335–41

Audley, B. G., Archer, B. L., Runswick, M. J. Ethylene production by *Hevea brasiliensis* tissues treated with latex yield-stimulating compounds, *Annals of Botany* **42**: 63–71

Bachman, R. J. 1980 Special purpose synthetic rubbers. In *Proceedings of the Twenty-sixth Assembly of the International Rubber Study Group, Kuala Lumpur*

Malaysia 29 September–4 October 1980, pp. 207–14

Baker, C. S. L., Gelling, I. R., Newell, R. 1985 Epoxidized natural rubber, *Rubber Chemistry and Technology* **58**: 67–85

Baptist, E. D. C. 1953 Improvements of yields in *Hevea brasiliensis, World Crops* **5**: 194–8

Baptist, E. D. C. 1961 Breeding for high yield and disease resistance in Hevea, *Proceedings of the Natural Rubber Research Conference Kuala Lumpur 1960*, pp. 430–45

Baptist, E. D. C., de Jonge, P. 1955 Stimulation of yield in *Hevea brasiliensis* (Parts I–III), *Journal of the Rubber Research Institute of Malaya* **14**: 355–406

Baptist, E. D. C., Jeevaratnam, A. J. 1952 A note of Guatemala grass as a mulch for rubber plants, *Quarterly Journal of the Rubber Research Institute of Ceylon* **35**(2): 26–31

Barker, P. W., Holt, E. G. 1938 Rubber Statistics 1900–37: production stocks, absorption, stocks and prices. *United States Department of Commerce Trade Promotion Series* No. **181**. Government Printer, Washington

Barker, P. W., Holt, E. G. 1939 Rubber history of the United States 1839–1939, *United States Department of Commerce Trade Promotion Series* No. **197**. Government Printer, Washington

Barker, P. W., Holt, E. G. 1940 Rubber: history, production and manufacture, *United States Department of Commerce Trade Promotion Series* No. **209**. Government Printer, Washington

Barlow, C. 1978 *The natural rubber industry. Its development, technology and economy in Malaysia.* Oxford University Press, Kuala Lumpur

Barlow, C. 1983 The natural rubber industry, *Planter, Kuala Lumpur* **59**: 252–67

Barlow, C., Chan, C. K. 1969 Towards an optimum size of rubber holding, *Journal of the Rubber Research Institute of Malaya* **21**: 613–653

Barlow, C., Lim, S. C. 1967 Effect of density of planting on growth, yield and economic exploitation of Hevea II. The effect on profit, *Journal of the Rubber Research Institute of Malaya* **20**: 44–51

Barlow, C., Lim, S. C., Thomas, P. O. 1966 Effect of planting systems on times of tapping and collection, *Journal of the Rubber Research Institute of Malaya* **19**: 205–13

Barlow, C., Ng, C. S. 1966 Budgeting on the merits of a shorter immature period, *Planters' Bulletin of the Rubber Research Institute of Malaya* No. **87**: 216–36

Barney, J. A. 1968 *Processing natural rubber – information on current practices.* Planting Manual No. 13 Rubber Research Institute of Malaya, Kuala Lumpur

Barrière, P. 1743 *Nouvelle relations de la France Equinoxiale.* Paris

Barry, R. G., Chorley, R. J. 1976 *Atmosphere, weather and climate*, 2nd edn. Methuen, London, p. 62

Basnayake, V. S. 1965 Safety tapping ladders and tapping knives, *Planters' Bulletin of the Rubber Research Institute of Malaya* No. **80**: 165–7

Bastos, T. X., Diniz, D. A. S. 1980 Microclima ribeirinho. Um controle do *Microcyclus ulei* em seringueira, *Boletim de Pesquisa, EMBRAPA, Belém* No. **13**

Basuki, R., Tobing, H. P. L., Siregar, M. 1978 Perkembangan penyadapan mikro pada tanaman karet [Micro-tapping of rubber; preliminary results], *Menara Perkebunan* **46**: 61–4

Bateman, L., Sekhar, B. C. 1966 Significance of PRI in raw and vulcanized natural rubber, *Journal of the Rubber Research Institute of Malaya* **19**: 133–40

Bauer, P. T. 1948 *The rubber industry. A study in competition and monopoly.* Longmans, Green, London

Baum, V. 1943 *The weeping wood.* Sun Dial Press, New York

Bealing, F. J. 1965 Role of rubber and other terpenoids in plant metabolism. In Mullins, L. (ed.) *Proceedings of the Natural Rubber Producers' Research Association Jubilee Conference, Cambridge 1964.* Maclaren, London, pp. 113–24

Bealing, F. J. 1969 Carbohydrate metabolism in *Hevea* latex: availability and utilization of substrates, *Journal of the Rubber Research Institute of Malaya* **21**: 445–55

Bealing, F. J. 1976 Quantitative aspects of latex metabolism: possible involvement of precursors other than sucrose in the biosynthesis of *Hevea* rubber, *Proceedings of the International Rubber Conference, Kuala Lumpur 1975*, vol. 2, pp. 543–65

Beaufils, E. R. 1954 Contribution to the study of mineral elements in field latex, *Proceedings of the Third Rubber Technology Conference, London 1953*, pp. 87–98

Beaufils, E. R. 1955 Mineral diagnosis of some '*Hevea brasiliensis*', *Archief voor de Rubbercultuur in Nederlandsch-Indië* **32**: 1–32

Beaufils, E. R. 1957 Research for rational exploitation of the *Hevea* using a physiological diagnosis based on the mineral analysis of various parts of the plant, *Fertilité* **3**: 27–38

Beinroth, F. H. 1975 Relationships between US Soil Taxonomy, the Brazilian Soil Classification System, and FAO/UNESCO Soil Units. In Bornemisza and Alvarado (eds) *Soil management in tropical America*. Soil Science Department, North Carolina State University, pp. 92–108

Bennett, D. A. 1980 Synthetic *cis*-polyisoprene – what now? In *Proceedings of the Twenty-sixth Assembly of the International Rubber Study Group, Kuala Lumpur Malaysia 29 September–4 October 1980*

Bevan, J. 1979 Credit 358-IND North Sumatra Smallholders Development Project. Full Supervision Report, 25 June

Bitancourt, A. A., Jenkins, A. E. 1956 Estudos sôbre as Miriangiales. VIII. A antracnose maculada da seringueira, causada par *Elisinoe*, *Arquivos do Instituto Biologico, Brasil* **23**: 41–66

Blair, J. A. S., Noor, N. M. 1981 Elephant barriers for crop defence in Peninsular Malaysia, *Planter, Kuala Lumpur* **57**: 289–312

Blencowe, J. W., Templeton, J. R. 1970 Establishing cocoa under rubber. In Blencowe, E. K., Blencowe, J. W. (eds) *Crop diversification in Malaysia*. Incorporated Society of Planters, Kuala Lumpur, pp. 286–96

Bloomfield, G. F. 1951 Studies in *Hevea* rubber IV. Characteristics of rubber in latex of untapped trees and in branches of trees in regular tapping, *Journal of the Rubber Research Institute of Malaya* **13**: 10–23

Boatman, S. G. 1966 Preliminary physiological studies on the promotion of latex flow by plant growth regulators, *Journal of the Rubber Research Institute of Malaya* **19**: 243–58

Bolle-Jones, E. W. 1954 Nutrition of *Hevea brasiliensis*. I. Experimental methods, *Journal of the Rubber Research Institute of Malaya* **14**: 183–208

Bolle-Jones, E. W. 1956 Visual symptoms of mineral deficiencies of *Hevea brasiliensis*, *Journal of the Rubber Research Institute of Malaya* **14**: 493–584

Bolle-Jones, E. W. 1957a A magnesium-manganese inter-relationship in the mineral nutrition of *Hevea brasiliensis*, *Journal of the Rubber Research Institute of Malaya* **15**: 22–8

Bolle-Jones, E. W. 1957b Molybdenum: effects on the growth and composition of *Hevea*, *Journal of the Rubber Research Institute of Malaya* **15**: 141–58

Bolton, J. 1960 The response of *Hevea* to fertilizers on a sandy latosol, *Journal of the Rubber Research Institute of Malaya* **16**: 178–90

Bolton, J. 1961 The effects of fertilizers on pH and the exchangeable cations of some Malayan soils, *Proceedings of the Natural Rubber Research Conference, Kuala Lumpur 1960*, pp. 70–80

Bolton, J. 1963 A study of the leaching of commonly used fertilizers on Malayan latosols used for the cultivation of *Hevea brasiliensis*. MSc thesis University of Leeds

Bolton, J. 1964a The response of immature *Hevea brasiliensis* to fertilizers in Malaya. I. Experiments on shale-derived soils, *Journal of the Rubber Research Institute of Malaya* **18**: 67–79

Bolton, J. 1964b The manuring and cultivation of *Hevea brasiliensis*, *Journal of the Science of Food and Agriculture* **1**: 1–8

Bolton, J., Shorrocks, V. M. 1961 The effect of magnesium limestone and other fertilizers on a mature planting of *Hevea brasiliensis*, *Journal of the Rubber Research Institute of Malaya* **17**: 31–9

Bonnemain, J. L. 1980 Micro-autoradiography as a tool for the recognition of phloem transport, *Bericht der Deutschen botanischen Gesellschaft* **93**: 99–107

Bonner, J. 1961 The biogenesis of rubber, *Proceedings of the Natural Rubber Research Conference, Kuala Lumpur 1960*, pp. 11–18

Booth, C., Holliday, P. 1973 *Sphaerostilbe repens. Commonwealth Mycological Institute Descriptions of Pathogenic Fungi and Bacteria* No. 391

Bos, H., McIndoe, K. G. 1965 Breeding Hevea for resistance against *Dothidella ulei*. P. Henn., *Journal of the Rubber Research Institute of Malaya* **19**: 98–107

Boulle, P. 1959 *Sacrilege in Malaya*. Secker and Warburg, London [*Le sacrilège Malais*. Rene Julliard, Paris]

Bridge, K. 1983 Plant growth regulator use in natural rubber. In Nickell, L. G. (ed.) *Plant growth regulating chemicals*. CRS Press, Florida

Brooks, F. T., Sharples, A. 1914 Pink disease. *Department of Agriculture Federated Malay States Bulletin* No. 21

Brookson, C. W. 1956 Importation and development of new strains of *Hevea brasiliensis* by Rubber Research Institute of Malaya, *Journal of the Rubber Research Institute of Malaya* **14**: 423–48

Broughton, W. J. 1977 Effect of various covers on soil fertility under *Hevea brasiliensis* Muell. Arg. and on growth of the tree, *Agro-Ecosystems* **3**: 147–70

Bryce, G., Campbell, L. E. 1917 On the mode of occurrence of latex vessels in *Hevea brasiliensis, Department of Agriculture Ceylon, Bulletin* **30**: 1

Bryce, G., Gadd, C. H. 1923 Yield and growth in *Hevea brasiliensis, Department of Agriculture Ceylon, Bulletin* **68**: 15

Buckler, E. J., Byrne, P. S., Schwammberger, R. G., Sykes, R. C. 1980 Future developments in general-purpose synthetic rubbers. In *Proceedings of the Twenty-sixth Assembly of the International Rubber Study Group, Kuala Lumpur Malaysia 29 September–4 October 1980*

Burkill, H. M. 1959 Large scale variety trials of *Hevea brasiliensis* on Malayan estates, *Journal of the Rubber Research Institute of Malaya* **16**: 1–37

Burkill, I. H. 1935 *Dictionary of the economic products of the Malay Peninsula*. Crown Agents for the Colonies, London

Buttery, B. R. 1961 Investigations into the relationship between stock and scion in *Hevea, Journal of the Rubber Research Institute of Malaya* **17**: 46–76

Buttery, B. R., Boatman, S. G. 1966 Manometric measurement of turgor pressures in laticiferous phloem tissues, *Journal of Experimental Botany* **17**: 283–96

Buttery, B. R., Boatman, S. G. 1967 Effects of tapping, wounding and growth regulators on turgor pressure in *Hevea brasiliensis, Journal of Experimental Botany* **18**: 644–59

Buttery, B. R., Boatman, S. G. 1976 Water deficits and flow of latex. In Kozlowski, T. T. (ed.) *Water deficits and plant growth* vol. 4 *Soil water measurement, plant responses, and breeding for drought resistance*. Academic Press, New York

Buttery, B. R., Westgarth, D. R. 1965 The effect of density of planting on the growth, yield and economic exploitation of *Hevea brasiliesis*. I. The effect on growth and yield, *Journal of the Rubber Research Institute of Malaya* **19**: 62–71

Campaignole, J., Bouthillou, J. 1955 Alliance clone-sujet, *Institut Recherches Caoutchouc Indochine Rapport 1954* p. 66

Campbell, D. S., Elliott, D. J. and Wheelans, M. A. 1978 Thermoplastic natural rubber blends, *NR Technology*, 21–31

Carpenter, J. B. 1951 Target leaf spot of the *Hevea* rubber tree in relation to host development, infection, defoliation and control, *United States Department of Agriculture Technical Bulletin* No. 1028

Carron, M. P., Enjalric, F. 1982 Studies on vegetative micro-propagation of *Hevea brasiliensis* by somatic embryogenesis and *in vitro* micro-cuttings. In *Plant tissue*

culture 1982. Japanese Association for Plant Tissue Culture, Tokyo, pp. 751–2

Cavallo, T. 1783 *The history and practice of aerostation.* London

Cervantes, V. 1794 Discurso pronunciado en el Real Jardin Botánico el 2 de Junio por el Cathedrático Don Vicente de Cervantes, *Suplemento a la Gazeta de Literatura* Mexico, 5 Noviembre 1794: 1–8

Ceulemans, R., Gabriels, R., Impens, I., Yoon, P. K., Leong, W., Ng, A. P. 1984 Comparative study of photosynthesis in several *Hevea brasiliensis* clones and *Hevea* species under tropical field conditions, *Tropical Agriculture, Trinidad* **61:** 273–5; 326

Chan, C. H. 1967 *The development of British Malaya 1896–1909*, 2nd edn. Oxford University Press, Oxford

Chan, C. K., Ng, C. S., Barlow, C. 1969 Results of 1964 sample survey of estates in West Malaysia, *Rubber Research Institute of Malaya Economics and Planning Division Document 61*

Chan, H. Y. 1972 Soil and leaf nutrient surveys for discriminatory fertilizer use in West Malaysia rubber holdings, *Proceedings of the Rubber Research Institute of Malaya Planters' Conference, Kuala Lumpur 1971*, pp. 201–13

Chan, H. Y. 1977 Soil classification. In Pushparajah, E., Amin, L. L. (eds) *Soils under Hevea and their management in Malaysia.* Rubber Research Institute of Malaysia, Kuala Lumpur, pp. 57–74

Chan, H. Y., Pushparajah, E. 1972 Productivity potentials of *Hevea* on West Malaysian soils: a preliminary assessment, *Proceedings of the Rubber Research Institute of Malaya Planters' Conference, Kuala Lumpur 1972*, pp. 97–125

Chan, H. Y., Soong, N. K., Woo, Y. K., Tan, K. H. 1972 Manuring in relation to soil series in West Malaysian mature rubber-growing plantations, *Proceedings of the Rubber Research Institute of Malaya Planters' Conference, Kuala Lumpur 1972*, pp. 127–139

Chan, S. K., Chew, P. S. 1983 Volatilisation losses of urea on various soils under oil palm, *Seminar on Fertilizers in Malaysian Agriculture, Malaysian Society for Soil Science and Universiti Pertanian, Serdang*

Chandapillai, M. M. 1969 Castor, a prospective intercrop in Malaysian plantations. In Turner, P. D. (ed.) *Progress in oil palm.* Incorporated Society of Planters, Kuala Lumpur, pp. 252–64

Chandrasekera, L. B. 1980 Crown budding with clone LCB 870, *Rubber Rubber Research Institute of Sri Lanka Bulletin* **15:** 24–7

Chang, A. K., Chan, E., Pushparajah, E., Leong, Y. S. 1977 Precision of field sampling intensities in nutrient surveys for two soils under *Hevea*, *Proceedings of the Conference on Chemistry and Fertility of Tropical Soils, Kuala Lumpur 1973.* Malaysian Society for Soil Science, pp. 25–37

Chang, A. K., Teoh, C. 1982 Commercial experience in the use of leaf analysis for diagnosing nutritional requirement of *Hevea*, *Proceedings of the Rubber Research Institute of Malaysia Planters' Conference, Kuala Lumpur 1981*, pp. 220–31

Chang, W. P., Pillai, N. M., Chin, P. S. 1969 Developments in polybag collection, *Planters' Bulletin of the Rubber Research Institute of Malaya* No. **104:** 165–72

Chapman, G. W. 1941 Leaf analysis and plant nutrition, *Soil Science* **52:** 63

Chapman, G. W. 1951 Plant hormones and yield in *Hevea brasiliensis. Journal of the Rubber Research Institute of Malaya* **13:** 167–76

Chee, K. H. (undated) *Una visita a Bahia (Brasil) para dar assistencia ao controle da 'queima da folha' da seringueira (Hevea brasiliensis)*, SUDHEVEA, Brazil

Chee, K. H. 1968 Patch canker of *Hevea brasiliensis* caused by *Phytophthora palmivora, Plant Disease Reporter* **52:** 132–3

Chee, K. H. 1969 Variability of *Phytophthora* species from *Hevea brasiliensis, Transactions of the British Mycological Society* **52:** 425–36

Chee, K. H. 1971 A new disease of *Hevea brasiliensis* caused by *Fusarium solani* and *Botryodiplodia theobromae, Plant Disease Reporter* **55:** 152–3

Chee, K. H. 1974 Hosts of *Phytophthora palmivora.* In Gregory, P. H. (ed.) *Phytophthora diseases of cocoa*, pp. 81–7

Chee, K. H. 1976a Assessing susceptibility of Hevea clones to *Microcyclus ulei*, *Annals of Applied Biology* **84**: 135–45

Chee, K. H. 1976b Micro-organisms associated with rubber (*Hevea brasiliensis* Mull. Arg.)., *Rubber Research Institute of Malaysia Agricultural Series Report* No. 4

Chee, K. H. 1984 Improved control of South American leaf blight of Hevea in Brazil, *Journal of Plant Protection in the Tropics* **1**: 1–7

Chee, K. H., Holliday, P. 1986 South American Leaf Blight of Hevea Rubber, *Malaysian Rubber Research and Development Board Monograph* No. 13

Chee, K. H., Wastie, R. L. 1980 The status and future prospects of rubber diseases in tropical America, *Review of Plant Pathology* **59**: 541–8

Chee, Y. K., Chin, T. V., Rashid, A. S. 1982a Testing of legume cover crop seeds, *Planters' Bulletin of the Rubber Research Institute of Malaysia* No. **170**: 3–9

Chee, Y. K., Lui, S., Chin, T. V. 1982b Establishment of legume cover crop on flat land, *Planters' Bulletin of the Rubber Research Institute of Malaysia* No. **177**: 119–23

Chee, Y. K., Tan, G. K. 1982 Pre-treatment of legume cover crop seeds, *Planters' Bulletin of the Rubber Research Institute of Malaysia* No. **170**: 10–13

Chen, K. Y., Amlir Aziz, Loke, K. M., Ong, C. T., Rama Rao, P. T. 1978 New features of SMR scheme, 1979. *Planters' Bulletin of the Rubber Research Institute of Malaysia* No. **156**: 89–96

Chevallier, M. H. 1984 Preliminary notes on genetic variability in the germplasm from the 1981 Amazonian prospection. In *Compte-rendu du colloque exploitation, physiologie et amélioration de l'*Hevea. Institut de Recherches sur le Caoutchouc en Afrique/Groupment d'Etudes et de Recherches pour le Développement de l'Agronomie Tropicale, Montpellier, France, pp. 453–61

Chilvers, L., Burley, T. 1984 Rubber: an assured place in Asia's future, *Asian Agribusiness* **1**(6)

Chin, H. F., Aziz, M, Ang, B. B., Hamzah, S. 1981 The effect of moisture and temperature on the ultrastructure and viability of seeds of *Hevea brasiliensis*, *Seed Science and Technology* **9**(2): 411–22

Chin, P. S. 1966 Versatility of the Heveacrumb process – application to oil extended and constant viscosity natural rubbers, *Planters' Bulletin of the Rubber Research Institute of Malaya* No. **86**: 111–25

Chin, P. S., O'Connell, J. 1969 Oil extension of natural rubber at the latex stage, *Journal of the Rubber Research Institute of Malaya* **22**: 91–103

Chittenden, A. E., Hawkes, A. J., Padden, A. R. 1973 An assessment of rubber wood from Liberia for particle board manufacture, *Tropical Products Institute*, Report L 35

Chong, F. C. 1981 The role of carbohydrates in the exploitation and latex flow of Hevea, *Journal of the Rubber Research Institute of Malaysia* **29**: 125–6

Chong, K. M., Pee, T. Y. 1976 Relative merits of replanting with different planting materials, *Proceedings of the Rubber Research Institute of Malaysia Planters' Conference, Kuala Lumpur 1976*, pp. 1–18

Chrestin, H., Bangratz, J., d'Auzac, J., Jacob, J. L. 1984 Role of the lutoidic tonoplast in the senescence and degeneration of the laticifers of *Hevea brasiliensis*, *Zeitschrift für Pflanzenphysiologie* **114**: 261–77

Chua, S. E. 1966 Studies on tissue culture of *Hevea brasiliensis*. I. Role of osmotic concentration, carbohydrates and pH values in induction of callus growth in plumule tissues from *Hevea* seedlings, *Journal of the Rubber Research Institute of Malaya* **19**: 272–6

Chua, S. E. 1967 Physiological changes in Hevea trees under intensive tapping, *Journal of the Rubber Research Institute of Malaya* **20**: 100–5

Chua, S. E. 1970 Physiology of foliar senescence and abscission in *Hevea brasiliensis*, *Rubber Research Institute of Malaya Archives Document* No. 63

Chuang, C. P. 1965 The problem of skilled tappers: a trial with 1200-tree tasks, *Planters' Bulletin of the Rubber Research Institute of Malaya* No. **80**: 168–9

Cockbain, E. G., Philpott, M. W. 1963 Colloidal properties of latex. In Bateman,

L. (ed.) *The chemistry and physics of rubber-like substances.* Maclaren, London, pp. 73–82

Collier, H. M., Chick, W. H. 1977 Problems and potential in the treatment of rubber factory and palm oil mill effluents, *Planter, Kuala Lumpur* **53:** 439–48

Collier, H. M., Lowe, J. S. 1969 Effect of fertilizer applications on latex properties, *Journal of the Rubber Research Institute of Malaya* **21:** 181–91

Collier, H. M., Morgan, F. A. 1969 Polybag collection trial on Bahau Estate, *Planters' Bulletin of the Rubber Research Institute of Malaya* No. **104:** 173–82

Collins, J. 1869 On India-rubber, its history, commerce and supply. Paper read at the 5th Annual Meeting. *Journal of the Royal Society of Arts* December 1869

Collins, J. 1872 *Report on the caoutchouc of commerce being information on the plants yielding it, their geographical distribution, climatic conditions and the possibility of their cultivation and acclimatization in India.* Allen, Stanford, King, Trübner, London [with map of species]

Combe, J. C., Gener, P. 1977a Influence de la famille du porte-greffe sur la croissance et la production des hévéas greffés, *Revue Générale du Caoutchouc et des Plastiques* **568:** 97–101

Combe, J. C., Gener, P. 1977b Effect of the stock family on the growth and production of grafted Heveas, *Journal of the Rubber Research Institute of Sri Lanka* **54:** 83–92

Compagnon, P. 1962 Mineral nutrition of Hevea, *Revue Générale du Caoutchouc et des Plastiques* **39:** 1105–35

Compagnon, P., Tixier, P. 1950 Sur une possibilité d'améliorer la production d'*Hevea brasiliensis* par l'apport d'oligo-éléments, *Revue Générale du Caoutchouc et des Plastiques* **27:** 525, 591, 663

Conduro Neto, J. M. H., Martins, E., Silva, H., da Cunha, R. L. M., Benchimol, R. L. 1983 Controle do mal das folhas da seringueira pela termonebulização do fungicida triadimefon, *FCAP Nota Prévia* No. 6

Constable, D. H. 1953 Manuring replanted rubber, 1938–1952. *Quarterly Circular of the Rubber Research Institute of Ceylon* **29:** 16

Cook, A. S., McMullen, A. I. 1951 The precoagulation of Hevea latex in wet weather, *Journal of the Rubber Research Institute of Malaya* **13:** 139–40

Cook, A. S., Sekhar, B. C. 1955 Volatile acids and the quality of concentrated natural latex, *Journal of the Rubber Research Institute of Malaya* **14:** 407–22

Corley, R. H. V. 1983 Potential productivity of tropical crops. *Experimental Agriculture* **19:** 217–37

Coupé, M., Lambert, C., Primot, L., d'Auzac, J. 1977 Cinétique d'action de l'acide 2-chloroéthylphosphonique sur les polyribosomes du latex d' *Hevea brasiliensis, Phytochemistry* **16:** 1133

d'Auzac, J., Jacob, J. L. 1969 Regulation of glycolysis in latex of Hevea brasiliensis, *Journal of the Rubber Research Institute of Malaya* **21:** 417–44

d'Auzac, J., Ribaillier, D. 1969 L'éthylène, nouvel agent stimulant de la production de latex chez l'*Hevea brasiliensis, Revue Générale du Caoutchouc et des Plastiques* **46:** 857–8

da Costa, L. 1968 The main tropical soils of Brazil. In *Approaches to soil classification. FAO World Soil Resources Report* No. 32, pp. 95–106

Dale-Rudwick, P. H., Hastie, D. C. 1970 The establishment of a ground cover under shade, *Planter, Kuala Lumpur* **46:** 379–82

Davidson, F. 1962 Replanting under shade, *Planters' Bulletin of the Rubber Research Institute of Malaya* No. **62:** 127–9

Dawkes, W. H. 1943 *The P and T lands. An agricultural romance of Anglo-Dutch enterprise.* Anglo-Dutch plantations of Java Ltd, London

De Abreu, J. M. 1982 Investigations on the rubber leaf caterpillar Erynnys ello in Bahia, Brazil, *Revista Theobroma* **12:** 85–99

de Barros, J. C. M., De Castro, A. M. G., Miranda, J., Schenkel, C. S. 1983 Development and present status of rubber cultivation in Brazil, *Proceedings of the*

Rubber Research Institute of Malaysia Planters' Conference, Kuala Lumpur 1983, pp. 18–30

de Camargo, A. P., Schmidt, N. C., Cardoso, R. M. G. 1976 South American leaf blight epidemics and rubber phenology in São Paulo, *Proceedings of the International Rubber Conference Kuala Lumpur 1975* vol. **3**: 251–65

de Fourcroy, A. F. 1791 *Annales de Chimie* **11**: 225

de Geus, J. G. 1973 *Fertilizer guide for the tropics and subtropics*, 2nd edn. Centre d' Etude de l'Azote, Zurich, p. 541

de Haan, van Aggelen-Bot, G. M. 1948 Die vorming van rubber bij *Hevea brasiliensis*, *Archief voor de Rubbercultuur in Nederlansch-Indië* **26**: 121–80

de Jonge, P. 1957 Stimulation of bark renewal of Hevea and its effect on yield of latex, *Journal of the Rubber Research Institute of Malaya* **15**: 53–71

de Jonge, P. 1969a Influence of depth of tapping on growth and yield of *Hevea brasiliensis*, *Journal of the Rubber Research Institute of Malaya* **21**: 348–52

de Jonge, P. 1969b Exploitation of Hevea, *Journal of the Rubber Research Institute of Malaya* **21**: 283–91

de Jonge, P., Tan, H. T. 1972 Chemara ethrel stimulation experiments: preliminary results, *Proceedings of the Rubber Research Institute of Malaya Planters' Conference, Kuala Lumpur 1971*, pp. 126–35

de Jonge, P., Westgarth, D. R. 1962 The effect of size of tapping task on the yield per tapper and yield per acre of *Hevea*, *Journal of the Rubber Research Institute of Malaya* **17**: 150–8

de la Condamine, C. M. 1745 *Relation abrégée d'un voyage fait dans l'intérieur de l'Amérique méridionale*. Lue à l'assemblé publique de l'Académie de Sciences, Paris

de la Neuville, A. J. 1723 Sur une poire faite de gomme qui sert aux Indiens de seringue, *Memoires de Trévoux* **1**: 527–8

de Sahagun, B. c. 1530 *General history of the things of New Spain* [Translated by Anderson, J. O. and Dibble, C. E., Santa Fé, New Mexico, 1950–51]

de Soyza, A. G. A., Samaranayaka, C., Abeywardene, V., Jayaratne, A. M. R., Wilberk, S. 1983 A survey of the incidence and pattern of distribution of the brown bast disease of *Hevea* in Sri Lanka, *Journal of the Rubber Research Intitute of Sri Lanka*, in press

De Torquemada, J. 1615 *De la monarchia Indiana* vol. **2**, pp. 83–180 [republished 1723]

de Vernou, P. 1981 Lutte contre *Eupatorium odoratum* en Cote d'Ivoire: cas de plantations d'hévéas villageois. *Compte Rendue de la ll*e Conférence de COLUMA 1981 tome 3, pp. 958–65

De Vries, O. 1926 Superieur plantmateriaal (zaailingen en oculaties), *De Bergcultures* **1**: 404

Dennis, R. W. G., Reid, D. A. 1957 Some marasmioid fungi allegedly parasitic on leaves and twigs in the tropics, *Kew Bulletin 1957*: 287–92

D'Hoore, J. L. 1964 Soil map of Africa – exploratory monograph, *CCTA Publication* 93, Lagos

Dickenson, P. B. 1965 The ultrastructure of the latex vessel of *Hevea brasiliensis*. In Mullins, L. (ed.) *Proceedings of the Natural Rubber Producers' Research Association Jubilee Conference Cambridge 1964*. Maclaren, London, pp. 52–66

Dickenson, P. B. 1969 Electron microscopical studies of the latex vessel system of *Hevea brasiliensis*, *Journal of the Rubber Research Institute of Malaya* **21**: 543–59

Dickenson, P. B., Sivakumaran, S., Abraham, P. D. 1976 Ethad and other new stimulants for *Hevea brasiliensis*, *Proceedings of the International Rubber Conference, Kuala Lumpur 1975* vol. **2** pp. 315–40

Dijkman, M. J. 1951 *Hevea. Thirty years of research in the Far East*. University of Miami Press.

Drabble, J. 1973 *Rubber in Malaya. The genesis of an industry*. Oxford University Press, Kuala Lumpur

Duckworth, I. H. 1964 Creamed latex concentrate,, *Planters' Bulletin of the Rubber Research Institute of Malaya* No. **74**: 111–22

Eaton, B. J. 1952 Wild and plantation rubber, gutta percha and balata. In Schidrowitz, P., Dawson, T. R. (eds) *History of the rubber industry*. Heffer, Cambridge

Edathil, T. T., Krishnankutty, V., Jacob, C. K. 1984 Thermal fogging: a new method for controlling powdery mildew disease of rubber in India, *Pesticides* **18** (5): 35–6

Edgar, A. T. 1958 *Manual of rubber planting (Malaya)*. Incorporated Society of Planters, Kuala Lumpur

Ellis, M. B., Holliday, P. 1972 *Drechslera heveae, Commonwealth Mycological Institute Descriptions of Pathogenic Fungi and Bacteria* No. 343

Emsley, J. 1977 Phosphate cycles, *Chemistry in Britain* **13**: 459–63

Enjalric, F., Carron, M. P. 1982 Microbouturage *in vitro* des jeunes plantes d'*Hevea brasiliensis*, *Comptes rendus des Séances de l'Académie des Sciences. III. Sciences de la Vie* **295** (3): 259–64

Eschbach, J. M. 1974 Some results of a rubber spacing trial, *Revue Générale du Caoutchouc et des Plastiques* **51** (4): 239–42

Eschbach, Y. M., Banchi, Y. 1985 Advantage of ethrel stimulation in association with reduced tapping intensity in the Ivory Coast, *Planter, Kuala Lumpur* **61**: 555–6

Eschbach, Y. M., Roussel, D., van de Sype, H., Jacob J. L., d'Auzac, J. 1984 Relationships between yield and clonal physiological characteristics of latex from *Hevea brasiliensis*, *Physiologie Végétale* **22**: 295–304

Faiz, A. A. 1979a General principles of weed control. In *Training manual on crop protection and weed control in rubber plantations*. Rubber Research Institute of Malaysia, Kuala Lumpur, pp. 102–10

Faiz, A. A. 1979b Weed control in rubber nurseries and leguminous cover crops. In *Training manual on crop protection and weed control in rubber plantations*. Rubber Research Institute of Malaysia, Kuala Lumpur, pp. 111–15

Faiz, A. A., Liu, S. 1982 Effects of some herbicide mixtures against several common weeds in rubber plantations. In *Proceedings of the International Conference on Plant Protection in the Tropics, Kuala Lumpur 1982*, pp. 531–41

Faizah, A. W. 1983a Towards an improved *Rhizobium* inoculant supply, *Planters' Bulletin of the Rubber Research Institute of Malaysia* No. **174:** 3–6

Faizah, A. W. 1983b Nitrogen fixatiton in legumes, *Planters' Bulletin of the Rubber Research Institute of Malaysia* No. **174:** 7–12

Fallows, J. C. 1961 The major elements in the foliage of *Hevea brasiliensis* and their inter-relation, *Proceedings of the Natural Rubber Research Conference, Kuala Lumpur 1960*, pp. 142–53

FAO-UNESCO 1974 In Dudal *et al.* (eds) *Soil map of the world* vol. 1. UNESCO, Paris, legend

Fauconnier, H. 1931 *The soul of Malaya*. Elkin Mathews and Marrot, London [*Malaisie*, Paris]

Feng, Y. Z., Wang, H. H., Zhang, J. H., Zhang, K. Y., Ma, W. J., Long, Y. M. 1982 Experimental and ecological studies on a mixed rubber and tea plantation, *Acta Botanica Sinica* **24**: 164–71

Fernando, D. M. 1966 An outline of the breeding, selection and propagation of rubber, *Quarterly Journal of the Rubber Research Institute of Ceylon* **42**: 9–12

Fernando, D. M. 1969 Breeding for multiple characters of economic importance in Hevea – preliminary assessment of recent selections, *Journal of the Rubber Research Institute of Malaya* **21**: 27–37

Fernando, D. M. 1977 Some aspects of *Hevea* breeding and selection. *Journal of the Rubber Research Institute of Sri Lanka* **54**: 17–32

Fernando, D. M., Liyanage, A. de S. 1976 Hevea breeding for leaf and panel disease resistance in Sri Lanka, *Proceedings of the International Rubber Conference, Kuala Lumpur 1975* vol. 2, pp. 236–46

Fernando, D. M., Tambiah, B. S. 1970a Sieve tube diameters and yield in *Hevea*

species: a preliminary study, *Quarterly Journal of the Rubber Research Institute of Ceylon* **46**: 88–92

Fernando, D. M., Tambiah, B. S. 1970b A study of the significance of latex in *Hevea* species, *Quarterly Journal of the Rubber Research Institute of Ceylon* **46**: 68–74

Ferrand, M. 1941 Observations sur les variations de la concentration du latex in situ par la micro méthode de la goutte de latex, *INEAC Serie Scientifique* No. **22**: 1

Ferwerda, F. P. 1969 Rubber. In Ferwerda, F. P., Wit, F. (eds) *Outlines of perennial crop breeding in the tropics.* Veenman en Zonen Wageningen, The Netherlands, pp. 427–58

Figart, D. M. 1925 The plantation rubber industry in the Middle East, *United States Department of Commerce Trade Promotion Series* No. 2. Government Printing Office, Washington

Fleming, T. 1965 Progressive reorganization of tapping tasks: tapping systems and field collection in Sumatra, *Planters' Bulletin of the Rubber Research Institute of Malaya* No. **80**: 170–6

Food and Agriculture Organization 1983 Indonesia: assistance to the tree-crop subsector investment planning. Summary report and working papers, *Investment Centre Working Paper* No. 101/83 TA. INS 41

Fournier, F., Tuong, C. C. 1961 The biosynthesis of rubber, *Rubber Chemistry and Technology* **34** (5): 1229–1305

Fox, R. A. 1961 White root disease of *Hevea brasiliensis*: the identity of the pathogen, *Proceedings of the Natural Rubber Research Conference, Kuala Lumpur 1960*, pp. 473–82

Fox, R. A. 1964 A report on a visit to Nigeria (9–30 May 1963) undertaken to make a preliminary study of root diseases of rubber, *Rubber Research Institute of Malaya Archives Document* No. 27

Fox, R. A. 1965 The role of biological eradication in root-disease control in replantings of *Hevea brasiliensis*. In Baker, K. F., Synder, W. C. (eds) *Ecology of soil-borne plant pathogens: prelude to biological control*, pp. 348–62

Fox, R. A. 1971 A comparison of methods of dispersal, survival, and parasitism in some fungi causing root diseases of tropical plantation crops. In Tousson, T. A., Bega, R. V., Nelson, P. H. (eds) *Root diseases and soil-borne pathogens*, pp. 179–87

Fox, R. A. 1977 The impact of ecological, cultural and biological factors on the strategy and costs of controlling root diseases in tropical plantation crops as exemplified by *Hevea brasiliensis*, *Journal of the Rubber Research Institute of Sri Lanka* **54**: 329–62

Freeman, R. 1954 Micro-gel in latex and sheet rubber, *Proceedings of the Third Rubber Technology Conference, London, June 1954*, pp. 3–12

Fresneau, F. 1755 Sur une résine élastique nouvellement découverte à Cayenne par M. Fresneau. Histoire et Mémoires de l'Académie pour l'année 1751, Paris, pp. 323–33

Frey-Wyssling, A. 1929 Microscopic investigation on the occurrence of resins in *Hevea* latex, *Archief voor de Rubbercultuur in Nederlandsch-Indië* **13** (7): 392–4

Frey-Wyssling, A. 1932 Investigations on the dilution-reaction and the movement of the latex of *Hevea brasiliensis* during tapping [in Dutch], *Archief voor de Rubbercultuur in Nederlandsch-Indië* **16**: 285

Frey-Wyssling, A. 1952 Latex flow. In *Deformation and flow in biological systems*. North Holland Publishing Co., Amsterdam, pp. 322–43

Fundação Instituto Brasileiro de Geografia e Estatistica 1980 *Anuário Estatistico do Brasil*, pp. 34–41

Gale, R. S. 1959 A survey of factors involved in an experimental study of the drying of sheet rubber, *Journal of the Rubber Research Institute of Malaya* **16**: 35–64

Gama, M. I. C., Kitajima, E. W., Avila, A. C., Lim, T. M. 1983 Um carlavirus em seringueira (*Hevea brasiliensis*), *Fitopatologia Brasileira* **8**: 621

Gan, L. T., Chew, D. K., Ho, C. Y., Wood, B. J. 1985 Ebor eaves – a new technique for rainguarding tapping panels, *Planter, Kuala Lumpur* **61**: 355–60

Garnier, R. 1922 *Ceylon Rubber Planters' Manual.* Times of Ceylon, Colombo, pp. 121–3

Gasparotto, L., Trindade, D. R., Martins, E., Silva, H. 1984 Doenças da seringueira, *Circular Técnica EMBRAPA* No. 4

Gener, P. 1966 Le greffage de l'*Hevea*: influence des stades de pousse foliare du greffon et du porte-greffe sur la réussite du greffage, *Rapport Technique Service Agronomique de l'Institut Recherches Caoutchouc de l' Afrique*

Ghandimathi, H., Yeang, H. Y. 1984 The low fruit set that follows conventional hand pollination in *Hevea brasiliensis*: insufficiency of pollen as a cause, *Journal of the Rubber Research Institute of Malaysia* **32**: 20–9

Ghani, M. N. A., Ong, S. H., Subramaniam, S. 1976 The production and utilization of rubber seeds and the approval of seed collection areas. In *Seed technology in the tropics.* Universiti Pertanian Malaysia, Serdang, pp. 237–45

Gilbert, N. E., Dodds, K. S., Subramaniam, S. 1973 Progress of breeding investigations with *Hevea brasiliensis* V. Analysis of data for earlier crosses, *Journal of the Rubber Research Institute of Malaya* **23**: 365–80

Goering, T. J. 1982 *Natural rubber. Sector Policy Paper.* International Bank for Reconstruction and Development, Washington

Gomez, J. B. 1977 Demonstration of latex coagulants in bark extracts of Hevea and their possible role in latex flow, *Journal of the Rubber Research Institute of Malaysia* **25**: 109–19

Gomez, J. B. 1982 Anatomy of *Hevea* and its influence on latex production, *Malaysian Rubber Research and Development Board Monogaph* No. 7

Gomez, J. B. 1983 Physiology of latex (rubber) production, *Malaysian Rubber Research and Development Board Monograph* No. 8

Gomez, J. B., Chen, K. T. 1967 Alignment of anatomical elements in the stem of *Hevea brasiliensis, Journal of the Rubber Research Institute of Malaya* **20**: 91–9

Gomez, J. B., Hamzah, S. bte 1980 Variation in leaf morphology and anatomy between clones of *Hevea, Journal of the Rubber Research Institute of Malaysia* **28**: 157–82

Gomez, J. B., Moir, G. F. J. 1979 The ultracytology of latex vessels in *Hevea brasiliensis, Malaysian Rubber Research and Development Board Monograph* No. 4

Gomez, J. B., Narayanan, R., Chen, K. T. 1972 Some structural factors affecting the productivity of *Hevea brasiliensis*. I. Quantitative determination of the laticiferous tissue, *Journal of the Rubber Research Institute of Malaya* **23**: 193–203

Gooding, E. G. B. 1952a Studies in the physiology of latex. II. Latex flow on tapping *Hevea brasiliensis*: associated changes in trunk diameter and latex concentration, *New Phytology* **51**: 11–29

Gooding, E. G. B. 1952b Studies in the physiology of latex. III. Effects of various factors on the concentration of latex of *Hevea brasiliensis, New Phytology* **51**: 139–53

Gorton, A. D. T. 1972 Effect of Ethrel stimulation on latex concentrate properties, *Proceedings of the Rubber Research Institute of Malaya Planters' Conference , Kuala Lumpur 1971*, pp. 234–54

Graham, D. J. 1964 New tunnel-type smokehouse, *Planters' Bulletin of the Rubber Research Institute of Malaya* No. **74**: 123–31

Graham, D. J., Morris, J. E. 1966 Manufacture of Heveacrumb. *Planters' Bulletin of the Rubber Research Institute of Malaya* No. **86**: 130–47

Graham, K. M. 1970 Pests and diseases of intercropped bananas in West Malaysia. In Blencowe, E. K., Blencowe, J. W. (eds) *Crop diversification in Malaysia.* Incorporated Society of Planters, Kuala Lumpur, pp. 237–44

Graham, P. H., Hubbel, D. H. 1975 Soil-plant-rhizobium interactions in tropical agriculture. In Bornemisza, E., Alvaredo, A. (eds) *Soil management in tropical America.* Soil Science Department, North Carolina State University, Raleigh, NC, USA, pp. 211–27

Grantham, J. 1924 Manurial experiments on *Hevea*, *Archief voor de Rubbercultuur in Nederlandsch-Indië* **8**: 501–6

Greenwood, J. M. F., Promdej, S. 1971 The effect of mulching on growth of *Hevea* rubber seedlings in a nursery, *Rubber Research Centre Thailand Report 1971*: 27–46

Greig, J. L. 1937 The use of chemicals for the eradication of lalang grass, *Malayan Agricultural Journal* **25**: 363–9

Grilli, E. R., Agostini, B. B., Hooft-Welvaars, M. J. 't 1980 The world rubber economy; structure, changes and prospects, *World Bank Staff Occasional Papers* No. 30, Johns Hopkins University Press

Grossart 1791 Sur les moyens de faire les instruments de gomme élastique avec les bouteilles qui viennent de Brésil, *Annales de Chimie* **11**: 143

Guha, M. M. 1969 Recent advances in fertilizer usage for rubber in Malaya, *Journal of the Rubber Research Institute of Malaya* **21**: 207–16

Guha, M. M. 1975 Fertilizer response from mature rubber in Liberia, *Proceedings of the International Rubber Conference, Kuala Lumpur 1975*, pp. 108–21

Guha, M. M., Narayanan, R. 1969 Variation in leaf nutrient content of *Hevea* with clone and age of leaf, *Journal of the Rubber Research Institute of Malaya* **21**: 225–39

Gunasekera, S. A., Sumanaweera, H. A. N. 1974 Control of blue-stain on rubber wood during the boron diffusion treatment, *Quarterly Journal of the Rubber Research Institute of Sri Lanka* **51**: 54–6

Gunn, J. S. 1960 *West African Institute for Oilpalm Research Annual Report 1959–60*, p. 52

Gunnery, H. 1935 Yield prediction in *Hevea*: a study of sieve tube structure in relation to latex yield, *Journal of the Rubber Research Institute of Malaya* **6**: 8–15

Habib, M., Rahman, R. A. 1978 Intercropping with upland rice in smallholder project, *Bulletin BPP, Medan* **9**: 115–24

Haines, W. B. 1932 On the effect of covers and cultivation methods on the growth of young rubber III, *Journal of the Rubber Research Institute of Malaya* **4**: 123–30

Haines, W. B. 1934 *The uses and control of natural undergrowth on rubber estates.* Rubber Research Institute of Malaya, Kuala Lumpur

Haines, W. B., Flint, C. F. 1931 The manuring of rubber. III. Résumé of present position with illustrative cases based on Malayan data, *Journal of the Rubber Research Institute of Malaya* **3**: 57–90

Haines, W. B., Guest, E. 1936 Recent experiments on manuring *Hevea* and their bearing on estate practice, *Empire Journal of Experimental Agriculture* **4**: 299–325

Hamzah, S. bte, Gomez, J. B. 1981a Ultrastructure of mineral deficient leaves of *Hevea*. III. Quantitative considerations, *Journal of the Rubber Research Institute of Malaysia* **29**: 15–23

Hamzah, S. bte, Gomez, J. B. 1981b Anatomy of bark renewal in normal puncture tapped trees, *Journal of the Rubber Research Institute of Malaysia* **29**: 86–95

Hamzah, S. bte, Gomez, J. B. 1982 Some structural factors affecting the productivity of *Hevea brasiliensis*. III. Correlation studies between structural factors and plugging, *Journal of the Rubber Research Institute of Malaysia* **30**: 148–60

Hamzah, S. bte, Mahmood, A. A., Sivanadyan, K., Gomez, J. B. 1976 Effects of mineral deficiencies on bark anatomy of *Hevea brasiliensis*. *Proceedings of the International Rubber Conference, Kuala Lumpur 1975*, vol. 2 pp. 165–80

Han, K. J., Maclean, R. J. 1983 Commercial evaluation of ultra-low volume technique for weed control on plantations in Johore, *Proceedings of the Rubber Research Institute of Malaysia Planters' Conference, Kuala Lumpur 1983*, pp. 245–60

Hancock, T. 1857 *Personal narrative of the origin and progress of the caoutchouc or India-rubber manufacture in England.* Longman, London

Hanower, P., Brzozowska, J. 1977 Les phénol-oxydases et la coagulation. Colloque sur la physiologie du latex de l'*Hevea brasiliensis*, *Travaux et documents de l'ORSTOM* **68**: 25

Hansell, J. R. F. (ed) 1981 Transmigration settlements in central Sumatra: identification, evaluation and physical planning, *Land Research Development Study* 33, Land Resources Development Centre, Surbiton

Hardenburg, W. E. 1912 *The Putamayo*. London

Hardjono, A., Angkapradipta, P. 1976 Discriminatory nutrition in Java and South Sumatra, *Symposium of the International Rubber Research and Development Board, Bogor, Indonesia, 8–9 November 1976*

Haridas, G. 1981 Selection, preparation and maintenance of nurseries. In *Training manual on soil management and nutrition of Hevea*. Rubber Research Institute of Malaysia, Kuala Lumpur, pp. 150–61

Haridas, G. 1984 The influence of irrigation on latex flow properties and yield of different *Hevea* cultivars, *Procedings of the International Conference on Soils and Nutrition of Perennial Crops, Kuala Lumpur August 1984*

Haridas, G., Sivanadyan, K., Tan, K. T., P'ng, T. C., Pushparajah, E. 1976 Interrelationship between nutrition and exploitation of *Hevea brasiliensis*, *Proceedings of the International Rubber Conference Kuala Lumpur 1975*, pp. 263–77

Haridasan, V. 1977 Utilization of rubber seeds in India, *Rubber Board Bulletin (India)* **14**: 19–24

Harper, R. S. 1973 Ground flora under rubber in Thailand, *PANS* **19**: 71–5

Harris, A. S., Siemonsma, J. S. 1975 Penilaian lanjutan beberapa klon harapan dan anjuran bahan tanaman karet di Sumatera Utara dewasa ini [Further evaluation of several promising rubber clones and rubber planting material recommendations in North Sumatra], *Menara Perkebunan* **43**: 227–33

Harris, E. M. 1979 Utilization of waste wood as fuel and chemical resource, *Planters' Bulletin of the Rubber Research Institute of Malaysia* No. **160**: 118–24

Hartley, C. W. S. 1977 *The oil palm*, 2nd edn. Longman, London

Hashim, I., P'ng, T. C. 1980 The micro-X tapping system, *Planters' Bulletin of the Rubber Research Institute of Malaysia* No. **164**: 112–23

Hawksworth, D. L. 1972 *Ustulina deusta, Commonwealth Mycological Institute Descriptions of Pathogenic Fungi and Bacteria* No. 360

Hebant, C., Devic, C., Fay, E. de C. 1981 Organisation fonctionelle du tissue producteur de l'*Hevea brasiliensis*, *Revue Générale du Caoutchouc et des Plastiques* **58**: 97–100

Hernandez, F. 1649 *Rerum medicarum Novae Hispaniae thesaurus*. Rome, pp. 50–1 [Report of a survey made in 1570–77]

Heubel, G. A. 1939 Voorloopige resultaten van eenige plantverband en uitdunningsproeven bij *Hevea* in Zuid-Sumatra, *De Bergcultures* **13**: 641–52; 682–95

Heusser, C. 1919 Over de voorplantings organen van *Hevea brasiliensis* Muel.-Arg. [On the propagating organs in *Hevea brasiliensis* Muel.-Arg.], *Archief voor de Rubbercultuur Nederlandsch-Indië* **3**: 455–514

Hilton, R. H. 1952 Bird's eye spot leaf disease of the *Hevea* tree caused by *Helminthosporium heveae* Petch., *Journal of the Rubber Research Institute of Malaya* **14**: 42–92

Hilton, R. H. 1958 Pink disease of *Hevea* caused by *Corticium salmonicolor* Berk. et Br., *Journal of the Rubber Research Institute of Malaya* **15**: 275–92

Ho, C. C., Subramaniam, A., Young, W. M. 1976 Lipids associated with the particles in Hevea latex, *Proceedings of the International Rubber Conference, Kuala Lumpur 1975*, vol. **2** pp. 441–56

Ho, C. Y. 1976 Clonal characters determining the yield of *Hevea brasiliensis*, *Proceedings of the International Rubber Conference, Kuala Lumpur 1975*, vol. **2**, pp. 27–38

Ho, C. Y. 1978 Conservation and utilisation of *Hevea* genetic resources, *SABRAO Workshop on Genetic Resources, Kuala Lumpur*, pp. 18–27

Ho, C. Y. 1979 *Contributions to improve the effectiveness of breeding, selection and planting recommendations of* Hevea brasiliensis. Thesis, Faculty of Agricultural Science, Ghent, Belgium

Ho, C. Y. 1986 Rubber, *Hevea brasiliensis*. In FAO, *Breeding for Durable Resistance in Perennial Crops*. FAO Plant Production and Protection Paper 70, pp. 85–114

Ho, C. Y., Chan, H. Y., Lim, T. M. 1974 Environmax planting recommendations – a new concept in choice of clones, *Proceedings of the Rubber Research Institute of Malaysia Planters' Conference, Kuala Lumpur 1974*, pp. 293–310

Ho, C. Y., Khoo, S. K., Meignanaratnam, K., Yoon, P. K. 1980 Potential new clones from mother tree selection, *Proceedings of the Rubber Research Institute of Malaysia Planters' Conference, Kuala Lumpur 1979*, pp. 201–18

Ho, C. Y., Leong, Y. S., Jeyathevan, V. 1973a Responses of clones and seedlings to stimulation in large scale variety trials. *Proceedings of the Rubber Research Institute of Malaysia Planters' Conference, Kuala Lumpur 1973*, pp. 101–21

Ho, C. Y., Narayanan, R., Chen, K. T. 1973b Clonal nursery studies in Hevea. I. Nursery yields and associated structural characteristics and their variation. *Journal of the Rubber Research Institute of Malaya* **23**: 305–16

Ho, C. Y., Ong, S. H. 1981 Potentials of wide crosses in Hevea breeding, *Third International Congress SABRAO, Kuala Lumpur*, pp. 447–69

Ho, C. Y., Paardekooper, E. C. 1965 Application of stimulants to the virgin bark in clone trials, *Planters' Bulletin of the Rubber Research Institute of Malaya* No. 80: 150–7

Ho, H. H., Liang, Z. R., Zhuang, W. J., Yu, Y. N. 1984 *Phytophthora* spp. from rubber tree plantations in Yunnan Province of China, *Mycopathologia* **86**: 121–4

Holliday, P. 1970 South American Leaf Blight (*Microcyclus ulei*) of *Hevea brasiliensis*. *Phytopathology Paper No. 12: 31*

Holt, E. G. 1943 Amazonian caucho. *United States Development Corporation Information Bulletin* No. 6

Homans, L. N. S., Dalfsen, J. W. van, Gils, G. E. van 1948 Complexity of fresh *Hevea* latex, *Nature, London* **161**: 177–8

Huang, Z, Zheng, X. 1983 Rubber cultivation in China, *Proceedings of the Rubber Research Institute of Malaysia Planters' Conference, Kuala Lumpur 1983*, pp. 31–44

Hunt, L. K. 1983 Response of selected clones to premature tapping, *Proceedings of the Rubber Research Institute of Malaysia Planters' Conference, Kuala Lumpur 1983*, pp. 159–70

Hurov, H. R. 1961 Green-bud strip budding of two to eight month old rubber seedlings, *Proceedings of the Natural Rubber Research Conference, Kuala Lumpur 1960*, pp. 419–29

Husin, S. bin M., Chin, H. F., Hor, Y. L. 1981 Fruit and seed development in *Hevea* (clone RRIM 600) in Malaysia, *Journal of the Rubber Research Institute of Malaysia* **29**: 101–9

Hutchison, F. W. 1958 Defoliation of *Hevea brasiliensis* by aerial spraying, *Journal of the Rubber Research Institute of Malaya* **15**: 241–74

Ibrahim, M. T. 1979 Herbicides. In *Training manual on crop protection and weed control in rubber plantations*. Rubber Research Institute of Malaysia, Kuala Lumpur, pp. 133–46

Ikram, A., Mahmud, A. W. 1984 Endomycorrhizal fungi in soils under rubber *Journal of the Rubber Research Institute of Malaysia* **32**: 198–206

Indonesia Department of Trade and Cooperatives 1981 *A study on the production potential of natural rubber in Indonesia*. DTC Jakarta

Inglis, B. 1973 *Roger Casement*. Hodder and Stoughton, London

Institut de Recherches sur le Caoutchouc en Afrique 1976 Etude de la résistance au vent: influence de la qualité du bois, *Rapport du Deuxième Semestre. Série Agronomie Physiologie*: 1–8

Institut de Recherches sur le Caoutchouc en Afrique 1983 *Annual Report*, pp. 61–2

International Fertilizer Development Center 1983 First full-scope project achieves goal, *International Development Center Report* **8**: 1–2

International Rubber Study Group 1974 *World rubber statistics handbook* vol.1

1946–1970. IRSG, London

International Rubber Study Group 1984 *World rubber statistics handbook* vol. 2 1965–1980. IRSG, London

Iyer, G. C., Chan, H. Y., Pushparajah, E. 1977 Computerised approach towards the diagnosis of fertilizer requirements in rubber, *Proceedings of the Conference on Classification and Management of Tropical Crops (CLAMATROPS), Kuala Lumpur, August 1977*, pp. 349–55

Jaafar, H. bin 1984 Effect of sodium dikegulac on the establishment of budded stumps and stumped buddings and multiplication of source bushes in *Hevea brasiliensis*, *Journal of the Rubber Research Institute of Malaysia* **32:** 73–81

Jackson, C. 1968 *Planters and speculators. Chinese and European agricultural enterprise in Malaya 1786–1921.* University of Malaya Press, Singapore

Jacob, J. L., Prevot, J. C., d'Auzac, J. 1983 Augmentation de la production de l'Hevea par l'éthylène. *Revue Générale du Caoutchouc et des Plastiques* **60:** 87–9

Jayaratnam, K., Nehru, C. R. 1984 White grubs and their management in rubber nursery, *Pesticides* **18:** 27–9

Jayaratne, A. H. R. 1982 Endomycorrhizas of rubber growing in soils in Sri Lanka, *Journal of the Rubber Research Institute of Sri Lanka* **60:** 47–57

Jayasekara, N. E. M., Samaranayake, P., Karunasekara, K. B. 1977 Initial studies on the nature of genotype-environment interaction in some Hevea cultivars, *Journal of the Rubber Research Institute of Sri Lanka* **54:** 33–42

Jayasekera, N. E. M., Senanayake, Y. D. A. 1971 A study of growth parameters in a population of nursery rootstock seedlings of *Hevea brasiliensis* cv. Tjir 1, *Quarterly Journal of the Rubber Research Institute of Ceylon* **48:** 66–81

Jeevaratnam, A. J. 1967 A note on boron toxicity in young replantings. *Rubber Research Institute Ceylon Bulletin* **2** (1, 2): 22–3

Jeevaratnam, A. J. 1969 Relative importance of fertilizer application during pre- and post-tapping phases of *Hevea*, *Journal of the Rubber Research Institute of Malaya* **21:** 175–80

Jeyasingham, T. 1974 Rubber wood is abundant and accessible; will it ever be successfully exploited? *Quarterly Journal of the Rubber Research Institute of Sri Lanka* **51:** 13–15

Joachim, A. W. R. 1955 The soils of Ceylon, *Tropical Agriculturist* **111**(3): 161–72

John, C. K., Mahmud , A. W., Mohamed, K., Pushparajah, E. 1977 Utilisation of wastes from natural rubber producing factories, *Planter, Kuala Lumpur* **53:** 449–67

John, K. P. 1964 Minor root diseases – a reappraisal, *Planters' Bulletin of the Rubber Research Institute of Malaysia* No. **75:** 244–8

Johnston, B. F., Kilby, P. 1975 *Agricultural and structural transformation in late-developing countries.* Oxford University Press, Oxford

Jollands, P., Turner, P. D., Kartika, D., Soebagyo, F. X. 1983 Use of the controlled droplet application (c.d.a.) technique for herbicide application, *Planter, Kuala Lumpur* **59:** 388–407

Joseph, K. T. 1970 The effect of phosphorus on nitrogen fixation by the cover crop *Pueraria phaseoloides* on a latosol, *Planter, Kuala Lumpur* **46:** 153–6

Jumpasut, P. 1985 Forecasts of rubber production and consumption to 2000. Paper tabled at *Proceedings of the Twenty-ninth Assembly, International Rubber Study Group, Abidjan, Ivory Coast, November 1985*

Junqueira, N. T. V, Gasparotto, L., Moraes, V. H. F., Martins e Silva, H., Lim, T. M. 1986 New diseases caused by virus, fungi and a bacterium on rubber from Brazil and their impact on international quarantine, *Proceedings of the International Rubber Conference, Kuala Lumpur 1985*, in press

Kalam, M. A., Amma, M. K., Punnoose, K. I., Potty, S. N. 1980 Effect of fertilizer application on growth and lead content of some important *Hevea* clones, *Rubber Board (India) Bulletin* **16:** 19–30

Kasasian, L. 1971 *Weed control in the tropics.* Leonard Hill, London

Kashyap, R. K, Verma, A. N., Bhanot, J. P. 1984 Termites of plantation crops, their damage and control, *Journal of Plantation Crops* **12**: 1–10

Kew Gardens Correspondence 1860 India Office Miscellaneous Reports 5. India economic products, *Cinchona*. (Markham correspondence on *Cinchona*)

Kew Gardens Correspondence 1876a India Office Miscelllaneous Reports Caoutchouc I. Wickham, H. A. The introduction of the India rubber tree to India (Description of Collection in Brazil. Undated)

Kew Gardens Correspondence 1876b India Office Miscellaneous Reports Caoutchouc I. (Letter dated 6 March 1876 from Wickham at Tapajos to Hooker on collection of *Hevea* seed)

Kew Gardens Correspondence 1876c India Office Miscellaneous Reports Caoutchouc I. (Letter dated 16 September 1876 from Ceylon Botanic Garden to Kew)

Kew Gardens Correspondence 1877a India Office Miscellaneous Reports Caoutchouc I. Cross, R. M. Report on plants and seeds of India-rubber trees of Para and Ceara and balsam of Capaiba, 29 March 1877

Kew Gardens Correspondence 1877b India Office Miscellaneous Reports Caoutchouc I. (Letter T. Thistleton Dyer to Sir Louis Mallet, India Office, 8 September 1877)

Kew Gardens Correspondence 1892 India Office Miscellaneous Reports Caoutchouc I. (Report from Ceylon Botanic Gardens to Kew)

Kew Gardens Correspondence 1907 Malaya Rubber (1852–1908). (Ridley's reports on 1905 tapping, made in 1907)

Khoo, S. K., Yoon, P. K., Meignanaratnam, K., Gopalan, A., Ho, C. Y. 1982 Early results of mother tree (ortet) selection, *Planters' Bulletin of the Rubber Research Institute of Malaysia* No. **171**: 33–49

Kira, T. 1978 Primary production of Pasoh forest – a synthesis, *Malayan Nature Journal* **30**: 291–7

Knorr, K. E. 1945 *World rubber and its regulation*. Stanford University Press, California

Köppen, W. S. 1923 *Die Klimate der Erde*. Walter der Gruyter, Berlin

Krisanasap. S. 1984 Rubber Research Institute of Thailand, personal communication

Landon, J. R. (ed) 1984 *Booker tropical soil manual*. Booker Agriculture International Ltd and Longman, London

Langford, M. H. 1953 *Hevea* diseases of the Amazon valley, *Boletím Tecnico do Instituto de Pesquisa Agropecuaria do Norte* No. 27

Langford, M. H., Osores, A. 1965 Enfermedades de jebe y recomendaciones para su control, *Boletim Técnico SIPA* **63**: 1–16

Lau, C. H. 1981 Fertilizer forms and characteristics. In *Training manual on soils, soil management and nutrition of* Hevea. Rubber Research Institute of Malaysia, Kuala Lumpur, pp. 162–74

Lau, C. H., Pushparajah, E., Yap, W. C. 1972 Evaluation of various soil-K indices in relation to nutrition, growth and yield of *Hevea brasiliensis*, *Proceedings of the Second ASEAN Soil Conference, Djakarta 17–19 July 1972*

Lau, C. H, Pushparajah, E., Yap, W. C. 1977 Evaluation of various soil-P indices of *Hevea*, *Proceedings of the Conference on Chemistry and Fertility of Tropical Soils, Kuala Lumpur 1973*, pp. 103–11

Laurence, J. C. 1931 *The world's struggle for rubber 1905–1931*. University of Minnesota

Leong, S. K., Leong, W., Yoon , P. K. 1982 Harvesting shoots for rubber extraction in *Hevea, Journal of the Rubber Research Institute of Malaysia* **30**: 117–22

Leong, S. K., Ooi, C. B., Yoon, P. K. 1976 Further development in the production of cuttings and clonal rootstocks in *Hevea, Proceedings National Plant Propagation Symposium, Kuala Lumpur 1976*, Rubber Research Institute of Malaysia, Kuala Lumpur, pp. 154–65

Leong, S. K., Yoon, P. K. 1979 Effects of bud types on early scion growth of *Hevea, Journal of the Rubber Research Institute of Malaysia* **27**: 1–7

Leong, T. T., Tan, H. T. 1978 Results of Chemara puncture-tapping trials, *Proceedings of the Rubber Research Institute of Malaysia Planters' Conference, Kuala Lumpur 1977*, pp. 111–31

Leong, W., Yip, E., Subramaniam, A., Loke, K. M., Yoon, P. K. 1986 Modification of Mooney viscosity and other rubber properties by crown budding. *Planters' Bulletin of the Rubber Research Institute of Malaysia* **186**: 29–37

Leong, W., Yoon, P. K. 1976 RRIM crown budding trials – progress report, *Proceedings of the Rubber Research Institute of Malaysia Planters' Conference, Kuala Lumpur 1976*, pp. 87–115

Leong, W., Yoon, P. K. 1978 Some properties of latex and raw rubber from three-part trees, *Proceedings of the International Rubber Research and Development Board Symposium, Kuala Lumpur 1978*

Leong, W., Yoon, P. K. 1982a Modification of crown development of *Hevea brasiliensis* by cultural practices. I. Pruning, *Journal of the Rubber Research Institute of Malaysia* **30**: 50–7

Leong, W., Yoon, P. K. 1982b, Modification of crown development of *Hevea brasiliensis* by cultural practices. II. Tree density, *Journal of the Rubber Research Institute of Malaysia* **30**: 123–30

Leong, W., Yoon, P. K. 1984 Effect of low and controlled pruning on growth and yield of *Hevea brasiliensis, Proceedings of the Rubber Research Institute of Malaysia Planters' Conference, Kuala Lumpur 1983*, pp. 261–85

Leong, W., Yoon, P. K. 1985 The role of cuttings and clonal rootstocks in *Hevea* cultivation, *Proceedings of the International Rubber Conference, Rubber Research Institute of Sri Lanka, 1984*

Levandowski, D. W. 1959 Propagation of clonal *Hevea brasiliensis* by cuttings, *Tropical Agriculture, Trinidad* **36**: 247–57

Leveque, J., Loyen, G., Dion, B. 1976 Commercial experience in tapping and processing polybag rubbers, *Proceedings of the International Rubber Conference, Kuala Lumpur 1975*, vol. 2, pp. 427–33

Lian, C. H., Cheam, S. T. 1974 The effects of Ethrel application on the optimum replanting ages of rubber and on the feasibility of maximising early yields in estates, *Malaysian Agricultural Research* **3**: 77–85

Liang, D. 1983 The effect of potash fertilizer on plantation crops in China, *Potash Review*, Subject 27 No. 5: 1–5

Lim, C. H., P'ng, T. C. 1984 Land application of rubber factory effluent on oil palm and rubber, *Proceedings of the International Conference on Soils and Nutrition of Perennial Crops, Kuala Lumpur, August 1984*, pp. 13–15

Lim, S. C. 1969 An agro-economic study of intercrops on rubber smallholdings, *Rubber Research Institute of Malaya Economics and Planning Division: Report No. 6*

Lim, S. C. 1972 Economic viability of future smallholdings, *Proceedings of the Rubber Research Institute of Malaysia Planters' Conference, Kuala Lumpur 1972*, pp. 17–33

Lim, S. C. 1976 Marketing of rubber by smallholders in Peninsular Malaysia, *Malaysian Rubber Research and Development Board, Kuala Lumpur, Monograph* No. 2

Lim, S. C. 1985 Malaysian smallholders' rubber: issues and approaches in further processing and manufacturing, *Malaysian Rubber Research and Development Board, Kuala Lumpur, Monograph* No. 10

Lim, S. C. 1986 Malaysian natural rubber production: trend to 2000, *International Rubber Study Group Ninety Sixth Group Meeting, London, June 1986*

Lim, S. C., Chong, K. M. 1973 Merits of estate participation and incentive wage system in public sector land development, *Proceedings of the Rubber Research Institute of Malaysia Planters' Conference, Kuala Lumpur 1973*, pp. 17–34

Lim, S. C., Ho, C. Y., Yoon, P. K. 1973 Economics of maximizing early yields and shorter immaturity, *Proceedings of the Rubber Research Institute of Malaysia Planters' Conference, Kuala Lumpur 1973*, pp. 1–16

Lim, S. C., Noor, G. 1974 Use of public funds on unsatisfactory land schemes, *Proceedings of the Rubber Research Institute of Malaysia Planters' Conference, Kuala Lumpur 1974*, pp. 1–14

Lim, T. M. 1971 Stem rot of *Hevea* caused by *Phellinus noxius*. In Wastie, R. L., Wood, B. J. (eds) *Crop protection in Malaysia*. Incorporated Society of Planters, Kuala Lumpur, pp. 221–7

Lim, T. M. 1982 Recent developments in the chemical control of rubber leaf diseases in Malaysia, *Proceedings of the International Conference on Plant Protection in the Tropics, Kuala Lumpur 1982*, pp. 219–31

Lim, T. M., Hashim, I. 1977 Folex – a promising rainfast defoliant for rubber, *Planters' Bulletin of the Rubber Research Institute of Malaysia* No. **148**: 3–9

Lim, T. M., Kadir, A. 1983 Triclopyr, an alternative arboricide to 2,4,5-T and sodium arsenite, *Planters' Bulletin of the Rubber Research Institute of Malaysia* No. **174**: 17–21

Lim, T. M., Kadir, A. A. S. A. 1978 Thermal fogging – a promising new method for controlling rubber leaf diseases. In Amin, L. L., Kadir, A. A. S. A., Soon, L. G., Singh, K. G., Tan, A. M., Varghese, G. (eds) *Proceedings of the Plant Protection Conference, Kuala Lumpur 1978*, 72–81

Lim, T. S. 1978 Nutrient uptake of clone RRIM 600 in relation to soil influence and fertilizer needs, *Proceedings of the Rubber Research Institute of Malaysia Planters' Conference, Kuala Lumpur 1977*, pp. 166–85

Ling, A, H., Mainstone, B. J. 1983 Effects of burning of rubber timber during land preparation on soil fertility and growth of *Theobroma cacao* and *Gliricidia maculata, Planter, Kuala Lumpur* **59**: 52–9

Liyanage, A. de S. 1977 Economics of white root disease control, *Bulletin of the Rubber Research Institute of Sri Lanka* **12**: 51–7

Liu, P. S. W. 1977 A supplement to a host list of plant diseases in Sabah, Malaysia, *Phytopathological Papers* No. 21

Liu, S. 1979 Weeds in rubber cultivation and their control. In *Training manual on crop protection and weed control in rubber plantations*. Rubber Research Institute of Malaysia, Kuala Lumpur, pp. 116–32

Lotfy, N., Paranjothy, K. 1978 Induction and control of flowering in Hevea, *Journal of the Rubber Research Institute of Malaysia* **26**: 123–34

Low, F. C. 1978 Distribution and concentration of major soluble carbohydrates in Hevea latex, *Journal of the Rubber Research Institute of Malaysia* **26**: 21–32

Low, F. C., Gomez, J. B., Hashim, I. 1983 Carbohydrate status of exploited Hevea II. Effect of micro-tapping on the carbohydrate content of latex, *Journal of the Rubber Research Institute of Malaysia* **31**: 27–34

Low, F. C., Yeang, H. Y. 1985 Effects of ethephon stimulation on latex invertase in *Hevea, Journal of the Rubber Research Institute of Malaysia* **33**: 37–47

Lowe, J. S. 1964 Copper sulphate as a yield stimulant for *Hevea brasiliensis* II. Techniques for application of copper sulphate, *Journal of the Rubber Research Institute of Malaya* **18**: 261–8

Lowe, J. S. 1969 The integration of livestock with rubber, *Planter, Kuala Lumpur* **45**: 18–21

Lukman 1979 Hubungan antara pembukaan sadapan dan stimulasi dengan produksi dan sifat sekunder klon AVROS 2037 [Relationships of size at opening and stimulation to the yield and secondary characteristics of clone AVROS 2037], *Bulletin Balai Penelitian Perkebunan Medan* **10**: 189–200

Lukman 1980a Pembukaan sadapan dan stimulasi sehebungan dengan besarnya lilit batang [Opening of tapping and stimulation in relation to girth size], *Bulletin Balai Penelitian Perkebunan Medan* **11**: 23–38

Lukman 1980b Penundaan pembukaan sadapan dari lilit batang 45 cm menjadi 50 cm pada klon GT 1 [Opening of rubber clone GT 1 delayed from 45 to 50 cm girth], *Bulletin Balai Penelitian Perkebunan Medan* **11**: 115–25

Lukman 1983 Revised international tapping notation for exploitation systems,

Journal of the Rubber Research Institute of Malaysia **31**: 130–40

Lustinec, J., Chai, K. C., Resing, W. L. 1966 L'aire drainée chez les jeunes arbres de l' *Hevea brasiliensis, Revue Générale du Caoutchouc et des Plastiques* **43** (10): 1343–54

Lustinec, J., Resing, W. L., Simmer, Y. 1968 Distinction des deux composants de l'aire drainée sur le tronc de l' *Hevea brasiliensis, Biologia Plantarum (Prague)* **10**: 284

Lustinec, J., Simmer, J., Resing, W. L. 1969 Trunk contraction of *Hevea brasiliensis* due to tapping, *Biologia Plantarum (Prague)* **11**: 236–44

Mass, J. G. J. A. 1934 Die selectie van *Hevea brasiliensis* bij s'lands caoutchoucbedrijf, *Archief voor de Rubbercultuur in Nederlandsch-Indië* **18**: 58–65

Macquer, P. J. 1770 *Histoire et mémoires de l'Académie pour 1768.* Paris, pp. 209–17

Mahmud, A. W. 1978 Phosphates for rubber and legume covers in Malaysia, *Proceedings of the International Rubber Research and Development Board Symposium, Kuala Lumpur 1978*, Part A, Section 3 pp. 1–11

Mainstone, B. J. 1963a Manuring of *Hevea*: effects of 'triple' superphosphate on transplanted stumps in Nigeria, *Empire Journal of Experimental Agriculture* **31**: 53–9

Mainstone, B. J. 1963b Manuring of *Hevea*. VI. Some long-term manuring effects, with special reference to phosphorus, on one of the Dunlop (Malaya) experiments, *Empire Journal of Experimental Agricultre* **31**: 175–85

Mainstone, B. J. 1963c Residual effects of ground cover and of continuity of nitrogen fertilizer treatments, applied prior to tapping on the yield and growth of *Hevea brasiliensis, Empire Journal of Experimental Agriculture* **31**: 213–25

Mainstone, B. J. 1963d The effects of nitrogen and phosphorus fertilizers on *Hevea brasiliensis* when applied after commencement of tapping, *Empire Journal of Experimental Agriculture* **31**: 226–42

Mainstone, B. J. 1969 Residual effects of ground cover and nitrogen fertilization of *Hevea* prior to tapping, *Journal of the Rubber Research Institute of Malaya* **21**: 113–25

Mainstone, B. J. 1970 Planting density – early effects on growth pattern of *Hevea brasiliensis, Journal of the Rubber Research Institute of Malaya* **23**: 56–67

Mainstone, B. J., Tan, K. S. 1964 Copper sulphate as a yield stimulant for *Hevea brasiliensis*. I. Experimental stimulation of 1931 budded rubber with 2, 4-D or 2,4,5-T in the presence and absence of copper sulphate injection, *Journal of the Rubber Research Institute of Malaya* **18**: 253–60

Majumder, S. K. 1964a Chromosome studies of some species of *Hevea, Journal of the Rubber Research Institute of Malaya* **18**: 269–73

Majumder, S. K. 1964b Studies of the germination of pollen of *Hevea brasiliensis in vivo* and on artificial media, *Journal of the Rubber Research Institute of Malaya* **18**: 185–93

Majumder, S. K. 1967 Male sterile clones in *Hevea brasiliensis, Canadian Journal of Botany* **45**: 145–51

Malaya, Department of Agriculture 1934 Bark consumption and bark reserves on rubber smallholdings in Malaya, *Economic Series* No. 4, Kuala Lumpur

Malaysian Rubber Research and Development Board 1983 *Malaysian natural rubber industry 1983–2000.* Report of the task force of experts. *MRRDB*, Kuala Lumpur

Markham, C. R. 1880 *Peruvian bark.* John Murray, London

Marty, G. 1966 La préparation des terrains, forestiers par empoisonnement des arbres aux Nouvelles Hebrides, *Oléagineux* **21**: 589–91

Mathews, G. A. 1982 Pesticide application in the tropics, *Proceedings of the International Conference on Plant Protection in the Tropics, Kuala Lumpur 1982*, pp. 671–9

McCulloch, G. C., Vanialingam, T. 1972 Highlands Research Unit Ethrel trials: preliminary results, *Proceedings of the Rubber Research Institute of Malaya Planters' Conference, Kuala Lumpur 1971*, pp. 118–25

McFadyean 1944 *The history of rubber regulation 1934–1943.* Allen and Unwin, London

McHale, T. R. 1964 Changing technology and shifts in the supply and demand for rubber: an analytical history, *Malayan Economic Review* **9**(2): 24–48

McMullen, A. I. 1959 Nucleotides of *Hevea brasiliensis* latex: a ribonucleoprotein component, *Biochemical Journal* **72**: 545

McMullen, A. I. 1962 Particulate ribonucleoprotein components of *Hevea brasiliensis* latex, *Biochemical Journal* **85**: 491–5

Melis, R. 1978 *Rapport de stage effectué a l'IRCA sur les cultures intercalaires 15 Juillet 1977–15 Janvier 1978.* University of Wageningen, The Netherlands

Middleton, K. R. 1961 Inconsistencies in the response of *Hevea brasiliensis* to phosphatic fertilizers in field trials and pot experiments with soil, *Proceedings of the Natural Rubber Research Conference, Kuala Lumpur 1960*, pp. 89–101

Middleton, K. R. M., Chin, T. T., Iyer, G. C. 1965 A comparison of rock phosphate with superphosphate, and of ammonium sulphate with sodium nitrate, as sources of phosphorus and nitrogen for rubber seedlings. II. Association with abnormal growth and effect on wood strength, *Journal of the Rubber Research Institute of Malaya* **19**: 108–19

Milanez, F. R. 1948 Segunda nota sobre laticiferos, *Lilloa* **16**: 193–6

Milford, G. F. J., Paardekooper, E. C., Ho, C. Y. 1969 Latex vessel plugging, its importance to yield and clonal behaviour, *Journal of the Rubber Research Institute of Malaya* **21**: 274–82

Mohamed, W. E. W. 1978 Utilization of ground vegetation in rubber plantation for animal rearing, *Proceedings of the Rubber Research Institute of Malaysia Planters' Conference, Kuala Lumpur 1977*, pp. 265–81

Mohamed, W. E. W., Chee, Y. K. 1976 Maximizing returns in immature rubber smallholdings, *Proceedings of the Rubber Research Institute of Malaysia Planters' Conference, Kuala Lumpur 1976*, pp. 34–43

Mohamed, W. E. W., Hamidy, M. Z. A. 1983 Performance of Dorset Horn crossbreds under rubber, *Proceedings of the Rubber Research Institute of Malaysia Planters' Conference, Kuala Lumpur 1983*, pp. 235–44

Mohinder Singh, Talibudeen, O. 1969 Thermodynamic assessment of the nutrient status of rubber-growing soils, *Journal of the Rubber Research Institute of Malaya* **21**: 240–9

Moir, G. F. J. 1959 Ultracentrifugation and staining of *Hevea* latex, *Nature, London* **161**: 177

Montenegro, A. 1908 *Album do Estado do Pará, oito anos do Governo (1901 a 1909)* [Album of Pará State, eight years of government (1901–09)]. Chaponet, Paris

Moore, A., Piddlestone, J.H. 1937 Subur-type smokehouses, *Journal of the Rubber Research Institute of Malaya* **7**: 147–64

Moraes, V. H. F. 1978 Mini-sangria da seringueira. Ensaios preliminares com o clone FX 25. *Pesquisa Agropecuaria Brasileira* **13**(1): 1–8

Morel, E. D. 1919 *Red rubber. The story of the rubber slave trade flourishing on the Congo in the year of grace 1906.* Fisher Unwin, London

Morgan-Jones, G. 1967 *Ceratocystis fimbriata, Commonwealth Mycological Institute Descriptions of Pathogenic Fungi and Bacteria* No. 141

Morris, J. E. 1964 Sole crepe, *Planters' Bulletin of the Rubber Research Institute of Malaya* No. **74**: 155–75

Morris, J. E., Nielsen, P. S. 1961 Properties of low-grade rubber (estate brown crepes), *Proceedings of the Natural Rubber Research Conference, Kuala Lumpur 1960*, pp. 626–35

Mulder, H. H. 1941 De nieuwe 'domper-kroon ontwikkelings methode' bij jonge rubber, *De Bergcultures* **15**: 472–4

Muzik, T. J., Cruzado, H. J. 1958 Transmission of juvenile rooting ability from seedlings to adults of *Hevea brasiliensis, Nature, London* **181**: 1288

Nadarajah, M. 1969 The collection and utilization of rubber seed in Ceylon, *Bulletin*

of the Rubber Research Institute of Ceylon **4**: 23–32

Nadarajah, M., Abeyasinjhe, A., Dayaratne, W. C., Tharmalingam, R. 1973 The potentialities of rubber seed collection and its utilization in Sri Lanka, *Bulletin of the Rubber Research Institute of Sri Lanka* **8**(1): 9–21

Nadarajapillai, N., Wijewantha, R. T. 1967 Productivity potentials of rubber seed, *Bulletin of the Rubber Research Institute of Ceylon* **2**: 8–17

Nair C. K. N. 1973 Mineral nutrition of *Hevea brasiliensis*, *Proceedings International Rubber Research and Development Symposium, Puneah, Indonesia.* 5 pp.

Nair, P. K. R., Varghese, P. T. 1980 Recent advances in the management of coconut-based systems in India. In Furtado, J. F. (ed.) *Tropical Ecology and Development*, part 1, pp. 569–80. International Society for Tropical Ecology, Kuala Lumpur, Malaysia

Nair, V. K. B., George, P. J., Amma, C. K. S. 1976 Breeding improved Hevea clones in India, *Proceedings of the International Rubber Conference, Kuala Lumpur 1975*, pp. 45–54

Najib, L., Paranjothy, K. 1978 Induction and control of flowering in *Hevea*, *Journal of the Rubber Research Institute of Malaysia* **26**: 123–34

Nandris, D., Tran, V. C., Geiger, J-P., Omont, H., Nicole, M. 1985 Remote sensing in plant diseases using infrared colour aerial photography; application trials in the Ivory Coast to root diseases of *Hevea brasiliensis*, *European Journal of Forest Pathology* **15**: 11–21

Napitupulu, L. A. 1977 Planting density experiment on rubber clone AVROS 2037, *Bulletin Balai Penelitian Perkebunan Medan* **8**: 99–104

Napper, R. P. N. 1932 A scheme of treatment for control of *Fomes lignosus* in young rubber areas, *Journal of the Rubber Research Institute of Malaya* **4**: 34–8

Narayanan, R., Gomez J. B., Chen, K. T. 1973 Some structural factors affecting the productivity of *Hevea brasiliensis*. II. Correlation studies between structural factors and yield, *Journal of the Rubber Research Institute of Malaya* **23**: 285–97

Narayanan, R., Ho, C. Y. 1970 Yield-girth relationship studies on *Hevea*, *Journal of the Rubber Research Institute of Malaya* **23**: 23–31

Narayanan, R., Ho, C. Y. 1973 Clonal nursery studies in *Hevea*. II. Relationships between yield and girth, *Journal of the Rubber Research Institute of Malaysia* **23**: 332–8

Narayanan, R., Ho, C. Y., Chen, K. T. 1974a Clonal nursery studies in *Hevea*. III. Correlation between yield, structural characters, latex constituents and plugging index, *Journal of the Rubber Research Institute of Malaysia* **24**: 1–14

Narayanan, R., Ho, C. Y., Subramaniam, S., Jeyathevan, V. 1974b Relative efficiency of simple lattice designs in clone trials of *Hevea*, *Journal of the Rubber Research Institute of Malaysia* **24**: 26–38

National Academy of Sciences 1977 Guayule: an alternative source of natural rubber. National Academy of Sciences, Washington

Newall, W. 1967 Clonal responses of flooding, *Planters' Bulletin of the Rubber Research Institute of Malaya* No. **90**: 112

Newsam, A. 1967 Clearing methods and root disease control, *Planters' Bulletin of the Rubber Research Institute of Malaya* No. **92**: 176–82

Ng, A. P. 1983 Performance of rootstocks, *Planters' Bulletin of the Rubber Research Institute of Malaysia* No. **175**: 56–63

Ng, A. P., Ho, C. Y., Sultan, M. O., Ooi, C. B., Lew, H. L., Yoon, P. K. 1982 Influence of six rootstocks on growth and yield of six scion clones of *Hevea brasiliensis*, *Proceedings of the Rubber Research Institute of Malaysia Planters' Conference, Kuala Lumpur 1981*, pp. 134–51

Ng, A. P., Sepien, A., Ooi, C. B., Leong, W., Lew, H. L., Yoon, P. K. 1979 Report on various aspects of yield, growth and economics of a density trial, *Proceedings of the Rubber Research Institute of Malaysia Planters' Conference, Kuala Lumpur 1979*, pp. 303–31

Ng, A. P., Yoon, P. K. 1982 Relationships between rootstock and scion in *Hevea*

brasiliensis. In *Abstracts: 21st International Horticultural Congress, Hamburg 1982*, vol. 1, Abstract No. 1318

Ng, C. S. 1967 Some aspects of estate replanting and new planting costs, *Planters' Bulletin of the Rubber Research Institute of Malaya* No. **92**: 164–75

Ng, C. S. 1984 International marketing of natural rubber: changes, challenges and prospects, *Malaysian Rubber Research and Development Board Monograph* No. 9

Ng, C. S., Ng, A. P., Yoon, P. K. 1972 Economics of early opening, *Proceedings of the Rubber Research Institute of Malaya Planters' Conferernce, Kuala Lumpur 1972*, pp. 34–57

Ng, E. K., Ng, C. S., Lee, C. K. 1969 Economic analysis of tapping experiments, *Journal of the Rubber Research Institute of Malaya* **21**: 360–87

Ng, C. S., Sekhar, B. C. 1977 International rubber markets: their role in the rubber trade, *Malaysian Rubber Research and Development Board Monograph* No. 3

Ng, K. Y. 1980 Grasshopper (*Varanga nigricornis*) control by aerial application in Malaysia, *Planter, Kuala Lumpur* **56**: 362–7

Nga, B. H., Subramaniam S. 1974 Variation in *Hevea brasiliensis*. I. Yield and girth data of the 1937 hand-pollinated seedlings, *Journal of the Rubber Research Institute of Malaysia* **24**: 69–74

Ninane, F., David, R. 1971 Problèmes relatifs aux fonctions mathématiques de l'écoulement du latex chez l'*Hevea brasiliensis*, *Revue Générale du Caoutchouc et des Plastiques* **48**(3): 285–9

Noordin, W. D. 1981 Distribution, properties and classification of soils under rubber. *Training manual on soils, management of soils and nutrition of* Hevea. Rubber Research Institute of Malaysia, Kuala Lumpur, pp. 24–44

Nor, A. bin M., Black, J. J., Chan, H. Y. 1982 A comparative feed-back study of two planting techniques: maxi stumps and soil core buddings, *Proceedings of the Rubber Research Institute of Malaysia Planters' Conference, Kuala Lumpur 1981*, pp. 152–9

Nor, A. M., Nayagam, J. 1982 *RRIM economic laboratory – a seven year experience.* Rubber Research Institute of Malaysia, Kuala Lumpur (mimeograph)

Nor, S. M. 1984 Heveawood – Timber of the future, *Planter, Kuala Lumpur* **60**: 370–81

Office of the Rubber Replanting Aid Fund and the Food and Agriculture Organization of the United Nations 1980 *Proposals for a second series of loans to support the accelerated rubber replanting programme in Thailand.* Working Paper AGO:THA/75/021, ORRAF/FAO Project, Rubber Research Centre, Hat Yai, Thailand (mimeograph)

Oldeman, L. R., Frère, M. 1982 A study of agroclimatology of the humid tropics of South-East Asia, *Secretariat of the World Meteorological Organization Geneva Technical Note* No. **179** WMO-No. 597

Omont, H. 1981 Some aspects of the mineral nutrition of young *Hevea* in the Ivory Coast, *Revue Générale du Caoutchouc et des Plastiques* No. 610: 87–94

Omont, H. 1982 Plantation d'*Hévéa* en zone climatique marginale, *Revue Générale du Caoutchouc et des Plastiques* No. **625**: 75–9

Ong, S. H. 1972 Flower induction in *Hevea*, *Proceedings of the Symposium on International Cooperation in Rubber Breeding*, Kuala Lumpur

Ong, S. H. 1979 *Cytotaxonomic investigation of the genus* Hevea. PhD thesis, University of Malaysia

Ong, S. H. 1981 Correlations between yield, girth and bark thickness of RRIM clone trials, *Journal of the Rubber Research Institute of Malaysia* **29**: 1–14

Ong, S. H., Ghani, M. N. bin A., Tan, A. M., Tan, H. 1983 New *Hevea* germplasm – its introduction and potential, *Proceedings of the Rubber Research Institute of Malaysia Planters' Conference, Kuala Lumpur 1983*, pp. 3–17

Ong, S. H., Naimah Ibrahim, Mohamed Nor, Abdul Ghani, Tan, H. 1984 Results of RRIM work on induced mutation and polyploid breeding. *In Compte Rendu du Colloque Exploitation, Physiologie et Amélioration de l'Hevea*: 383–99. Institut de

Recherches sur le Caoutchouc en Afrique/Groupement d'Etudes et de Recherches pour le Développement de l'Agronomie Tropicale, Montpellier, France.

Ong, S. H., Subramaniam, S. 1973 Mutation breeding in *Hevea brasiliensis*. In *Induced mutations in vegetatively propagated plants*. International Atomic Energy Agency, Vienna, pp. 117–27

Ong, S. H., Sultan, M. O., Khoo, S. K. 1984 Promotion plot trials (first series): second report, *Planters' Bulletin of the Rubber Research Institute of Malaysia* No. 178: 22–5

Ooi, C. B., Leong, S. K., Yoon, P. K. 1976 Vegetative propagation techniques in *Hevea*, *Proceedings of the National Plant Propagation Symposium 1976*, Rubber Research Institute of Malaysia, Kuala Lumpur, pp. 116–31

Ooi, C. B., Leong, S. K., Yoon, P. K. 1978 Production of advanced scion/stock plants as polybag planting materials, *Proceedings of the Rubber Research Institute of Malaysia Planters' Conference, Kuala Lumpur 1977*, pp. 3–20

Ooi, J-B. 1963 *Land, people and economy in Malaya*. Longman, London

Ostendorf, F. W. 1948 Twee proeven met meervoudige *Hevea* oculaties, *Archief voor de Rubbercultuur in Nederlandsch-Indië* **26**: 1–19

Othman, R. B., Paranjothy, K. 1980a Induced flowering in young Hevea buddings, *Journal of the Rubber Research Institute of Malaysia* **28**: 149–56

Othman, R. B., Paranjothy, K. 1980b Isolation of *Hevea* protoplasts, *Journal of the Rubber Research Institute of Malaysia* **28**: 61–6

Owen, G. 1951 A provisional classification of Malayan soils, *Journal of the Rubber Research Institute of Malaya* **13**: 20–42

Owen, G. 1953 Determination of available nutrients in Malayan soils, *Journal of the Rubber Research Institute of Malaya* **14**: 109–20

Owen, G., Westgarth, D. R., Iyer, G. C. 1957 Manuring *Hevea*: effects of fertilizers on growth and yield of mature rubber trees, *Journal of the Rubber Research Institute of Malaya* **15**: 29–52

Paardekooper, E. C. 1954 Resultaten van twee onderstamproeven bij *Hevea*, *De Bergcultures* **23**: 551–6

Paardekooper, E. C. 1964 A note on the effect of changes in exploitation methods on profit per acre, *Rubber Research Institute of Malaya Confidential Report* No. 19

Paardekooper, E. C. 1968 Report on tapping and stimulation experiments, *Rubber Research Centre of Thailand, Hat Yai, Report* No. 12/68

Paardekooper, E. C. 1970 Second report on latex flow investigations, *Rubber Research Centre of Thailand, Hat Yai, Report* No. 12/70

Paardekooper, E. C. 1971 Report on Ethrel stimulation experiments 1970/71, *Rubber Research Centre of Thailand, Hat Yai, Report* No. 18/71

Paardekooper, E. C., Langlois, S. J. C., Sookmark, S. 1976 Influence of tapping intensity and stimulation on yield, girth and latex constitution, *Proceedings of the International Rubber Conference, Kuala Lumpur 1975*, vol. 2, pp. 290–314

Paardekooper, E. C., Newall, W. 1977 Considerations of density in *Hevea* plantations, *Planter, Kuala Lumpur* **53**: 143–56

Paardekooper, E. C., Samosorn, S. 1969 Clonal variation in latex flow pattern, *Journal of the Rubber Research Institute of Malaya* **21**: 264–73

Paardekooper, E. C., Sookmark, S. 1969 Diurnal variation in latex yield and dry rubber content, in relation to saturation deficit of air, *Journal of the Rubber Research Institute of Malaya* **21**: 341–7

Pakianathan, S. W. 1967 Determination of osmolarity of small latex samples by vapour pressure osmometer, *Journal of the Rubber Research Institute of Malaya* **20**: 23–6

Pakianathan, S. W. 1977 Some factors affecting yield reponse to stimulation with (2-chloroethyl)-phosphonic acid, *Journal of the Rubber Research Institute of Malaysia* **25**: 50–60

Pakianathan, S. W., Boatman, S. G., Taysum, D. H. 1966 Particle aggregation following dilution of Hevea latex: a possible mechanism for the closure of latex vessels after tapping, *Journal of the Rubber Research Institute of Malaya* **19:** 259–71

Pakianathan, S. W., Hamzah, S. bte, Sivakumaran, S., Gomez, J. B. 1982 Physiological and anatomical investigations on long-term ethephon-stimulated trees, *Journal of the Rubber Research Institute of Malaysia* **30:** 63–79

Pakianathan, S. W., Jaafar, H., Ghani, A. 1978 Practical uses of plant hormones in controlling latex flow and plant growth, *Planters' •Bulletin of the Rubber Research Institute of Malaysia* No. **155:** 61–9

Pakianathan, S. W., Milford, G. F. J. 1973 Changes in the bottom fraction contents of latex during flow in *Hevea brasiliensis, Journal of the Rubber Research Institute of Malaya* **23:** 391–400

Pakianathan, S. W., Tharmalingam, C. 1982 A technique for improved field planting of *Hevea* budded stumps for smallholdings, *Planters' Bulletin of the Rubber Research Institute of Malaysia* No. **172:** 79–84

Pakianathan, S. W., Wain, R. L. 1976 Effects of endogenous and exogenous growth regulators on some growth processes in *Hevea brasiliensis, Proceedings of the International Rubber Conference, Kuala Lumpur 1975*, vol. **2**, pp. 109–42

Pakianathan, S. W., Wain, R. L., Ng, E. K. 1976 Studies on displacement area on tapping in mature Hevea trees, *Proceedings of the International Rubber Conference, Kuala Lumpur 1975*, vol. **2**, pp. 225–46

Pakianathan, S. W., Wong, T. K., Jaafar, H. 1980 Use of indolebutyric acid on budded stumps to aid earlier root initiation and growth, *Proceedings of the Rubber Research Institute of Malaysia Planters' Conference, Kuala Lumpur 1979*, pp. 273–96

Pan, H-S. 1983 Adaptabilities of four rubber clones to the higher latitude and elevation areas of Yunnan Province, *Symposium of the International Rubber Research and Development Board Beijing Peoples' Republic of China 12–14 May*

Panichapong, S. 1984 Discussant-2 Problems associated with the application of soil taxonomy in Thailand, *Proceedings of the International Workshop on Soils, Townsville Australia 12–16 September 1983.* Australian Centre for International Agricultural Research, pp. 31–4

Paranjothy, K., Ghandimathi, H. 1976 Tissue and organ culture of *Hevea, Proceedings of the International Rubber Conference, Kuala Lumpur 1975*, vol. **2**, pp. 59–84

Paranjothy, K., Gomez, J. B., Yeang, H. Y. 1976 Physiological aspects of brown bast development, *Proceedings of the International Rubber Conference, Kuala Lumpur 1975*, vol. **2**, pp. 181–202

Paranjothy, K., Yeang, H. Y. 1978 A consideration of the nature and control of brown bast, *Proceedings of the Rubber Research Institute of Malaysia Planters' Conference, Kuala Lumpur 1977*, pp. 74–90

Parker, C. 1972 The *Mikania* problem, *PANS* **18:** 312–15

Parkin, J. 1900 Observations on latex and its functions, *Annals of Botany* **14:** 193–214

Parkin, J. 1910 The right use of acetic acid in the coagulation of *Hevea* latex, *India Rubber Journal* **40:** 752

Parliamentary Papers 1912–13 lxviii (Cd 6266) *Correspondence on the Putamayo.* [The Putamayo Blue Book]

Peacock, F. 1981 Rural poverty and development in West Malaysia, *Journal of Development Administration*: 639–52

Pee T. Y. 1977 Social returns from rubber research in Peninsular Malaysia. *Ph.D. Thesis, Michigan University USA*

Pee, T. Y., Khoo, S. K. 1976 Diffusion and impact of high yielding materials on rubber production in Malaysia, *Proceedings of the International Rubber Conference, Kuala Lumpur 1975*, Vol. **3**, pp. 327–349

Peel, J. D. 1959–60 Pulp and paper manufacture in Malaya. I, *Malayan Forester* **22:** 62, 69; **23:** 5, 147, 255

Pegler, D. N., Gibson, I. A. S. 1972 *Armillariella mellea, Commonwealth Mycological Institute Descriptions of Pathogenic Fungi and Bacteria* No. 321

Pegler, D. N., Waterston, J. M. 1968 *Phellinus noxius, Commonwealth Mycological Institute Descriptions of Pathogenic Fungi and Bacteria* No. 195

Peries, O. S. 1979 Studies on the relationship between weather and incidence of leaf diseases of *Hevea*, *Planter, Kuala Lumpur* **55**: 158–69

Petch, T. 1911 *The physiology and diseases of* Hevea brasiliensis. Dulau, London

Petch, T. 1914 Notes on the history of the plantation rubber industry in the East, *Annals of the Royal Botanic Gardens Peradeniya* **5**(8): 433–520, 1911–14

Petch, T. 1921 *The diseases and pests of the rubber tree.* Macmillan, London

Philpott, M. W., Westgarth, D. R. 1953 Stability and mineral composition of *Hevea* latex, *Journal of the Rubber Research Institute of Malaya* **14**: 133–48

Pichel, R. J. 1956 Les pourridiées de l'Hévéa dans la cuvette congolaise, *Publication de l'Institut National pour l'Etude Agronomique du Congo Belge. Série Technique* No. 49

Pillai, K. R. 1977 A note on the application of mixed factory effluent to mature rubber, *Planter, Kuala Lumpur* **53**: 468–9

Pillai, K. R. 1978 A review of chemical weed control in rubber and legumes, *Planter, Kuala Lumpur* **54**: 669–80

Pillay, R. N. R. (ed.) 1980 *Handbook of natural rubber production in India.* Rubber Research Institute of India, Rubber Board, Kottayam

Pinheiro, E., Pinheiro, F. S. V., Alves, R. M. 1981 Comportamento de alguns clones de *Hevea*, em Açailândia, na regiao pre-Amazonica Maranhense (Dados preliminares). In EMBRAPA/FCAP *3° Seminário Nacional da Seringueira Belém Brasil 1980*

Pires, J. M. 1981 Notas de herbario 1, *Boletim do Museo Paraense Emilio Goeldi, Nova Serie: Botanica, Belém* **52**: 4–8

Plessix, J. du 1968 Essais de mise en evidence de l'intérêt d'une étude de développement pour l'amélioration de l'*Hevea*, *Rapport de Recherche, Institut de Recherche Caoutchouc Afrique 1968*, pp. 168–74

P'ng, T. C., Leong, W., Abraham, P. D. 1973 A new method of applying novel stimulants, *Proceedings of the Rubber Research Institute of Malaysia Planters' Conference, Kuala Lumpur 1973*, pp. 122–32

P'ng, T. C., Yoon, P. K., Rahman, K., Goh, C B., Chiah, H. S. 1976 Controlled upward tapping, *Proceedings of the Rubber Research Institute of Malaysia Planters' Conference, Kuala Lumpur 1976*, pp. 177–207

Polhamus, L. G. 1957 Rubber content of miscellaneous plants, *United States Department of Agriculture Production Research Report* No. 10

Polhamus, L. G. 1962 *Rubber, botany, production and utilization.* Leonard Hill, London

Polinière, J-P., van Brandt, H. 1969 Pot culture techniques for study of *Hevea* nutrition, *Journal of the Rubber Research Institute of Malaya* **21**: 250–7

Populer, C. 1972 Les épidémies de l'oidium de l'hévéa et la phénologie de son hôte dans le monde, *Publication de l'Institut National pour l'Etude Agronomique du Congo. Série Scientifique* No. 115

Porritt, B. D. 1926 The early history of the rubber industry, *India Rubber Journal* **71**: 405, 444, 483

Potty, S. N., Kalam, A., Mathew, M., Ananth, K. C. 1968 Effect of mulching nurseries with different mulching materials on the growth of rubber seedlings, *Rubber Board Bulletin (India)* **10**: 83–90

Prawit, W., Varunee, B., Jirakorn, K., Samer, S., Slearmlarp, W. 1983 Three years' experience on the trial establishment of *Hevea* plantation in semi-arid area, *Symposium of the International Rubber Research and Development Board Beijing Peoples' Republic of China 12–14 May*

Prawit, W., Waiwit, B., Templeton, J. K. 1975 Girth growth of rubber and interrow management, *Rubber Research Centre, Hat Yai, Thailand, Document* No. 56

Premakumari, D., Annamma, Y., Bhaskaran Nair, V. K., 1979 Clonal variability for stomatal characters and its application in *Hevea* breeding and selection, *Indian Journal of Agricultural Science* **49**: 411–17

Premakumari, D., Parikar, A. O. N., Sobhana, S. 1985 Occurrence of interxylary phloem in *Hevea brasiliensis, Annals of Botany* **55**: 275–7

Premakumari, D., Sherief, P. M., Sethuraj, M. R. 1980 Lutoid stability and rubber particle stability as factors influencing yield during drought in rubber, *Journal of Plantation Crops* **8**: 43–7

Primot, L., Tupy, J. 1976 Sur l'exploitation de l'Hevea par microsaignée. *Revue Générale du Caoutchouc et des Plastiques* **53**: 77–9

Pujarniscle, S. 1968 Caractère lysosomal des lutoides du latex d'*Hevea brasiliensis, Physiologie Végétale* **6**: 27–46

Pujarniscle, S., Ribaillier, D. 1966 Etude préliminaire sur les lutoides du latex et leur possibilité d'intervention dans le biosynthèse du caoutchouc, *Revue Générale du Caoutchouc et des Plastiques* **43**: 226–28

Pujarniscle, S., Ribaillier, D., d'Auzac, J. 1970 Du rôle des lutoides dans l'écoulement du latex chez l'*Hevea brasiliensis.* I, II, *Revue Générale du Caoutchouc et des Plastiques* **47**: 1001–3, 1317–21

Punnoose, K. I., Potty, S. N., Mathew, M., George, C. M. 1976 Responses of *Hevea brasiliensis* to fertilizers in South India, *Proceedings of the International Rubber Conference, Kuala Lumpur 1975*, vol. 3, pp. 84–103

Purseglove, J. W. 1968 *Tropical crops. Dicotyledons.* Longman, London

Pushparajah, E. 1969 Response in growth and yield of *Hevea brasiliensis* to fertilizer application on a Rengam series soil, *Journal of the Rubber Research Institute of Malaya* **21**:: 165–74

Pushparajah, E. 1975 Recent developments in the nutrition of *Hevea* in West Malaysia, *Planter, Kuala Lumpur* **51**: 177–90

Pushparajah, E. 1979 Fertilizer standards, *Planter, Kuala Lumpur* **55**: 40–7

Pushparajah, E. 1983a Manuring of rubber (*Hevea*): current global practices and future trends, *Symposium of the International Rubber Research and Development Board Beijing Peoples' Republic of China 12–14 May*

Pushparajah, E. 1983b Problems and potentials for establishing *Hevea* under difficult environment conditions, *Planter, Kuala Lumpur* **59**: 242–51

Pushparajah, E. 1984 Discussant-1 Nutrient availability in acid soils of the tropics following clearing and cultivation with plantation crops, *Proceedings of the International Workshop on Soils. 12–16 September 1983, Townsville, Australia.* Australian Centre for International Agricultural Research, pp. 47–51

Pushparajah, E., Chan, H. Y. 1973 Optimizing of land use for perennial crops in West Malaysia, *Proceedings of the Symposium on National Utilization of Land Resources Serdang Malaysia*, pp. 7–22

Pushparajah, E., Chan, H. Y., Sivanadyan, K. 1983 Recent developments for reduced fertilizer applications in *Hevea, Proceedings of the Rubber Research Institute of Malaysia Planters' Conference, Kuala Lumpur 1983*, pp. 313–27

Pushparajah, E., Chellapah, K. 1969 Manuring of rubber in relation to covers, *Journal of the Rubber Research Institute of Malaya* **21**: 126–39

Pushparajah, E., Guha, M. M. 1969 Fertilizer response in *Hevea brasiliensis* in relation to soil type and leaf nutrition studies, *Transactions of the Ninth International Congress on Soil Science Adelaide 1968*, pp. 85–92

Pushparajah, E., Haridas, G. 1977 Developments in reduction of immaturity period of *Hevea* in Peninsular Malaysia, *Journal of the Rubber Research Institute of Sri Lanka* **54**: 93–105

Pushparajah, E., Haridas, G., Soong, N. K., Zeid, P. 1975 Utilization of effluent and bowl sludge from natural rubber processing factories, *Proceedings of the Symposium on Agronomy of Industrial Wastes, Kuala Lumpur 1975*, pp. 170–7

Pushparajah, E., Mahmud, A. W., Lau, C. H. 1977a Residual effect of applied phosphates on performance of *Hevea brasiliensis* and *Pueraria phaseoloides, Journal*

of the Rubber Research Institute of Malaysia **25:** 101–8

Pushparajah, E., Ng, S. K., Ratnasingham, K. 1977b Leaching losses of nitrogen, potassium and magnesium on Peninsular Malaysian soil, *Proceedings of the Conference on Chemistry and Fertility of Tropical Soils, Kuala Lumpur 1973*, pp. 121–9

Pushparajah, E., Sivanadyan, K., P'ng, T. C., Ng, E. K. 1972 Nutritional requirements of *Hevea brasiliensis* in relation to stimulation, *Proceedings of the Rubber Research Institute of Malaya Planters' Conference, Kuala Lumpur 1971*, pp. 189–200

Pushparajah, E., Sivanadyan, K., Subramaniam, A., Tan, K. T. 1976a Influence of fertilizers on nutrient content, flow and properties of *Hevea* latex, *Proceedings of the International Rubber Conference, Kuala Lumpur 1975*, vol. 3, pp. 122–31

Pushparajah, E., Soong, N. K., Yew, F. K., Zainal, E. 1976b Effect of fertilizers on soils under *Hevea*, *Proceedings of the International Rubber Conference, Kuala Lumpur 1975*, vol. 3, p. 37

Pushparajah, E., Tan, K. H., Chin, S. L. 1982 Nitrogenous fertilizers for *Hevea* cultivation, *Proceedings of the Rubber Research Institute of Malaysia Planters' Conference, Kuala Lumpur 1981*. pp. 203–19

Pushparajah, E., Tan, K. H., Soong, N. K. 1977 Influence of covers and fertilizers and management on soil. In *Soils under* Hevea *and their management in Peninsular Malaysia*. Rubber Research Institute of Malaysia, p. 83

Pushparajah, E., Tan, K. T. 1972 Factors influencing leaf nutrient levels in rubber, *Proceedings of the Rubber Research Institute of Malaya Planters' Conference, Kuala Lumpur 1972*, pp. 140–54

Pushparajah, E., Tan, S. K. 1970 Tapioca as an intercrop in rubber. In Blencowe, E. K., Blencowe, J. W., (eds) *Crop Diversification in Malaysia*. Incorporated Society of Planters, Kuala Lumpur, pp. 128–38

Pushparajah, E., Wahab, M. A. W. 1978 Manuring in relation to covers, *Proceedings of the Rubber Research Institute of Malaysia Planters' Conference, Kuala Lumpur 1977*, pp. 150–65

Pushparajah, E., Yew, F. K. 1977 Management of soils. In *Soils under* Hevea *and their management in Peninsular Malaysia*. Rubber Research Institute of Malaysia, pp. 94–117

Pyke, E. E. 1932 *Second Annual Report on cacao Research*. Imperial College of Tropical Agriculture, Trinidad

Pyke, E. E. 1941 Trunk diameter of trees of *Hevea brasiliensis*: experiments with a new dendrometer, *Nature, London* **148:** 51–2

Radhakrishna Pillay, P. N. 1977 Aerial spraying against abnormal leaf fall disease of rubber in India, *Planters' Bulletin of the Rubber Research Institute of Malaysia* No. **148:** 10–14

Radjino, A. J. 1969 Effect of *Oidium* and *Dothidella* resistant crowns on growth and yield of *Hevea brasiliensis, Journal of the Rubber Research Institute of Malaya* **21:** 56–63

Rajalakshmy, V. K., Radhakrishna Pillay, P. N. 1978 *Poria* root disease of rubber in India, *Indian Phytopathology* **31:** 199–202

Rajaratnam, J. A. 1979 Aerial application of fertilizer, *Planter, Kuala Lumpur* **55:** 12–24

Rama Rao, J. S., Nayagam, J., Yeoh, C. H. 1982 *An Evaluation of the seven-picul smokehouses*. Rubber Research Institute of Malaysia, Kuala Lumpur (mimeograph)

Ramakrishnan, T. S. 1963 Patch canker or bark canker caused by *Phytophthora palmivora* Butl. and *Pythium vexans* de Bary, *Rubber Board (India) Bulletin* **7:** 11–13

Ramakrishnan, T. S., Radhakrishna Pillay, P. N. 1962 Report on the first year of trials in increasing yield from existing plantations of rubber, *Rubber Board (India) Bulletin* **6:** 9

Rao, A. N. 1963 Reticulate cuticle on leaf epidermis in *Hevea, Nature, London* **197**: 1125

Rao, B. S. 1961 Pollination of *Hevea* in Malaya, *Journal of the Rubber Research Institute of Malaya* **17**: 14–18

Rao, B. S. 1963 Pests of leguminous covers in Malaya and their control, *Planters' Bulletin of the Rubber Research Institute of Malaya* No. **68**: 182–6

Rao, B. S. 1964a Root-knot nematodes of leguminous covers in rubber plantations, *Journal of the Rubber Research Institute of Malaya* **18**: 146–50

Rao, B. S. 1964b The use of light traps to control the cockchafer *Lachnosterna bidentata* Burm. in Malayan rubber plantations, *Journal of the Rubber Research Institute of Malaya* **18**: 243–52

Rao, B. S. 1965 *Pests of Hevea plantations in Malaya.* Rubber Research Institute of Malaya, Kuala Lumpur

Rao, B. S. 1969 Cockchafers attacking rubber in West Malaysia and their integrated control, *FAO Plant Protection Bulletin* **17**: 52–5

Rao, B. S. 1973 Some observations on South American leaf blight in South America, *Planter, Kuala Lumpur* **49**: 2–9

Rao, B. S. 1975 *Maladies of Hevea in Malaysia.* Rubber Research Institute of Malaysia, Kuala Lumpur

Rao, B. S., Azaldin, M. Y. 1973 Progress towards recommending artificial defoliation for avoiding secondary leaf fall, *Proceedings of the Rubber Research Institute of Malaysia Planters' Conference, Kuala Lumpur 1973*, pp. 267–80

Reed, M. E. D. 1974 Report on a survey of intercropping in immature rubber. Part II. Preliminary Tables, FAO/UNDP Rubber Development Working Paper. *Rubber Research Centre, Hat Yai, Thailand, Document* No. 48

Reed, M. E. D., Sumana 1976 Economic aspects of intercropping in immature rubber and oil palm in Indonesia, *UNDP/FAO Project INS/72/004 Note* No. 131

Ribaillier, D. E. 1970 Importance des lutoïdes dans l'écoulement du latex: action de la stimulation, *Revue Générale du Caoutchouc et des Plastiques* **47**: 305–10

Ribaillier, D. 1972 Quelques aspects du rôle des lutoïdes dans la physiologie et l'écoulement du latex d'*Hevea brasiliensis.* Thèse, Faculté des Sciences Université d'Abidjan

Ribaillier, D., d'Auzac, J. 1970 Nouvelles perspectives de stimulation, hormonale de la production chez l'*Hevea brasiliensis, Revue Générale du Caoutchouc et des Plastiques* **47**: 433–9

Ribeiro, R. P. 1980 Production planning for the natural rubber industry of Brazil. *Proceedings of the Twenty-sixth Assembly of the International Rubber Study Group, Kuala Lumpur Malaysia 29 September–4 October 1980*, pp. 177–82

Riches, J. B., Gooding, E. G. B. 1952 Studies in the physiology of latex. I. Latex flow on tapping – theoretical considerations, *New Phytology* **51**: 1–10

Ridley, H. N. 1897 Rubber cultivation, *Agricultural Bulletin of the Malay Peninsula*

Ridley, H. N. 1910 Historical notes on the rubber industry, *Agricultural Bulletin, Federated Malay States* **6**: 202–3

Ridley, H. N. 1912 *The story of the rubber industry in Malaya.* Malay States Development Agency, London

Ridley, H. N. 1955 Evolution of the rubber industry. In *Proceedings of the Institution of the Rubber Industry* vol 2. Heffer, Cambridge

Riepma, P. 1965 A comparison of the pre-emergence herbicides atratone, prometone, prometryne and simazine applied in planting strips of *Hevea brasiliensis. Journal of the Rubber Research Institute of Malaya* **19**: 74–80

Riggenbach, A. 1966 Observations on root diseases of *Hevea* in West Africa and the East, *Tropical Agriculture, Trinidad* **43**: 53–8

Rogers, T. H., Peterson, A. L. 1976 Control of South American leaf blight on a plantation scale in Brazil, *Proceedings of the International Rubber Conference, Kuala Lumpur 1975*, vol. 3, pp. 266–77

Romano, R., Rao, S. 1983 Desfolhamento quimico em seringueira na Bahia. *Pesquisa Agropecuaria Brasileira* **18**: 507–14

Rosenquist, E. A. 1961 Manuring of rubber in relation to wind damage, *Proceedings of the Natural Rubber Research Conference Kuala Lumpur 1960*, pp. 81–8

Rosenquist, E. A., McEwen J. F., Shepherd R. 1976 Field data retrieval by clip cards and its practical implications, *Proceedings of the International Rubber Conference, Kuala Lumpur 1975*, vol. 3, pp. 173–96

Ross, J. M. 1964 Summary of breeding carried out at the RRIM during the period 1928–1963, *Research Archives of the Rubber Research Institute of Malaya Document* No. 28

Ross, J. M., Brookson, C. W. 1966 Progress of breeding investigations with *Hevea brasiliensis*. III. Further data on the crosses made in the years 1937–41, *Journal of the Rubber Research Institute of Malaya* **19**: 158–172

Ross, J. M., Dinsmore, C. S. 1972 Firestone Plantation's experience with Ethrel yield stimulant, *Proceedings of the Rubber Research Institute of Malaya Planters' Conference, Kuala Lumpur 1971*, pp. 91–103

Royal Botanic Gardens, Kew 1878 Report on the progress and condition of the Royal Gardens at Kew during the year 1877, *Reports and documents 1784–1884*. Part 47

Royal Botanic Gardens, Kew 1898 Para rubber, *Kew Bulletin* No. 142

Royal Botanic Gardens, Kew 1914 The introduction of Para rubber to Buitenzorg, *Kew Bulletin* No. 4 162–5 (Correspondence D. Prain and P. J. S. Cramer)

Rubber Industry Smallholders' Development Authority 1983 *TRIDELTA project – RISDA's strategy to consolidate extension activities in the smallholders's sector*. RISDA, Kuala Lumpur (mimeograph)

Rubber Research Institute of Ceylon 1959 Minor pests of rubber plantations, *Quarterly Journal of the Rubber Research Institute of Ceylon* **35**: 86–92

Rubber Research Institute of India 1981 Rain guarding, *Rubber Board Bulletin (India)* **16**(3): 26–9

Rubber Research Institute of India 1983 New 'head light' for tapping rubber trees early morning, *Rubber Board Bulletin (India)* **19**(3): pp. 6–7

Rubber Research Institute of Malaya 1932 The forestry system of rubber planting, *Journal of the Rubber Research Institute of Malaya* **4**: 54

Rubber Research Institute of Malaya 1940 RRI tunnel-type smokehouse, *Planters' Bulletin of the Rubber Research Institute of Malaya* (old series) No. **14**: 9–10

Rubber Research Institute of Malaya 1952a Report on the technical classification of natural rubber in Malaya, *Journal of the Rubber Research Institute of Malaya* **14**: 1–9

Rubber Research Institute of Malaya 1952b *Application of the RRI bleaching process to the preparation of white latex crepe*. Rubber Research Institute of Malaya Circular No. 36

Rubber Research Institute of Malaya 1953 Air dried sheet, *Planters' Bulletin of the Rubber Research Institute of Malaya* No. **7**: 107–9

Rubber Research Institute of Malaya 1954a Mycorrhiza in *Hevea*, *Planters' Bulletin of the Rubber Research Institute of Malaya* No. **12**:57–8

Rubber Research Institute of Malaya 1954b White latex crepes, *Planters' Bulletin of the Rubber Research Institute of Malaya* No. **10**: 7–9

Rubber Research Institute of Malaya 1956 Mulching, *Planters' Bulletin of the Rubber Research Institute of Malaya* No. **26**:90–2

Rubber Research Institute of Malaya 1957 Superior processing rubber, *Planters' Bulletin of the Rubber Research Institute of Malaya* No. **32**: 83–7

Rubber Research Institute of Malaya 1958 Rooting habit, *Planters' Bulletin of the Rubber Research Institute of Malaya* No. **39**: 120–8

Rubber Research Institute of Malaya 1959a Effect of high watertable on growth of taproot, *Planters' Bulletin of the Rubber Research Institute of Malaya* No. **41**: 47–8

Rubber Research Institute of Malaya 1959b Wind damage, *Planters' Bulletin of the Rubber Research Institute of Malaya* No. **43**: 79–93

Rubber Research Institute of Malaya 1960 A crown budding experiment, *Planters' Bulletin of the Rubber Research Institute of Malaya* No. **49:** 84–7

Rubber Research Institute of Malaya 1961 Cover plants, manuring and wind damage, *Planters' Bulletin of the Rubber Research Institute of Malaya* No. **57:** 183–9

Rubber Research Institute of Malaya 1962a Thrips, *Planters' Bulletin of the Rubber Research Institute of Malaya* No. **59:** 47–9

Rubber Research Institute of Malaya 1962b Defects in sheet rubber, *Planters' Bulletin of the Rubber Research Institute of Malaya* No. **59:** 32–47; No. **60:** 57–60; No. **63:** 163–72

Rubber Research Institute of Malaya 1963 Identification of plants on Malayan rubber estates: Plates 17–24, Grasses, *Planters' Bulletin of the Rubber Research Institute of Malaya* No. **67:**88–99

Rubber Research Institute of Malaya 1964a *Annual Report 1963*, p. 64

Rubber Research Institute of Malaya 1964b Flood damage, *Planters' Bulletin of the Rubber Research Institute of Malaya* No. **72:** 61–3

Rubber Research Institute of Malaya 1964c *Annual Report 1963*, pp. 20–1

Rubber Research Institute of Malaya 1964d Warm blooded animals, *Planters' Bulletin of the Rubber Research Institute of Malaya* No. **70:** 10–18

Rubber Research Institute of Malaya 1965a Seed gardens for estates, *Planters' Bulletin of the Rubber Research Institute of Malaya* No. **81:** 257–60

Rubber Research Institute of Malaya 1965b Two minor leaf diseases, *Planters' Bulletin of the Rubber Research Institute of Malaya* No. **81:** 261–3

Rubber Research Institute of Malaya 1965c Standard Malayan Rubber, *Planters' Bulletin of the Rubber Research Institute of Malaya* No. **78:** 75–98

Rubber Research Institute of Malaya 1966 Termites, *Planters' Bulletin of the Rubber Research Institute of Malaya* No. **84:** 67–71

Rubber Research Institute of Malaya 1967a Pruning to prevent wind damage, *Planters' Bulletin of the Rubber Research Institute of Malaya* No. **91:** 147–57

Rubber Research Institute of Malaya 1967b Snails and slugs, *Planters' Bulletin of the Rubber Research Institute of Malaya* No. **93:** 284–7

Rubber Research Institute of Malaya 1968 Cockchafers, *Planters' Bulletin of the Rubber Research Institute of Malaya* No. **97:** 102–9

Rubber Research Institute of Malaya 1970 Mites, *Planters' Bulletin of the Rubber Research Institute of Malaya* No. **106:** 3–6

Rubber Research Institute of Malaya 1972a Phosphorus: its role in rubber cultivation, *Planters' Bulletin of the Rubber Research Institute of Malaya* No. **120:** 82–91

Rubber Research Institute of Malaya 1972b Cover management in rubber, *Planters' Bulletin of the Rubber Research Institute of Malaya* No. **122:** 170–80

Rubber Research Institute of Malaya 1972c Banana and tapioca as intercrops in immature rubber, *Planters' Bulletin of the Rubber Research Institute of Malaya* No. **123:** 203–12

Rubber Research Institute of Malaya 1973 RRIM clonal seedling trials: final report, *Planters' Bulletin of the Rubber Research Institute of Malaya* No. **127:** 115–26

Rubber Research Institute of Malaysia 1974a Collecting, handling and planting propagation materials of Para rubber, *Hevea brasiliensis*, *Planters' Bulletin of the Rubber Research Institute of Malaya* No. **132:** 98–103

Rubber Research Institute of Malaysia 1974b *Reduction of immature period of rubber*. Mimeographed report

Rubber Research Institute of Malaysia 1974c Towards a wider use of rubber wood, *Planters' Bulletin of the Rubber Research Institute of Malaysia* No. **135:** 181–94

Rubber Research Institute of Malaysia 1974d Root diseases. Part II: Control, *Planters' Bulletin of the Rubber Research Institute of Malaysia* No. **134:** 157–64

Rubber Research Institute of Malaysia 1975a *Nursery techniques for rubber plant propagation. Agricultural Series*, Report No. 2

Rubber Research Institute of Malaysia 1975b Enviromax planting recommendations, *Planters' Bulletin of the Rubber Research Institute of Malaysia* No. **137:** 27–50

Rubber Research Institute of Malaysia 1975c *Corynespora* leaf spot, *Planters' Bulletin of the Rubber Research Institute of Malaysia* No. **139:** 84–6

Rubber Research Institute of Malaysia 1976a Branch induction of young trees, *Planters' Bulletin of the Rubber Research Institute of Malaysia* No. **147:** 149–58

Rubber Research Institute of Malaysia 1976b *Annual Report 1975*, p. 81

Rubber Research Institute of Malaysia 1977a *Annual Report 1976*, p. 67

Rubber Research Institute of Malaysia 1977b *Mucuna cochinensis* – a potential short-term legume cover plant, *Planters' Bulletin of the Rubber Research Institute of Malaysia* No. **150:** 78–82

Rubber Research Institute of Malaysia 1977c A soil suitability technical grouping system for *Hevea*, *Planters' Bulletin of the Rubber Research Institute of Malaysia* No. **152:** 135–46

Rubber Research Institute of Malaysia 1977d Rubber wood – an underexploited resource, *Planters' Bulletin of the Rubber Research Institute of Malaysia* No. **149:** 54–60

Rubber Research Institute of Malaysia 1978a *Annual Report 1977*, p. 67

Rubber Research Institute of Malaysia 1978b Sources of nitrogen, *Annual Report 1977*, p. 113

Rubber Research Institute of Malaysia 1979a *Annual Report 1978*, p. 74

Rubber Research Institute of Malaysia 1979b Influence of nutrients on latex flow and properties, *Annual Report 1978*, p. 125

Rubber Research Institute of Malaysia 1980a Puncture tapping, an overview, *Planters' Bulletin of the Rubber Research Institute of Malaysia* No. **164:** 101–112

Rubber Research Institute of Malaysia 1980b *RRIM training manual on natural rubber processing.* Rubber Research Institute of Malaysia, Kuala Lumpur

Rubber Research Institute of Malaysia 1981a *Annual Report 1980*, p. 31

Rubber Research Institute of Malaysia 1981b Crop protection, *Annual Report 1980*, pp. 38–41

Rubber Research Institute of Malaysia 1982a *Annual Report 1981*, p. 39

Rubber Research Institute of Malaysia 1982b *Annual Report 1981*, pp. 52–3

Rubber Research Institute of Malaysia 1982c Crop protection, *Annual Report 1981*, pp. 44–7

Rubber Research Institute of Malaysia 1983a *Annual Report 1982*, p. 39

Rubber Research Institute of Malaysia 1983b *Annual Report 1982*, p. 30

Rubber Research Institute of Malaysia 1983c RRIM planting recommendations 1983–85, *Planters' Bulletin of the Rubber Research Institute of Malaysia* No. **175:** 37–55

Rubber Research Institute of Malaysia 1984a *Annual Report 1983*, p. 38

Rubber Research Institute of Malaysia 1984b *Annual Report 1983*, p. 64

Rubber Research Institute of Malaysia 1984c *Annual Report 1983*, p. 31

Rubber Research Institute of Malaysia 1984d Screening of commercial herbicides, *Annual Report*, p. 53

Rubber Research Institute of Malaysia 1984e *Annual Report 1983*, p. 50

Rubber Research Institute of Malaysia 1985a *Annual Report 1984*, p. 19

Rubber Research Institute of Malaysia 1985b *Annual Report 1984*, pp. 26–7

Rubber Research Institute of Malaysia 1985c *Annual Report 1983*, p. 27

Rubber Research Institute of Sri Lanka 1978 *Rubber Research Institute of Sri Lanka Annual Review*

Rubber Research Institute of Thailand 1983 *Unpublished survey data. Rubber Research Institute of Thailand* Department of Agriculture, Bangkok

Rubber Statistical Bulletin: various issues from 1946

Ryden, J. C. 1984 Fertilizers for grassland, *Chemistry and Industry* **18:** 652–7

Saccas, A. M. 1959 Une grave maladie des Hévéas des Terres Rouges en Oubangi-Chari, *Agronomie Tropicale* **14:** 409–59

Saengruksowong, C., Dansagoonpan, S., Thammarat, C. 1983 Rubber planting in the North-Eastern and Northern Regions of Thailand, *Symposium of the International Rubber Research and Development Board Beijing Peoples' Republic of China 12–14 May*

Samaranayake, C. 1975 Selection of rootstocks, *Rubber Research Institute of Sri Lanka Bulletin* **10**: 35–7

Samaranayake, C., de Soyza, A. G. A. 1984 Studies on exploitation of Hevea clones in Sri Lanka, *Proceedings of the International Rubber Conference of the Rubber Research Institute of Sri Lanka 1984:* in press

Samaranayake, C., Gunaratne, R. B. 1977 The use of leaf buds and scale buds in the vegetative propagation of *Hevea, Journal of the Rubber Research Institute of Sri Lanka* **54**: 65–9

Samosorn, S., Creencia, R. P., Wasuwat, S. 1978 Study on the yield, sucrose level of latex and other important characteristics of *Hevea brasiliensis, Thai Journal of Agricultural Science* **11**: 171–81, 183–92, 193–207

Samsidar binte Hamzah, Gomez J B 1982 Some structural factors affecting the productivity of *Hevea brasiliensis.* III Correlations between structural factors and plugging. *Journal of the Rubber Research Institute of Malaysia* **30**: 148–60

Samsuddin, Z., Impens, I. 1978a Water vapour and carbon dioxide diffusion resistances of four *Hevea brasiliensis* clonal seedlings, *Experimental Agriculture* **14**: 173–7

Samsuddin, Z., Impens, I. 1978b Comparative net photosynthesis of four *Hevea brasiliensis* clonal seedlings, *Experimental Agriculture* **14**: 337–40

Samsuddin, Z., Impens, I. 1979a Relationship between leaf age and some carbon dioxide exchange characteristics of four *Hevea brasiliensis* clones, *Photosynthetica* **13**: 208–10

Samsuddin, Z., Impens, I. 1979b The development of photosynthetic rate with leaf age in four *Hevea brasiliensis* clonal seedlings, *Photosynthetica* **13**: 267–70

Samsuddin, Z., Impens, I. 1979c Photosynthetic rates and diffusion resistances of seven *Hevea brasiliensis* clones, *Biologia Plantarum* **21**: 154–6

Samsuddin, Z., Khir, M., Impens, I. 1978 Development of the leaf blade class concept for the characterization of *Hevea brasiliensis* leaf age, *Journal of the Rubber Research Institute of Malaysia* **26**: 1–5

Samsuddin, Z., Tan, H., Yoon, P. K. 1985 Variations, heritabilities and correlations of photosynthetic rates, yield and vigour in young *Hevea brasiliensis* progenies, *Proceedings International Rubber Conference, Kuala Lumpur 1985*

Samsuddin, Z., Tan, H., Yoon, P. K. 1987 Correlation studies on photosynthetic rates, girth and yield in *Hevea brasiliensis, Journal of Natural Rubber Research* **2**: 46–54

Sanchez, J. T., Rivera, D. M. 1980 Planning for future production of natural rubber in Mexico, *Proceedings of the Twenty-sixth Assembly of the International Rubber Study Group, Kuala Lumpur Malaysia 29 September–4 October 1980,* pp. 183–206

Sachez, P. A. 1976 *Properties and management of soils in the tropics.* John Wiley, New York, p. 52

Satchuthananthavale, R., Irugalbandara, Z. E. 1972 Propagation of callus from *Hevea* anthers, *Quarterly Journal of the Rubber Research Institute of Ceylon* **49**: 65–8

Schidrowitz, P., Dawson, T. R. (eds) 1952 *History of the rubber industry.* Heffer, Cambridge

Schmöle, J. F. 1926 Manuring experiments on rubber, *Archief voor de Rubbercultuur in Nederlandsch- Indië* **10**: 233–301

Schmöle, J. F. 1940 De invloed van den onderstam op de productie van oculaties, *Archief voor de Rubbercultuur in Nederlandsch-Indië* **24**: 305–14

Schmöle, J. F. 1941 *Hevea brasiliensis* and *Hevea spruceana* hybrids as stocks for bud-grafts, *Archief voor de Rubbercultuur in Nederlandsch-Indië* **25**: 159–67

Schultes, R. E. 1956 The Amazonian Indian and the evolution of *Hevea* and related genera, *Journal of the Arnold Arboretum* **37**: 123–47

Schultes, R. E. 1970 The history of taxonomic studies in *Hevea, Botanical Review* **36**: 197–211

Schultes, R. E. 1977a The Odyssey of the cultivated rubber tree, *Endeavour* New Series **1**(34): 133–8

Schultes, R. E. 1977b Wild Hevea: an untapped source of germplasm, *Journal of the Rubber Research Institute of Sri Lanka* **54**: 227–57

Schurer, H. 1955 Bicentenary of the first publication on rubber, *Rubber Journal* January 1: 4–8

Schurer, H. 1956 The discovery of the rubber tree. Achievements of the 18th Century explorations in America, Africa and Asia, *Rubber Journal* February 4: 132–7 and May 12: 552–8

Schurer, H. 1957a Aspects of early rubber technology, *Rubber Journal* September 7: 306–15

Schurer, H. 1957b Rubber, a magic substance of ancient America, *Rubber Journal* April 7: 543–9

Schurer, H. 1958 The Spanish discovery of rubber. A story through three centuries, *Rubber Journal and International Plastics* August 23: 269–87

Schurer, H. 1959a Inventions of the first rubber articles. Developments from early technology, *Rubber Journal and International Plastics* Oct 3: 279–82

Schurer, H. 1959b Inventions of the first rubber articles. Developments from early technology, *Rubber Journal and International Plastics* Oct 10: 321–4

Schurer, H. 1960a Early knowledge of rubber. Part 1 The latex enigma, *Rubber Journal and International Plastics* August 27: 304–8

Schurer, H. 1960b Early knowledge of rubber. Part 2 The naming of latex, *Rubber Journal and International Plastics* September 3: 337–40

Schurtz, W. L., Hargis, O. D., Marbut, C. F., Manifold, C. B. 1925 Rubber production in the Amazon Valley, *United States Department of Commerce Trade Promotion Series* No. 23. Government Printer, Washington

Schweizer, J. 1938 Over den wederzijdschen invloed van boven-en onderstam bij *Hevea, De Bergcultures* **12**: 773–8

Schweizer, J. 1940 Verdere gegevens over het pagger system bij *Hevea, De Bergcultures* **14**: 1658–72

Schweizer, J. 1949 Hevea latex as a biological substance, *Archief voor de Rubbercultuur in Nederlansch-Indië* **26**: 345–97

Sekhar, B. C. 1961 Inhibition of hardening in natural rubber, *Proceedings of the Natural Rubber Research Conference, Kuala Lumpur 1960*, pp. 512–19

Sekhar, B. C., Chin, P. S., Graham, D. J., O'Connell, J. 1965 Heveacrumb, *Rubber Developments* **18**: 78–84

Sekhar, B. C., Nielsen, P. S. 1961 New types of superior processing rubbers, *Proceedings of the Natural Rubber Research Conference, Kuala Lumpur 1960*, pp. 572–86

Sekhar, B. C., Pee, T. Y. 1980 Natural rubber potentials for the future, *Malaysian Rubber Research and Development Board Monograph* No. 6, Kuala Lumpur

Senanayake, Y. D., Jayasekera, N. E. M., Samaranayake, P. 1975 Growth of nursery rootstock seedlings of *Hevea brasiliensis* cv. Tjir 1, *Quarterly Journal of the Rubber Research Institute of Sri Lanka* **52**: 29–37

Senanayake, Y. D., Samaranayake, P. 1970 Intraspecific variation in stomatal density in *Hevea brasiliensis, Quarterly Journal of the Rubber Research Institute of Ceylon* **46**: 61–5

Senechal, V. 1984 *Etude sur le* Colletotrichum gloeosporioides *à HEVECAM*, Rapport d'Activité Phytopathologie, Janvier–Septembre 1984, HEVECAM Plantations, Cameroon

Sepien, A. bin 1980 Economics of high planting density on rubber smallholdings,

Proceedings of the Rubber Research Institute of Malaysia Planters' Conference, Kuala Lumpur 1979, pp. 82–92

Sepien, A. bin, Lim, F. H. 1981 Economics of various tapping systems with stimulation; a review, *Planters' Bulletin of the Rubber Research Institute of Malaysia* No. **166:** 3–14

Sergeant, C. J. 1967 The soil core method of transplanting young rubber as practised on Kirby Estate, *Planters' Bulletin of the Rubber Research Institute of Malaya* No. **92:** 256–63

Seth, A. K. 1969 Use of 'Gramoxone' for *Mikania cordata* control on oil palm and rubber plantations, *Planter, Kuala Lumpur* **45:** 34–40

Seth, A. K. 1977 Integrated weed control in tropical plantations. In Fryer, J., Matsunaka, S. (eds) *Integrated control of weeds*. University of Tokyo Press, Tokyo, pp. 69–87

Sethuraj, M. R. 1968 Studies on the physiological aspects of rubber production I. Theoretical considerations and preliminary observations, *Rubber Board Bulletin (India)* **9**(4): 47–53

Sethuraj, M. R. 1981 Yield components in Hevea brasiliensis – theoretical considerations, *Plant, Cell and Environment* **4:** 81–3

Sethuraj, M. R., George, M. J. 1975 Seasonal changes in the effectiveness of Ethrel, 2, 4-D and NAA in the stimulation of latex flow in *Hevea brasiliensis, Indian Journal of Plant Physiology* **18:** 163–8

Sethuraj, M. R., George, M. J., Sulochanamma, S. 1976 Physiological studies on yield stimulation of *Hevea brasiliensis, Proceedings of the International Rubber Conference, Kuala Lumpur 1975*, vol. **2**, pp. 280–9

Sethuraj, M. R., Nair, N. U., George, M. J., Mani, K. T. 1977 Physiology of latex flow in *Hevea brasiliensis* as influenced by intensive tapping, *Journal of the Rubber Research Institute of Sri Lanka* **54:** 221–6

Sethuraj, M. R., Subronto, Sulochanamma, S., Subbarayalu, G. 1978 Two indices to quantify latex-flow characteristics in *Hevea brasiliensis, Indian Journal of Agricultural Sciences* **48:** 521–4

Sharp, C. C. T. 1940 Progress of breeding investigations with *Hevea brasiliensis* I. The Pilmoor crosses 1928–1931 series, *Journal of the Rubber Research Institute of Malaya* **10:** 34–66

Sharp, C. C. T. 1951 Progress of breeding investigations with *Hevea brasiliensis* II. The crosses made in the years 1937–1941, *Journal of the Rubber Research Institute of Malaya* **13:** 73–99

Sharples, A. 1936 *Diseases and pests of the rubber tree*. Macmillan, London

Shepherd, R. 1967 A study of the comparative merits of different planting techniques, *Planters' Bulletin of the Rubber Research Institute of Malaya* No. **92:** 214–20

Shepherd, R. 1969a Aspects of Hevea breeding and selection investigations undertaken on Prang Besar Estate, *Planters' Bulletin of the Rubber Research Institute of Malaya* No. **104:** 206–19

Shepherd, R. 1969b Induction of polyploidy in *Hevea brasiliensis*; preliminary observations on trials conducted at Prang Besar Estate, *Planters' Bulletin of the Rubber Research Institute of Malaya* No. **104:** 248–56

Shepherd, R., Teoh, C. H., Lim, K. P. 1974 Responses in a PB 5/51 planting trial, *Proceedings of the Rubber Research Institute of Malaysia Planters' Conference Kuala Lumpur 1974*, pp. 148–59

Sherief, P. M., Sethuraj, M. R. 1978 The role of lipids and proteins in the mechanism of latex vessel plugging in *Hevea brasiliensis, Physiologia Plantarum* **42:** 351–3

Shorrocks, V. M. 1964 *Mineral deficiencies in* Hevea *and associated cover plants*, Rubber Research Institute of Malaya, Kuala Lumpur

Shorrocks, V. M. 1965a Mineral nutrition, growth and nutrient cycle of *Hevea brasiliensis*. I. Growth and nutrient content, *Journal of the Rubber Research Institute of Malaya* **19:** 32–47

Shorrocks, V. M. 1965b Leaf analysis as a guide to the nutrition, of *Hevea brasi-*

liensis. VI. Variation in the leaf nutrient composition with age of leaf and with time, *Journal of the Rubber Research Institute of Malaya* **19**: 1–8

Shorrocks, V. M. 1965c Mineral nutrition, growth and nutrient cycle of *Hevea brasiliensis*. II. Nutrient cycle and fertilizer requirements, *Journal of the Rubber Research Institute of Malaya* **19**: 48–61

Shorrocks, V. M. 1965d Mineral nutrition, growth and nutrient cycle of *Hevea brasiliensis*. IV. Clonal variation in girth with reference to shoot dry weight and nutrient requirements, *Journal of the Rubber Research Institute of Malaya* **19**: 93–7

Shorrocks, V. M., Templeton, J. K., Iyer, G. C. 1965 Mineral nutrition, growth and nutrient cycle of *Hevea brasiliensis*. III. The relationship between girth and shoot dry weight, *Journal of the Rubber Research Institute of Malaya* **19**: 85–92

Shorrocks, V. M., Watson, G. A. 1961 Managanese deficiency in *Hevea*: the effect of soil application of manganese sulphate on the manganese status of the tree, *Journal of the Rubber Research Institute of Malaya* **17**: 19–30

Siahaan, M. M. 1980 Percobaan sadap tusuk pada tanaman karet muda [Puncture tapping on young rubber], *Bulletin Balai Penelitian Perkebunan Medan* **11**: 85–91

Silva, C. G. 1969 Provisional classification of rubber soils of Ceylon and their relationship to Malayan soils, *Journal of the Rubber Research Institute of Malaya* **21**: 217–24

Silva, C. G. 1975 The fertilizer requirements of smallholdings in Sri Lanka, *Bulletin of the Rubber Research Institute of Sri Lanka* **10**: 31–4

Silva, C. G. 1976 Discriminatory fertilizer recommendations for rubber in Sri Lanka, *Proceedings of the International Rubber Conference, Kuala Lumpur 1975*, vol. **3**, pp. 132–44

Silva, C. G., de Silva, M. U. J. 1971 A survey of wind damage in rubber plantations. *Bulletin of Rubber Research Institute of Ceylon* **67**: 1–11

Silva, C. G., Perera, A. M. A. 1971 A study of the urease activity on the rubber soils of Ceylon. *Quarterly Journal of the Rubber Research Institute of Ceylon* **47**: 30–6

Simandjuntak, M. T. 1974 Penyadapan ke atas dapat menaikkan hasil dan menurukan biaya secara nyata [Upward tapping to increase yield and reduce costs], *Menara Perkebunan* **42**(6): 295–304

Simmonds, N. W. 1969 Genetical bases of plant breeding, *Journal of the Rubber Research Institute of Malaya* **21**: 1–10

Simmonds, N. W. 1979 *Principles of Crop Improvement*. Longman, London

Simmonds, N. W. 1982 Some ideas on botanical research on rubber, *Tropical Agriculture, Trinidad* **59**: 2–8

Simmonds, N. W 1983 Strategy of disease resistance breeding, *FAO Plant Protection Bulletin* **31**: 2–10

Simmonds, N. W. 1985 Perspectives on the evolutionary history of tree crops. In Cannell, M. G. R., Jackson, J. E., Gordon, J. C. (eds) *Attributes of trees as crop plants*. Institute of Terrestrial Ecology, Huntingdon, England, pp. 3–12

Simmonds, N. W. 1986a The strategy of rubber breeding, *Proceedings of the International Rubber Research Conference, Kuala Lumpur 1985*, **3**: 115–26

Simmonds, N. W. 1986b Theoretical aspects of synthetic/polycross populations of rubber seedlings, *Journal of Natural Rubber Research* **1**: 1–15

Singh, K. G. 1978 Regional cooperation in pest control through legislation. In Amin, L. L., Kadir, A. A. S. A., Soon, L. G., Singh, K. G., Tan, A. M., Varghese, G. (eds) *Proceedings of the Plant Protection Conference, Kuala Lumpur 1978*, pp. 251–5

Sivakumaran, S. 1982 Ethephon stimulation – minimum time after application for response, *Planters' Bulletin of the Rubber Research Institute of Malaysia* No. **172**: 85–8

Sivakumaran, S. 1983 Ethephon stimulation – inter-relationship between concentration, dosage and frequency of application. *Planters' Bulletin of the Rubber Research Institute of Malaysia* No. **174**: 22–5

Sivakumaran, S. 1984 Prospects of puncture tapping in specific areas, *Planters' Bulletin of the Rubber Research Institute of Malaysia* No. **179**: 59–64

Sivakumaran, S., Gomez, J. B. 1980 Puncture tapping – an overview. *Planters' Bulletin of the Rubber Research Institute of Malaysia* No. **164**: 101–11

Sivakumaran, S., Pakianathan, S. W. 1983 Studies on dryness. I. A simple and rapid method of inducing dryness in Hevea trees, *Journal of the Rubber Research Institute of Malaysia* **31**: 88–101

Sivakumaran, S., Pakianathan, S. W., Abraham, P. D. 1982 Long-term ethephon stimulation. II. Effect of continuous ethephon stimulation with low frequency tapping systems. *Journal of the Rubber Research Institute of Malaysia* **30**: 174–96

Sivakumaran, S., Pakianathan, S. W., Abraham, P. D. 1983 Long-term ethephon stimulation. III. Effect of continuous ethephon stimulation with short-cut panel-changing systems, *Journal of the Rubber Research Institute of Malaysia* **31**: 151–74

Sivakumaran, S., Pakianathan, S. W., Abraham, P. D. 1984 Continuous yield stimulation – plausible cause for yield decline, *Journal of the Rubber Research Institute of Malaysia* **32**: 119–43

Sivakumaran, S., Pakianathan, S. W., Gomez, J. B. 1981 Long-term ethephon stimulation. I. Effect of continuous ethephon stimulation with half spiral alternate daily tapping, *Journal of the Rubber Research Institute of Malaysia* **29**: 57–85

Sivanadyan, K. 1983 Manuring of *Hevea*: recent evidence and a possible new outlook, *Proceedings of the Rubber Research Institute of Malaysia Planters' Conference, Kuala Lumpur 1983*, pp. 286–312

Sivanadyan, K., Haridas, G., Pushparajah, E. 1976 Reduced immaturity period of *Hevea brasiliensis*, *Proceedings of the International Rubber Conference, Kuala Lumpur 1975*, vol. **3** pp. 147–57

Sivanadyan, K., P'ng, T. C., Pushparajah, E. 1972 Nutrition of *Hevea brasiliensis* in relation to Ethrel stimulation, *Proceedings of the Rubber Research Institute of Malaya Planters' Conference, Kuala Lumpur 1972*, pp. 83–96

Sivanadyan, K., Said, M. bin M., Woo, Y. K., Soong, N. K., Pushparajah, E. 1973 Agronomic practices towards reducing the period of immaturity, *Proceedings of the Rubber Research Institute of Malaysia Planters Conference, Kuala Lumpur 1973*, pp. 226–42

Sivanesan, A., Holliday, P. 1976 Oidium heveae, *Commonwealth Mycological Institute Descriptions of Pathogenic Fungi and Bacteria* No. 508

Sly, J. M. A., Tinker, P. B. H. 1962 An assessment of burning in the establishment of oil palm plantations in Southern Nigeria, *Tropical Agriculture, Trinidad* **39**: 271–80

Smee, L. 1964 Insect pests of *Hevea brasiliensis* in the territory of Papua and New Guinea: their habits and control, *Papua and New Guinea Agricultural Journal* **17**: 21–8

Smit, H. P. 1980 Prospective demand for rubber, In *Proceedings of the Twenty-sixth Assembly of the International Study Group Kuala Lumpur Malaysia 29 September–4 October 1980*

Smit, H. P. 1982 The world rubber economy to the year 2000, its prospects and the implications of production policies on market conditions for natural rubber. Thesis, Amsterdam University

Smit, H. P. 1984 *Forecasts for the world rubber economy to the year 2000.* MacMillan, London, and Globe Book Services Basingstoke

Smit, H. P. 1985 The outlook for world rubber demand. Paper presented to the International Rubber Study Group Assembly, Abidjan, 1985

Soedarsan, A., Budiman, S. 1969 Suatu metodik penggumpalan lateks karet-rakjat sehitungan dengan pembuatan karet remah (crumb rubber) dalam rangka peningkatan mutu karet-rakjat [A method of smallholders' latex collection in relation to crumb-rubber processing], *Menara Perkebunan* **38** (3/4): 1–4

Soedarsan, A., Budiman, S. 1970 Penggumpalan lateks dalam kantong plastik-besar diperkebunan rakjat Djambi [Latex coagulation in large plastic bags in smallholders' rubber in Djambi], *Menara Perkebunan* **39** (1/2): 2–6

Soepadmo, B. 1975 *Colletotrichum gloeosporioides* sebagai penyebab penyakit gugur daun pada karet. *Menara Perkebunan* **43**: 299–302

Soepadmo, B. 1981 The effect of time of cover crop establishment on root disease incidence in the replanting of *Hevea* [in Indonesian], *Menara Perkebunan* **49**: 129–33

Soepadmo, B., Pawirosoemardjo, S. 1977 Kepekaan repuluh macam klon karet terhadap *Colletotrichum gloeosporioides* condisi rumah kaca. *Menara Perkebunan* **45**: 133–5

Sookmark, S., Langlois, S. J. C. 1974 High level tapping and stimulation, *Rubber Research Centre of Thailand, Hat Yai*, Document No. 39

Soong, N. K. 1973 Effects of nitrogenous fertilizers on growth of rubber seedlings and leaching losses of nutrients, *Journal of the Rubber Research Institute of Malaya* **23**: 356–64

Soong, N. K. 1976 Feeder root development of *Hevea brasiliensis* in relation to clones and environment, *Journal of the Rubber Research Institute of Malaysia* **24**: 283–98

Soong, N. K. 1981 Discriminatory fertilizer use for *Hevea*. In *Training manual on soils, soil management and nutrition of* Hevea. Rubber Research Institute of Malaysia, pp. 203–10

Soong, N. K., Yap, W. C. 1976 Effect of cover management on physical properties of rubber-growing soils, *Journal of the Rubber Research Institute of Malaysia* **24**: 145–59

Soong, N. K., Yeoh, C. S., Chin, L. S., Haridas, G. 1976 Natural rubber encapsulated fertilizers for controlled nutrient release, *Proceedings of the Rubber Research Institute of Malaysia Planters' Conference, Kuala Lumpur 1976*, pp. 63–74

Sorensen 1942 Crown budding for healthy *Hevea*, *Agriculture in the Americas* **11**: 191–3

Southorn, W. A. 1961 Microscopy of *Hevea* latex, *Proceedings of the Natural Rubber Research Conference, Kuala Lumpur 1960*, pp. 766–76

Southorn, W. A. 1963 In *Rubber Research Institute of Malaya Annual Report 1962*, p. 82

Southorn, W. A. 1967 Local changes in bark dimensions of *Hevea brasiliensis* very close to the tapping cut, *Journal of the Rubber Research Institute of Malaya* **20**: 36–43

Southorn, W. A. 1968a Latex flow studies. I. Electron microscopy of *Hevea brasiliensis* in the region of the tapping cut, *Journal of the Rubber Research Institute of Malaya* **20**: 176–86

Southorn, W. A. 1968b Latex flow studies. IV. Thixotropy due to lutoids in fresh latex demonstrated by a microviscometer of new design, *Journal of the Rubber Research Institute of Malaya* **20**: 226–35

Southorn, W. A. 1969 Latex collection in disposable plastic bags and the use of expanded plastic rainguards, *Planters' Bulletin of the Rubber Research Institute of Malaya* No. **104**: 156–64

Southorn, W. A., Edwin, E. E. 1968 Latex flow studies. II. Influence of lutoids on the stability and flow of *Hevea* latex, *Journal of the Rubber Research Institute of Malaya* **20**: 187–200

Southorn, W. A., Gomez, J. B. 1970 Latex flow studies. VII. Influence of length of tapping cut on latex flow pattern, *Journal of the Rubber Research Institute of Malaya* **23**: 15–22

Southorn, W. A., Yip, E. 1968a Latex flow studies. III. Electrostatic considerations in the colloidal stability of fresh *Hevea* latex, *Journal of the Rubber Research Institute of Malaya* **20**: 201–15

Southorn, W. A., Yip, E. 1968b Latex flow studies. V. Rheology of fresh *Hevea latex* flow in capillaries, *Journal of the Rubber Research Institute of Malaya* **20**: 236–47

Spears, J. S. 1980 Can farming and forestry co-exist in the tropics? *Unasylva* **33**: 2–12

Spencer 1939 On the nature of the blocking of the laticiferous system at the leaf base of *Hevea brasiliensis, Annals of Botany, London (NS)* **3**: 231–5

Stahel, G. 1947 A new method of rooting cuttings of *Hevea* and other trees, *Tropical Agriculture, Trinidad* **24**: 4–6

Steinmann, A. 1925 *De ziekten en plagen van* Hevea brasiliensis *in Nederlandsch-Indië* [Diseases and pests of *Hevea brasiliensis* in the Netherlands Indies], Archipel Drukkerij, Buitenzorg

Stern, H. J. 1982 History. In Blow, C. M., Hepburn, C. (eds) *Rubber technology and manufacture.* Butterworth, London

Stevenson, J. A., Imle, E. P. 1945 *Periconia* blight of *Hevea, Mycologia* **37**: 576–81

Steyaert, R. L. 1975 *Ganoderma philippii, Commonwealth Mycological Institute Descriptions of Pathogenic Fungi and Bacteria* No. 446

Stosic, D. D., Kaykay, J. M. 1981 Rubber seeds as animal feed in Liberia, *World Animal Review* **39**: 29–39

Strahler, A. N. 1969 *Physical geography*, 3rd edn. Wiley, New York

Subramaniam, A. 1972 Effects of ethrel stimulation on raw rubber properties, *Proceedings of the Rubber Research Institute of Malaya Planters' Conference, Kuala Lumpur 1971*, pp. 255–62

Subramaniam, A. 1976 Molecular weight and other properties of natural rubber: a study of clonal variations, *Proceedings of the International Rubber Conference, Kuala Lumpur 1975*, vol. **4**, pp. 3–10

Subramaniam, S. 1969 Performance of recent introductions of Hevea in Malaysia, *Journal of the Rubber Research Institute of Malaya* **21**: 11–18

Subramaniam, S. 1970 Performance of Dothidella resistant Hevea clones in Malaysia, *Journal of the Rubber Research Institute of Malaya* **23**: 39–46

Subramaniam, S. 1973 Recent trends in the breeding of *Hevea, Indian Journal of · Genetics and Plant Breeding* **34A**: 132–40

Subramaniam, S., Ong, S. H. 1974 Conservation of the gene pool in *Hevea, Indian Journal of Genetics and Plant Breeding* **34A**: 33–6

Subronto 1982 Studi aliran lateks dengan sadap tusuk [Latex flow study in puncture tapped trees], *Bulletin Balai Penelitian Perkebunan Medan* **13**: 99–109

Subronto, Harris, S. A. 1977 Indeks aliran sebagai parameter fisiologis penduga produksi lateks [Flow index as physiological parameter for predicting the yield], *Bulletin Balai Penelitian Perkebunan Medan* **8**: 433–41

Subronto, Napitupulu, L. A. 1978 Pengujian klon dengan menggunakan parameter fisiologis untuk menaksir kemampuan produksi [Screening clones by physiological parameters for predicting yield potential], *Bulletin Balai Penelitian Perkebunan Medan* **9**: 171–83

Sultan, M. O. 1973 Assessment of some new clones for the future, *Proceedings of the Rubber Research Institute of Malaysia Planters' Conference, Kuala Lumpur 1973*, pp. 281–99

Sumarno-Kertowardjono, Sube, A., Taha Surdia, N. M., Amin-Tjasadihardja 1976 Beberapa pengaruch stimulasi Ethrel techadays sifat lateks alam dan Karet Alam [Some influences of Ethrel stimulation on properties of natural rubber and latex], *Menara Perkebunan* **44**: 75–81

Sys, C. 1975 Report on the *ad hoc* expert consultation on land evaluation, *FAO World Soil Resources Report* No. 45, Rome, pp. 59–79

Talib, A. bin. Raveendran, N., Wong, S. P., Pushparajah, E. 1984 Manganese toxicity symptoms in *Hevea, Proceedings of the International Conference on Soils and Nutrition of Perennial Crops, Kuala Lumpur 13–15 August 1984*

Tan, A. G., Sujan, A. 1981 Rubber wood for furniture manufacture, *Planter, Kuala Lumpur* **57**: 649–55

Tan, A. G., Sujan, A., Chong, K. F., Tam, M. K. 1979 Bio-deterioration of rubber wood and control measures, *Planters' Bulletin of the Rubber Research Institute of Malaysia* No. **160**: 106–17

Tan, A. G., Sujan, A., Khoo, T. C. 1980 Rubber wood for parquet manufacture, *Planters' Bulletin of the Rubber Research Institute of Malaysia* No. **163:** 81–7

Tan, A. M. 1979 *Phytophthora* diseases of rubber in Peninsular Malaysia, *Planters' Bulletin of the Rubber Research Institute of Malaysia* No. **158:** 11–19

Tan, A. M. 1983 A new fungicide for the control of black stripe, *Planters' Bulletin of the Rubber Research Institute of Malaysia* No. **174:** 13–16

Tan, H. 1977 Estimates of general combining ability in Hevea breeding at the RRIM I. Phases II and IIIA, *Theoretical and Applied Genetics* **50:** 29–34

Tan, H. 1978a Estimates of parental combining abilities in rubber based on young seedling progeny, *Euphytica* **27:** 817–23

Tan, H. 1978b Assessment of parental performance for yield in *Hevea* breeding, *Euphytica* **27:** 521–28

Tan, H. 1979a Heritabilities of six biometrical characters of single pair mating families in *Hevea brasiliensis, Journal of the Rubber Research Institute of Malaysia* **27:** 127–31

Tan, H. 1979b A biometrical approach to study crown-trunk relationships in *Hevea, Journal of the Rubber Research Institute of Malaysia* **27:** 79–91

Tan, H. 1981 Estimates of genetic parameters and their implications in *Hevea* breeding, *Proceedings of the Third International Congress, SABRAO Kuala Lumpur*, pp. 439–49

Tan, H. 1987 Current status of *Hevea* breeding. In Campbell, A. I., Abbott, A. J., Atkin, R. K. (eds) *Improvement of vegetatively propagated plants*. Academic Press, London [8th Long Ashton Symposium, 1982]

Tan, H., Mukherjee, T. K., Subramaniam, S. 1975 Estimates of genetic parameters of certain characters in *Hevea brasiliensis, Theoretical and Applied Genetics* **46:** 181–90

Tan, H., Ong, S. H., Sultan, M. O., Khoo, S. K. 1984 Accelerated approach to *Hevea* selection, *Proceedings of the Natural Rubber Conference Salvador, Brazil*

Tan, H., Subramaniam, S. 1976 A five-parent diallel cross analysis for certain characters of young *Hevea* seedlings, *Proceedings of the International Rubber Conference, Kuala Lumpur 1975*, vol. **2**, pp. 13–26

Tan, H. T., Jap, K. T., Ismail, P. L. 1961 *Psophocarpus palustris*: an ideal ground cover for oil palm and rubber, *Proceedings of the Natural Rubber Research Conference, Kuala Lumpur 1960*, pp. 312–19

Tan, H. T., Leong, T. T. 1976 Chemara crown budding trials, *Proceedings of the Rubber Research Institute of Malaysia Planters' Conference, Kuala Lumpur 1976*, pp. 116–28

Tan, H. T., Menon, C. M. 1973 Further progress in Chemara Ethrel experiments, *Proceedings of the Rubber Research Institute of Malaya Planters' Conference, Kuala Lumpur 1973*, pp. 152–78

Tan, H. T., Pillai, K. R. 1979 New chemical techniques in the establishment, maintenance and rehabilitation of leguminous covers, *Proceedings of the Rubber Research Institute of Malaysia Planters' Conference, Kuala Lumpur 1979*, pp. 135–49

Tan, H. T., Pillai, K. R., Barry, D. J. 1976 Possible utilization of rubber factory effluent on crop land, *Proceedings of the International Rubber Conference, Kuala Lumpur 1975*, vol. **3**, pp. 158–72

Tan, H. T., Yeow, K. H., Chandapillai, M. M. 1969 Possibilities of intercropping, *Planter, Kuala Lumpur* **45:** 8–17

Tan, K. H. 1982 Studies on nitrogen in Malaysian soils. II. Urea hydrolysis and transformations, *Journal of the Rubber Research Institute of Malaysia* **30:** 19–31

Tan, K. H., Bremner, J. M. 1982 Studies on nitrogen in Malaysian soils. III. Denitrification, *Journal of the Rubber Research Institute of Malaysia* **30:** 102–9

Tan, K. J., Chan, H. Y., Chiah, H. S. 1983 Observations on early stimulation under estate practise: preliminary results, *Proceedings of the Rubber Research Institute of Malaysia Planters' Conference, Kuala Lumpur 1983*, pp. 171–92

Tan, K. H., Pushparajah, E., Shepherd, R., Teoh, C. H. 1976 *Calopogonium caeruleum*, a shade-tolerant leguminous cover for rubber, *Proceedings of the Rubber Research Institute of Malaysia Planters' Conference, Kuala Lumpur 1976*, pp. 45–62

Tata, S. J., Beintema, J. J., Balabaskaran, S. 1983 The lysozyme of *Hevea brasiliensis* latex: isolation, purification, enzyme kinetics and a partial amino-acid sequence, *Journal of the Rubber Research Institute of Malaysia* **31:** 35–48

Tata, S. J., Boyce, A. N., Archer, B. L., Audley, B. G. 1976 Lysozymes: major components of the sedimentable phase of *Hevea brasiliensis* latex, *Journal of the Rubber Research Institute of Malaysia* **24:** 233–6

Tayeb, D. M. 1978 Agricultural utilization of rubber factory effluent: a review of its potential, *Proceedings of the International Rubber Research and Development Board Symposium, Kuala Lumpur*

Tayeb, D. M, Zaid, I., Zim, M. K. 1982 Land-disposal of rubber effluent: soil-plant system as a pollutant remover, *Proceedings of the Rubber Research Institute of Malaysia Planters' Conference, Kuala Lumpur 1981*, pp. 364–79

Tayeb, D. M., Zim, K. M., Zaid, I. 1979 Land-disposal of rubber factory effluent: its effects on soil properties and performance of rubber and oil palm, *Proceedings of the Rubber Research Institute of Malaysia Planters' Conference, Kuala Lumpur 1979*, pp. 436–57

Taysum, D. H. 1961 Effect of ethylene oxide on the tapping of *Hevea brasiliensis*, *Nature, London* **191:** 1319–20

Templeton, J. K. 1968 Growth studies in *Hevea brasiliensis*. I. Growth analysis up to seven years after budgrafting, *Journal of the Rubber Research Institute of Malaya* **20:** 136–46

Templeton, J. K. 1969a Partition of assimilates, *Journal of the Rubber Research Institute of Malaya* **21:** 259–63

Templeton, J. K. 1969b Where lies the yield summit for Hevea? *Planters' Bulletin of the Rubber Research Institute of Malaya* No. **104:** 220–5

Templeton, J. K. 1970 Don't intercrop with unimproved types of castor. In Blencowe, E. K., Blencowe, J. W. (eds) *Crop diversification in Malaysia*. Incorporated Society of Planters, Kuala Lumpur, pp. 99–106

Templeton, J. K. 1978 *Natural rubber – organizations and research in producing countries*. International Agricultural Development Service, New York

Templeton, J. K., Shepherd, R. 1967 Some technical aspects of green budding, *Planters' Bulletin of the Rubber Research Institute of Malaya* No. **92:** 190–7

Teoh, C. H., Adham, A., Reid, W. M. 1979 Critical aspects of legume establishment and maintenance, *Proceedings of the Rubber Research Institute of Malaysia Planters' Conference, Kuala Lumpur 1979*, pp. 252–71

Teoh, C. H., Toh, P. Y., Chong, C. F., Evans, R. C. 1978 Recent developments in the use of herbicides on estates. *Proceedings of the Rubber Research Institute of Malaysia Planters' Conference, Kuala Lumpur 1977*, pp. 91–110

Teoh, C. H., Toh, P. Y., Khairudin, H. 1982 Chemical control of *Asystasia intrusa* (B1.), *Clidemia hirta* (Don.) and *Elattoriopsis curtisii* (Bak.) in rubber and oil palm plantations, *Proceedings of the International Conference on Plant Protection in the Tropics, Kuala Lumpur 1982*, pp. 497–510

Teoh, K. S. 1972 A novel method of rubber propagation, *Proceedings of the Rubber Research Institute of Malaysia Planters' Conference, Kuala Lumpur 1972*, pp. 59–71

Thainugul, W., Sinthurahas, S. 1982 The feasibility of rubber growing areas in Thailand, *First International Symposium on Soil, Geology and Landforms – Impact on Land Use Planning in Developing Countries, Bangkok 1–3 April*

Thiagalingam, K., Grimme 1976 The evaluation of the K status of some Malaysian soils by means of electro-ultra-filtration, *Planter, Kuala Lumpur* **52:** 83–9

Thomas, P. O. 1980 Production planning in the Malaysian natural rubber industry to the year 2000. In *Proceedings of the Twenty-sixth Assembly of the International*

Rubber Study Group Kuala Lumpur Malaysia 29 September–4 October 1980

Thompson, A. 1924 A preliminary note on a new bark disease of *Hevea, Malayan Agricultural Journal* **12:** 163–4

Ti, T., Pee, T. Y., Pushparajah, E. 1972 Economic analyses of cover policies and fertilizer use in rubber cultivation, *Proceedings of the Rubber Research Institute of Malaya Planters' Conference, Kuala Lumpur 1971*, pp. 214–33

Tinker, P. B., Leigh, R. A. 1984 Nutrient uptake by plants – efficiency and control, *Proceedings of the International Conference of Soils and Nutrition Perennial Crops, Kuala Lumpur 13–15 August 1984*

Tinley, G. H. 1961 Effect of ferric dimethyl dithiocarbamate on rooting of cuttings of *Hevea brasiliensis, Nature, London* **191:** 1217–18

Tinley, G. H., Garner, R. J. 1960 Developments in the propagation of clones of *Hevea brasiliensis* by cuttings, *Nature, London* **186:** 407

Tonnelier, M., Primot, L., Trancard, J., Omont, H. 1979 La saignée par piqures. Bilan provisoire et perspectives d'avenir, *Revue Générale du Caoutchouc et des Plastiques* **56** (594): 71–6

Toruan, N. L., Suryatmana, N. 1977 Tissue culture of *Hevea brasiliensis, Menara Perkebunan* **45:** 17–21

Tran, V. C. 1982 Lutte contre le *Fomes*: nouvelle méthode d'étude, *Caoutchouc et Plastiques* No. 617/8

Tugiman, S., Said, I. 1983 *The need for an effective extension service in the smallholder sector.* Confidential Report Rubber Research Institute of Malaysia, Kuala Lumpur (mimeograph)

Tupy, J. 1969 Nucleic acids in latex and production of rubber in *Hevea brasiliensis, Journal of the Rubber Research Institute of Malaya* **21:** 468–76

Tupy, J. 1973a Possible mode of action of potassium on latex production, *Symposium of the International Rubber Research and Development Board, Puncak, Indonesia 2–4 July 1973*

Tupy, J. 1973b The level and distribution pattern of latex sucrose along the trunk of *Hevea brasiliensis* as affected by the sink region induced by latex tapping, *Physiologie Végétale* **11:** 1–11

Tupy, J. 1973c The sucrose mobilizing effect of auxins in *Hevea brasiliensis, Physiologie Végétale* **11:** 13–23

Typy, J. 1973d The activity of latex invertase and latex production in *Hevea brasiliensis, Physiologie Végétale* **11:** 633–41

Tupy, J. 1973e The regulation of invertase activity in the latex of *Hevea brasiliensis, Journal of Experimental Botany* **24:** 516–24

Tupy, J. 1973f Influence de la stimulation hormonale de la production sur la teneur en saccharose du latex d'*Hevea brasiliensis, Revue Générale du Caoutchouc et des Plastiques* **50:** 311–14

Tupy, J. 1973g Possibilité d'exploitation de l'Hévéa par micro-saignée, *Revue Générale du Caoutchouc et des Plastiques* **50:** 7–8

Tupy, J. 1985 Some aspects of sucrose transport and utilization in latex producing bark of *Hevea brasiliensis, Biologia Plantarum (Prague)* **27:** 51–64

Tupy, J., Primot, L. 1976 Control of carbohydrate metabolism by ethylene in latex vessels of *Hevea brasiliensis* in relation to rubber production, *Biologia Plantarum (Prague)* **18:** 373–82

Tupy, J., Primot, L. 1982 Sucrose synthetase in the latex of *Hevea brasiliensis, Journal of Experimental Botany* **33:** 988–95

Tupy, J., Resing, W. L. 1968 Anaerobic respiration in latex of *Hevea brasiliensis*: substrate and limiting factors, *Biologia Plantarum* **10:** 72–80

Turner, P. D., Gillbanks, R. A. 1974 *Oil palm cultivation and management.* Incorporated Society of Planters, Kuala Lumpur

Turner, P. D., Myint, U-H. 1980 Rubber diseases in Burma, *FAO Plant Protection Bulletin* **28:** 85–91

Udomsakdhi, B., Munsakul, S., Sthapitanonda, K. 1974 Potential value of rubber seed, *Thai Journal of Agricultural Science* **7:** 259–71

United States Department of Agriculture 1975 Soil taxonomy. *United States Department of Agriculture Soil Conservation Service Agricultural Handbook* No. 436. Government Printer, Washington

United States Department of Commerce 1952 *Materials survey: rubber.* Government Printer, Washington

Van Dienst, A. 1979 Factors affecting the availability of potassium in soils, International Potash Institute Potassium research – reviews and trends, *Proceedings of the 11th Congres Bern, Switzerland 4–8 September 1978*, pp. 75–97

Verhaar, G. 1973 Processing of natural rubber, *FAO Agricultural Series Bulletin* No. 20. Food and Agriculture Organization of the United Nations, Rome

Viégas, R. M. F., Pinheiro, F. S. V., da Cunha, R. L. M. 1980 Consorciação seringueira y pimenta-do-reino. Resultados preliminares [Association between rubber and vine pepper. Preliminary results]. In EMBRAPA/FCAP *3° Seminario Nacional da Seringueira, Belém, Brasil 1980*

Vimal, O. P. 1981 The uses of rubber seed, *Planters' Chronicle* July: 333–6

Vollema, J. S. 1931 Manuring experiments with *Hevea* in West Java, *Archief voor de Rubbercultuur in Nederlandsch-Indië* **15:** 481–574

Von Saher, H., Verhaar, G. 1979 One hundred years of *Hevea brasiliensis* in Indonesia 1876–1976 [in Indonesian, English summary], *Menara Perkebunaan* **47:** 57–69

Wahab, M. H. A., Dolmat, M., Isa, Z. 1979 Bowl sludge – a potential fertilizer. *Planters' Bulletin of the Rubber Research Institute of Malaysia* No. **159:** 41–53

Waidyanatha, U. P. de S., Angammana, D. K. 1981 Early exploitation of Hevea rubber trees by puncture and short-cut tappings, *Experimental Agriculture* **17:** 303–9

Waidyanatha, U. P. de S., Pathiratne, L. S. S. 1971 Studies on latex flow patterns and plugging indices of clones, *Quarterly Journal of the Rubber Research Institute of Ceylon* **48:** 47–55

Waidyanatha, U. P. de S., Wijesinghe, D. S., Stauss, R. 1984 Zero-grazed pasture under immature *Hevea* rubber: productivity of some grasses and grass-legume mixtures and their competition with *Hevea*, *Tropical Grasslands* **18:** 16–4

Wallace, G. L. 1952 Chapter 30, Statistical and economic outline. In Schidrowitz, P., Dawson, T. R. (eds) *History of the rubber industry.* Heffer, Cambridge

Warmke, H. E. 1951 Studies on pollination of *Hevea brasiliensis* in Puerto Rico, *Science* **113:** 646–8

Warmke, H. E. 1952 Studies on natural pollination of *Hevea brasiliensis* in Brazil, *Science* **116:** 474–8

Warriar, S. M. 1969 Cover plant trials, *Journal of the Rubber Research Institute of Malaya* **21:** 158–64

Wastie, R. L. 1965 The occurrence of an *Endogone* type of endotrophic mycorrhiza in *Hevea brasiliensis*, *Transactions of the British Mycological Society* **48:** 167–8

Wastie, R. L. 1972 Factors affecting secondary leaf fall of *Hevea* in Malaysia, *Journal of the Rubber Research Institute of Malaya* **23:** 232–47

Wastie, R. L. 1973a Nursery screening of *Hevea* for resistance to *Gloeosporium* leaf disease, *Journal of the Rubber Research Institute of Malaya* **23:** 339–50

Wastie, R. L. 1973b Influence of weather on the incidence of *Phytophthora* leaf fall of *Hevea brasiliensis* in Malaysia, *Journal of the Rubber Research Institute of Malaya* **23:** 381–90

Wastie, R. L. 1975 Diseases of rubber and their control, *PANS* **21:** 268–87

Wastie, R. L., Mainstone, B. J. 1969 Economics of controlling secondary leaf fall of Hevea caused by *Oidium heveae* Steinm., *Journal of the Rubber Research Institute of Malaya* **21:** 64–72

Watson, G. A. 1957a Cover plants in rubber cultivation, *Journal of the Rubber Research Institute of Malaya* **15:** 2–18

Watson, G. A. 1957b Nitrogen fixation by *Centrosema pubescens*, *Journal of the*

Rubber Research Institute of Malaya **15**: 168–74

Watson, G. A. 1960 Interactions of lime and molybdate in the nutrition of *Centrosema pubescens* and *Pueraria phaseoloides, Journal of the Rubber Research Institute of Malaya* **16**: 126–38

Watson, G. A. 1961 Cover plants and the soil nutrient cycle in *Hevea* cultivation, *Proceedings of the Natural Rubber Research Conference,* Kuala Lumpur 1960, pp. 352–61

Watson, G. A. 1962 Rubber cultivation in a diversified agriculture: notes on its relegation to the poorer soils of Malaya, North Borneo and Sarawak, *Sarawak Museum Journal* **10**: 590–7

Watson, G. A. 1963 Cover plants and tree growth, Part II. Leguminous creeping covers and manuring, *Planters' Bulletin of the Rubber Research Institute of Malaya* No. **68**: 172–6

Watson, G. A. 1964 Maintenance of soil fertility in the permanent cultivation of *Hevea brasiliensis* in Malaya, *Outlook on Agriculture* **4**: 103–109

Watson, G. A. 1980 A study of tree crop farming systems in the lowland humid tropics, *World Bank AGR Technical Note* No. 2, vols **1,2**. Washington

Watson, G. A. 1983 Development of mixed tree and food crop systems in the humid tropics; a response to population pressure and deforestation, *Experimental Agriculture* **19**: 311–32

Watson, G. A., Chin, T. T., Wong, P. W. 1962 Loss of ammonia by volatilisation from surface dressings of urea in rubber cultivation, *Journal of the Rubber Research Institute of Malaya* **17**: 77–90

Watson, G. A., Narayanan, R. 1965 Effect of fertilizer on seed production by *Hevea brasiliensis, Journal of the Rubber Research Institute of Malaya* **19**: 22–31

Watson, G. A., Wong, P. W., Narayanan, R. 1963 Effects of cover plants on soil nutrient status and on growth of *Hevea.* II. The influence of applications of rock phosphate, basic slag and magnesium limestone on the nutrient content of leguminous cover plants, *Journal of the Rubber Research Institute of Malaya* **18**: 28–37

Watson, G. A., Wong, P. W., Narayanan, R. 1964a Effects of cover plants on soil nutrient status and on growth of *Hevea.* III. A comparison of leguminous creepers with grasses and *Mikania cordata, Journal of the Rubber Research Institute of Malaya* **18**: 80–95

Watson, G. A., Wong, P. W., Narayanan, R. 1964b Effects of cover plants on soil nutrient status and on growth of *Hevea.* IV. Leguminous creepers compared with grasses, *Mikania cordata* and mixed indigenous covers on four soil types, *Journal of the Rubber Research Institute of Malaya* **18**: 123–45

Watson, G. A., Wong, P. W., Narayanan, R. 1964c Effects of cover plants on soil nutrient status and on growth of *Hevea.* V. Loss of nitrate-nitrogen and of cations under bare soil conditions. A progress report on results from a small-scale field trial, *Journal of the Rubber Research Institute of Malaya* **18**: 161–74

Watson, I. 1965 The economic evaluation of tapping systems, *Planters' Bulletin of the Rubber Research Institute of Malaya* No. **80**: 236–44

Watson, P. J. 1984 World natural rubber consumption 1975–82 by type and grade. In *Proceedings of the Twenty-eighth Assembly of the International Rubber Study Group, London, 18–22 June 1984,* pp. 104–19

Watson, P. J. 1985 *Proceedings of the Twenty-ninth Assembly of the International Rubber Study Group,* Abidjan, Ivory Coast, 25–29 November 1985

Weinstein, B. 1983 *Amazon rubber boom 1850–1920.* Stanford University Press, Stanford, California, U.S.A.

Weir, J. R. 1926 A pathological survey of the Para rubber tree (*Hevea brasiliensis*) in the Amazon valley, *United States Department of Agriculture Bulletin* No. 1380

Wheeler, L. C. 1978 *Hevea* (rubber) seeds for human food, *Bulletin of the Rubber Research Institute of Sri Lanka* **13**: 17–21

Whitby, S. 1919 Variation in *Hevea brasiliensis, Annals of Botany* **33**: 313–21

Whitford, H. N. 1929 *Estate and native plantation rubber in the Middle East 1928.* Rubber Manufacturers' Association, New York

Whitford, H. N. 1930 *Estate and native plantation rubber in the Middle East 1929.* Rubber Manufacturers' Association, New York

Whitford, H. N. 1931 *Estate and native plantation rubber in the Middle East 1930.* Rubber Manufacturers' Association, New York

Whitford, H. N. 1932 *Fourth report on plantation rubber in the Middle East.* Rubber Manufacturers' Association, New York

Whitford, H. N. 1934 *Fifth report on plantation rubber in the Middle East.* Rubber Manufacturers' Association, New York

Whitford, H. N., Anthony, A. 1926 Rubber production in Africa, *United States Department of Commerce Trade Promotion Series* No. 34. Government Printer, Washington

Wickham, H. A. 1908 *Notes on the plantation, cultivation and curing of Para Indian rubber* (Hevea brasiliensis) *with an account of its introduction from the West to the Eastern Tropics.* Kegan Paul, London

Wiersum, L. K. 1955 Observations on the rooting of *Hevea* cuttings, *Archief voor de Rubbercultuur in Nederlandsch-Indië* **32**: 213–15

Wilson, H. M., Street, H. E. 1975 The growth, anatomy and morphogenetic potential of callus and cell suspension culture of *Hevea brasiliensis, Annals of Botany* **39**: 671–82

Winder, J. A. 1976 Ecology and control of *Erinnys ello* and *E. alope*, important insect pests in the New World, *PANS* **22**: 449–66

Wong, C. B., Chan, H. Y., Selvarajah, N. 1977 Influence of common parent rocks, *Proceedings of the Conference on Chemistry and Fertility of Tropical Soils, Kuala Lumpur 1973*, pp. 192–98

Wong, P. W. 1964 Evidence for the presence of growth inhibitory substances in *Mikania cordata* (Burm. f.) B L Robinson, *Journal of the Rubber Research Institute of Malaya* **18**: 231–42

Wong, P. W. 1966 Weed control under partial shade with Weedazol TL, sodium chlorate and 2,4-D, *Planters' Bulletin of the Rubber Research Institute of Malaya* No. **87**: 191–6

Woodroffe, J. F., Hammel-Smith, H. 1915 *The rubber industry of the Amazon and how its supremacy can be maintained.* John Bale, Sons and Dan elsson, London

Woodruff, W. 1958 *The rise of the British rubber industry during the 19th Century.* Liverpool University Press

Wren, W. G 1947 African wild rubber. In *Proceedings of the Eleventh International Congress on Pure and Applied Chemistry 1947*, pp. 373–82

Wright, H. 1912 Hevea brasiliensis *or Para rubber: botany, cultivation, chemistry and diseases*, 4th edn. Maclaren, London

Wycherley, P. R. 1959 The Singapore Botanic Gardens and rubber in Malaya, *Gardens' Bulletin, Singapore* **17**(2): 175

Wycherley, P. R. 1963a Variation in the performance of *Hevea* in Malaya, *Journal of Tropical Geography* **17**: 143

Wycherley, P. R 1963b The range of cover plants, *Planters' Bulletin of the Rubber Research Institute of Malaya* No. **68**: 117–22

Wycherley, P. R. 1964 The cultivation and improvement of the plantation rubber crop, *Rubber Research Institute of Malaya Archives Document* No. 29

Wycherley, P. R. 1967 Rainfall in Malaysia, *Rubber Research Institute of Malaya Planting Manual* No. 12

Wycherley, P. R 1968a Introduction of *Hevea* to the Orient, *Planter, Kuala Lumpur* **44**: 1–11

Wycherley, P. R 1968b Introduction of *Hevea* to the Orient, *Planter, Kuala Lumpur* **44**: 127–37

Wycherley, P. R. 1969 Breeding of *Hevea, Journal of the Rubber Research Institute of Malaya* **21**: 38–55

Wycherley, P. R 1971 *Hevea* seed (I); (II); (III); *Planter, Kuala Lumpur* **47**: 291–8; 345–50; 405–10

Wycherley, P. R 1975 *Hevea*: Long flow, adverse partition and storm losses, *Planter, Kuala Lumpur* **51**: 6–13

Wycherley, P. R 1976a Rubber. In Simmonds, N. W. (ed) *Evolution of crop plants*. Longman, London pp. 77–80

Wycherley, P. R. 1976b Tapping and partition, *Journal of the Rubber Research Institute of Malaysia* **24**: 169–94

Wycherley, P. R., Buttery, B. R., Templeton, J. K. 1962 Trunk snap in *Hevea brasiliensis, Proceedings of the 16th International Horticultural Congress, Brussels 1962*, vol. 4, pp. 474–80

Wycherley, P. R., Chandapillai, M. M. 1969 Effects of cover plants, *Journal of the Rubber Research Institute of Malaya* **21**: 140–57

Yahampath, C. 1968 Growth rate of PB 86 on different rootstocks, *Quarterly Journal of the Rubber Research Institute of Ceylon* **44**: 27–34

Yeang, H. Y., David, M. N. 1984 Quantification of latex vessel plugging by the 'intensity of plugging', *Journal of the Rubber Research Institute of Malaysia* **32**: 164–9

Yeang, H. Y., Paranajothy, K. 1982 Some primary determinants of seasonal variation in clone RRIM 623, *Journal of the Rubber Research Institute of Malaysia* **30**: 131–47

Yee, Y. L. 1983 Effects of yield stimulation on profitability and rubber production hypersurface in the estate sector, *Journal of the Rubber Research Institute of Malaysia* **31**: 5–26

Yee, Y. L., Lim, F. H. 1982 Financial performance of rubber estates in 1979, *Planters' Bulletin of the Rubber Research Institute of Malaysia* No. **171**: 50–63

Yeoh, C. H., Lee, K. A., Lim, H. H., Phang, A. K. 1980 Comparison of chemical and manual weedings in rubber nursery, *Planters' Bulletin of the Rubber Research Institute of Malaysia* No. **162**: 28–35

Yeoh, C. H., Pushparajah, E. 1976 Chemical control of *Imperata cylindrica* in Malaysia, *Proceedings of the Rubber Research Institute of Malaysia Planters' Conference, Kuala Lumpur 1976*, pp. 250–74

Yeoh, C. S., Soong, N. K. 1977 Natural rubber-based slow release fertilizers, *Journal of the Rubber Research Institute of Malaysia* **21**: 1–8

Yeow K. H., Yeop A. K. 1983 The present status of effluent utilisation in Malaysia. *Proceedings of the Rubber Research Institute of Malaysia Planters' Conference 1983*, pp. 347–68

Yescombe, E. R. 1976 *Plastics and rubber. World sources of information.* Applied Science Publishers, London

Yip, E., Gomez, J. B. 1980 Factors influencing the colloidal stability of fresh clonal *Hevea* latices as determined by the Aerosol OT test, *Journal of the Rubber Research Institute of Malaysia* **28**: 86–106

Yip, E., Gomez, J. B. 1984 Characteristics of cell sap of *Hevea* and its influence on cessation of latex flow, *Journal of the Rubber Research Institute of Malaysia* **32**: 1–19

Yip, E., Southorn, W. A. 1968 Latex flow studies. VI. Effects of high pressure gradients on flow of fresh *Hevea* latex in narrow bore capillaries, *Journal of the Rubber Research Institute of Malaya* **20**: 248–56

Yip, E., Southorn, W. A. 1973 Latex flow studies. VIII. Rheology of fresh latex from *Hevea* collected over successive intervals from tapping, *Journal of the Rubber Research Institute of Malaya* **23**: 277–84

Yip, E., Southorn, W. A. 1974 Latex flow studies. IX. Effects of application of yield stimulants on rheology of *Hevea* latex, *Journal of the Rubber Research Institute of Malaysia* **24**: 103–10

Yogaratnam, N. 1971 Weed control under *Hevea* in Ceylon with herbicides based on

MSMA, *Quarterly Journal of the Rubber Research Institute of Ceylon* **48**: 147–59

Yogaratnam, N., Perera, M. A. 1981 Urea as a nitrogen fertilizer in Sir Lanka. II. Agronomic investigations, *Journal of the Rubber Research Institute of Sri Lanka* **59**: 20–30

Yogaratnam, N., Weerasuriya, S. M. 1984 Fertilizer responses in mature *Hevea* under Sri Lankan conditions, *Proceedings of the International Conference on Soils and Nutrition of Perennial Crops, Kuala Lumpur August 1984*, pp. 13–15

Yoon, P. K. 1967 RRIM crown budding trials, *Planters' Bulletin of the Rubber Research Institute of Malaya* No. **92**: 240–9

Yoon, P.K. 1972 Further progress in crown budding, *Proceedings of the Rubber Research Institute of Malaya Planters' Conference, Kuala Lumpur 1971*, pp. 143–53

Yoon, P. K. 1973 Horticultural manipulations towards shorter immaturity period, *Proceedings of the Rubber Research Institute of Malaysia Planters' Conference, Kuala Lumpur 1973*, pp. 203–25

Yoon, P. K., Leong, S. K. 1976 Induction of pseudo-taproots in cuttings and production of clonal rootstocks in *Hevea, Proceedings of the International Rubber Conference, Kuala Lumpur 1975*, vol. **2**, pp. 85–108

Yoon, P. K., Leong, W., Ghouse, M. 1976 An approach to modify branching habits – its effects and potentials, *Proceedings of the Rubber Research Institute of Malaysia Planters' Conference, Kuala Lumpur 1976*, pp. 143–76

Yoon, P. K., Ooi, C. B. 1976 Deep planting of propagated materials of *Hevea* – its effects and potentials, *Proceedings of the National Plant Propagation Symposium, Kuala Lumpur 1976*, pp. 273–302. Rubber Research Institute of Malaysia, Kuala Lumpur

Young, A. 1976 *Tropical soils and soil survey*. Cambridge University Press, p. 235

Zainuddin, R. N., Lim, T. M. 1979 Fogging as a method of controlling bird's eye spot, *Planters' Bulletin of the Rubber Research Institute of Malaysia* No. **158**: 3–5

Index